ELECTRONIC STRUCTURE, CORRELATION EFECTS AND PHYSICAL PROPERTIES OF *d*- AND *f*-METALS AND THEIR COMPOUNDS

ELECTRONIC STRUCTURE, CORRELATION EFECTS AND PHYSICAL PROPERTIES OF *d*- AND *f*-METALS AND THEIR COMPOUNDS

V.Yu.Irkhin and Yu.P.Irkhin

Viva Books

New Delhi | Mumbai | Chennai | Kolkata | Bangalore | Hyderabad | Kochi | Guwahati

First Indian Edition 2010

VIVA BOOKS PRIVATE LIMITED

- 4737/23 Ansari Road, Daryaganj, New Delhi 110 002
  E-mail: vivadelhi@vivagroupindia.net, Tel. 42242200
- Plot No. 76, Service Industries, Shirvane, Sector 1, Nerul, Navi Mumbai 400 706
  E-mail: vivamumbai@vivagroupindia.net, Tel. 27721273, 27721274
- Jamals Fazal Chambers, 26 Greams Road, Chennai 600 006
  E-mail: vivachennai@vivagroupindia.net, Tel. 28290304, 28294241
- B-103, Jindal Towers, 21/1A/3 Darga Road, Kolkata 700 017
  E-mail: vivakolkata@vivagroupindia.net, Tel. 22836381, 22816713
- 7, Sovereign Park Apartments, 56-58, K. R. Road, Basavanagudi, Bangalore 560 004
  E-mail: vivabangalore@vivagroupindia.net, Tel. 26607409, 26607410
- 101-102 Mughal Marc Aptt., 3-4-637 to 641, Narayanguda, Hyderabad 500 029
  E-mail: vivahyderabad@vivagroupindia.net, Tel. 27564481, 27564482
- First Floor, Beevi Towers, SRM Road, Kaloor, Kochi 682 018, Kerala
  E-mail: vivakochi@vivagroupindia.net, Tel: 0484-2403055, 2403056
- 232, GNB Road, Beside UCO Bank, Silpukhuri, Guwahati 781 003
  E-mail: vivaguwahati@vivagroupindia.net, Tel: 0361-2666386

Published by arrangement with

Cambridge International Science Publishing
7, Meadow Walk, Great Abington
Cambridge CB1 6AZ, UK

ISBN 978-81-309-1235-6

Published by Vinod Vasishtha for Viva Books Private Limited, 4737/23, Ansari Road, New Delhi - 110 002. Printed and bound by Anand Sons, Delhi - 110 092.

PREFACE

There exist several reasons to pick out the physics of transition group metals as a separate branch of solid state physics. The first one is the variety of properties of transition metals (TM) and their compounds, which are not only much more complicated, but have some important peculiarities in comparison with simple metals. In particular, ferromagnetic ordering takes place practically in TM and their compounds only. The second reason is the importance of investigating TM both from theoretical and practical point of view. TM yield an example of a strongly interacting many-electron system, which does not enable one to introduce in a simple way an effective one-electron potential. Thus we deal with the full many-particle quantum problem which requires application of the all modern theoretical physics methods.

The term 'transition metals' has two meanings. In the narrow sense of the word, TM are elements with partly occupied 3*d*, 4*d* and 5*d*-shells, which form large periods in the Mendeleev periodic table. Sometimes this term is applied to all the elements with partly filled inner electron shells (transition-group elements), including 4*f* (rare earths, RE) and 5*f* (actinides). As a rule, we shall use the notion 'transition metals' in the narrow sense, but all the classes of transition-group elements will be discussed in the book.

We have 24 TM, 13 RE and 8 actinides, so that about one half of elements belong to TM in the broad sense (at the same time, there exist only 25 simple metals). Among TM we find most practically important metals which exhibit maximum strength, melting temperature etc. The most well-known example is Fe: up to now we live in the iron age. More and more applications find rare earths. For example, the intermetallic compound $SmCo_5$ yields a basis for best permanent magnets. Actinidies are widely used in nuclear energetics.

A rather large number of books is available which treat metal physics. They are devoted mainly to simple metals and contain usually some separate paragraphs concerning TM. True, detailed monograps [15–17] discuss properties of rare-earth metals and their compounds. At the same time, similar books on *d*-metals seem to be absent. This is probably explained by that systematization of corresponding large material, which may be found in original and review papers on particular topics, is a more difficult problem. However, in our opinion, it is useful to collect most important results on TM physics and discuss some general regularities.

Let us list some distinctive physical properties of TM and their compounds:

(*i*) large binding energies (high strength and melting temperature)

(*ii*) large electronic specific heat (heavy electron masses)

(*iii*) strong magnetism: large paramagnetic susceptibility and sometimes ferro- or antiferromagnetic ordering

(*iv*) superconductivity, often with high T_C

(*v*) anomalous transport properties (in particular, extraordinary halvanomagnetic effects).

The book is devoted to consideration of these non-trivial physical properties, especial attention being paid to connection with the electronic structure. (The latter term includes both the properties of partially filled *d*- and *f*-shells and the anomalies of band structure.) We do not pretend to consider all the variety of TM compounds and alloys, but try to illustrate some interesting physical phenomena, which are not pronounced for elemental metals, by some bright examples.

The book contains, where possible, a simple physical discussion of a number of problems. At the same time, we use widely in last three Chapters such methods of theoretical physics as the second quantization, atomic representation, Green's function approach. These methods permit to apply microscopic many-electron models which describe systems with strong interelectron correlations. Besides traditional questions of the solid state physics, we treat some modern topics, e.g. magnetism of highly correlated and low-dimensional electron systems, anomalous properties of exotic rare earth and actinide systems (Kondo lattices, heavy-fermion compounds), formation of unusual quantum states with non-trivial excitation spectrum etc. More difficult mathematical aspects of these topics are considered in Appendices A–P. Inclusion of numerous rather long Appendices makes the 'topological' structure of the book somewhat non-trivial and non-traditional. However, such a structure reflects many-sided connections which exist between various branches of TM physics.

At present the TM theory is far from completeness, and a number of important problems are still not resolved. Therefore the treatment of some TM properties may seem to be not so clear and logical as that of corresponding issues of general solid state physics in classical textbooks and monographs [1–14]. However, we believe that description of the modern complicated situation in the TM physics is justified since this may excite the interest in the unsolved questions and stimulate further researches.

The plan of the book is as follows. In the introductory Chapter 1 we treat atomic aspects of TM physics which, unlike the case of simple metals, are rather important since d- and especially f-states retain in a large measure atomic features. Detailed mathematical consideration of some related questions is given in Appendices A–C. In particular, we review applications of the Racah's angular momentum formalism and Hubbard's many-electron operator representation in the solid state theory, which are seldom discussed in the literature on the metal theory.

Chapter 2 considers the electronic structure of TM from the 'band' side. We review briefly methods of band structure calculations, including the density functional approach, with especial attention to TM peculiarities. Besides that, we discuss some simple model approaches to the band spectrum and related experimental (especially spectral) data. We consider also theoretical and experimental results concerning the Fermi surfaces. In Chapter 3 we discuss thermodynamical properties of TM: cohesive energy and related properties, stability of crystal structures, and specific heat, electronic contributions being treated in details.

Chapter 4 deals with magnetic properties. Here we discuss various theoretical models describing highly correlated d- and f-electrons. These models permit all-round consideration of the complicated metallic magnetism problem which includes the atomic ('localized') and band ('itinerant') aspects of d-electron behaviour. A large number of Appendices, related to this Chapter (D–K), demonstrate concrete practical applications of the many-electron models, mainly within the simple method of double-time retarded Green's functions.

Chapter 5 is devoted to transport phenomena in TM which demonstrate a number of peculiarities in comparison with simple metals, e.g., occurrence of spontaneous (anomalous) effects. Quantitative treatment of these effects is performed with the use of density-matrix approach in the operator form (Appendix M).

Finally, in Chapter 6 we treat some questions of the anomalous f-compound theory. In particular, we discuss various mechanisms for occurence of 'heavy' electron mass and the problem of competition between the Kondo effect and magnetic interactions. We consider also model descriptions of the electronic structure in two-dimensional highly-correlated systems including copper-oxide high-T_c superconductors.

We intended to consider in the book all main properties of TM and corresponding theoretical concepts. However, the choice and volume of the material in different Chapters are determined to some extent by scientific interests of the authors. In particular, we pay a great attention to the theory of magnetism and transport phenomena, but discuss in less details lattice properties, and almost do not touch superconductivity (the latter topic develops now extensively and is widely discussed in modern review and monograph literature). We list once more some examples of non-traditional questions which are treated in the book: the influence of density of states singularities on electron properties; many-electron description of strong itinerant magnetism; the problem of quenching and unquenching of orbital magnetic magnetic momenta in solids and their significance for magnetic anisotropy; microscopic theory of anomalous transport phenomena in ferromagnets.

The book is partly based on the lection course of the transition metal physics, which was read for a number of years at the Ural State University.

Except for numerical estimations, we use often in formulas the system of units with $e = k_B = \hbar = 1$. We hope that the book may be of interest for researchers working in solid state physics and for beginners, both for theorists and experimentators. Some parts of the material may be used in lecture courses for students.

The authors are grateful to A.Zarubin, A. Ignatenko, P. Igoshev, A.Nasibullin and A.Katanin for their help in preparing the electronic version of the book.

CONTENTS

Chapter 1

INTRODUCTION

1.1 PARTIALLY FILLED ATOMIC SHELLS AND ELECTRON LOCALIZATION IN TRANSITION METALS

The 'hydrogen-like' scheme of energy levels depending on one-electron principal and orbital quantum numbers n and l corresponds to consecutive filling of atomic shells with increasing n and l. (The 'accidental' degeneracy of the levels with different l for a given n in the purely Coulomb potential $1/r$, which is connected with a dynamical symmetry, is lifted for many-electron atoms where the potential r-dependence is modified.) However, in the periodic table this sequence is violated several times. This results in the formation of large periods of transition elements with partially filled $d(f)$-shells and peculiar physical properties.

To first time, such a situation occurs in the fourth period where filling of $4s$-states starts in potassium, the $3d$-shell remaining empty. This tendency holds also in calcium (configuration of valence electrons is $4s^2$), and filling of $3d$-shell starts only for the next element, i.e.scandium (configuration $3d4s$). This element opens the $3d$ transition group (iron group). The filling of the $3d$-shell goes in a not quite regular way (Table 1.1). So, chromium has the atomic configuration $3d^14s^2$ (instead of $3d4s^2$), and copper $3d^{10}4s$ (instead of $3d^94s^2$). One can see that a tendency to formation of configurations d^0, d^5 and d^{10} exists. The stability of these configurations is apparently connected with that they correspond to zero summary orbital momentum of all the d-electrons, i.e. to a spherically symmetric electron density. It should be noted that in some cases copper demonstrates a considerable contribution of d^9s configuration, which is connected with an appreciable s–d hybridization. Therefore, we shall discuss Cu side by side with transition metals.

A situation, similar to $3d$ metals, takes place for $4d$, $5d$, $4f$ and $5f$ transition metals groups (Tables 1.1, 1.2). Thus, the filling of d- and f-shells is delayed, and, after it begins, the electrons from the

Table 1.1. Atomic configurations and ground terms for the free atoms and M^{2+} ions, and third ionization potentials (eV) in *d*-series.

3d	Sc	Ti	V	Cr	Mn	Fe	Co	Ni	Cu
4d	Y	Zr	Nb	Mo	Tc	Ru	Rh	Pd	Ag
5d	La	Nb	Ta	W	Re	Os	Ir	Pt	Au
3d	d^1s^2	d^2s^2	d^3s^2	d^5s	d^5s^2	d^6s^2	d^7s^2	d^8s^2	$d^{10}s$
4d	d^1s^2	d^2s^2	d^4s	d^5s	d^5s^2	d^7s	d^8s	d^{10}	$d^{10}s$
5d	d^1s^2	d^2s^2	d^3s^2	d^4s	d^5s^2	d^6s^2	d^7s^2	d^9s	$d^{10}s$
3d	$^2D_{3/2}$	$^3F_2$	$^4F_{3/2}$	$^7S_3$	$^6S_{5/2}$	$^5D_4$	$^4F_{9/2}$	$^3F_4$	$^2S_{1/2}$
4d	$^2D_{3/2}$	$^2F_2$	$^6D_{1/2}$	$^7S_3$	$^6D_{9/2}$	$^5F_5$	$^4F_{9/2}$	$^1S_0$	$^2S_{1/2}$
5d	$^2D_{3/2}$	$^3F_2$	$^4F_{3/2}$	$^5D_0$	$^6S_{5/2}$	$^5D_4$	$^4F_{9/2}$	$^4D_3$	$^2S_{1/2}$
M^{2+}	d^1	d^2	d^3	d^4	d^5	d^6	d^7	d^8	d^9
3d, 4d, 5d	$^2D_{3/2}$	$^3F_2$	$^4F_{3/2}$	$^5D_0$	$^6S_{5/2}$	$^5D_4$	$^4F_{9/2}$	$^3F_4$	$^2D_{5/2}$
3d	24.75	27.47	29.31	30.95	33.69	30.64	33.49	35.16	36.83
4d	20.51	22.98	25.04	27.13	31.9	28.46	31.05	32.92	34.82
5d	19.18	28.1	22.3	24.1	26	25	27	28.5	30.5

Table 1.2. Atomic configurations and ground terms for free atoms and R^{3+} ions of rare earths and actinides. Rare earths and heavy actinides (starting from Am) are characterized by the most stable valence of 3+. Beside that, the 4+ state is possible for Ce, Tb and Pr, and the 2+ state for Nd, Sm, Eu, Tm and Yb. Light actinides exhibit a large variety of valence states.

	La	Ce	Pr	Nd	Pm	Sm	Eu
R	f^0d	f^1d	f^3	f^4	f^5	f^6	f^7
	$D^2_{3/2}$	H^3_4	$I^4_{9/2}$	I^5_4	$H^6_{5/2}$	F^7_0	$S^8_{7/2}$
R^{3+}	f^0	f^1	f^2	f^3	f^4	f^5	f^6
	$^1S_0$	$^2F_{5/2}$	$^3H_4$	$^4I_{9/2}$	$^5I_4$	$^6H_{5/2}$	$^7F_0$
	Gd	Tb	Dy	Ho	Er	Tm	Yb
R	f^7d	f^8d	f^{10}	f^{11}	f^{12}	f^{13}	f^{14}
	$^9D_2$	$^8H_{17/2}$	$^5I_8$	$^4I_{15/2}$	$^3H_6$	$^2F_{7/2}$	$^1S_0$
R^{3+}	f^7	f^8	f^9	f^{10}	f^{11}	f^{12}	f^{13}
	$^8S_{7/2}$	$^7F_6$	$^6H_{15/2}$	$^5I_8$	$^4I_{15/2}$	$^3H_6$	$^2F_{7/2}$

	Ac	Th	Pa	U	Np	Pu	Am	Cm	Bk	Cf
R	f^0d^1	f^0d^2	f^2d^1	f^3d^1	f^5d^0	f^6d^0	f^7d^1	f^7d^1	f^8d^1	$f^{10}d^0$
	$^2D_{3/2}$	$^3F_2$	$^4K_{11/2}$	$^2L_6$	$^6H_{5/2}$	$^7F_0$	$^8S_{7/2}$	$^9D_2$	$^8H_{17/2}$	$^5J_8$
R^{3+}	f^0	f^1	f^2	f^3	f^4	f^5	f^6	f^7	f^8	f^9
	$^1S_0$	$^2F_{5/2}$	$^3H_4$	$^4I_{9/2}$	$^5I_4$	$^6H_{5/2}$	$^7F_0$	$^8S_{7/2}$	$^7F_6$	$^6H_{15/2}$

$4s$, $4p$ shells with higher energies do not pass into 'free' $d(f)$-states. These phenomena are connected with inapplicability of the simple one-electron picture, which is based on the hydrogen atom theory, for many-electron atoms.

Consider the radial atomic potential $V_l(r)$, 'felt' by electrons. This is an effective potential which is obtained after averaging two-particle interaction among electrons and including the rotational motion energy (centrifugal potential):

$$V_l(r) = -\frac{Z(r)}{r} + \frac{l(l+1)}{r^2}, \tag{1.1}$$

where $Z(r)$ is an effective nuclear charge, which depends on the electron coordinate r (measured in Bohr radia). For the hydrogen atom we have $Z(r)=1$ and $V_l(r)$ has the usual form with a minimum at $r_0=l(l+1)$ (Fig.1.1). This picture holds in hydrogen-like atoms where the condition $Z(r)$ = const is satisfied to high accuracy. However, in more complicated situations the dependence $Z(r)$ becomes important because of non-uniform screening of nuclear potential. It is this dependence that may lead to an anomalous form of the function $V_l(r)$ for a large centrifugal term (i.e. sufficiently large values of l). In particular, for

$$Z(r) = \begin{cases} A/r, & r_1 < r < r_2, \\ \text{const}, & r < r_1, \quad r > r_2, \end{cases} \tag{1.2}$$

$V_l(r)$ may have two minima separated by a potential barrier. Concrete calculations demonstrate that $V(r)$ may even become positive in an interval $[r_1,r_2]$. The results for Ba and La are shown in Fig.1.2. In fact, the interval $[r_1,r_2]$ corresponds to the position of orbits of $5s$- and $5p$-electrons which screen strongly the nucleus. In other words, strong repulsion between $4f$ and s, p-electrons occurs in the

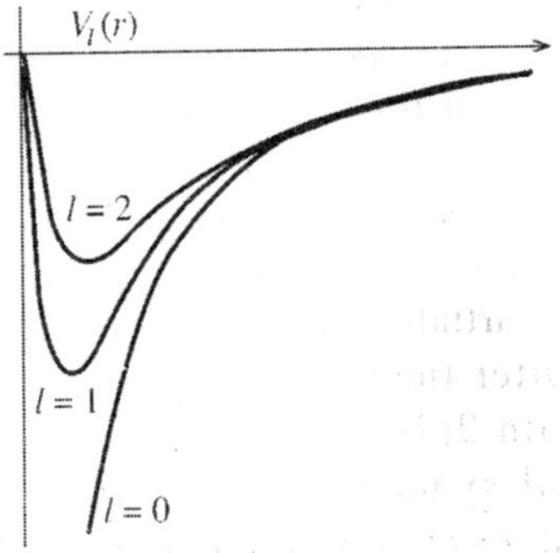

Fig.1.1. The effective potential for a hydrogen-like atom.

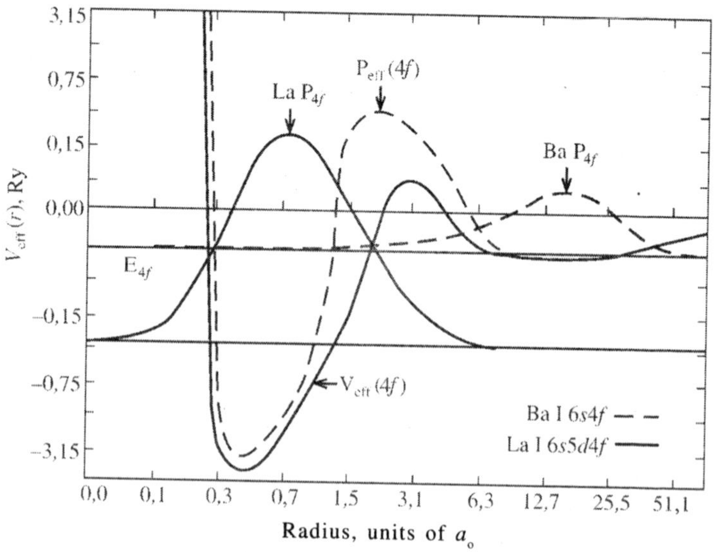

Fig.1.2. The effective potential V_{eff} and radial wavefunctions $P_{4f}(r)$ for BaI (broken lines) and LaI (solid lines) [18].

region of localization of the latter.

In the presence of two minima, the electron density may be concentrated in any one, depending on the form of $Z(r)$. The energy of the corresponding states is, generally speaking, considerably different, as one can see from Fig.1.2. At passing from Ba to La, the lanthanide collapse of 4*f*-states, i.e. a sharp decrease of the 4*f*-shell radius takes place. Further, for cerium 4*f*-states become more energetically favourable than 5*s*, 5*p* and 6*s* ones, the maximum of 4*f*-electron density passes to the first minimum and filling of 4*f*-levels starts.

Thus the main reason of irregular filling of atomic levels are interelectron interaction and appreciable value of the orbital energy at $l \neq 0$. These effect become important from $l=2$ and are strongly pronounced for rare earths ($l=3$). In Fig.1.3 the data on energy levels of external electrons and their density distribution are shown for the gadolinium atom.

Now we treat the question about true capacity of $d(f)$-states which remain in transition elements, from the point of view of the simple one-electron theory, partially unfilled (despite filling of higher shells). According to the latter theory, each shell with the orbital quantum number l may contain $2(2l+1)$ electrons, their energies being equal because of spherical symmetry. However, this is is not true when interelectron correlations are taken into account. This may be performed in the many-configuration approximation [20] where a Hartree–

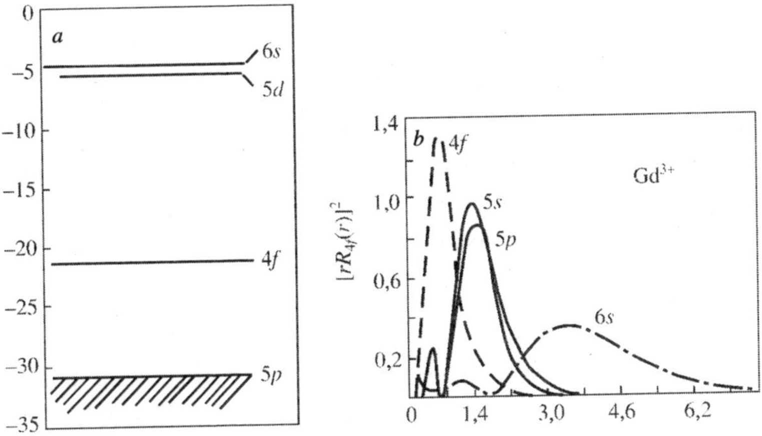

Fig.1.3. (a) energy levels (eV) of external atomic electrons in the Gd atom [19] (b) radial distribution of the Hartree–Fock charge density for $4f$-, $5s$-, $5p$- and $6s$-electrons of the Gd^{3+} ion [16].

Fock-type potential, which depends on electron configuration, is introduced.

We present a simple illustration of this effect. Write down the energy of n electrons in a l-shell (the principal quantum number is dropped for brevity)

$$E_l^n = n\varepsilon_l + \frac{1}{2}n(n-1)Q_l, \tag{1.3}$$

where ε_l is the sum of the kinetic energy and additive part of the potential energy, Q_l is the interelectron repulsion for the shell. The energies of the configurations l^{n+1} and $l^n l'$ (e.g., $d(f)^{n+1}$ and $d(f)^n s$) read

$$\begin{aligned} E_l^{n+1} &= (n+1)\varepsilon_l + \frac{1}{2}n(n+1)Q_l, \\ E_l^n &= n\varepsilon_l + \varepsilon_{l'} + \frac{1}{2}n(n-1)Q_l + nQ_{ll'}, \end{aligned} \tag{1.4}$$

with $Q_{ll'}$ being the repulsion between l and l'-electrons. One can see that the configuration l has the higher energy provided that

$$Q_l - Q_{ll'} > (\varepsilon_{l'} - \varepsilon_l)/n. \tag{1.5}$$

In transition elements, the intrashell repulsion is much stronger than the intershell one, $Q_{sd(f)} \ll Q_{d(f)}$, and the difference $\varepsilon_s - \varepsilon_{d(f)}$ is not too large. Therefore the maximum possible filling of the $d(f)$–shell is not energetically favourable.

The atomic picture of $4f$-electrons which are well localized holds in metals. A possible exception is given by cerium, which opens the 4*f*-series, so that the *f*-electron lies apparently near the centrifugal potential barrier. The γ–α transition in the metallic Ce is assumed now to be connected with the tunneling of the *f*-electron through the barrier and its appreciable delocalization (see 6.5). A similar situation occurs in Sm, Nd and Pr under high pressure of order of 1Mbar [21]. Besides that, in some rare earth compounds *f*-electrons become partially delocalized because of hybridization mixing with conduction electrons.

The description of *d*-states in solids is rather difficult. Unlike 4*f*-electrons, they demonstrate both 'localized' and itinerant features. Such a behaviour is determined by the corresponding atomic potential, which differs essentially from that for *s*, *p*-electrons. The presence of a potential barrier results in an appreciable localization of *d*-states and decrease of overlap between *d*-functions at different lattice sites. However, the density of *d*-electrons still lies partially outside the potential barrier. As a result, their kinetic energy turns out to be considerable, and the corresponding bandwidth values are comparable with those for *s*, *p*-electrons.

The second factor, which determines the *d*-electron spectrum, is the Coulomb repulsion between them. This depends strongly on the number of *d*-electrons inside the barrier. Therefore the effective one-electron energies and wavefunctions should be rather sensitive to the many-electron configuration $d^{n\pm1}s^{1\mp1}$, and the degeneracy is lifted:

$$\varepsilon_d = \varepsilon_d(d^n), \qquad \varepsilon_d(d^n) < E_F < \varepsilon_d(d^{n+1}) \simeq \varepsilon_d(d^n) + Q \tag{1.6}$$

(the Fermi energy E_F determines the band filling). In Appendix C we discuss these effects in more detail by using many-electron Hamiltonian and angular momentum theory.

Owing to increase of the *d*-shell radius, the localization degree of *d*-electrons decreases when we move in the periodic Table both from right to left and from up to below. Thus 3*d*-electrons are considerably more localized than 4*d*- and especially 5*d*-electrons. This explains the fact that magnetic ordering, which is connected with existence of pronounced local moments, occurs for 3*d*-metals only. A similar difference in the localization degree takes place between 4*f*-electrons in rare earths and 5*f*-electrons in actinides. In light actinides (from Th to Pu) 5*f*-electrons form wide energy bands and may be considered as itinerant, and in actinides with higher atomic numbers they are localized. However, the delocalization in Am, Cm, Bc and

Cf may occur under pressure [22, 23].

1.2 ATOMIC AND BAND APPROACHES IN THE TRANSITION ELEMENT THEORY

Most important peculiarity of transition metals from the theoretical point of view is an important role of electron correlations. There exist a number of approaches to the problem of treating many-electron (ME) systems with strong correlations. The first one was the self-consistent field (Hartree–Fock) approximation at solving the Schroedinger equation for ME atoms, which yielded satisfactory quantitative results. The Hartree–Fock method allows to take into account ME atomic terms, but its full version requires the solution of a complicated system of integro-differential non-linear equations [20], so that its direct generalization on solids (systems of large number of atoms) is hardly possible. Main successes of the solid state theory were connected with first-principle one-electron band structure calculations. Modern versions of this approach, which are based on the spin-density functional method (2.3), enable one to obtain a precise description of ground state characteristics [24].

At the same time, the band theory is wittingly insufficient for strongly localized *f*-states, and also at treating some physical phenomena, e.g. magnetism (especially at finite temperatures) and metal-insulator (Mott-Hubbard) transition [25]. Such oversimplified versions of the Hartree-Fock method, as the Stoner mean-field theory of itinerant magnetism, turned out to be not too successful. Later, some shortcomings of the Stoner theory were improved by semi-phenomenological spin-fluctuation theories [26], which, however, do not take into account in most cases correlation effects in the ground state. To treat electron correlations, various perturbation approaches in the electron-electron interaction (e.g., diagram techniques) were used. Besides that, the Fermi-liquid theory was proposed to describe systems with strong electron interactions [27]. However, this theory is violated in the cases where the correlations result in a reconstruction of the ground state, e.g. for systems with a Mott–Hubbard gap [25].

Another approach to the problem of electron correlations was proposed by Hubbard [28–31] who considered the simplest microscopic model including the strong on-site Coulomb repulsion. Starting from the atomic limit, Hubbard carried out a decoupling of Green's functions and obtained an interpolation solution, describing both atomic and band limits for *s*-states [28] and the simplest model of degenerate band [29]. In the subsequent paper [30] this solution was improved to describe the metal-insulator transition. In the paper [31] Hubbard proposed a general formalism of ME X-operators (atomic representation) which enables one to take into account intraatomic interactions in the zeroth-order approximation. This formalism for real atomic configurations is considered in Appendix A.

It is clear from the physical point of view that interelectron correlations are most important for the electrons of the same atomic shell (the equivalent electrons). The modern form of the theory of atomic spectra is based on the Racah formalism for angular momenta (see, e.g., [20]). In some cases, calculations may be considerably simplified by using the irreducible tensor operator technique in the second quantization representation [32] (related questions are considered in Appendix B). This powerful mathematical technique introduces the representation of many-electron quantum numbers $\Gamma=\{SL\mu M\}$ instead of one-electron ones, $\gamma=\{lm\sigma\}$,

$$\mathbf{S}=\sum_i \mathbf{s}_i, \qquad \mathbf{L}=\sum_i \mathbf{l}_i \tag{1.7}$$

being the total angular spin and orbital momenta and μ and M their projections. Then numerous possible combinations of γ-sets for a partially occupied shell are replaced by the sets of Γ. The total number of ME state is the same, but the energy degeneracy is lifted, so that in most physical problems one can retain only the lowest ME term. According to Hund's rules, this term corresponds to maximum L and S. Within this approach, the problem of electrostatic interaction in the system is reduced to calculating a few Slater integrals $F^{(p)}$ (Appendix C). These integrals may be calculated with the use of atomic wavefunctions [33] or determined from experimental data on energies of atomic terms (see [34]).

Information on the energy of the Hund atomic terms may be obtained from the values of atomic ionization potentials. Table 1.1 presents the values of third ionization potentials (of M^{2+} ions) in d-rows, which characterize the binding of d-electrons (first and second ionization potentials correspond as a rule to moving away of s-electrons from the d^ns^2 configuration). One can see that with increasing atomic number the binding in the d-rows increases. Such a dependence explains the corresponding increase of d-electron localization in solids, which is discussed in previous Section.

In solids the values of the Slater integrals become modified. In the recent paper [35] the lowering $\Delta F^{(p)}/F^{(0)}$, which is due to the change in correlation effects (in particular, in interconfiguration interaction) in a crystal, was calculated. At p=2, the values of this quantity of 0.22 and 0.16 were obtained for $Ti^{2+}(d^2)$ and Ni^{2+} (d^8) ions, respectively.

Besides the band approach, the ME approach with inclusion of the atomic spectroscopy mathematical technique should be very useful for the development of quantitative theory of solids. Such an approach is applicable in the case of strong electron localization and especially effective for equivalent electron shells of *d*- and *f*-ions with a complicated term structure. It is important for the solid state theory that the second quantization representation (unlike the standard atomic theory) enables one to consider processes with changing electron number in a shell or an atom (Appendix C). An account of band energy dependence on many-electron quantum numbers corresponds to the degenerate Hubbard model (Appendices C,H).

The localization degree of 4*f*-electrons is sufficient to justify using the atomic picture. On the other hand, for *d*- and 5*f*-electrons the crystal field effects result in at least partial quenching of orbital momenta and destruction of atomic terms. However, atomic structure is important for local effects (e.g., in the case of *d*-metal impurities). Atomic description may be useful also at considering some physical phenomena in periodic crystals (e.g., strong magnetism with pronounced local magnetic moments, formation of an insulating state in *d*-compounds).

An interesting many-electron term effect for a pure metal is the 6eV satellite in the XPS spectrum of nickel which may be attributed to the multiplet structure of the d^8-configuration (see also 2.6). Another example is provided by Auger spectra for *d*-states which exhibit clearly the term structure. This is due to large localization of *d*-electrons in the presence of highly correlated two-hole atomic states which occur after the core hole decay. The Coulomb interaction of *d*-holes U in Cu is so large that the two-hole final state is split off from the one-hole band states. Therefore one can distinguish in the Cu spectrum (Fig.1.4) all the terms of the d^8 configuration.

A somewhat less pronounced multiplet structure of the *d*-configuration is observed in the Auger spectrum of metallic nickel. The value of U in Ni is of order of the one-hole bandwidth and the two-hole states mix strongly with band states (the initial state is a mixture of d^9 and d^{10} configurations). Figure 1.5 shows an interpretation of the L_3VV spectrum in terms of the multiplet structure corresponding to the main peak ^1G with

$$U(^1G) = F^{(0)} + \frac{4}{49}F^{(2)} + \frac{1}{441}F^{(4)}. \tag{1.8}$$

The comparison of XPS and L_3VV spectra for nickel and its compounds

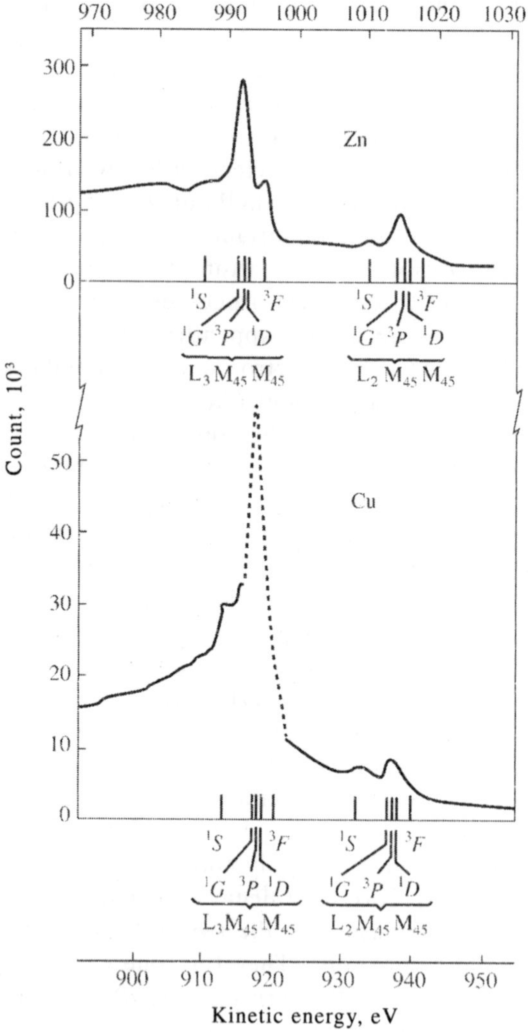

Fig.1.4. Experimental $L_{2,3}M_{45}M_{45}$ Auger spectra from copper and zinc. Atomic multiplet structures are shown as lines under spectra [36].

[37] permitted to estimate the value of U and determine its dependence on crystal surrounding. Experimental energies of the XPS satellite correlate with the energy of the L_3VV peak in all the cases. A fit of the Slater integral yields

$$F^{(0)} = 1.7, \qquad F^{(2)} = 9.6, \qquad F^{(4)} = 6.4,$$

(in eV) which leads to

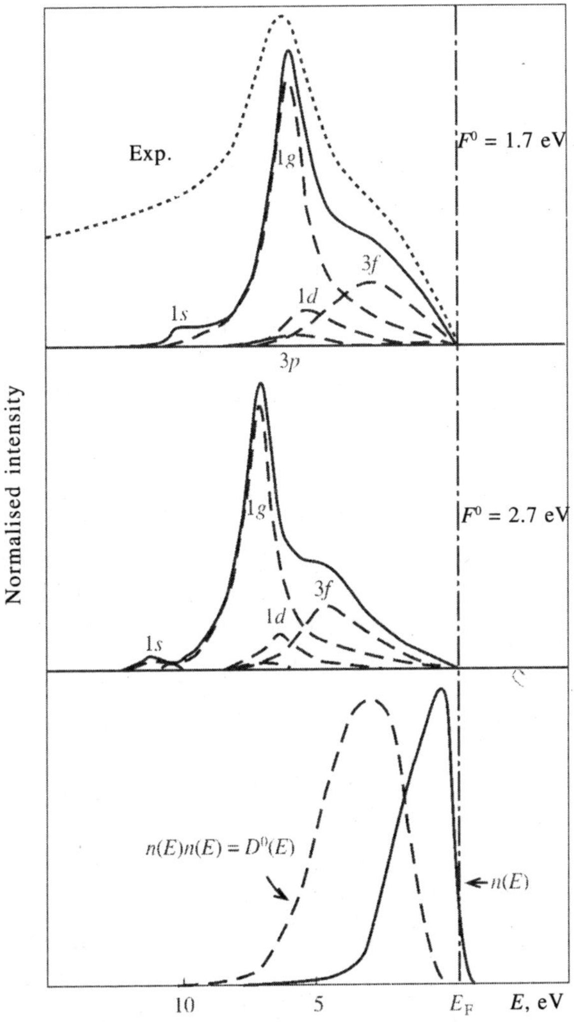

Fig.1.5. Effect of multiplets on the L_3VV Auger spectra of nickel. Lower panel: XPS VB after subtraction of satellite $N(E)$ and its self-fold $D^0(E)$. Middle and top panels: experimental spectrum and calculated contributions from individual d^8 terms with different values of Slater integrals [37].

$$U(^1S) = 6.3, \qquad U(^1G) = 2.5, \qquad U(^3P) = 7.85,$$
$$U(^1D) = 1.6, \qquad U(^3F) \simeq 0.$$

A considerable decrease in the value of F in comparison with atomic one demonstrates strong screening of the Coulomb interaction by

Table 1.3. Probabilities of Auger transitions in Cu (10^{-4} a.u.) [38].

Term	^1S	^1G	^3P	^1D	^3F
Multiplicity	1	9	9	5	21
$L_3M_{45}M_{45}$	1.15	2.88	1.39	7.40	16.18
$L_2M_{45}M_{45}$	0.58	21.43	0.69	3.69	8.10

conduction electrons. The overlap of the atomic ^3F state with band-like states may influence appreciably electronic properties of nickel.

Besides the energy characteristics, the transition probabilities in spectral measurements are also of interest for investigating the ME term structure features in the solid state. The probabilities of the Auger processes $L_{2,3}M_{45}M_{45}$ for the d^8 multiplet of Cu were calculated in [38] with the use of modified wavefunctions of continuum Auger electrons (using the plane waves yielded a drastic disagreement with experimental data). The results are presented in Table 1.3. One can see that the probabilities are not proportional to term multiplicities, which indicates a considerable change of atomic states in the metal.

The values of Slater integrals with account of Coulomb interaction screening by extra-atomic electrons in two-hole states, which were obtained in [38], are

$$F^{(2)} = 10.0, \qquad F^{(4)} = 5.6.$$

They are appreciably suppressed in comparison with the atomic values

$$F^{(2)} = 11.65, \qquad F^{(4)} = 7.18.$$

1.3 CRYSTAL FIELD AND ORBITAL MOMENTA IN SOLIDS

As discussed above, electrostatic interactions in free transition-element atoms result in occurrence of partially filled $d(f)$-shells, their states being characterized by many-electron quantum numbers. Now we consider the role of extra interactions which are present in crystals containing such atoms. In particular, we discuss the problem of role of orbital magnetic momenta in solids.

The orbital momentum L is one of principal atomic characteristics. This quantum number determines the ground state of the atom according to the second Hund's rule and the general picture of excited terms. Existence of orbital magnetic momenta (OMM) in ME atomic configurations should result in a number of important effects in solids. In particular, one might expect the occurrence of the orbital magnetism with large magnetic momenta and strong anisotropy. Such a situation does take place in rare-earth systems [16, 39] where magnetism is determined by the total angular momentum $\mathbf{J} = \mathbf{L}+\mathbf{S}$. Orbital momenta play also an important role in actinides because of strong spin-orbital coupling.

In most d-metals and their compounds, the contribution of OMM to physical properties is considerably smaller (e.g., for Ni the investigation of L absorption edge [40] yielded l=0.05μ_B). This is explained by the fact that the d-electrons are influenced by the crystal field (CF) considerably stronger than well-localized 4f-electrons. The ME levels of d-ions are split by CF, so that OMM become quenched: $\langle \mathbf{L} \rangle = 0$ in not too strong magnetic fields. However, other interactions in the crystal unquench them partially, and (even small) OMM that arise are important for the anisotropy of physical properties due to coupling to the lattice. In particular, strong anisotropy of paramagnetic susceptibility owing to the orbital contribution takes place in a number of hexagonal transition metals (see a detailed discussion in Sect.4.2).

In the problem of quenching of orbital momenta, three basic interactions should be taken into account: the Coulomb interaction V_Q, the spin-orbit coupling V_{so} and the CF potential V_{cf}. There occur three cases. For a weak crystal field where

$$V_{cf} \ll V_{so} \ll V_Q, \tag{1.9}$$

we have the coupling scheme

$$\sum_i \mathbf{s}_i = \mathbf{S}, \qquad \sum_i \mathbf{l}_i = \mathbf{L}, \qquad \mathbf{L}+\mathbf{S} = \mathbf{J}. \tag{1.10}$$

Then OMM are unquenched (the whole quantum number set is $\{SLJM\}$). The coupling between $\mathbf{L}$ and $\mathbf{S}$ is rigid, and CF orients the total momentum $\mathbf{J}$ in some crystallographic directions, which results in a large magnetic anisotropy. The weak field approximation is valid for rare-earth ions where 4f-shells are strongly screened from V_{cf} by s, p, d shells, and the value of V_{so} is appreciable due to the large nuclear charge Z.

For the intermediate crystal field

$$V_{so} \ll V_{cf} \ll V_Q, \tag{1.11}$$

L and **S** still persist, but the momentum **J** is not formed because of CF splitting. Then OMM are oriented in easy crystal directions and spin momenta are coupled to them by a weak spin-orbit coupling. At inclusion of an external magnetic field H in a hard direction, the deviation of spin momentum from the easy direction will be proportional to the ratio $\mu_B H/V_{cf}$. The orbital contribution to magnetization will be negligible because of smallness of the ratio $\mu_B LH/V_{cf}$ in any realistic fields. The intermediate-field situation apparently takes place for d-impurities in some salts [41] and for some non-metallic d-compounds. This case is compatible with orbital ordering, which exists in a number of d-systems [42].

Finally, for the strong crystal field

$$V_{so} \ll V_Q \ll V_{cf} \tag{1.12}$$

the Russel–Saunders coupling between d-electrons is totally destroyed. Then their energy levels are to be considered as one-electron ones which are split by CF. The Coulomb interaction should be now treated as a perturbation for the states with a new symmetry, and orbital momenta are transformed into quasimomenta which are determined by the corresponding irreducible representations of the point group. The strong field situation is usually assumed to take place for transition metals and their alloys. However, the estimation $V_{cf} \sim V_Q$ seems to be more realistic in such systems. This leads to difficulties at developing a consistent quantitative theory.

The CF theory is well developed for paramagnetic ions in non-metallic crystals where a general qualitative analysis of the d-shell structure can be made on the basis of the local point group, corresponding to a given site. General expansion of the CF potential in spherical functions has the form

$$V_{cf} = \sum_{\lambda\mu} A_\lambda^\mu \langle r^\lambda \rangle Y_{\lambda\mu}(\theta, \phi). \tag{1.13}$$

For a l-shell, it is sufficient to retain in (1.13) the terms with $\lambda \le 2l$ only; λ is even in the presence of inversion symmetry. The crystal potential can be also expanded in cubic harmonics V_λ^μ. For $\lambda=2$

$$V_2^0 = 3z^2 - r^2, \qquad V_2^2 = x^2 - y^2,$$

$$V_2^1 = xz, \qquad V_2^{-1} = yz, \qquad V_2^{-2} = xy. \tag{1.14}$$

In many cases, it is convenient to use the method of Stevens equivalent momentum operators which have the same matrix elements as spherical functions. For example,

$$V_2^0 = \alpha_J \langle r^2 \rangle [3(J^z)^2 - J(J+1)], \qquad V_2^2 = \frac{1}{2}\alpha_J \langle r^2 \rangle [(J^+)^2 + (J^-)^2],$$

$$V_2^1 = V_2^2 = \frac{1}{2}\alpha_J \langle r^2 \rangle (J^x J^z + J^z J^x),$$

where α_J (and similar factors β_J, γ_J for λ=4.6) are the Stevens proportionality coefficients which depend on the configuration l^n [16, 41, 43]. The Stevens operator representation holds beyond the multiplet with a given J (or L) only. Of course, the operator (1.5), unlike (1.7), has also matrix elements between different terms. The method of equivalent operators in problems with changing J or L is more complicated (see [41]).

Consider concrete examples of d^n configurations in the crystal field. Since various situations may take place for d-electrons in CF, constructing interpolation schemes is useful [43].

The picture of the two-hole spectrum for the Ni^{2+} ion (configuration d^8) is shown in Fig.1.6. We use Bethe's notations $^{2S+1}\Gamma_i$ for irreducible representations of the cubic group: $\Gamma_1(A_{1g})$ and $\Gamma_3(E_g)$ are one-dimensional, $\Gamma_4(T_{1g})$ is two-dimensional, Γ_4 (T_{1g}) and $\Gamma_5(T_{2g})$ are three-dimensional. An arbitrary position of one-electron levels is chosen; an analogue of the Hund rule takes place for the effective orbital momentum in CF [41]. One can see that the initial orbital momentum L=3 is quenched in intermediate crystal field provided that the lowest state is the orbital singlet $^3\Gamma_2$. For three-dimensional 'Hund' representations $^3\Gamma_4$ and $^3\Gamma_5$, quenching is absent. In the strong CF, electron orbital momenta are quenched provided that the lowest one-electron state is e_g (which corresponds to metallic nickel). We see that, a regrouping of e_g and t_{2g} states takes place at increasing Coulomb interaction, which may result in a new ground state with unquenched OMM.

The ion Fe^{2+} (configuration d^6) possesses the lowest term ^5D in the intermediate CF and the configuration $(e_g)^4(t_{2g})^2$ in the strong field (the case of the configuration $(t_{2g})^6$ is trivial). The corresponding interpolation scheme is shown in Fig.1.7. We see that in the intermediate CF quenching takes place only provided that the level $^3\Gamma_3$ is lower.

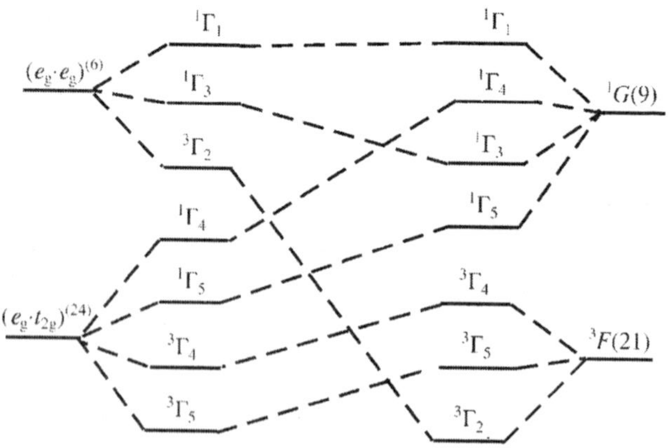

Fig.1.6. Interpolation scheme of levels for $Ni^{2+}(d^8)$ ion in the crystal field. The numbers in brackets stand for degeneracy of a state.

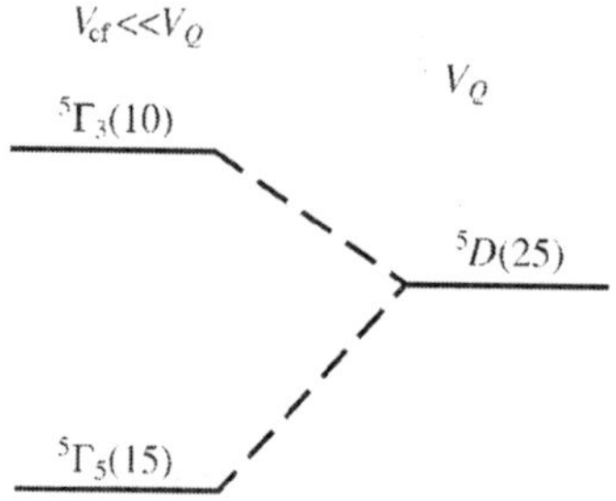

Fig.1.7. Fragment of level scheme for $Fe^{2+}(d^6)$ ion. Position of levels in the ground state corresponds to unquenched orbital momenta for both strong and intermediate CF. Unlike Fig.1.6, interpolation is not shown, since not all corresponding levels are included.

In the strong CF, the state $^3\Gamma_5$, corresponding to three-dimensional t_{2g} representation, has an unquenched orbital quasimomentum.

The examples considered demonstrate that, generally speaking, in the case of a high lattice symmetry a local CF results in only partial degeneracy lift which cannot provide total quenching of OMM. Even in the case where such a quenching does take place, OMM may be partially unquenched by the Coulomb interaction which has off-diagonal matrix elements between different point group irreducible representations, both for intermediate and strong CF cases, e.g., the matrix elements of the type $\langle e_g|Q|t_{2g}\rangle$ unquench OMM in the state e_g. On the other hand, in the case of intermediate CF, off-diagonal matrix elements of the crystal field itself may mix different terms and unquench OMM of the ground state term.

Thus the local CF potential does not explain almost total quenching of OMM in transition metals and their alloys. The quenching mechanism owing to the periodic lattice potential will be considered in Sect.4.8.1. This potential results in an appreciable kinetic energy of *d*-electrons. The formation of rather wide energy *d*-bands leads also to destruction of localized spin momenta and suppression of magnetism in most *d*-metals.

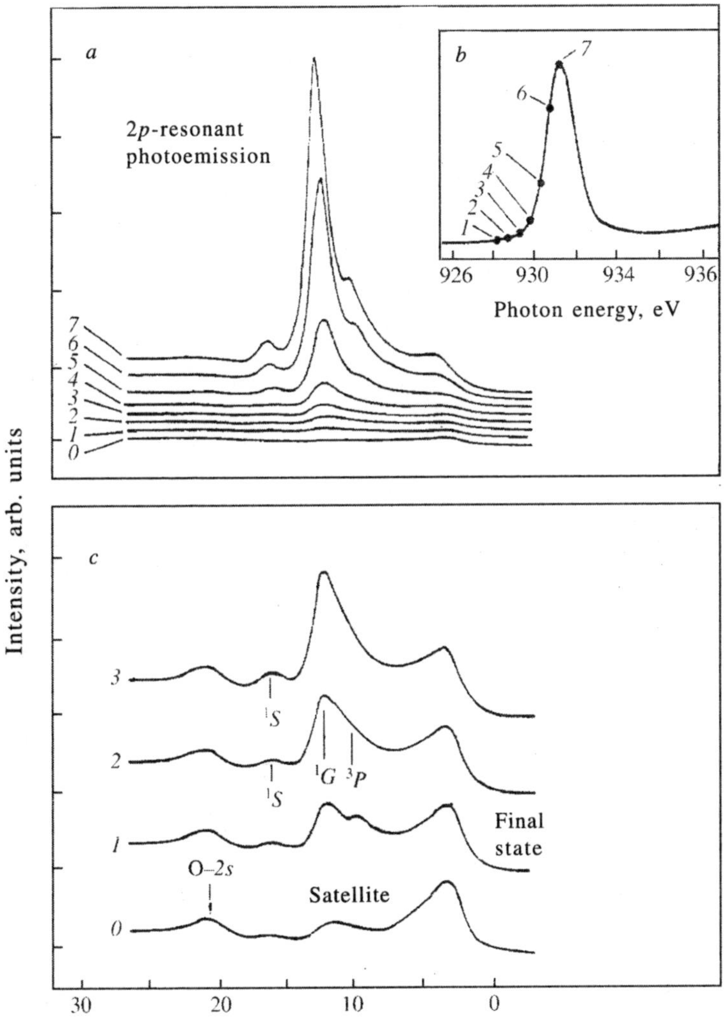

Fig.1.8. Resonance photoemission experiment on CuO (a) spectra with the photon energies sweeping to the point of the $2p$ resonance (b), the resonance intensity (c) and the number of spectra on the extended scale [47].

The role of OMM effects in highly-correlated d-compounds is debatable. Often the description of the electronic structure of Mott insulators (e.g, transition metal oxides and sulphides) is performed within the density functional approach, which does not take into account ME term structure [44, 45]. At the same time, there exist attempts of interpretation of their optical properties within a picture including spectroscopic terms with account of CF splitting [46, 13]. Detailed investigations of optical and X-ray spectra (Sect.2.5), which contain information on selection rules in L, seem to be useful for solving these problems. It should be noted that the orbital contributions may be different for the configurations $s(p)d^{n-1}$ and d^n which are mixed by the crystal field.

The resonance photoemission spectrum of CuO is shown in Fig.1.8. The resonance intensity allows the identification of the various atomic multiplets of the d^8 final state. Figures 1.9 and 1.10 [45] show a comparison of the XPS valence band spectra of MnO and NiO with the results of cluster calculations. The parameters of cluster calculations are shown in diagrams. One can see from Fig.1.10 the strong hybridization between the d^7 and d^8L^{-1} final state.

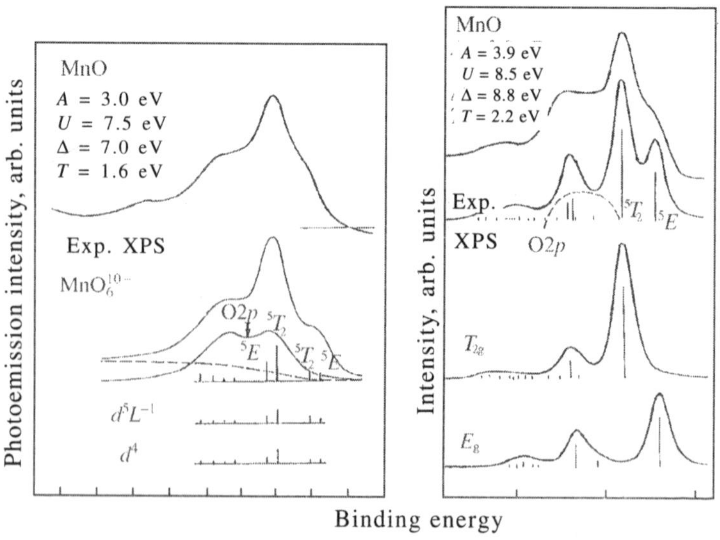

Fig.1.9. XPS-valence band spectrum of MnO and analysis of the spectra by cluster calculations of two groups [48,49].

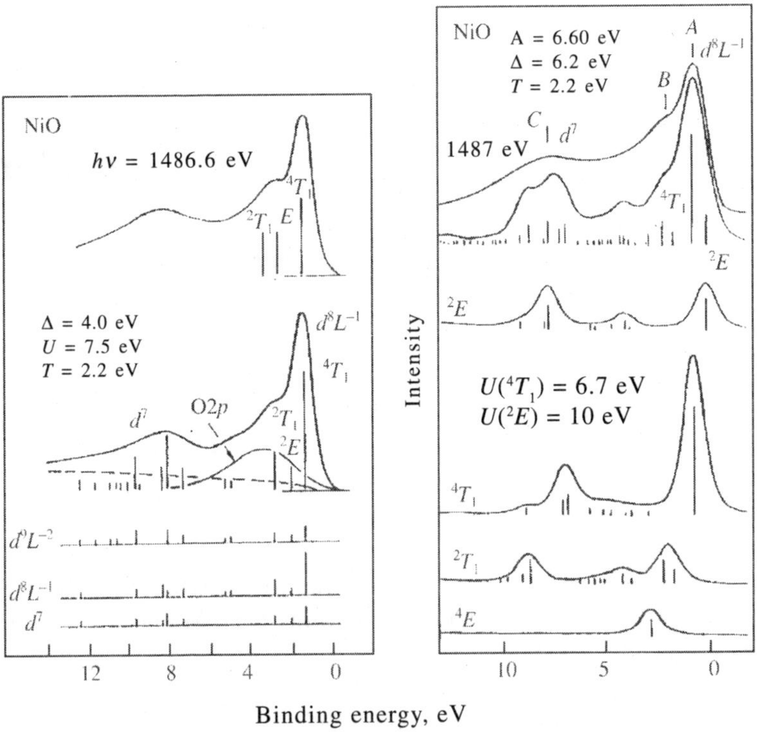

Fig.1.10. Comparison of experimental XPS valence band spectra of NiO with the results of cluster calculations [50, 51].

Chapter 2

BAND THEORY

Band theory is now a large and rather independent branch of solid state physics, which uses the whole variety of modern computational methods. The corresponding problems are discussed in detail in a number of monographs and reviews [12, 13, 52–57]. Nevertheless, we include this material in the book with two main purposes (i) to consider in a simple form main band calculation methods bearing in mind their physical basis (ii) to concretize the advantages and drawbacks of various methods as applied to transition metals. Further we discuss the results of band calculations for TM and compare them with experimental data.

Band theory deals with description of electron spectrum in a regular lattice with a periodic crystal potential,

$$V(\mathbf{r}) = V(\mathbf{r}+\mathbf{R}) = \sum_{\mathbf{R}} \upsilon(\mathbf{r}-\mathbf{R}),$$

where $\upsilon(\mathbf{r})$ is the potential of an ion, **R** are the lattice sites. According to the Bloch theorem, one-electron states in a periodic potential may be chosen in the form

$$\psi_{n\mathbf{k}}(\mathbf{r}) = u_{n\mathbf{k}}(\mathbf{r})e^{i\mathbf{k}\mathbf{r}}, \tag{2.1}$$

They are classified by the quasimomentum $u_{n\mathbf{k}}(\mathbf{r})$ and form a set of energy bands with the index n.

At considering the band spectrum, two opposite approaches are possible. The first one starts from the picture of isolated atoms and uses the crystal potential of neighbour atoms as a perturbation, the intraatomic potential being taken into account in the zeroth-order approximation. The simplest method of band calculations is the tight-binding method which uses one-electron atomic wave functions $\phi_\gamma(\mathbf{r})$. The corresponding wavefunctions of a crystal are written as

$$\phi_{\gamma\mathbf{k}}(\mathbf{r}) = \sum_{\mathbf{R}} \phi_\gamma(\mathbf{r}-\mathbf{R})e^{i\mathbf{k}\mathbf{R}}. \tag{2.2}$$

The band spectrum in this approximation contains a set of bands originating from atomic levels. In the case of the s-band for the simple cubic lattice in the nearest-neighbour approximation we have

$$E_{\mathbf{k}} = \varepsilon_s + 2\beta(\cos k_x + \cos k_y + \cos k_z), \tag{2.3}$$

where

$$\beta = \int \phi^*(\mathbf{r}) v(\mathbf{r}) \phi(\mathbf{r} + \mathbf{a}) d\mathbf{r}. \tag{2.4}$$

is the transfer integral. The width of the energy bands is determined by the overlap of atomic wavefunctions at neighbour lattice sites and decreases rapidly for inner shells. As a rule, the bands, which originate from different levels, overlap considerably. Taking into account off-diagonal matrix elements and combining the atomic functions with $\gamma \neq \gamma'$ we come to the method of linear combination of atomic orbitals (LCAO). This method was widely used in early band calculations. A version of this method was developed by Slater and Coster [58]. An important drawback of the LCAO method is non-orthogonality of atomic functions at different lattice sites (see Apendix C) and absence of delocalized states with positive energies in the basis. Thus it is difficult to describe weakly localized electron states in metals within this method.

The opposite limiting case corresponds to the picture of nearly free electrons with a large kinetic energy, so that the whole periodic crystal potential may be considered as a perturbation which results in formation of gaps (forbidden bands) in the energy spectrum. Contrary to the tight-binding approximation, the strength of crystal potential determines the widths of gaps rather than of electron bands.

Free electrons have the wavefunctions and the energy spectrum

$$\psi_{\mathbf{k}}(\mathbf{r}) = e^{i\mathbf{k}\mathbf{r}} \equiv |\mathbf{k}\rangle, \qquad E_{\mathbf{k}} = \frac{\hbar^2 k^2}{2m}. \tag{2.5}$$

The matrix elements of the periodic lattice potential are given by

$$\langle \mathbf{k} | V | \mathbf{k}' \rangle = \sum_{\mathbf{g}} V_{\mathbf{g}} \delta_{\mathbf{k}-\mathbf{k}',\mathbf{g}}, \tag{2.6}$$

where $\mathbf{g}$ are reciprocal lattice vectors which are defined by

$$\mathbf{gR} = 2\pi n, \quad n = 0, \pm 1, \ldots \tag{2.7}$$

For most values of $\mathbf{k}$, corrections to the spectrum may be calculated to second order of perturbation theory

$$\Delta E_{\mathbf{k}} = \sum_{\mathbf{g}} \frac{|V_{\mathbf{g}}|^2}{E_{\mathbf{k}} - E_{\mathbf{k}-\mathbf{g}}}. \tag{2.8}$$

However, the denominators of the terms in (2.8) with $E_{\mathbf{k}} = E_{\mathbf{k}-\mathbf{g}}$, i.e.

$$2(\mathbf{kg}) = g^2, \tag{2.9}$$

vanish and we have to calculate the spectrum with the use of the Brillouin–Wigner perturbation theory which yields

$$\begin{vmatrix} E - E_{\mathbf{k}} & V_{\mathbf{g}}^* \\ V_{\mathbf{g}} & E - E_{\mathbf{k-g}} \end{vmatrix} = 0, \tag{2.10}$$

or

$$E = \frac{1}{2}(E_{\mathbf{k}} + E_{\mathbf{k-g}}) \pm \left[\frac{1}{4}(E_{\mathbf{k}} - E_{\mathbf{k-g}})^2 + |V_{\mathbf{g}}|^2\right]^{1/2}. \tag{2.11}$$

Thus we obtain the splitting of the spectrum with formation of the energy gap

$$\Delta E_{\mathbf{g}} = 2|V_{\mathbf{g}}|. \tag{2.12}$$

The equation (2.9) just determines the boundaries of the Brillouin zone where the gaps occur.

The nearly-free electron approximation is applicable provided that the crystal potential is small in comparison with the bandwidth W. Since the matrix elements of crystal potential are of order of a few eV, this condition may be satisfied for external s- and p-electrons with $W \sim 10$ eV. On the other hand, d-electrons form more narrow bands and disturb considerably the conduction band because of s–d hybridization. Thus this approximation describes satisfactorily the conduction electron states in simple (in particular, alcaline) metals only.

In most cases both above pictures are insufficient to obtain a quantitative description of the electron spectrum. Especially difficult is this problem for d-electrons in transition metals, which are characterized by an intermediate degree of localization.

It should be noted that the shortcomings of the simplest band calculation methods occur because of the approximate solution of the problem in any real calculations. Using an infinite full set either of localized or extended wavefunctions would in principle provide an exact result (of course, in the one-electron approximation only). However, the use of a physically reasonable basis enables one to obtain much better results for a finite set. Evidently, the true wavefunctions should have an atomic character in the region near the lattice site and are close to plane waves in the outer space. Therefore modern methods of band calculations (e.g., the methods of augmented and orthogonalized plane waves), which are discussed briefly in the next sections, use combinations of atomic functions and plane waves.

The second difficulty of the band theory is connected with the problem of electron correlations, which are large for transition metals. The Coulomb interaction among electrons, which has a two-particle nature and prevents reducing the many-electron problem to the one-electron one. In principle, the possibility of such a reducing is provided by the Hohenberg–Kohn theorem [59]. This theorem guarantees the existence of an unique density functional which yields the exact ground state energy. However, explicit constructing of this functional is a very complicated problem. In the situation, where correlations are not too strong, the Coulomb interaction may be taken into account by introducing a self-consistent potential which depends on electron density. These questions are treated in Sect.2.3.

The correlation effects are especially important for narrow *d*- and *f*-bands. The strong intrasite Coulomb repulsion may lead to splitting of one-electron bands into many-electron subbands, and the usual band description (even in the tight-binding approach) is inapplicable. In particular, in the case of one conduction electron per atom the Mott–Hubbard transition takes place with increasing interatomic distances, so that electrons become localized at lattice sites. In such a situation, we have to construct the correlation Hubbard subbands with the use of the atomic statistics and many-electron quantum numbers (Appendices C, H). Development of first-principle band calculation methods in the many-electron representation is an exciting problem. Such calculations might be based on a modification of the eigenfunctions of a spherically symmetric potential, which are used, e.g., in APW and OPW methods, by expressing them in terms of many-electron functions of the atomic problem. In particular, the dependence of one-electron radial functions on many-electron quantum numbers S, L occurs in the full Hartree–Fock approximation (see Sect.2.3). Picking out the functions, which correspond to lowest (Hund) atomic terms should modify considerably the results of band calculations.

2.1 ORTHOGONALIZED PLANE WAVE METHOD AND PSEUDOPOTENTIAL

The main shortcoming of the nearly-free electron method is a poor description of strong oscillations of electron wavefunctions near atomic cores in terms of not too large a set of plane waves. This shortcoming is removed in the orthogonalized plane wave method (OPW) where the wavefunctions are chosen to be orthogonal with respect to core states:

$$\chi_{\mathbf{k}}(\mathbf{r}) = e^{i\mathbf{kr}} + \sum_c b_c \phi^c_{\mathbf{k}}(\mathbf{r}). \tag{2.13}$$

where $\phi^c_{\mathbf{k}}(\mathbf{r})$ are the tight-binding Bloch functions originating from the core levels. The condition

$$\int \chi_{\mathbf{k}}(\mathbf{r})\phi^c_{\mathbf{k}}(\mathbf{r})d\mathbf{r} \equiv \langle \chi | c \rangle = 0 \tag{2.14}$$

yields

$$b_c = -\int \phi^c_{\mathbf{k}}(\mathbf{r})e^{-i\mathbf{kr}}\, d\mathbf{r} \equiv -\langle \mathbf{k} | c \rangle. \tag{2.15}$$

The orthogonalized plane wave (2.13) oscillates rapidly near atomic nucleus and is close to the plane wave between the atoms. The trial wavefunction of the crystal is chosen as a superposition of OPW:

$$\psi_{\mathbf{k}} = \sum_{\mathbf{g}} \alpha_{\mathbf{k}-\mathbf{g}} \chi_{\mathbf{k}-\mathbf{g}} \tag{2.16}$$

(**g** are reciprocal lattice vectors). The coefficients α are determined from the Schrödinger equation. The matrix elements of the crystal potential for OPW (unlike those for plane waves) turn out to be small, so that it is sufficient to take in the expansion (2.16) not too many terms. It should be noted that the OPW basis is overfilled so that the increasing the number of functions in the set does not necessarily increase the accuracy of calculations.

Further development of the OPW method led to the idea of introducing a weak pseudopotential which permits (unlike the real crystal potential) the use of perturbation theory. To this end one introduces the pseudowavefunction

$$\tilde{\psi}_{\mathbf{k}} = \sum_{\mathbf{g}} \alpha_{\mathbf{k}-\mathbf{g}} e^{i(\mathbf{k}-\mathbf{g})\mathbf{r}} = \psi_{\mathbf{k}} + \sum_c \langle c | \tilde{\psi} \rangle \phi^c_{\mathbf{k}}, \tag{2.17}$$

which coincides with the true wavefunction outside the core region but does not exhibit oscillations, which are due to strong core level potential, within it. Substituting (2.17) into the usual Schroedinger equation $H\psi = E\psi$ we get

$$H\tilde{\psi}_{\mathbf{k}} + \sum_c (E - E_c)\langle c | \tilde{\psi} \rangle \phi^c_{\mathbf{k}} = E\tilde{\psi}_{\mathbf{k}}, \tag{2.18}$$

which may be represented in the form of a new Schrödinger equation

$$\left(-\frac{\hbar^2}{2m}\Delta + W \right)\tilde{\psi}_{\mathbf{k}} = E\tilde{\psi}_{\mathbf{k}}. \tag{2.19}$$

where the non-local energy-dependent pseudopotential operator W is defined by

$$W = V(\mathbf{r}) + V^R; \tag{2.20}$$

$$V^R\tilde{\psi}_{\mathbf{k}}(\mathbf{r})=\sum_c (E-E_c)\langle c|\tilde{\psi}\rangle\phi^c_{\mathbf{k}}. \tag{2.21}$$

For $E > E_c$, the matrix elements of the potential V^R

$$\int\tilde{\psi}^*_{\mathbf{k}}(\mathbf{r})V^R\tilde{\psi}_{\mathbf{k}}(\mathbf{r})d\mathbf{r}=\sum_c (E-E_c)|\langle c|\tilde{\psi}\rangle|^2 \tag{2.22}$$

are positive. At the same time, the matrix elements of the true periodic potential V are negative and large in the absolute value in the core region. Thus the diagonal matrix elements of W turn out to be positive and small because of partial cancelation of two terms in (2.20). In particular, the pseudopotential does not result, unlike the real crystal potential, in formation of bound states, corresponding to core levels. Therefore the pseudopotential method describes conduction electron states only.

Generally speaking, the pseudopotential approach is not simpler than the original OPW method since it requires introducing a non-local potential function which is not defined in a unique way. Therefore one uses frequently empirical or model local pseudopotentials which are fitted to experimental data, rather than first-principle ones. Perturbation theory with the use of such potentials enables one to describe the whole variety of properties of simple metals [55].

The standard form of the pseudopotential theory is inapplicable both in noble and transition metals because d-electrons are much stronger localized than s, p-electrons. To describe d-electron states one introduces into (2.17) the contribution from functions [9]

$$|\psi\rangle=|\tilde{\psi}\rangle-\sum_c\langle c|\tilde{\psi}\rangle|c\rangle+\sum_d a_d|d\rangle, \tag{2.23}$$

where, as well as for simple metals, the pseudowavefunction $\tilde{\psi}$ is a linear combination of plane waves. One has to take into account the difference between potentials of free ion and of ion in a crystal,

$$\delta v(\mathbf{r})=v^{\mathrm{ion}}(r)-v(\mathbf{r}).$$

Define the operator V^{h} by

$$V^{\mathrm{h}}|d\rangle=\delta v|d\rangle-|d\rangle\langle d|\delta v|d\rangle. \tag{2.24}$$

Then we obtain

$$a_d=-\langle d|\tilde{\psi}\rangle+\frac{\langle d|V^{\mathrm{h}}|\tilde{\psi}\rangle}{E_d-E},$$

and the Schrödinger equation for the pseudowavefunction takes the form

$$\left(-\frac{\hbar^2}{2m}\Delta+W\right)|\tilde{\psi}\rangle+\sum_d\frac{V^{\rm h}|d\rangle\langle d|V^{\rm h}|\tilde{\psi}\rangle}{E-E_d}=E|\tilde{\psi}\rangle, \qquad (2.25)$$

where the pseudopotential operator of the transition metal is given by

$$W|\tilde{\psi}\rangle=V|\tilde{\psi}\rangle+\sum_{i=c,d}(E-E_i)|i\rangle\langle i|\tilde{\psi}\rangle+\sum_d(|d\rangle\langle d|V^{\rm h}|\tilde{\psi}\rangle+V^{\rm h}|d\rangle\langle d|\tilde{\psi}\rangle). \qquad (2.26)$$

The most important peculiarity of Eq.(2.25) is the presence of the terms, which have a resonance denominator $E–E_{\rm d}$. The physical origin of this is as follows. As discussed in Sect.1.1, the atomic potential for d-electrons may have a centrifugal barrier where $v(r)$ is positive. Therefore there exist bound d-levels merged in the continuous spectrum. The strong scattering of continuous spectrum electron by localized levels takes place with the phase shift of $\pi/2$. Such a situation may be also described by the $s–d$ hybridization model. The corresponding energy spectrum is strongly distorted and should be calculated by exact diagonalization which yields

$$E_{\mathbf{k}}^{1,2}=\frac{1}{2}\left(\frac{\hbar^2k^2}{2m}+E_d\right)\pm\left[\frac{1}{4}\left(\frac{\hbar^2k^2}{2m}-E_d\right)^2+|V_{\mathbf{k}}^h|^2\right]^{1/2}. \qquad (2.27)$$

Far above the resonance the hybridization potential may be considered, as well as the usual pseudopotential, within the framework of perturbation theory. This permits a satisfactory description within the modified pseudopotential approach [60] of noble metals where the d-resonance, although crossing conduction band, lies well below $E_{\rm F}$.

In transition metals the resonance lies near the Fermi energy and perturbation theory is inapplicable. Thus one has to pick out the singularity, corresponding to the resonance. A method of completely orthogonalized plane waves [61], application of which to transition metals is discussed in [57, 62], gives in principle a possibility to eliminate this difficulty. However, this method is rather complicated and goes far beyond the original concept of the OPW method. Generally, such approaches require in a sense exact diagonalization of a matrix and lose in fact the main advantage of the pseudopotential idea – applicability of perturbation theory. More recent applications of the pseudopotential method for band calculations of transition metals are considered in [63, 64].

There exist a number of model pseudopotential forms, in particular including explicitly the d-resonance. However, they do not provide as a rule sufficiently satisfactory description of TM. Recently a model pseudopotential was successfully applied to explain properties of iridium [65]. The reasons of the good fit to experimental data for this particular d-metal are not quite clear.

2.2 AUGMENTED PLANE WAVE (APW) AND KORRINGA–KOHN–ROSTOKER (KKR) METHODS

The idea of other modern band calculation methods is to describe electron states in different space regions in a proper way using different bases. To this end, one builds around each site an atomic sphere with the volume which is somewhat smaller than that of the lattice cell, so that spheres centered at different sites do not overlap. The crystal potential at each site is taken in the so-called muffin-tin (MT) form

$$v_{\mathrm{MT}}(\mathbf{r}) = \begin{cases} v(r), & r < r_{\mathrm{MT}}, \\ v_c, & r > r_{\mathrm{MT}}. \end{cases} \tag{2.28}$$

Inside the sphere the potential is supposed to be spherically symmetric, and the corresponding wavefunctions for a given energy may be expanded in spherical harmonics

$$\phi(\mathbf{r}) = \sum_{lm} C_{lm} R_l(r, E) Y_{lm}(\hat{\mathbf{r}}), \quad r < r_{\mathrm{MT}} \tag{2.29}$$

(R_l are the solutions of the corresponding radial Schrödinger equation) which permits to simplify greatly the calculations due to using the angular momentum technique.

Since beyond the spheres the MT potential is constant, the basis wavefunctions $\phi(\mathbf{r})$ may be chosen in the form of plane waves. Using the expansion (C.28) we may join continuously the solutions at the sphere boundaries and determine the coefficients $C_{lm} = C_{lm}(\hat{\mathbf{k}})$

$$C_{lm} = i^l (2l+1) \frac{j_l(kr_{\mathrm{MT}})}{R_l(r_{\mathrm{MT}}, E)} Y^*_{lm}(\hat{\mathbf{k}}). \tag{2.30}$$

Then the function (2.29) is called the augmented plane wave (APW). The wavefunction of the crystal are searched as a linear combination of APW's with the same energy:

$$\psi_{\mathbf{k}}(\mathbf{r}) = \sum_{\mathbf{g}} \alpha_{\mathbf{k-g}} \phi_{\mathbf{k-g}}(\mathbf{r}). \tag{2.31}$$

Substituting (2.31) into the Schrödinger equation one obtains the system of linear equations. Poles of its determinant yield the dispersion law $E(\mathbf{k})$ which contains the crystal potential in a complicated way. To obtain accurate results, it is sufficient to restrict oneself by not too large number of terms in (2.31). Practically, one uses up to 100 APW's, and for a larger set the solution is as a rule stabilized.

The application of the APW method to d-bands requires an account of a larger number of APW's in comparison with, s,p-bands. However, such band calculations were performed for all the transition d-metals [53, 54]. For $4d$- and $5d$-metals, relativistic effects (in particular, spin-orbital interaction) are important. The corresponding version of the APW method (the RAPW method) was developed by Loucks [66].

The Green's function method, developed by Korringa, Kohn and Rostoker (KKR), uses, as well as APW method, the expansion in spherical functions inside the MT spheres (2.28). At the same time, the wavefunctions outside them are constructed not from plane waves, but also from spherical waves which are scattered from other sites. Formally, this is achieved by using the integral form of the Schrödinger equation ($\kappa = |E|^{1/2}$)

$$\psi(\mathbf{r}) = -\frac{1}{4\pi}\int \frac{\exp(i\kappa|\mathbf{r}-\mathbf{r}'|)}{|\mathbf{r}-\mathbf{r}'|}V(\mathbf{r}')\psi(\mathbf{r}')d\mathbf{r}'. \tag{2.32}$$

Using the Bloch theorem condition

$$\psi_{\mathbf{k}}(\mathbf{r}+\mathbf{R}) = \psi_{\mathbf{k}}(\mathbf{r})e^{i\mathbf{k}\mathbf{R}},$$

one reduces (2.32) to the form

$$\psi_{\mathbf{k}}(\mathbf{r}) = -\frac{1}{4\pi}\int G_{\mathbf{k}}(\mathbf{r}-\mathbf{r}', E)v(\mathbf{r}')\psi_{\mathbf{k}}(\mathbf{r}')d\mathbf{r}', \tag{2.33}$$

where the integration goes over one lattice cell and

$$G_{\mathbf{k}}(\mathbf{r}, E) = \sum_{\mathbf{R}} \frac{\exp(i\kappa|\mathbf{r}-\mathbf{R}|)}{|\mathbf{r}-\mathbf{R}|}e^{i\mathbf{k}\mathbf{R}}. \tag{2.34}$$

is the lattice Green's function.

The functional describing the system and corresponding to (2.33) has the form

$$\Lambda = \int|\psi_{\mathbf{k}}(\mathbf{r})|^2 d\mathbf{r} + \frac{1}{4\pi}\iint \psi^*_{\mathbf{k}}(\mathbf{r})v(\mathbf{r})G_{\mathbf{k}}(\mathbf{r}-\mathbf{r}', E)v(\mathbf{r}')\psi_{\mathbf{k}}(\mathbf{r}')d\mathbf{r}d\mathbf{r}'. \tag{2.35}$$

Substituting (2.29) into (2.35) yields [52]

$$\Lambda = \sum_{lm,l'm'} (A_{lm,l l'm'} + \kappa\delta_{ll'}\delta_{mm'}\cot\eta_l)C_{lm}C_{l'm'}. \tag{2.36}$$

Here η_l is the l-dependent phase shift owing to the MT-potential,

$$\cot \eta_l = \frac{\kappa r_{MT} n'_l(\kappa r_{MT}) - D_l(E) n_l(\kappa r_{MT})}{\kappa r_{MT} j'_l(\kappa r_{MT}) - D_l(E) j_l(\kappa r_{MT})}, \tag{2.37}$$

where j_i and n_l are the spherical Bessel and Neumann functions,

$$D_l(E) = r_{MT} R'_l(r_{MT}, E)/R_l(r_{MT}, E). \tag{2.38}$$

are the logarithmic derivatives of the radial wave function. The energy spectrum is determined by zeros of the determinant of the matrix Λ. The structural constants A are given by

$$A_{lm,l'm'} = 4\pi\kappa \sum_{\mathbf{R}\neq 0} e^{i\mathbf{kR}} \sum_{l''m''} [n_{l''}(\kappa R) - ij_{l''}(\kappa R)] Y^*_{l''m''}(\hat{\mathbf{R}}) \tilde{C}^{l'm'}_{lm,l'',m''} \tag{2.39}$$

(the quantities $\tilde{C}$ are defined in (C.9)). These constants do not depend on crystal potential and may be calculated once for all for a given lattice.

The advantage of the KKR method in comparison with the APW one is the decoupling of structural and atomic factors. For the same lattice potentials, the KKR and APW methods yield usually close results. The intimate relation of these methods is discussed in [52].

The main difficulty of the KKR method is the energy dependence of the structural constants. However, this dependence may be eliminated if we use so called atomic sphere approximation (ASA) where the volume of the sphere in the MT potential (2.28) is put to be equal to the lattice cell volume, $r_{MT} = s$, so that the volume of the intersphere region vanishes. Then one puts for simplicity the kinetic energy beyond the atomic spheres, $E-v_c$ to be zero, and the ASA–KKR equations take a very simple form [56]

$$\det | S_{lm,l'm'}(\mathbf{k}) - P_l(E)\delta_{ll'}\delta_{mm'} | = 0, \tag{2.40}$$

where the potential function is given by

$$P_l(E) = 2(2l+1)\frac{D_l(E)+l+1}{D_l(E)-l}, \tag{2.41}$$

and the structural constants are energy independent,

$$S_{lm,l'm'}(\mathbf{k}) = (4\pi)^{1/2}(-1)^{m'+1}(-i)^{\lambda}\frac{(2l'+1)(2l+1)}{(2\lambda+1)}\times$$

$$\times\frac{(\lambda+\mu)!(\lambda-\mu)!}{(l'+m')!(l'-m')!(l+m)!(l-m)!}\sum_{\mathbf{R}\neq 0} e^{i\mathbf{kR}} Y^*_{\lambda\mu}(\hat{\mathbf{R}})(s/R)^{\lambda+1}, \tag{2.42}$$

where

$$\lambda = l + l', \qquad \mu = m - m'.$$

The wavefunctions corresponding to the MT-potential with zero kinetic energy $E-\upsilon_c$ may be chosen in the form

$$\phi_{lm}(\mathbf{r},E)=i^l Y_{lm}(\hat{\mathbf{r}})\phi_l(r,E), \tag{2.43}$$

$$\phi_l(r,E)=R_l(r,E)\begin{cases}1, & r<s,\\ \dfrac{D_l+l+1}{2l+1}\left(\dfrac{r}{s}\right)^l+\dfrac{l-D_l}{2l+1}\left(\dfrac{r}{s}\right)^{-l-1}, & r>s.\end{cases}$$

where the coefficients are determined from the condition of joining the tail continuously and smoothly at the boundary of the MT-sphere. It is convenient to subtract the term which describes the diverging spherical wave. Then we obtain the so-called MT-orbitals

$$\chi_l(r,E)=\phi_l(r,E)\begin{cases}\dfrac{\phi_l(r,E)}{\phi_l(s,E)}-\dfrac{D_l+l+1}{2l+1}\left(\dfrac{r}{s}\right)^l, & r<s,\\ \dfrac{l-D_l}{2l+1}\left(\dfrac{r}{s}\right)^{-l-1}, & r>s.\end{cases} \tag{2.44}$$

Condition of cancelation of tails of the MT orbitals, which originate from other sites, leads again to the KKR equations.

A modification of LCAO method which uses instead of atomic functions the MT-orbitals is called the LCMTO method, the linear combinations of MT-orbitals being constructed similar to (2.1). Substituting these trial functions into the Schrödinger equation we obtain the LCMTO equations [67].

In the general APW, KKR and LCMTO methods the matrix elements are functions of energy. Therefore, at calculating eigenvalues one has to compute the determinants in each point of **k**-space for a large number values of E (of order of 100) which costs much time. To simplify the calculation procedure Andersen [67] proposed to expand the radial wavefunctions at some energy value to linear terms in E. Then the Hamiltonian and overlap matrices do not depend on energy. The error owing to the linearization does not as a rule exceed that owing to the inaccuracy in the crystal potential (in particular, owing to the MT-approximation). The accuracy may be increased by account of higher-order terms in the expansion. The linear methods (LMTO and LAPW) permitted to perform band calculations of a large number of transition metal compounds with complicated crystal structures [57].

Using the MT-orbital basis gives a possibility to carry out band calculations beyond MT-approximation for the potential itself. Such an approach (the full-potential LMTO method) enables one to considerably the accuracy of calculations. Very good re... .re achieved also within the framework of the full-potential LAPW method [68].

2.3 THE HARTREE-FOCK-SLATER AND DENSITY FUNCTIONAL APPROACHES TO THE PROBLEM OF ELECTRON CORRELATIONS

The above consideration of the electron spectrum was performed within the one-electron approximation. Main purpose of the calculation methods discussed was joining of atomic-like solutions near the lattice sites with wavefunctions in the outer space where the crystal potential $V(\mathbf{r})$ differs strongly from atomic one. The Coulomb interaction among conduction electrons was assumed to be included in $V(\mathbf{r})$. Such an inclusion should be performed in a self-consistent way. In the simplest approximation, this may be achieved by introducing into the Schrödinger equation the averaged Coulomb potential,

$$\varepsilon_i \psi_i(\mathbf{r}) = \left[-\frac{\hbar^2}{2m}\Delta + V(\mathbf{r}) + V_C(\mathbf{r}) \right] \psi_i(\mathbf{r}); \tag{2.45}$$

$$V_C(\mathbf{r}) = -e \int \frac{\rho(\mathbf{r}')}{|\mathbf{r}-\mathbf{r}'|} d\mathbf{r}', \qquad \rho(\mathbf{r}) = -e \sum_i{}' |\psi_i(\mathbf{r})|^2, \tag{2.46}$$

where the sum goes over occupied electron states and integration includes summation over spin coordinate.

The Hartree equations (2.45) do not include the exchange interaction which results from antisymmetry of electron wavefunctions. Substituting the antisymmetrized product of one-electron wavefunctions (Slater determinant) (A.2) in the Schrödinger equation for the many-electron system with the Hamiltonian (C.1) and minimizing the average energy

$$\mathcal{E} = \langle \Psi | H | \Psi \rangle / \langle \Psi | \Psi \rangle,$$

we obtain the Hartree–Fock equations

$$\varepsilon_i \psi_i(\mathbf{r}) = \left[-\frac{\hbar^2}{2m}\Delta + V(\mathbf{r}) + V_C(\mathbf{r}) \right] \psi_i(\mathbf{r}) - \\ - e^2 \sum_j{}' \psi_j(\mathbf{r}) \int \frac{\psi_j^*(\mathbf{r}')\psi_i(\mathbf{r}')}{|\mathbf{r}-\mathbf{r}'|} \delta_{\sigma_i \sigma_j} d\mathbf{r}', \tag{2.47}$$

where ε_i are the Lagrange multipliers. The exchange term (last term in the left-hand side of (2.47)) has a non-local form, i.e. is not reduced to multiplication of the wavefunction ψ_i by a potential function. It should be noted that the terms, describing the electron self-interaction ($i = j$) are mutually cancelled in the Coulomb and exchange contributions to (2.47). Due to orthogonality of spin wave functions, the exchange interaction occurs between electron with parallel spins only.

Parameters of the Hartree–Fock equation depend, generally speaking, not only on one-electron quantum numbers, but on the whole set of many-electron quantum numbers in a given state. For example, in the many-electron atom, the coefficients at the Slater parameters (except for F^0) in the multipole expansion of the Coulomb interaction (C.19) depend explicitly on the many-electron SL-term. This results in the dependence of the radial wavefuctions on the quantum numbers S,L. The corresponding expression for the parameters ψ_i may be represented as [20]

$$\varepsilon_l = E(l^n SL) - \sum_{S'L'} \left(G^{SL}_{S'L'}\right)^2 \tilde{E}(l^{n-1}S'L'). \tag{2.48}$$

where $\tilde{E}$ is the energy of the 'frozen' ion with the same radial functions as the n-electron atom, G are the fractional parentage coefficients. Evidently, $\tilde{E}$ exceeds the energy of the true ion, which is calculated from the Hartree–Fock equation. In systems with large number of electrons N the differences E–$\tilde{E}$ may be neglected since they give corrections of order of $1/N$ only [9]. The quantity (2.48) differs from the energy, measured in spectral experiments

$$\varepsilon_l(SL, S'L') = E(l^n SL) - \sum_{S'L'} E(l^{n-1}S'L'), \tag{2.49}$$

which corresponds to a concrete transition $SL \rightarrow S'L'$ (experiments with not too high resolution yield the Lorentz broadening which differs from (2.48)).

In the band theory the dependence on ME quantum numbers is usually neglected. Then the parameters ε_i have the meaning of one-electron energies, $-\varepsilon_i$ being equal to the ionization energy of the corresponding state in the crystal (the Koopmans theorem [9]).

Obviously, the Koopmans theorem is inapplicable for partially filled shells in transition metals where the energy of a level depends essentially on filling of other states (see Sect.1.1). The self-consistent field method, which considers one-electron levels only, does not take into account such correlation effects. Nevertheless, simple averaged exchange-correlation potentials, expressed in terms of electron density, are widely used to calculate electron structure of partially filled d-bands in transition metals. When applied to free atoms, this approach yields atomic levels with a non-integer filling. Although not quite physically correct, such a picture permits a satisfactory description of atomic and molecular spectra [69]. To calculate the energy of optical transitions, one takes the self-consistent potential calculated in the so-called transition state which corresponds to the level filling of 1/2.

The solution of the Schrödinger equation with a non-local potential resulting from the exchange interaction is a very complicated problem. The case of free electron may be considered analytically. Performing the integration for plane waves we obtain from (2.46)

$$\varepsilon_{\mathbf{k}}\psi_{\mathbf{k}}(\mathbf{r}) = \left[\frac{\hbar^2 k^2}{2m} + V_{\mathrm{x}}(k)\right]\psi_{\mathbf{k}}(\mathbf{r}) \tag{2.50}$$

with the local exchange potential

$$V_{\mathrm{x}}(k) = -8\left(\frac{3}{8\pi}\rho\right)^{1/3} F(k/k_{\mathrm{F}}), \tag{2.51}$$

where $\rho = N/V$ is the electron density,

$$F(z) = \frac{1}{2} + \frac{1-z^2}{4z}\ln\left|\frac{1+z}{1-z}\right|. \tag{2.52}$$

The logarithmic singularity in $F(z)$, and, consequently, in the electron spectrum results in a non-physical behaviour of the density of states near the Fermi level, $N(E \to E_{\mathrm{F}}) \to 0$. This singularity is due to the long-range character of the Coulomb interaction and should be in fact removed by correlation effects which result in screening. Thus the Hartree–Fock approximation turns out to be insufficient to describe the free electron gas because of the strong k-dependence. To avoid this difficulty Slater [69] proposed to use the potential which describes the motion of an electron in an averaged exchange field.

The exchange potential for a given state may be represented in the form

$$V_{\mathrm{x}i}(\mathbf{r}) = -e^2 {\sum_j}' \delta_{\sigma_i\sigma_j} \int \left(\frac{\psi_i^*(\mathbf{r})\psi_j^*(\mathbf{r}')\psi_j(\mathbf{r})\psi_i(\mathbf{r}')}{|\mathbf{r}-\mathbf{r}'|} \Big/ [\psi_i^*(\mathbf{r})\psi_i(\mathbf{r})] \right) d\mathbf{r}'. \tag{2.53}$$

The expression (2.53) may be interpreted as the potential of a charge with a value of e which is removed from the hole surrounding the i-th electron. We may write for the averaged potential

$$V_{\mathrm{x}}(\mathbf{r}) = -e^2 \int {\sum_{ij}}' \delta_{\sigma_i\sigma_j} \frac{\psi_i^*(\mathbf{r})\psi_j^*(\mathbf{r}')\psi_j(\mathbf{r})\psi_i(\mathbf{r}')}{|\mathbf{r}-\mathbf{r}'|} d\mathbf{r}' \Big/ {\sum_i}' \psi_i^*(\mathbf{r})\psi_i(\mathbf{r}). \tag{2.54}$$

Averaging the free-electron gas potential (2.51) over k we obtain

$$\bar{F} = \int_0^1 F(z) z^2 dz \bigg/ \int_0^1 z^2 dz = \frac{3}{4}; \tag{2.55}$$

$$V_{\mathrm{x}} = -6\left(\frac{3}{8\pi}\rho\right)^{1/3}. \tag{2.56}$$

Starting from the expression (2.56), Slater introduced the local exchange-correlation potential with the same dependence on electron density

$$V_{\mathrm{xS}}(\mathbf{r}) = -6\left(\frac{3}{8\pi}\rho(\mathbf{r})\right)^{1/3}. \tag{2.57}$$

An alternative approach to the problem of constructing exchange-correlation potential [70] uses the calculation of the total energy. Carrying out the integration of one-electron energy for the electron gas one obtains

$$\mathcal{E} = 2\sum_{k<k_{\mathrm{F}}}\left[\frac{\hbar^2 k^2}{2m} + \frac{1}{2}V_{\mathrm{x}}(k)\right] =$$

$$= \frac{e^2}{2a_{\mathrm{B}}}\left[\frac{3}{5}(k_{\mathrm{F}}a_{\mathrm{B}})^2 - \frac{3}{2\pi}k_{\mathrm{F}}a_{\mathrm{B}}\right] = \left[\frac{2.21}{(r_s/a_{\mathrm{B}})^2} - \frac{0.916}{r_s/a_{\mathrm{B}}}\right]\mathrm{Ry}, \tag{2.58}$$

where $\mathrm{Ry} = e^2/2a_{\mathrm{B}} = 13.6$ eV,

$$r_s = \left(\frac{3}{4\pi\rho}\right)^{1/3} \tag{2.59}$$

is the radius of the sphere with the volume which corresponds to one electron for a given uniform density ρ. When measured in the units of Bohr radius a_{B}, r_s is of order of the ratio of potential energy of an electron to its mean kinetic energy. The second summand in (2.58), which comes from the exchange interaction, is comparable with the first one for realistic metallic values $r_s = 2$–6. In the high-density limit the problem of the electron gas may be treated more rigorously by expanding ε in r_s [12]:

$$\mathcal{E} = \left[\frac{2.21}{(r_s/a_{\mathrm{B}})^2} - \frac{0.916}{r_s/a_{\mathrm{B}}} + 0.622\ln\frac{r_s}{a_{\mathrm{B}}} - 0.096 + O\left(\frac{r_s}{a_{\mathrm{B}}}\right)\right]\mathrm{Ry}. \tag{2.60}$$

The corrections in (2.60) describe correlation effects beyond the Hartree-Fock approximation.

Taking into account the exchange energy of the uniform gas per particle and neglecting correlations we obtain the Gaspar–Kohn–

Sham exchange potential [70]

$$V_{\mathrm{GKS}}(\mathbf{r}) = -4\left(\frac{3}{8\pi}\rho(\mathbf{r})\right)^{1/3}. \tag{2.61}$$

This potential differs from the Slater approximation (2.57) by a factor of 2/3. This fact is due to that the operations of varying the total energy and of statistical averaging do not commute. The expression (2.60) may be also obtained from the potential (2.51) by putting $k = k_F$ $(F = 1/2)$.

A more general exchange-correlation potential (xα-potential) may be written as

$$V_{x\alpha}(\mathbf{r}) = V_{xS}(\mathbf{r}). \tag{2.62}$$

The optimization of the parameter $\alpha < 1$ enables one to take into account effects of electron correlations. The potential (2.61) was successfully applied for calculations of electron structure of large number of solids [69,57]. However, the reasons of this success were not clear until the density functional method was developed. The results of xα-calculations for pure metals turn out to be not too sensitive to the choose of α. Earlier calculations with $\alpha = 1$ yielded results, which agreed satisfactory with experimental data and are close to those for optimal α. This is apparently due to that the error in the total energy for $\alpha = 1$ is partially compensated by increasing the depth of the potential well. However, the correct choose of α is important for metallic and insulating compounds. For transition metals the atomic densities are usually taken for $d^{n-1}s$ configuration in the 3d- and 4d-series and for the $d^{n-2}s^2$ configuration in the 5d-series, the optimal values of α lying in the interval 0.65÷0.8.

An important step in the theory of electron correlations in solids was made by Hohenberg and Kohn [59] who proved the general theorem, according to which the ground state energy is an unique functional of electron density. Kohn and Sham [70] represented this density functional (DF) in the form

$$\mathcal{E}[\rho(\mathbf{r})] = \int \rho(\mathbf{r})V_{\mathrm{ext}}(\mathbf{r})d\mathbf{r} + \frac{1}{2}\iint\left(\frac{\rho(\mathbf{r})\rho(\mathbf{r}')}{|\mathbf{r}-\mathbf{r}'|} + T[\rho(\mathbf{r})] + \mathcal{E}_{\mathrm{xc}}[\rho(\mathbf{r})]\right)d\mathbf{r}d\mathbf{r}'. \tag{2.63}$$

where V_{ext} is the external potential of lattice ions, T is the kinetic energy of non-interacting electrons with the density ρ, the functional $\varepsilon_{\mathrm{xc}}$ corresponds to exchange and correlation contributions. The DF approach may be generalized to consider spin-polarized systems.

Representing the electron density in the form (2.46) and varying (2.63) with respect to ψ_i we obtain

$$\varepsilon_i\psi_i(\mathbf{r}) = \left[-\frac{\hbar^2}{2m}\Delta + V_{\mathrm{ext}}(\mathbf{r}) + V_{\mathrm{C}}(\mathbf{r}) + V_{\mathrm{xc}}(\mathbf{r})\right]\psi_i(\mathbf{r}), \tag{2.64}$$

where the Lagrange multipliers ψ_i correspond to the quasiparticle energies, the exchange-correlation potential V_{xc} is defined by varying $\varepsilon_{\mathrm{xc}}$ with respect to density,

$$V_{\mathrm{xc}}(\mathbf{r}) = \delta\mathcal{E}_{\mathrm{xc}}[\rho(\mathbf{r})]/\delta\rho(\mathbf{r}). \tag{2.65}$$

Thus the many-electron problem is reduced to the one-particle one with the local exchange-correlation potential. However, it should be stressed that the DF approach guarantees the correct value of the ground state energy only and describes exactly the distribution of charge and spin densities and related physical quantitities. Here belong elastic characteristics and exchange parameters, which are expressed in terms of corresponding energy differences, saturation magnetic moment etc. On the other hand, the Kohn–Sham quasiparticles do not coincide with true electrons, so that the wavefunctions ψ_i do not have a direct physical meaning. Therefore the possibility of calculation of the whole excitation spectrum is, generally speaking, not justified. Especially doubtful is the description of the states far from E_{F} and of finite temperature behaviour. Nevertheless, the DF method is widely used to consider various physical properties.

The true functional (2.63) is of course unknown, and one has to use some approximations. The formally exact expression for $\varepsilon_{\mathrm{xc}}$ may be represented in the form of the Coulomb interaction of an electron with the exchange–correlation hole which surrounds it:

$$\mathcal{E}_{\mathrm{xc}}[\rho(\mathbf{r})] = \frac{1}{2}\iint \rho(\mathbf{r})\frac{\rho_{\mathrm{xc}}(\mathbf{r},\mathbf{r}'-\mathbf{r})}{|\mathbf{r}-\mathbf{r}'|}d\mathbf{r}d\mathbf{r}'. \tag{2.66}$$

The corresponding electron density may be obtained by integration over the coupling constant λ [71,72]

$$\rho_{\mathrm{xc}}(\mathbf{r},\mathbf{r}'-\mathbf{r}) = \rho(\mathbf{r}')\int_0^1[g(\mathbf{r},\mathbf{r}',\lambda)-1]d\lambda, \tag{2.67}$$

where g is the pair correlation function. Since the hole contains one electron, the charge density should satisfy the sum rule

$$\int\rho_{\mathrm{xc}}(\mathbf{r},\mathbf{r}'-\mathbf{r})d\mathbf{r}' = -e. \tag{2.68}$$

The functional $\varepsilon_{\mathrm{xc}}$ depends in fact on the spherically averaged charge density, which follows from the isotropic nature of the Coulomb interaction [72]. Indeed, the variable substitution in (2.66) allows to perform integration over angles of the vector **r**−**r**':

$$\mathcal{E}_{xc} = \frac{1}{2}\iint \rho(\mathbf{r}) R^2 \frac{\bar{\rho}_{xc}(r,R)}{R} d\mathbf{r}\, dR,$$

$$\bar{\rho}_{xc}(r,R) = \int \rho_{xc}(\mathbf{r},R)\, d\Omega. \tag{2.69}$$

Thus approximate expressions for ε_{xc} can give exact results even if the detailed description of ρ_{xc} (in particular, of its non-spherical part) is inaccurate.

Further we have to concretize the form of the correlation function. In the local density approximation (LDA), which is used as a rule in band calculations, ρ_{xc} is taken in the same form, as for uniform electron density with the replacement $\rho \to \rho(r)$ and the corresponding correlation function g_0:

$$\rho_{xc}(\mathbf{r},\mathbf{r}'-\mathbf{r}) = \rho(\mathbf{r}) \int_0^1 [g_0(|\mathbf{r}-\mathbf{r}'|,\lambda,\rho(r)) - 1]\, d\lambda. \tag{2.70}$$

Substituting (2.70) into (2.69) we obtain the LDA functional, which has the local form

$$\mathcal{E}_{xc}[\rho] = \int \rho(\mathbf{r}) \varepsilon_{xc}(\rho)\, d\mathbf{r}, \tag{2.71}$$

where ε_{xc} is the exchange and correlation energy per electron for the interacting uniform electron gas. This quantity was calculated numerically to a high accuracy and various interpolation analytical formulae are available. Therefore the LDF approach enables one to construct various local exchange-correlation potentials. The simplest exchange potential is given by the xα-expression (2.61). This expression may be generalized as

$$V_{xc}(\mathbf{r}) = \beta(r_s/a_B) V_{GKS}(\mathbf{r}). \tag{2.72}$$

Thus the xα-method becomes justified in the density functional theory, the parameter α being dependent of electron density. The value of α changes from 2/3 to 0.85 as r_s increases from zero (high-density limit) to values which are typical for metals in interatomic region (r_s/a_B = 4). An example of an interpolation approximation is the Hedin–Lundqvist potential [73] which corresponds to

$$\beta_{HL}(z) = 1 + 0.0316 z \ln(1 + 24.3/z). \tag{2.73}$$

The functional of local spin density (LSD) approximation which is applied to consider magnetic transition metals reads

$$\mathcal{E}_{xc}[\rho_\uparrow,\rho_\downarrow] = \int \rho(\mathbf{r}) \varepsilon_{xc}(\rho_\uparrow(\mathbf{r}),\rho_\downarrow(\mathbf{r}))\, d\mathbf{r}. \tag{2.74}$$

The simplest LSDA von Barth–Hedin potential [74] has in the first order in spin polarization $m(\mathbf{r}) = n_\uparrow(\mathbf{r}) - n_\downarrow(\mathbf{r})$ the form

$$V_{\mathrm{xc}}^{\sigma}(\mathbf{r}) = \frac{1}{3}\sigma\delta(n)\frac{m(\mathbf{r})}{n(\mathbf{r})}V_{\mathrm{GKS}}(\mathbf{r}), \tag{2.75}$$

where $n(\mathbf{r}) = n_\uparrow(\mathbf{r}) + n_\downarrow(\mathbf{r})$ and the parameter δ takes into account correlation effects. In the high-density limit $\delta = 1$, and $\delta = 0.55$ for $r_s = 4$.

The LD approximation oversimplifies greatly the general DF approach which is based on the Hohenberg–Kohn theorem. In particular, LDA does not take into account correctly the dependence of $\rho_{\mathrm{xc}}(\mathbf{r}, \mathbf{r}')$ on $\rho(\mathbf{r})$ in the whole space (compare (2.67) with (2.70)). However, it provides good results in the case of a slow varying density $\rho(\mathbf{r})$. Besides that, this approximation works well in the high-density limit since it includes correctly the kinetic energy of the non-interacting system.

In the original papers [70], attempts were made to take into account non-uniform corrections to LDA by expanding ρ_{xc} with respect to grad $\rho(\mathbf{r})$. However, such an expansion yields small contributions only and does not hold in the case of strongly non-uniform electron systems (e.g., for 4*f*-electrons in rare earths and narrow *d*-bands), and is therefore, strictly speaking, not justified for TM. Other modifications of LDA are discussed in the review [72]. The problem of account of multiplet structure in the density functional approach is treated in [75,76].

One of drawbacks of LDA is the inexact (unlike the full Hartree-Fock approximation) cancelation of the unphysical interaction of an electron with itself. In this connection, the so-called self-interaction correction (SIC) approach [72] was developed. The SIC functional within the LSDA reads

$$\mathcal{E}_{\mathrm{SIC}} = \mathcal{E}_{\mathrm{LSD}}[\rho_\uparrow(\mathbf{r}), \rho_\downarrow(\mathbf{r})] - \sum_{i\sigma}\delta\mathcal{E}_{i\sigma}. \tag{2.76}$$

The subtracted SI correction for the orbital $i\sigma$ has the form

$$\delta\mathcal{E}_{i\sigma} = \frac{1}{2}\iint\frac{\rho_{i\sigma}(\mathbf{r})\rho_{i\sigma}(\mathbf{r}')}{|\mathbf{r}-\mathbf{r}'|}d\mathbf{r}d\mathbf{r}' + \mathcal{E}_{\mathrm{xc}}[\rho_{i\sigma}, 0], \tag{2.77}$$

where the first term is the self-interaction energy, and the second term is the LSD approximation to the exchange-correlation energy of a fully spin-polarized system with the density $\rho_{i\sigma}(\mathbf{r})$. The functional (2.76) is exact for a system with one electron. The SI corrections are negligible for plane-wave-like states due to smallness of $\rho_{i\sigma}(\mathbf{r})$.

However, they may be important for description of strongly localized states. The SIC approach was successfully applied to model systems with Hubbard correlations and to transition-metal oxides [72, 77, 100].

2.4 DISCUSSION OF BAND CALCULATION RESULTS

Band structure calculations yield the whole picture of electron spectrum of a given substance. The corresponding dependences $E(\mathbf{k})$ and densities of states may be found in [24, 78]. The calculations of [24] are performed for cubic symmetry only. Besides that, the 5*d*-metals are omitted since they require relativistic methods. In the handbook [78], real crystal structure for all the transition metals was taken into account. The Slater–Coster parametrization of the band spectrum was applied, the basis containing t_{2g} and e_g states being used even for metals with the hexagonal structure. Bibliography on earlier band structure calculations is given in Sect.2.7 (see Table 2.6). Some examples of densities of states for transition metals are shown in Figs.2.1–2.7.

The real accuracy of the detailed and complicated information on the electron spectrum turns out often to be in fact insufficient. Therefore it is instructive to consider some general characteristics of band spectra (classification and widths of energy bands, position of the Fermi level, the value of the density of states $N(E_F)$) within simple model concepts.

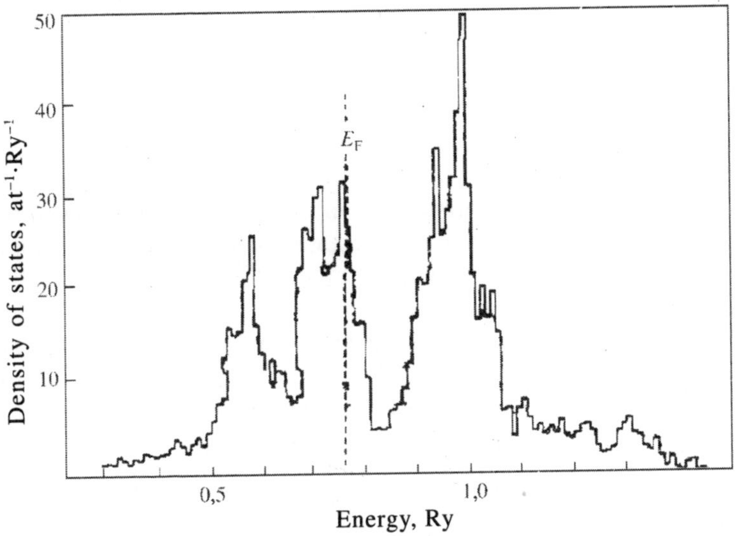

Fig. 2.1. Total density of states for titanium [159].

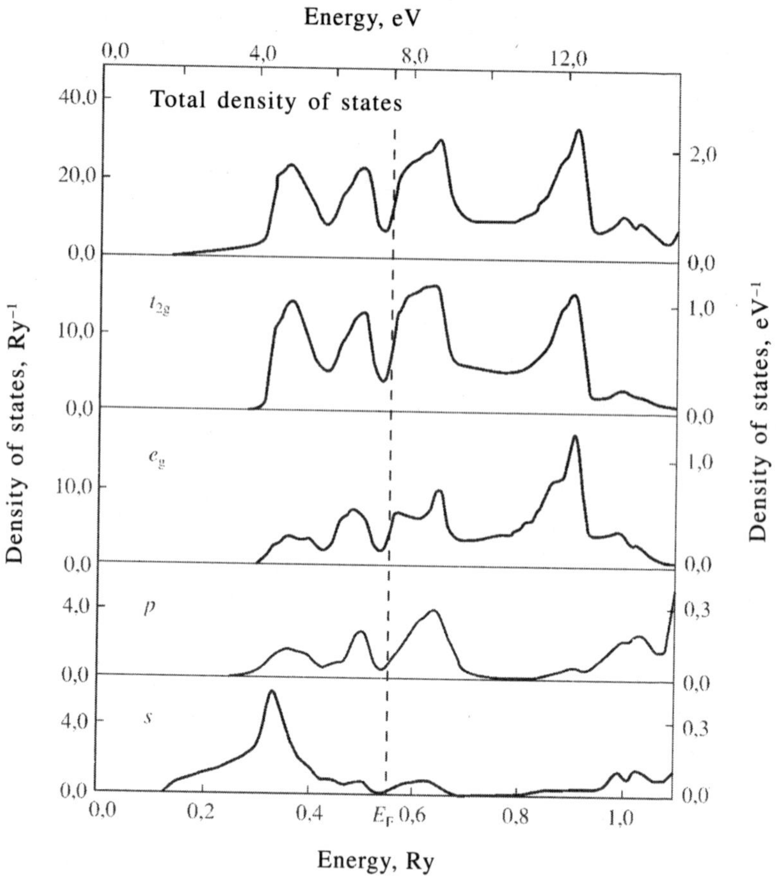

Fig. 2.2. Total and partial densities of states for zirconium [78].

The KKR method gives a possibility to separate the structural and dynamical aspects of the band structure. Using the atomic sphere approximation, which is discussed in Sect.2.2, enables one to obtain so-called canonical bands which depend on the crystal structure only. These bands are obtained by neglecting off-diagonal structural constants with $l \neq l'$ in the KKR equations (2.40), considering the potential functions $P_l(E)$ as independent variables and diagonalizing the matrix $S_{lm,l'm'}(\mathbf{k})$ for each value of l. Then for the set n, l we get $2l$+1 non-hybridized energy bands $E_{ni}(\mathbf{k})$ defined by

$$S_{li}(\mathbf{k}) = P_l(E), \quad i = 1, 2, \ldots, 2l+1. \tag{2.78}$$

The canonical bands $S_l(\mathbf{k})$ for the bcc, fcc and hcp lattices are shown in Figs.2.8–2.10, and the corresponding densities of states in Fig.2.11. Because of the infinite range of s-type MT-orbitals, a pure canonical

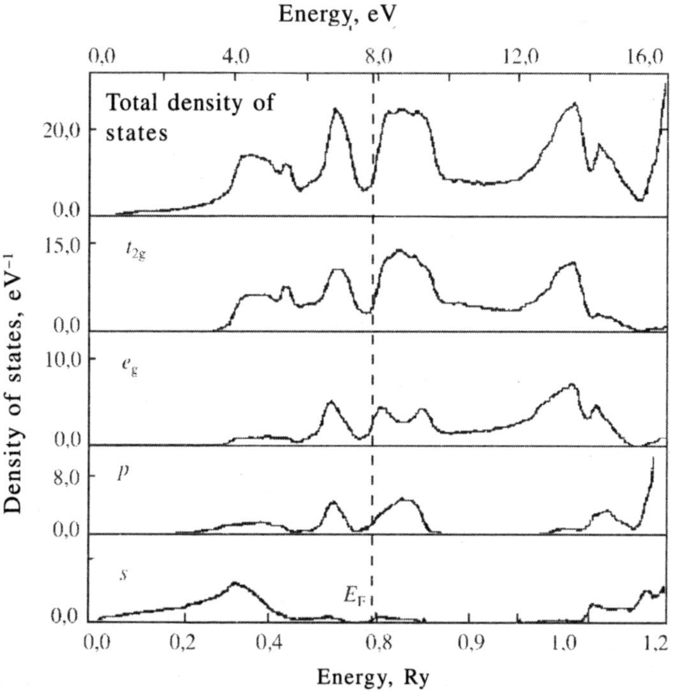

Fig. 2.3. Total and partial densities of states for hafnium [78].

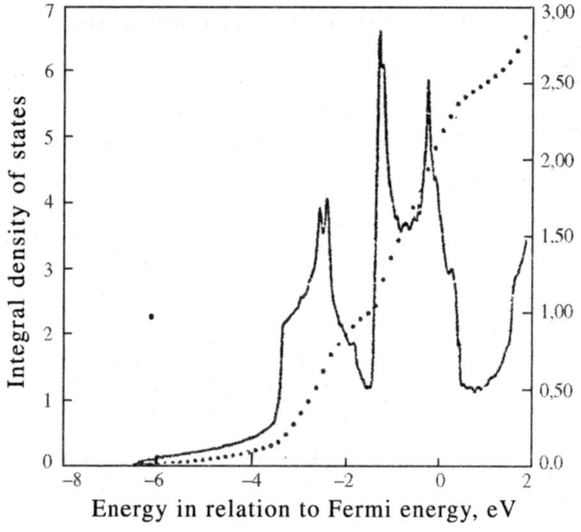

Fig.2.4. Total and partial densities of states for vanadium [78].

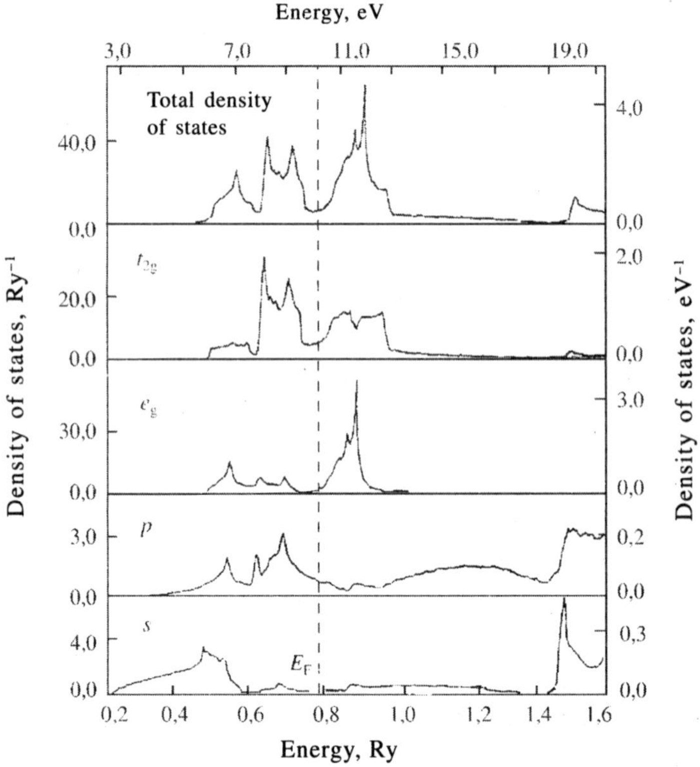

Fig. 2.5. Total and partial densities of states for chromium [78].

s-band diverges at the centre of the Brillouin zone,

$$S_s(\mathbf{k}) \to -6(ks)^{-2} + \text{const}.$$

Therefore one has to use in Figs.2.8–2.10 the free-electron-like scale $[1-(2/\pi)^2 S_s]^{-1}$. The width of a canonical band is estimated from the second moment:

$$\tilde{W}_l = (12 S_l^2)^{1/2},$$

$$S_l^2 = \frac{1}{2l+1} \sum_{i=1}^{2l+1} \sum_{\mathbf{k}} S_{li}^2(\mathbf{k}) =$$

$$= 2^{l+2}(2l+1)\frac{(4l-1)!!}{l!(2l)!} \sum_{\mathbf{R}\neq 0} (s/R)^{2(2l+1)} \qquad (2.79)$$

and depends only on the number of atoms in coordination spheres. For the bcc, fcc and ideal hcp lattices one has respectively

$$\tilde{W}_p = 18.8, \quad 18.7, \quad 18.6;$$

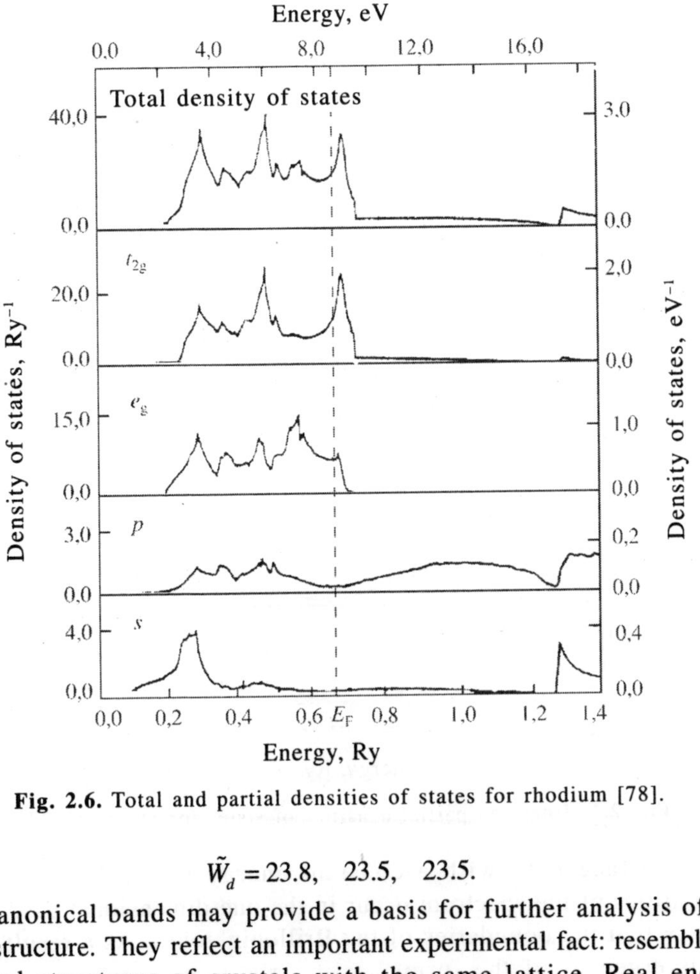

Fig. 2.6. Total and partial densities of states for rhodium [78].

$$\tilde{W}_d = 23.8, \quad 23.5, \quad 23.5.$$

The canonical bands may provide a basis for further analysis of the band structure. They reflect an important experimental fact: resemblance of band structures of crystals with the same lattice. Real energy bands for various elements are obtained from the canonical bands by specifying the functions $P_l(E)$. The hybridization of bands with different l does not influence strongly the spectrum provided that the bands do not cross each other. However, for crossing s,p,d-bands in transition metals the hybridization may lead to qualitative changes.

Consider the problem of the shape of energy bands, in particular the origin of peaks in the density of states $N(E)$. As follows from the formula

$$N(E) = \int \delta(E - E_{\mathbf{k}})d\mathbf{k} = \int \frac{dS}{|\operatorname{grad}_{\mathbf{k}} E_{\mathbf{k}}|}, \tag{2.80}$$

a peak of $N(E)$ corresponds to a nearly flat region (e.g., extremum)

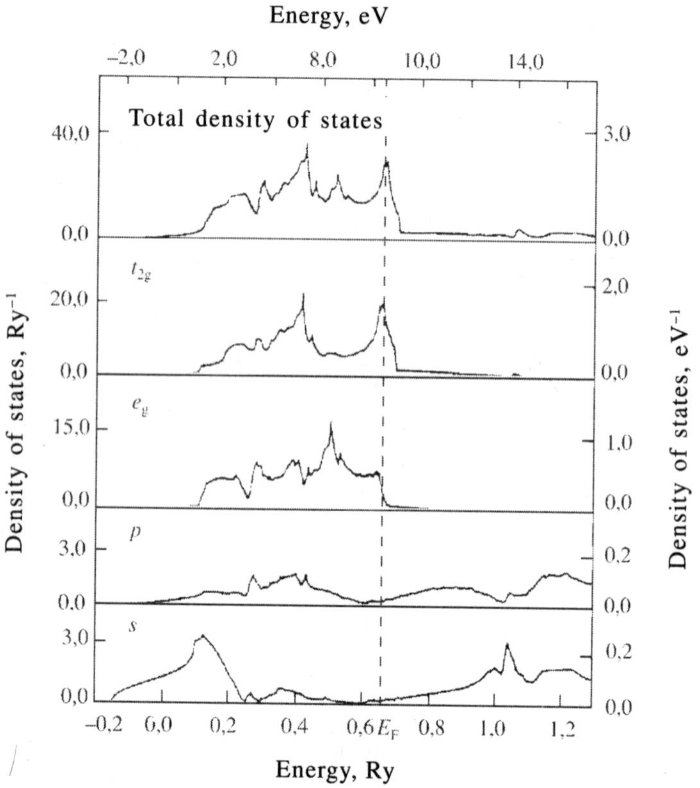

Fig. 2.7. Total and partial densities of states for platinum [78].

of the surface $E(\mathbf{k})$ with a constant energy in the **k**-space. Such peculiarities of the spectrum occur in the one-dimensional case where $dE/dk = 0$ at the boundaries of the Brillouin zones. As a result, DOS has divergences of the form

$$N(E) = |E - E_b|^{-1/2} . \tag{2.81}$$

According to the Van Hove theorem, in the general situation, DOS singularities (divergence of $N(E)$ or $N'(E)$) occur at some points E_c by topological reasons. In the two-dimensional case, these singularities are logarithmic. The $N(E)$ pictures demonstrating the Van Hove singularities were obtained for simple cubic, fcc and bcc lattices for an s-band in the nearest-neighbour approximation [79] (Figs.2.12, 2.13). The divergence of DOS at the band centre in the bcc lattice and the logarithmic singularity in the fcc lattice at the band bottom are removed when one includes next-nearest neighbours (Fig.2.14). Thus in the three-dimensional case the usual Van Hove singularities

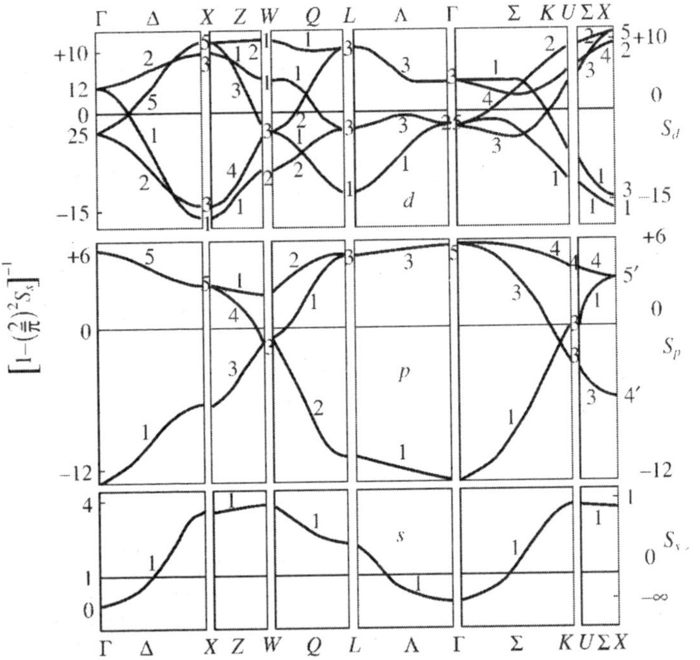

Fig. 2.8. The non-hybridized canonical bands for the fcc lattice [56].

have as a rule the one-sided square-root form

$$\delta N(E) = |E - E_c|^{1/2}\,\theta(\pm(E - E_c)) \tag{2.82}$$

and are too weak for explaining sharp DOS peaks. However, the peaks may form at merging of Van Hove singularities along some lines in the **k**-space. As demonstrated in [81], this leads to two-sided logarithmic singularities

$$\delta N(E) = -\ln|E - E_c|. \tag{2.83}$$

As follows from analysis of calculations [24,78], the merging along the P–N line in the bcc lattice results in formation of the 'giant' Van Hove singularities (2.83) in Li, V, Cr, Fe, Ba [81]. A similar situation takes place for fcc Ca and Sr.

Some peaks in the canonical DOS's may be identified with the Van Hove singularities. Despite a general similarity, the DOS, obtained in realistic band calculations, contain a more number of peaks than canonical DOS's. The additional peaks in calculated DOS pictures, which take into account off-diagonal matrix elements (hybridization) between s,p,d-bands, may be connected with splitting of a peak at the band intersection.

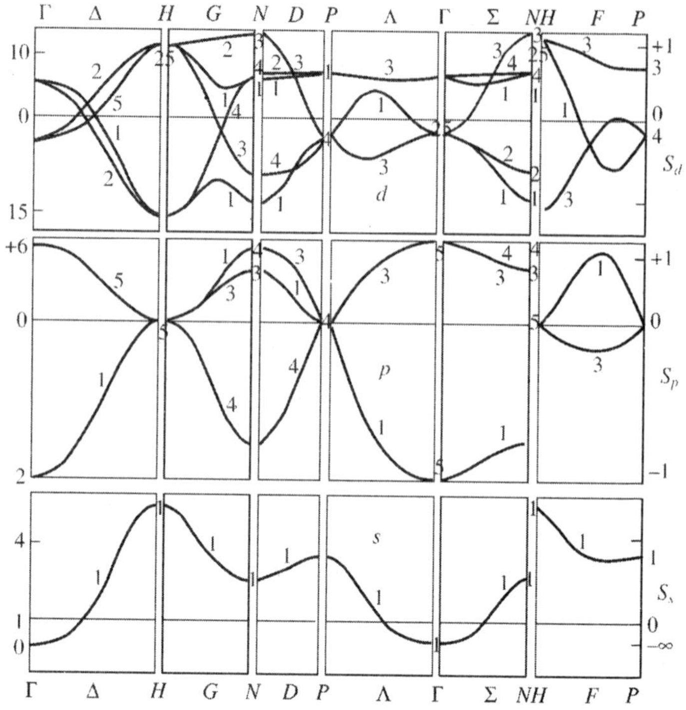

Fig. 2.9. The non-hybridized canonical bands for the bcc lattice [56].

A comparison of canonical bands with real band calculations which take into account hybridization is performed in the review [56]. The effect of hybridization turns out to depend strongly on the mutual position of the bands. As an example, Fig.2.15 shows the density of *d*-states for Nb. The effect of *p–d* hybridization is weak since *p*- and *d*-band are well separated. The *s–d* hybridization effect is more appreciable since *s*- and *d*-bands overlap (as well as in other transition metals). However, the *s*-band is influenced stronger than the degenerate *d*-band with a large electron capacity. Another typical example is provided by molybdenum where the splitting of some peaks occurs due to *s–d* hybridization (Fig.2.16) [82].

Now we discuss the position and width of energy bands. These characteristics of band structure may be investigated either by direct using results of numerical calculations or by parametrization within simple models which neglect unimportant details. The latter approach permits to obtain a more clear physical picture and to treat the problem analytically.

The simplest model of transition metals is the Friedel model [83] which considers their electron system as containing free *s*-electrons

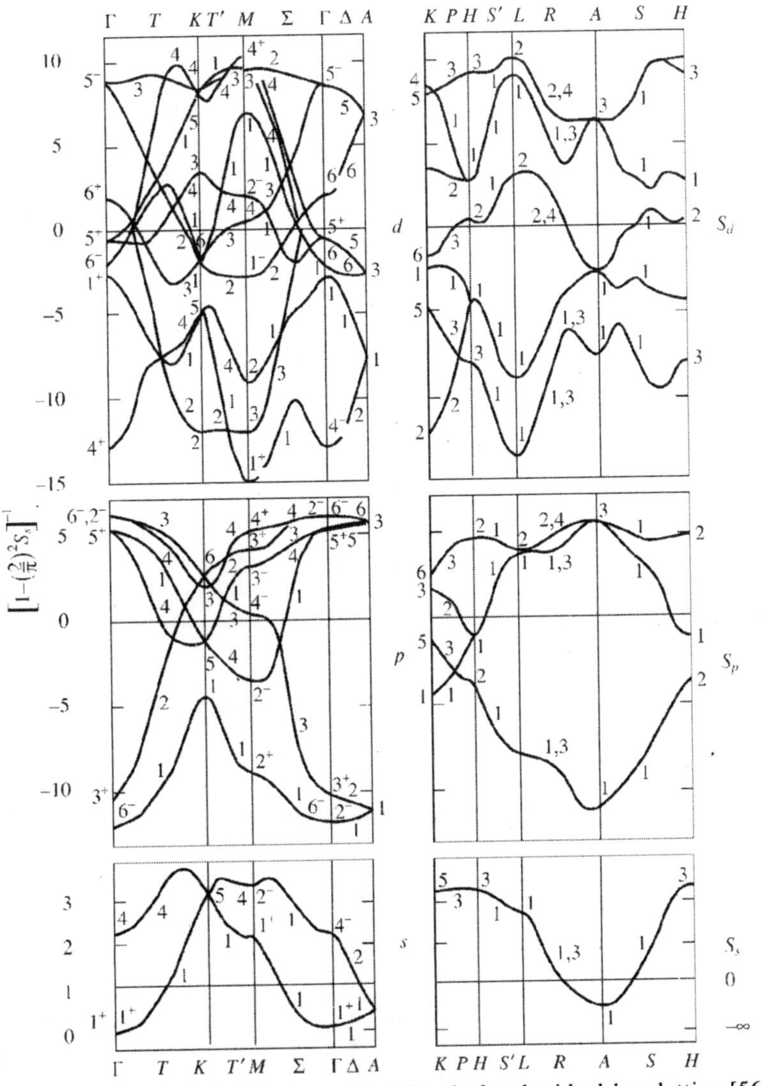

Fig. 2.10. The non-hybridized canonical bands for the ideal hcp lattice [56].

and the narrow d-band with the constant density of states of $10/W_d$. The lowering of electron energy at formation of the d-band from the level is given by the expression

$$\delta E_{\text{band}} = 5W_d\left[-\frac{n_d}{10}+\left(\frac{n_d}{10}\right)^2\right]. \tag{2.84}$$

A model description of band spectrum was proposed by Harrison [13]. Using the set of one s-function and five d-functions in the form

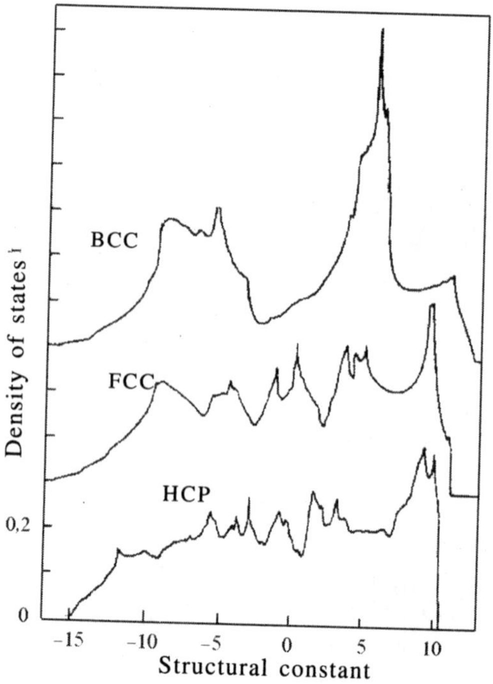

Fig. 2.11. Densities of states (per spin) for the non-hybridized canonical bands shown in Figs.2.8–2.10.

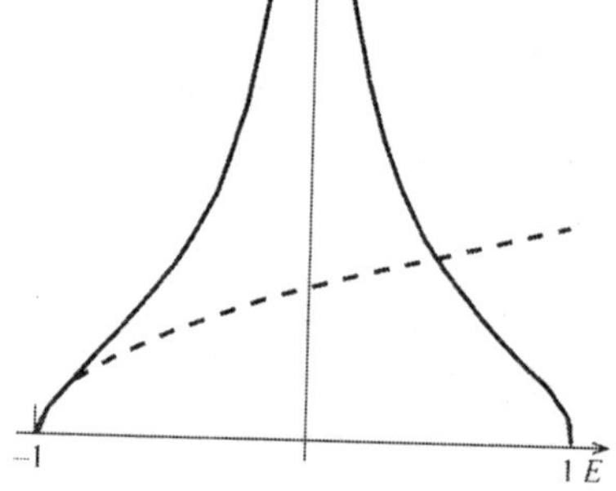

Fig. 2.12. The density of states for the *s*-band in the bcc lattice in the nearest-neighbour approximation [79].

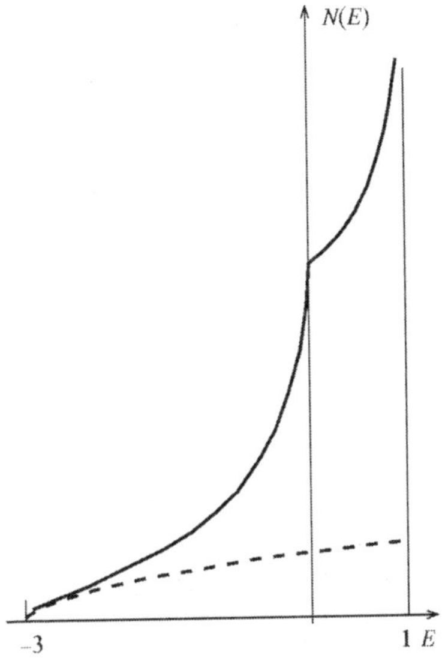

Fig. 2.13. The density of states for the s-band in the fcc lattice in the nearest-neighbour approximation [79].

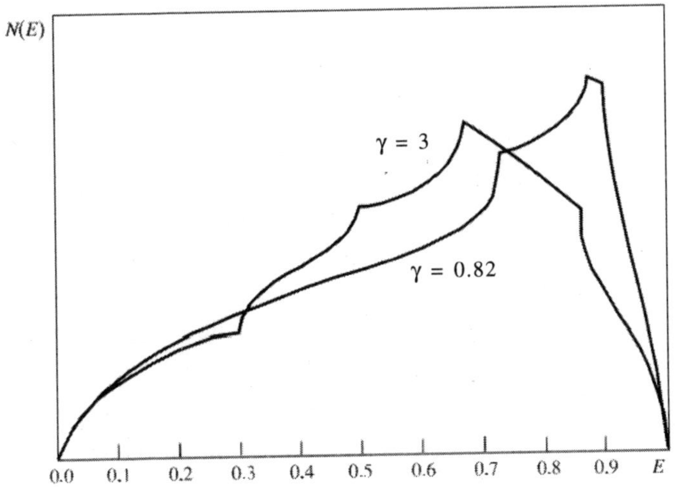

Fig. 2.14. The density of states for s-band in the fcc lattice with the ratio of next nearest and nearest-neighbour transfer integrals $\gamma = t_2/t_1 = 0.8$ and 3 [80].

of MT orbitals (2.44) he reduced the potential matix elements to two parameters – the width of d-band W_d and its position with respect to the s-band bottom E_d.

As follows from (2.44), the R-dependence of the intersite matrix

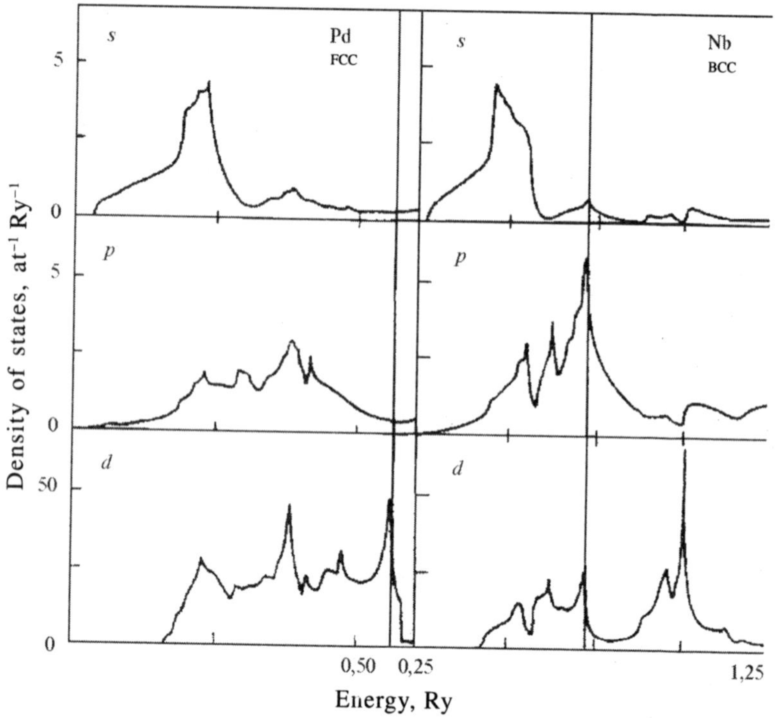

Fig. 2.15. *s*-, *p*- and *d*-projected densities of states for fcc palladium and bcc niobium [56].

elements of the potential has the form

$$V_{ll'm} = C_{ll'm} = (r_{\mathrm{MT}} / R)^{l+l'+1}, \tag{2.85}$$

where l, l' are the orbital quantum numbers and $m = \sigma, \pi, \delta$ are determined by the angle between the direction of the orbital and the vector which connects the atoms. In particular,

$$V_{ss\sigma} \sim R^{-1}, \qquad V_{sp\sigma} \sim R^{-2}, \qquad V_{pp\sigma,\pi} \sim R^{-3}, \qquad V_{ddm} \sim R^{-5} \tag{2.86}$$

(Note that in the free-electron approximation all the matrix elements are proportional to R^{-2} which results in a divergence after volume integration). In an explicit form one may write down

$$V_{ddm} = \eta_{ddm} \frac{\hbar^2 r_d^3}{mR^5}. \tag{2.87}$$

where the d-state radius r_d is a characteristics of the element, η are dimensionless constants which are tabulated in [13] (see also consideration of f-elements in paper [84]). The values of r_d are obtained

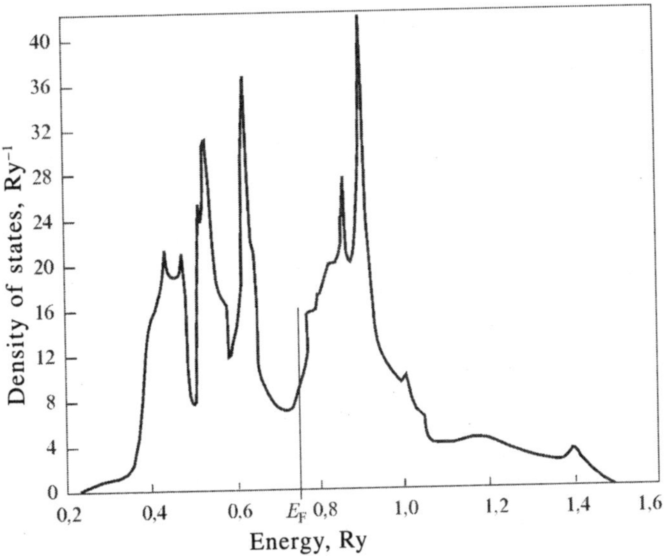

Fig. 2.16. Density of states of molybdenum for the lowest six bands [82].

from comparison with the first-principle band calculations. The quantities W_d are obtained in terms of the matrix elements with the use of standard formulas. For example, for the bcc lattice with the account of the nearest and next-nearest interactions,

$$W_d = -\frac{8}{3}V^{(1)}_{dd\sigma} + \frac{32}{9}V^{(1)}_{dd\pi} - 3V^{(2)}_{dd\sigma} + 4V^{(2)}_{dd\pi}.$$

The values of E_d and W_d in the d-rows, together with the results of band calculations [78] and experimental data [85], are presented in the Table 2.1. One can see that E_d and W_d increase from Sc to V and decrease from V to Cu.

In the pseudopotential theory, one can derive also the relation between the effective masses of s- and d-electrons [13]

$$\frac{m_s}{m_d} = \left(1 + 2.91\frac{m}{m_d}\right)^{-1}. \tag{2.88}$$

A number of theoretical models were proposed to describe the electronic structure of ferromagnetic 3d-metals Fe, Co and Ni [85–88]. These models are based on the idea of separating of localized and itinerant d-states with different symmetry (e.g., of t_{2g} and e_g states).

Now we consider some general regularities in the results of band calculations. These results agree qualitatively with above-discussed simple model notions. In agreement with the idea of canonical bands,

Table 2.1. Calculated characteristics of band structure for *d*-metals: the number of *d*-electrons n_d, position of the Fermi level E_d (eV) and density of states $N(E_F)$ (eV^{-1} at^{-1}) [78], the model parameters [13]: average energy E_d (relative to the *s*-band bottom, eV) and the bandwidth W_d (eV); experimental data on W_d from PES and IPES [40] are presented for 3*d*-metals only.

3d 4d 5d	Sc Y La	Ti Zr Hb	V Nb Ta	Cr Mo W	Mn Tc Re	Fe Ru Os	Co Rh Ir	Ni Pd Pt	Cu Ag Au
n_d	1.76 1.68 –	2.90 2.96 2.69	3.98 4.10 3.78	4.96 5.07 4.73	5.98 6.23 5.73	$4.6^{\uparrow}$ $2.34_{\downarrow}$ 7.24 6.70	$4.8^{\uparrow}$ $3.06_{\downarrow}$ 7.99 7.65	$4.82^{\uparrow}$ $4.15_{\downarrow}$ 8.96 8.74	9.91 10.01 9.89
E_F	5.82 5.39 –	8.03 7.42 7.87	9.26 9.06 9.31	10.6 10.6 11.3	10.3 10.3 10.9	10.2 10.1 11.6	9.53 8.61 10.4	8.71 7.06 8.68	7.90 6.30 7.32
$N(E_F)$	1.76 1.69 –	0.54 0.53 0.41	1.61 1.20 1.04	0.65 0.52 0.36	2.72 0.91 0.70	$0.85^{\uparrow}$ $0.24_{\downarrow}$ 0.81 0.65	$0.14^{\uparrow}$ $0.66_{\downarrow}$ 1.32 0.90	$0.15^{\uparrow}$ $1.54_{\downarrow}$ 2.31 1.67	0.15 0.06 0.18
E_d	7.05 6.75 –	7.76 7.17 9.12	8.13 7.29 9.50	8.01 7.12 9.45	7.91 6.67 8.99	7.64 6.02 8.38	7.36 5.08 7.35	6.91 4.52 6.51	5.90 2.49 5.18
W_d	5.13 6.59 –	6.08 8.37 9.56	6.77 9.72 11.2	6.56 9.98 11.4	5.60 9.42 11.0	4.82 8.44 10.3	4.35 6.89 8.71	3.78 5.40 7.00	2.80 3.63 5.28
W_e^{exp} (3d)	6.2	6.6	6.8	6.5	8.5	8.5	6.9	5.4	2.9

the DOS pictures for elements of one column in the periodic table demonstrate a similarity. This is illustrated by comparison of DOS for hcp metals Ti, Zr and Hf (Figs.2.1–2.3). A similar situation takes place for Sc and Y (hcp), V, Nb and Ta (bcc), Cr, Mo and W (bcc).

In Tc and Re (hcp) the heights of the DOS peaks (especially of the lower ones) differ appreciably. The same difference occurs between Ru and Os (hcp), Rh and Ir (fcc), Pd and Pt (fcc). The canonical band idea works also for elements of different columns with the same lattice, the position of the Fermi level being changed roughly according the rigid band model. This may be illustrated by comparison DOS of V and Cr (bcc), Rh and Pt (fcc).

The values of partial DOS of *s,p,d*-type, which are given in

Table 2.2. Partial *s*, *p*, *d* densities of states at the Fermi level and the number of *d*-electrons in 3*d*-metals [78].

	Sc	Ti	V	Cr	Mn	$Fe^{\uparrow}$	$Fe_{\downarrow}$	$Co^{\uparrow}$	$Co_{\downarrow}$	$Ni^{\uparrow}$	$Ni_{\downarrow}$
s	0.34	0.07	0.43	0.08	0.19	0.22	0.02	0.36	0.06	0.28	0.16
p	6.06	0.80	2.46	0.72	0.31	0.14	0.04	0.23	0.44	0.24	0.09
t_{2g}	13.70	4.61	18.0	6.52	17.9	10.5	2.80	0.65	5.54	1.25	13.07
e_g	10.12	2.69	3.95	2.31	19.1	1.07	0.41	1.22	3.48	0.74	7.94
n_g	1.76	2.9	3.98	4.96	5.98	4.6	2.34	4.8	3.06	4.82	4.15

Table 2.2, demonstrate that *d*-states in transition metals are dominating ones. They make up about 70–90% of the total DOS at the Fermi level. One can see also from Table 2.2 that large and small values of $N_d(E_F)$ alternate when passing to a neighbour element. This rule holds even irrespective of the crystal structure and demonstrates an important role of atomic configurations. The regularity is to some extent violated in the end of *d*-periods. This may be explained by ferromagnetism of Fe, Co, and Ni and violation of regularity in *d*-shell filling. It should be also noted that the numbers of *d*-electrons per ion in metals exceed by about unity the atomic values [78].

To end this section, we discuss the problem of accuracy of band calculations. This problem is connected with a number of approximations: choose of initial atomic configuration, the density functional approach for states which are far from E_F, the local approximation, the form of crystal potential (including exchange–correlation contribution), the calculation method used, an account of relativistic effects etc. The role of these approximations may be clarified by comparing results of various calculations. Such a comparison is performed in numerous

Table 2.3. Experimental (angle-resolved photoemission) and calculated energy on the $\Gamma\Delta L$ axis in palladium (in eV relative to the Fermi level) [90].

k-point, band number	Exper. [90]	RAPW [91]	RAPW [92]	HFS [93]	Pseudopot. [94]
$\Gamma_{2,3,4}$	−2.55 ±0.15	−2.79	−2.49	−2.59	−2.56
$\Gamma_{2,3,}$	−2.4± 0.2	−2.98	−2.62	−2.70	−2.66
$\Gamma_{5,6}$	−1.15± 0.1	−1.17	–	−2.70	−2.66
Γ_4	−0.4± 0.2	−0.14	–	−0.06	−0.09
Γ_5	0.1± 0.1	+0.05	–	−0.06	−0.09
Γ_7	+7.7± 0.3	+7.30	–	–	–
Γ_7	+18.4± 0.5	+17.71	–	–	–
Γ_8	+21.7±0.5	+21.65	–	–	–

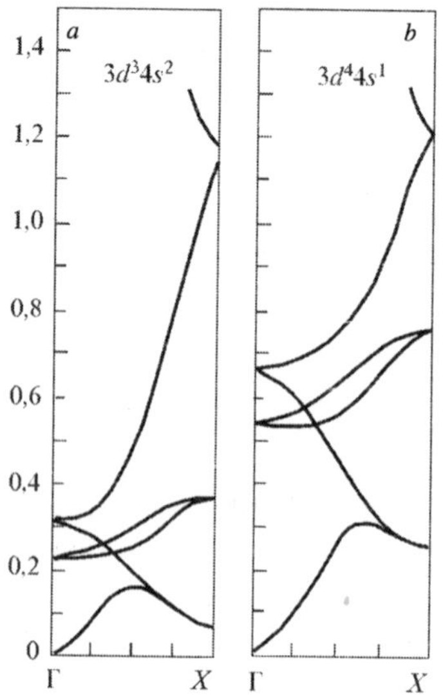

Fig. 2.17. Energy bands of vanadium, calculated for two possible choices of the crystal potential [83].

review papers and monographs (see e.g. [57]). For example, influence of initial atomic configuration is illustrated by band calculations of vanadium (Fig.2.17). The width of the *s*-band differs by two times for the configurations $3d^34s^2$ and $3d^44s^1$. The Table 2.3 shows results of band energy calculations at some Brillouin zone points in Pd by different methods [91–94] and corresponding experimental results [92].

We may conclude that the accuracy of energy determination makes up about 0.1–0.3eV. The disagreement with experimental data may be about 0.5eV. At comparing the calculated band spectrum with angle-resolved emission and de Haas–van Alphen effect data, one has often to shift it by such values. In temperature units, the corresponding uncertainty is of the order of 10^3 K which makes difficult precise calculations of thermodynamic properties within the band approach.

2.5 EXPERIMENTAL INVESTIGATIONS OF BAND STRUCTURE: SPECTRAL DATA

A number of quantities, which are obtained in band calculations (e.g., the position of energy bands and the shape of the Fermi surface, the electronic structure near the Fermi level) may be analyzed and compared with results of experimental investigations.

There exist a large number of spectral methods for investigating electron structure. They may be divided in emission and absorption spectroscopy methods. The first ones enable one to obtain data on the filled part of the band (below E_F), and the second on the empty part (above E_F). Both the optical and X-ray regions may be investigated. Let us list main spectral methods and corresponding notations [40]:

- PES – Photoelectron (Photoemission) Spectroscopy
- XPS – X-ray Photoelectron Spectroscopy
- UPS – Ultraviolet Photoelectron Spectroscopy
- IPES – Inverse Photoemission Spectroscopy
- AES – Auger Electron Spectroscopy
- BIS – Bremsstrahlung Isochromat Spectroscopy
- EELS – Electron Energy Loss Spectroscopy
- MXD – Magnetic X-ray Dichroism
- MXS – Magnetic X-ray Scattering
- SEC – Secondary Electron Spectroscopy
- XAS – X-ray Absorption Spectroscopy
- XANES – X-ray Absorption Near Edge Spectroscopy
- EXAFS – Extended X-ray Absorption Fine Structure

When the spectroscopies are performed angle-resolved and spin-polarized, acronyms AR and SP are added.

An advantage of the X-ray spectroscopy is the simple separation electron states with different angular moments. By virtue of the selection rule, $\Delta l = \pm 1$, *K*-spectra (transitions from 1*s*-states of inner core levels of an atom) contain information on *p*-states of higher bands, and *L*-spectra (transitions from 2*p*-states) on *s* and *d*-states. Due to large difference in the energies of *K*- and *L*-shells, these spectra are wel! separated. This simplifies the comparison with band calculations which also yield separately *s,p,d*-contributions. The optical part of the spectrum corresponds to transitions between external overlapping energy bands or inside them. Therefore the qualitative analysis of the optical spectra is much more complicated. We do not discuss in detail these spectra of transition metals where the interpretation is especially difficult (see [95]).

On the other hand, the resolution of X-ray spectra in the energy measurement is smaller than for optical spectra because of the large intrinsic width of internal levels (of order of 1eV, see the handbook [96]). The situation is somewhat better only for soft X-ray spectra (e.g., for the N-spectra the width of the internal $4f$-level makes up about 0.15eV). Besides that, there exist the contribution to level width which is due to finite lifetime of electrons and holes. This contribution increases with increasing distance from the Fermi energy:

$$\Gamma_{\mathbf{k}} \sim (E_{\mathbf{k}} - E_{\mathrm{F}})^2. \tag{2.89}$$

Provided that the density of states in a band contains a symmetric maximum which does not overlap with contributions of other bands, its position may be determined with a satisfactory accuracy (about 0.1 eV). However, such a determination is prevented in the presence of overlapping bands. Usually comparison of X-ray data with calculated spectra is performed by smoothing out the band structure with account of the level width.

Occupied states are investigated in PES and ultraviolet PS methods. The former yields, in particular, information on K, L, M, N and O-spectra of $3d$, $4d$ and $5d$-metals. Very promising is the angle resolved photoemission method which yields information on not only the density of state, but also the spectrum $E(\mathbf{k})$. Methods of X-ray spectroscopy for empty states include XAS, XANES, BIS, IPS. The relation between

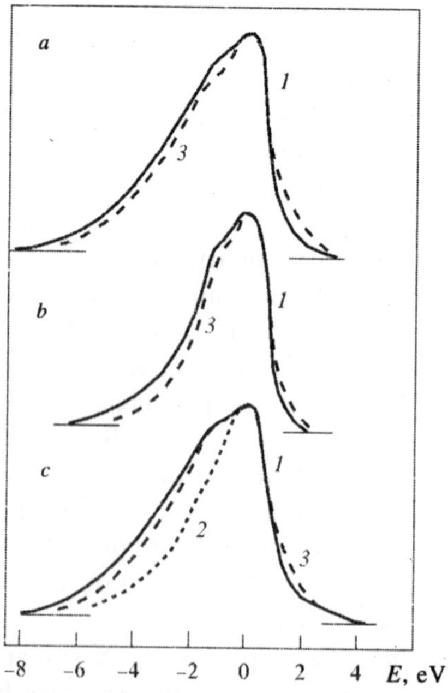

Fig. 2.18. X-ray emission spectra of vanadium [57] (a) K-band (b) L_{III}-band (c) M_{III} -band; 1 are the experimental data, 2 and 3 are the results of calculations with different perturbation operators.

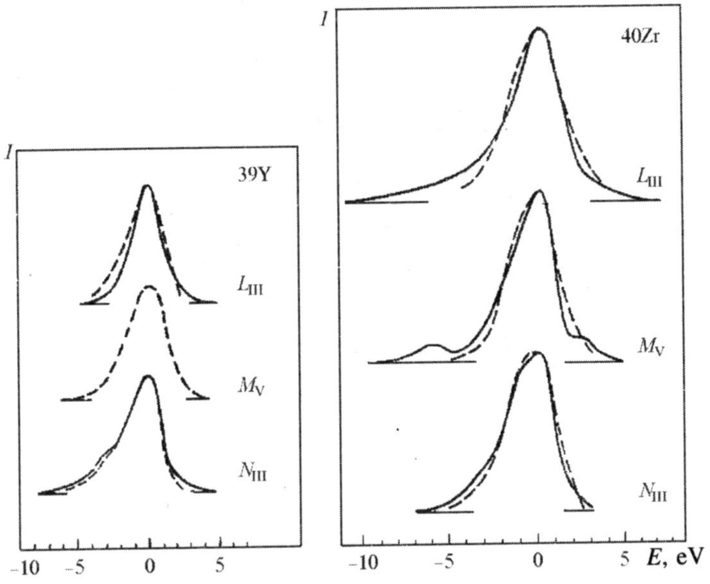

Fig. 2.19. X-ray emission spectra of Y and Zr; solid line shows experimental data, and broken line the results of calculations [57].

the IPES and BIS methods is roughly the same as between the ultraviolet PS and X-ray PS ones.

Large widths of internal levels in *K*, *L* and *M* PES-spectra prevent as a rule do resolve the density of state structure in the conduction band. The comparison on *K*, *L* and *M*-spectra of vanadium with results of theoretical calculations are shown in Fig.2.18. One can see that the complicated DOS structure [78] becomes smeared. At the same time, total width of the spectrum coincides roughly with the calculated bandwidth after account of the level broadening. So, the width is 8 eV for the *K*-spectrum and 6 eV for the *L* spectrum, the calculated bandwidth being equal to 4 eV.

The second typical example is the spectrum of zirconium (Fig.2.19). Here the L_{III} spectrum (4*d*–2*p* transitions) has the width of 3.4 eV, and the M_V spectrum (3*d*–4*d* transitions) the width of 2.9 eV. High-energy maximum corresponds to the M_{IV} spectrum, and low-energy maximum at 6–7 eV has probably a satellite origin. Although the width of the internal 4*p*-level is considerably smaller than that of the 2*p*-level, the experimental broadening is approximately the same for the L_{III} and N_{III} spectra. Possibly, this is due to larger contributions from the *s*–4*p* transitions to the N_{III} spectrum.

For next elements of 3*d*- and 4*d*-series, the widths of internal

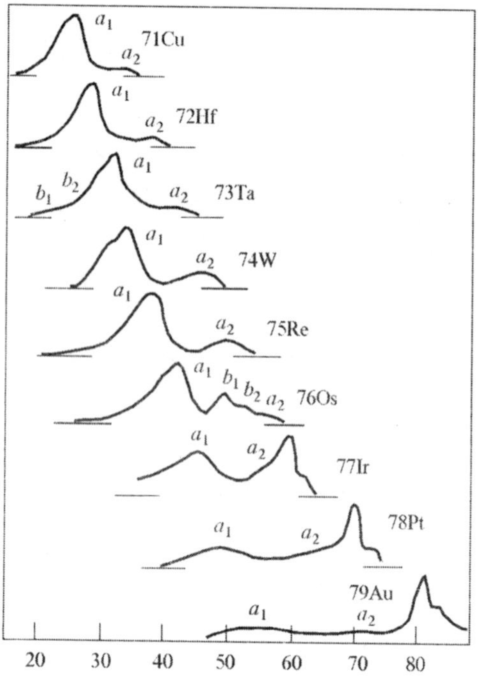

Fig. 2.20. Experimental X-ray emission bands of 5*d*-metals [57].

levels for *K*, *L*, *M* spectra increase so that their quality becomes worse. The situation is more favorable for the 5*d* series where the spectra $N_{VI,VII}$ (5*d*–4*f*) and $O_{II,III}$ (5*d*–5*p*, 5*s*–5*p*) with a small internal level width occur. Because of the closeness of 4*f* and 5*d* levels, four lines lie in the interval of the order of 10 eV. Nevertheless, because of small widths of *f* and 5*p* levels, these lines are clearly resolved in the beginning of the 5*d* series, emission bands being rather contrast and intensive. In the end of the series, the intensity of lines decreases and their structure becomes unclear. For the $N_{VI,VII}$ bands an opposite tendency takes place (Fig.2.20).

An important role in the spectra of 5*d*-metals belongs to the spin-orbital interaction. According to the selection rules, the transitions into internal doublets $f_{5/2}$, $f_{7/2}$ go from $d_{3/2}$ and $d_{5/2}$-levels of the conduction band. Analysis of experimental data demonstrates that the distributions of $d_{3/2}$ and $d_{5/2}$-states in the conduction band are different. Fig.2.21 shows the $N_{VI,VII}$-spectra of iridium and platinum with resolution of N_{VI} (i.e. $d_{3/2}$) and N_{VII} ($d_{5/2}$) states. The main maximum b_2 is connected with states of both types. The high-energy maximum b_3 is due to $d_{3/2}$-states, and the low-energy one b_1 to the

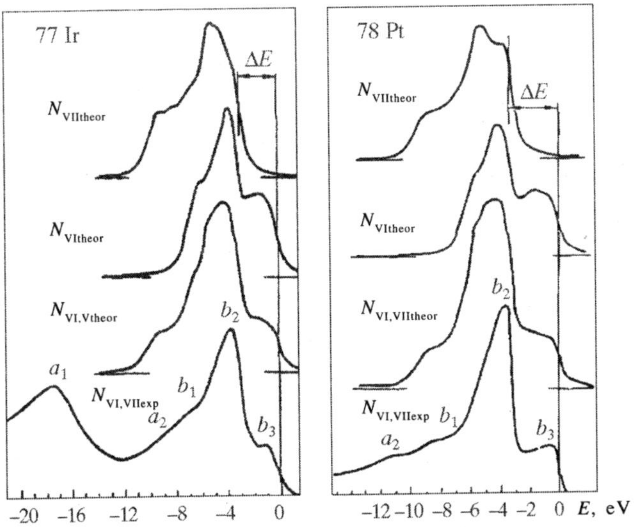

Fig. 2.21. $N_{VI,VII}$ emission spectra of Ir (a) and Pt (b). Lower curves are experimental data [57].

$d_{5/2}$-states. Thus, the $d_{5/2}$-states dominate at the d-band bottom and the $d_{3/2}$-states near the Fermi level. It is interesting that the spectra of Ir and Pt are satisfactory explained by atomic calculations which indicates a considerable localization of 5d-electrons in these metals.

A more complicated situation occurs for the $N_{\text{VI,VII}}$-spectra in the middle of the 5d-series (Ta, W, Re, Os), see [57]. The experimental width of the O_{III}-spectrum turns out to be larger than the calculated one. This is connected with extra interactions and many-electron effects (e.g., the influence of the vacancy in the O_{III}-shell).

The ARP method takes into account not only energy, but also quasimomentum of an electron in the conduction band. Consequently, it enables one to determine experimentally the dispersion law $E(\mathbf{k})$ for various branches, which gives a possibility to calculate the density of states to high accuracy. This method was applied to investigate Cu [97], Ag [98], Au, Pt [99], Pd [90], Ir [57].

The results for the spectrum of Cu in the $\langle 211 \rangle$ direction [97] agree well with the free electron model. Himpsel and Eastmann [90] measured $E(\mathbf{k})$ for Pd in the $\langle 111 \rangle$ direction to accuracy of 0.1–0.2 eV, the accuracy in determining $\mathbf{k}$ being about 5% of $k_{\max}$ (Table 2.3). One can see that agreement is not too good. A similar situation takes place as for Pt and Au [99].

Although investigating the occupied part of the conduction band is more important for the electron properties of a metal, data for

Table 2.4. Positions of density of states peaks, and widths of empty and occupied part of the band $W^{\pm}$ for 3d and 4d metals (in eV) [101].

3d 4d	Sc Y	Ti Zr	V Nb	Cr Mo	Mn Tc	Fe Ru	Co Rh	Ni Pd
E^{exp}_{peak}	1.8, 3.8	1.0, 2.8	2.3	1.0	1.4	1.6	0.5	0.3
	2.0, 5.4	0.9, 4.4	3.5	2.0	–	1.2	0.4	0.2
E^{theor}_{peak}	1.4, 3.5	0.8, 3.1	2.5	1.2		1.6	0.5	0.3
	1.3, 4.6	0.9, 4.6	3.4	1.8	–	1.3	0.5	0.1
W^{-}_{theor}	1.7	3.1	3.2	4.5	–	5.3	5.5	5.0
	2.0	3.1	3.5	5.4	–	6.2	5.8	4.8
W^{+}_{theor}	4.0	3.65	4.0	2.5	–	2.6	1.15	0.4
	5.3	5.3	5.6	3.9	–	1.95	1.2	0.4
W^{+}_{exp}	4.4-4.7	3.2-3.8	3.4-3.8	1.8-2.3	3.5	2.9-3.5	1.4	0.4
	6.1	5.2-5.7	5.2-5.9	3.9-4.5		1.95	1.1	0.4
W^{+}_{theor} – $-W^{+}_{theor}$	–0.5	0.0	+0.4	+0.5	–	–0.6	–0.25	0.0
	–0.8	–0.1	0	–0.3	–	0.0	0.1	0.0

the empty part of the band from absorption spectra are also useful from the point of view of comparison with band calculations.

The BIS data for 3*d* and 4*d*-metals were obtained and compared with band calculations in the paper [102] (see Table 2.4). The total resolution was about 0.7eV, and the lifetime broadening at E_F about 0.25eV. On the whole, the band shapes obtained are close to those for canonical bands. The agreement with calculated $N(E)$ turns out to be better than for XPS spectra which are strongly distorted in the beginning of periods [103]. Besides that, BIS spectra do not exhibit satellites which correspond to the d^{n+1} configuration. The *d*-bandwidth increases up to the middle of periods by about 25%. The *s–d* hybridization is important near the band bottom and is most strong for hcp metals. Peaks near E_F, which correspond to the empty *d*-band, are followed by a structureless plateau and at 7–10 eV by a step, which is interpreted as the van Hove singularity. The peaks (*d*-band tails) at E_F are observed even in Cu and Ag. Two peaks are observed for Sc, Ti, V and Zr, and one peak in other metals, which is in agreement with band calculations. Strong asymmetry of the peak is observed in molybdenum in agreement with theoretical predictions. The position of 1.6eV peak in Fe corresponds to DOS in ferromagnetic rather than paramagnetic state (Figs.2.22, 2.23). The relative *d*-band BIS intensity in the 3*d*-series was derived and demonstrate to have a maximum in the middle of the series.

Calculated spin-resolved densities of states of ferromagnetic transition metals Fe and Ni [24] are shown in Figs.2.24, 2.25. Of great interest

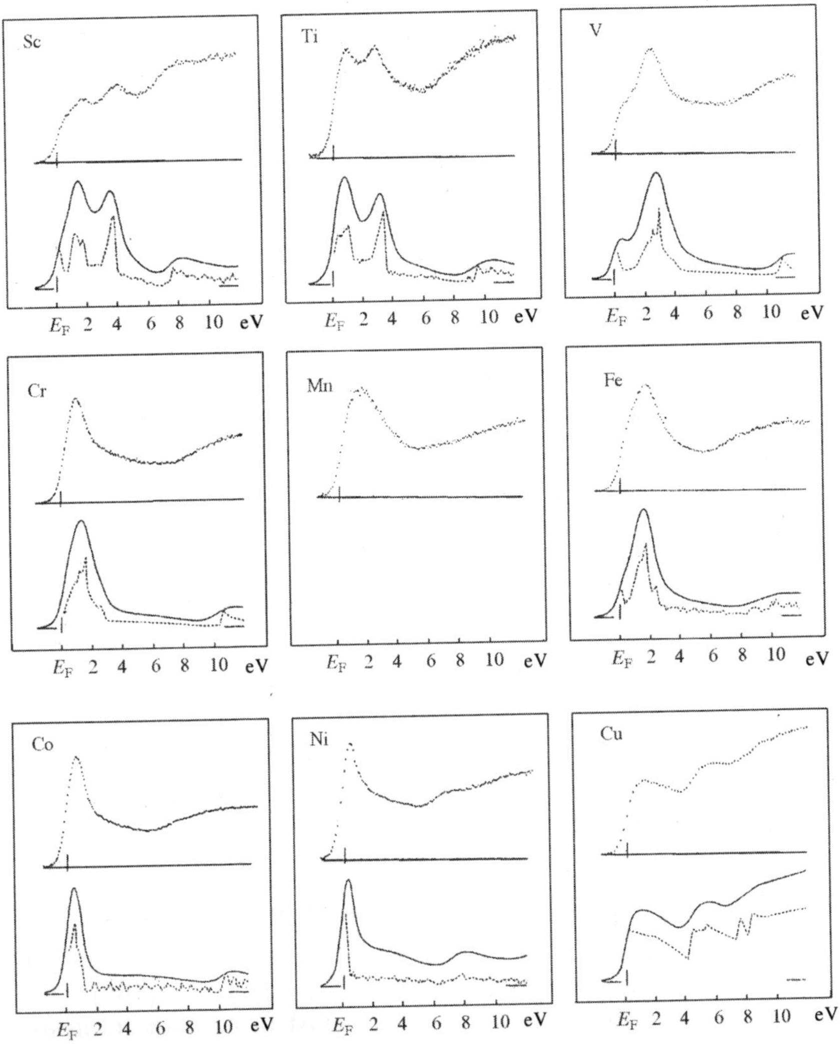

Fig. 2.22. Comparison of BIS and density of states curves for 3*d*-metals and Cu. The dashed curves correspond to the non-broadened DOS, the solid lines to the DOS broadened with Gaussian and Lorentz broadening to simulate instrumental and lifetime broadening respectively. The upper dotted curve is the measured BIS spectrum [102].

for the theory of both the electron structure and magnetism is the investigation of spin splitting Δ. In principle, such investigations may be carried out within the standard PES and IPES methods. However, these methods provide only a rough estimations of averaged values of Δ. According to band calculations, Δ depends considerably on E and $\mathbf{k}$, so that angle resolved methods are required for obtaining detailed

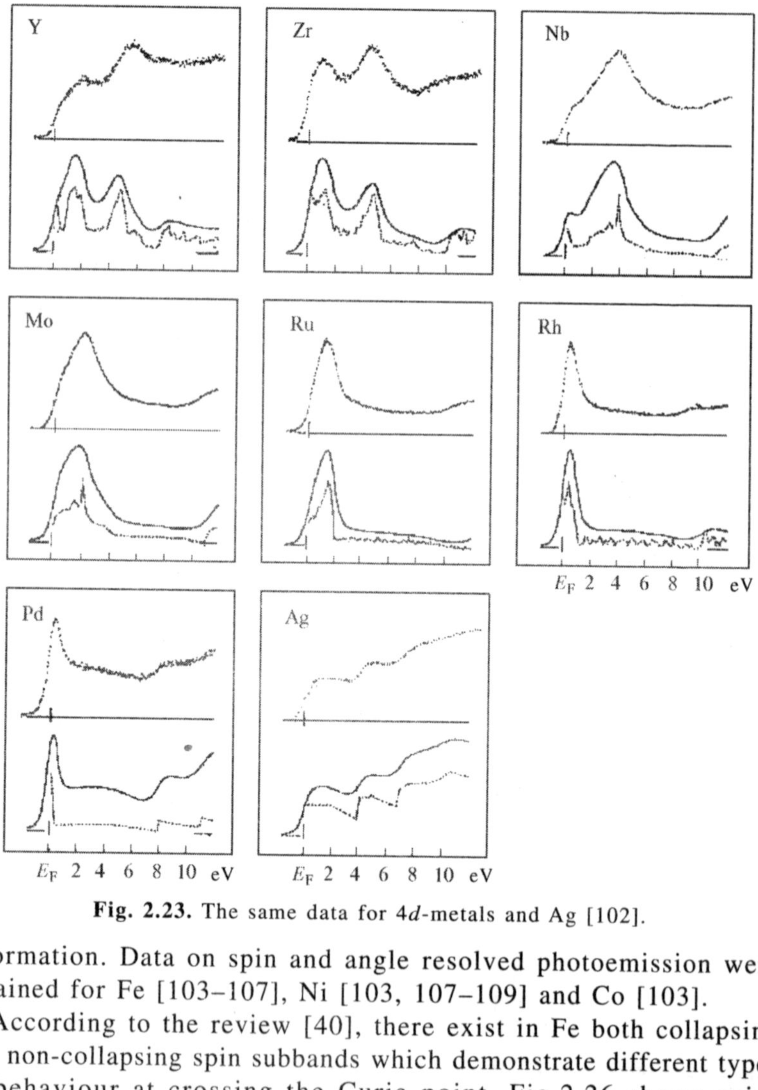

Fig. 2.23. The same data for 4*d*-metals and Ag [102].

information. Data on spin and angle resolved photoemission were obtained for Fe [103–107], Ni [103, 107–109] and Co [103].

According to the review [40], there exist in Fe both collapsing and non-collapsing spin subbands which demonstrate different types of behaviour at crossing the Curie point. Fig.2.26 shows spin-polarized angle-resolved spectra of Fe in the (100) direction near the Γ-point at $h\nu$ = 60 eV. These data demonstrate the energy and temperature dependence of Δ.

The ultraviolet photoemission spectrum of Fe was investigated by Pessa et al [110]. The observed peaks at 0.58 and 2.4 eV below the Fermi energy may be interpreted within the calculated density of states of ferromagnetic iron provided that one shifts the latter by 0.5eV to higher energies. The value of $N(E_F)$ for Fe obtained in [110] coincides approximately with the results of band calculations

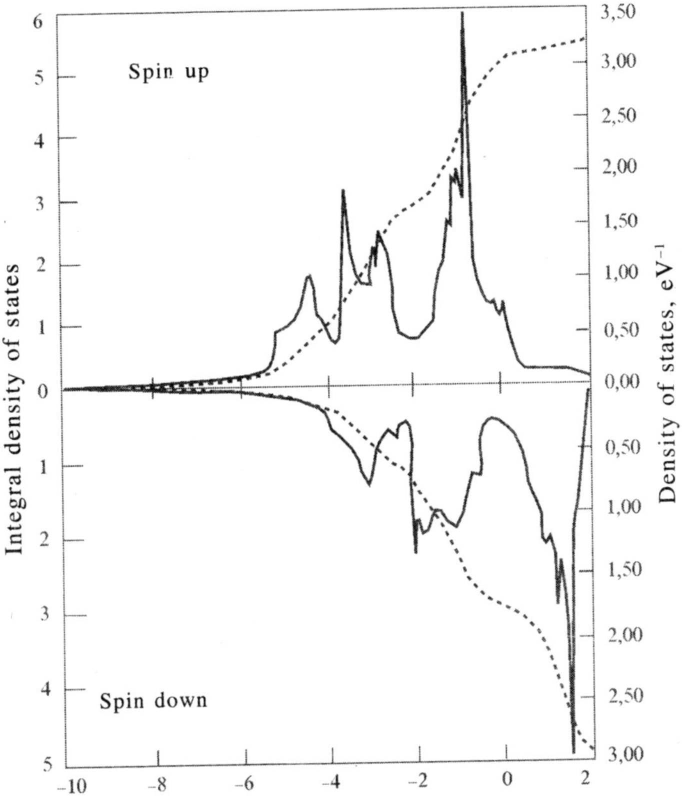

Fig. 2.24. Calculated spin up and down densities of states of ferromagnetic iron [24].

[194], but is considerably smaller than that determined from electronic specific heat [266] and paramagnetic susceptibility (see the detailed discussion in the corresponding chapters). It should be noted that an error of order 0.1eV in the energy measurement may influence strongly the $N(E_F)$ value. Comparison of results [110] with the spin-polarized spectra [111] confirms that the 0.58 eV peak belongs to the majority spin subband. The energy dependence of spin polarization is not explained by the theoretical density of states. The experimental estimation of spin splitting yields the value not more than 1.9 eV which is considerably smaller than theoretical one (2.35 eV). No shift of the spectrum at the Curie point is observed which may indicate retaining of strong short-range order above T_C.

The d-bandwidth of Ni from the ARP data [108] makes up 3.4 eV at the point L, the theoretical value being about 4.5 eV. The measured value of the exchange splitting is 0.31 eV at 293 K (the

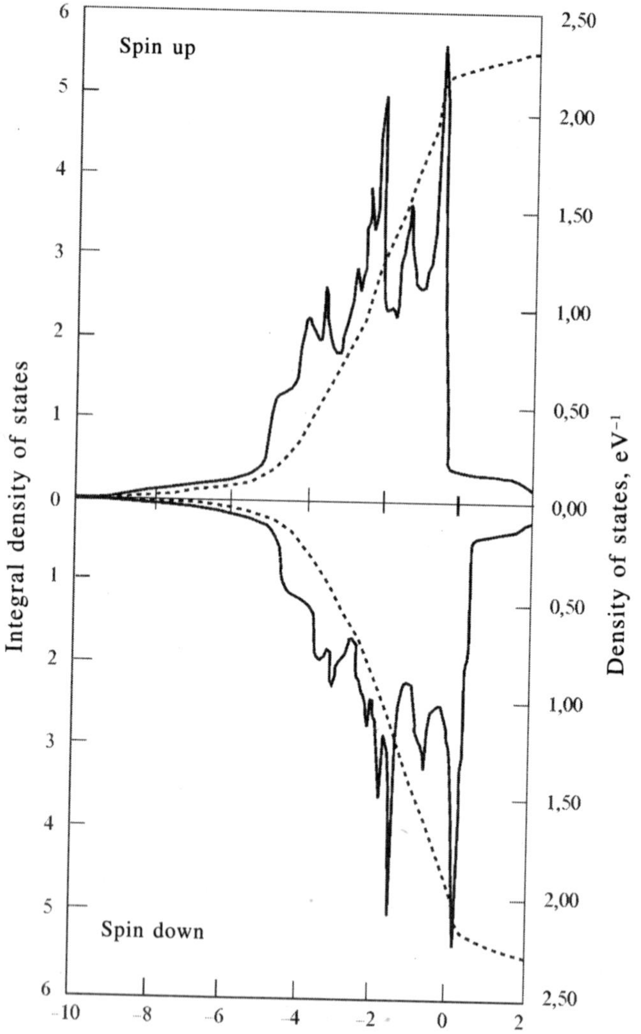

Fig. 2.25. Calculated spin up and down densities of states of ferromagnetic nickel [24].

theoretical value is 0.7 eV) and about 0.2 eV above the Curie point (Fig.2.27). Evidence of the existence of spontaneous spin splitting above T_C in Ni was obtained by the positron annihilation technique [112].

Spin-resolved photoemission data on the short-range magnetic order above T_C in Co were obtained in [113]. The problem of spontaneous spin splitting in iron group metals may also be investigated in optical experiments [114].

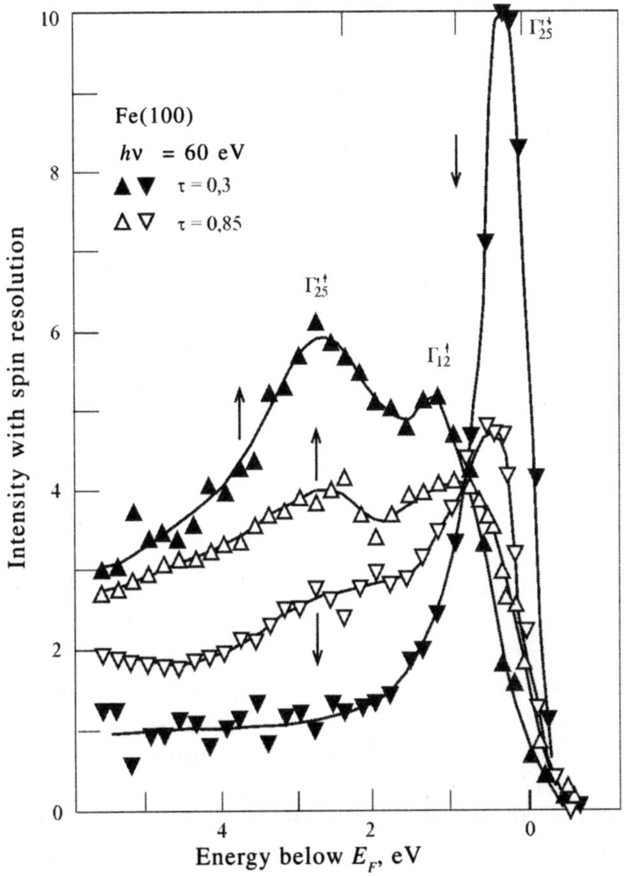

Fig. 2.26. Spin and angle-resolved distribution curves form Fe (100) at 60 eV photon energy for two different temperatures (T/T_C = 0.3 and 0.85) [105]. The arrows refer to the spontaneous magnetization direction.

Recently, the Magnetic X-Ray Dichroism (MXD) method was applied to determine the conduction electron spin polarization [40]. The L_{23} absorption edge ($2p \rightarrow 3d$ transitions) was investigated for Ni, and the asymmetry in photon absorption with different polarization permitted to find both spin and orbital momenta which turned out to be $s = 0.52\mu_B$ and $l = 0.05\mu_B$. These values are close to theoretical ones, which are obtained with for many-electron effects [115]. According to the latter paper, the ground state of the Ni ion is a superposition of the $3d^{10}$, $3d^9$ and $3d^8$-states with the weights of 15–20%, 60–70% and 15–20% respectively.

Consider the question about the origin of low-energy satellites in spectra of some d-metals. Such satellites are observed in X-ray

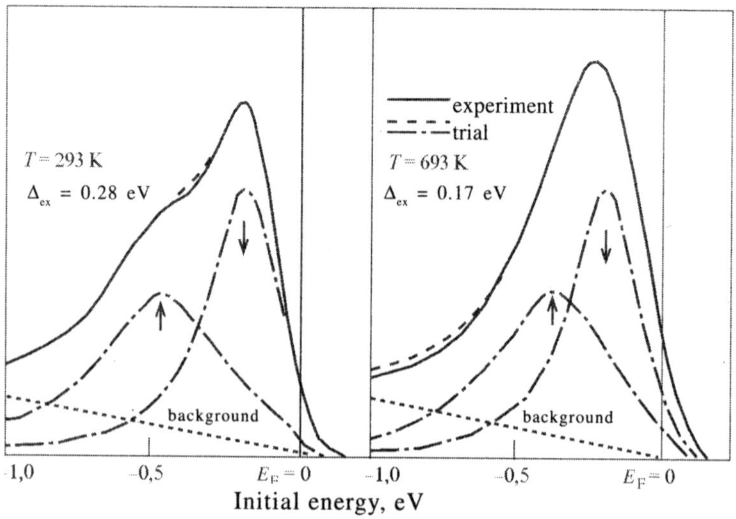

Fig. 2.27. Angle-resolved energy distribution curve showing temperature-dependent exchange splitting $\delta E_{ex} = 1.09\Delta_{ex}$ in Ni [108]. According to the experimental conditions, only the uppermost d-band is observed.

emission spectra of Ni (at 6 eV below E_F, see, e.g., [116]) and Zr [57]. There exist mechanisms of satellite formation which are connected with vacancies at excited atoms, Auger processes, shake up potential changes etc. (see [57]). We discuss more interesting many-electron mechanisms which are specific for transition metals.

Liebsh [117] treated the Coulomb interaction between two d-holes in Ni. Introducing atomic Slater integrals $F^{(p)}$ ($p = 0, 2, 4$), he obtained in the low density approximation additional peaks in the spectrum and compared them with experimental data. Besides the 6 eV satellite, the theory yields the second satellite which lies lower by 2 eV. The occurrence of satellites resulted in narrowing of the d-band since the total number of states is constant. The exchange splitting obtained turned out to be different for e_g and t_{2g} states (0.37 and 0.21 eV), the average value 0.3 eV being in satisfactory agreement with experiment. Penn [118] performed calculations within the degenerate Hubbard model. The fitted value of the Hubbard parameter was $U = 2$ eV.

The authors of the paper [119] treated this problem exactly at restricting to four point in the Brillouin zone, which is equivalent to consideration of a four-atom cluster. An agreement with experimental data was obtained for $U = 4.3$ eV and the ratio of the Hund and Coulomb parameters $J/U = 1/7$.

A more consistent consideration of the satellite formation problem should include many-electron terms (the satellite corresponds to the multiplet structure of the configuration d^8). The corresponding data of the Auger spectroscopy which contain more information on many-electron effects are discussed in Sect.1.2.

The ARP investigations of γ-cerium [120] yielded two dispersionless density of states peaks near E_F which did not obey the one-electron selection rule. They were attributed to atomic-like $4f$-states.

To describe high-energy spectroscopy data for cerium and its compounds, Gunnarsson and Schoenhammer [121] proposed a model, based on the one-impurity Anderson Hamiltonian and including effects of d–f hybridization. Further this model was applied to consider spectra of actinides and Ti (see [40]).

2.6 BAND CALCULATIONS OF RARE EARTHS AND ACTINIDES

One-electron band calculations of rare earths are performed by the same methods as for $5d$-metals, the role of relativistic effects becoming here still more important because of strong spin-orbital interaction. Since the $4f$-shell lies well below the conduction band, one may assume that the picture of this band is similar for all the $4f$-elements (especially for heavy rare earths). This is confirmed by data of photoemission spectroscopy from the $6s$–$5d$ band [122]. However, attempts of description in the one-electron approach of highly-correlated f-electron themselves meet with serious problems.

Results of band structure calculations of various RE metals by non-relativistic and relativistic augmented plane wave method (APW and RAPW) are presented in the review [19] (see also the discussion of electronic structure for heavy rare earths [15]). For most RE metals (except for Ce, Sm, Eu, Tm and Yb) the conclusions are as follows. The $4f$-band lies by about 5–10 eV below the conduction ($5s$,$6sp$) band and has the width about 0.05eV. The $5d$-band moves up relative to the conduction band with increase of the atomic number from La to Lu, the d-electron density of states at E_F increasing from 1.9 to 2.5 $eV^{-1}\cdot atom^{-1}$.

For elements with unstable valence Ce, Eu and Yb, especial attention was paid to investigation of $4f$-level position ε_f. A strong sensitivity to x was found in APW calculations of Ce in $4f^{2-x}5d^x6s^2$ (x = 0.5–2.0) configurations [123]. The $4f$-level is considerably lower than $5d$–$6s$ band for x = 0, but rapidly becomes higher and broader with increasing x. The calculated in [124] distance $E_F-\varepsilon_f$ equals to 0.36 and 0.24 Ry, and $4f$-bandwidth to 0.01 and 0.02 Ry for

γ- and α-cerium, respectively.

The full calculation of band structure of Ce in the linear RAPW method with account of correction to MT-approximation was performed by Pickett et al [125]. The width of $4f$-band is about 1 eV and increases by about 60% at the γ–α transition. The occupation numbers of f-electron (about 1.1) changes weakly under presure. This leads to the conclusion that the γ–α transition is accompanied by delocalization of $4f$-electrons rather than by promotion into sd-band.

Relativistic APW calculations for Yb in $4f^{14}6s^2$ configuration [126] yielded unexpectedly the $4f$-level position which is 0.1 Ry higher than the $6s$ conduction band (but still lower than E_F by 0.2–0.3 Ry). A small gap near E_F was obtained in contradiction with experimental situation. In the self-consistent calculations by Koelling (see [127]), $E_F \simeq \varepsilon f$ and the gap in the spectrum vanishes.

APW calculation of $4f$-band in thulium [128] with the account of $5s6d$–$4f$-hybridization yielded $\varepsilon_{5d} - \varepsilon_{4f}$ = 0.69 Ry, $\varepsilon_{6s} - \varepsilon_{4f}$= 1.27 Ry and the f-bandwidth of the order of 10^{-2} eV.

According to the first calculation of ferromagnetic gadolinium with the account of $4f$-states within the self-consistent RAPW method by Harmon [127], the f-level for spin down states (with the width about 0.03 Ry) lies slightly above E_F because of large spin splitting. This should lead to giant values of $N(E_F)$, and, consequently, of electronic specific heat, which drastically contradicts the experimental data. Similar conclusions were obtained in [129–131] for Tb, Dy and Gd. These results demonstrate the inadequacy of usual band calculations within local-spin-density approximation for $4f$-states. This problem is discussed in recent papers [132] with application with gadolinium.

In a number of works, attempts to take into account correlation effects were made. Authors of papers [133–134] obtained a picture of $4f$- and conduction-electron states with account of correlation, screening and relaxation. The f-level energies for the ground and excited configurations f^n and f^{n+1} were determined by

$$\Delta_- = E(f^n) - E(f^{n-1}), \qquad \Delta_+ = E(f^{n+1}) - E(f^n), \tag{2.90}$$

Calculations were performed within non-relativistic [133] and relativistic [134] approximations.

The crystal potential in [133] was constructed from renormalized Hartree–Fock functions for the configurations $4f^n5d^{m-1}\,6s^1$ and $4f^{n\pm1}5d^{m-1\mp1}6s^1$ (m = 2,3). The renormalization (transfer of atomic wavefunction tails inside the Wigner–Seitz cell) resulted in a strong change of density for $5d$, $6s$ electrons and, because of screening effects, in a considerable (by about 0.05 Ry) increase of one-electron f-energies. The values of Δ_+ were corrected by introducing correlation effects from comparison with spectral data (the correlation energy was assumed to be the same in the cases of free atom and crystal). Although the correlation energy in each state may be large, their difference

$$\xi^{+}=E_{\text{corr}}(f^{n}d^{2}s)-E_{\text{corr}}(f^{n+1}ds) \quad (2.91)$$

makes up about 0.1 Ry only. The results of calculation of the quantity ξ^{+} are shown in Fig.2.28.

The edges of 5*d*- and 6*s*-bands were determined from the zero of the wavefunction and its derivative at the Wigner–Seitz cell boundary. The results turned out to be somewhat different from those of standard calculations. The 4*f*-bandwidth made up 0.03 Ry for γ-Ce, 0.01 Ry for Pr and Nd and smaller than 0.005 Ry for other rare earths.

One of important drawbacks of the calculation [133] was using of *f*-level energies ε_f which are averaged over many- electron terms. At the same time, spectral data, which were used, corresponded to transitions between ground states with definite *S* and *L*. Since the difference in energies of terms reaches 5eV, the error in the correlation energy may be large. A detailed calculation of the term structure for the f^{n-1} configuration was performed in [135]. The corresponding line spectrum should be observed in experiments with high resolution. As follows from equation (2.48), the averaged energy $\varepsilon_4 f$ may differ considerably from the ionization energy with the final state corresponding to the Hund term. The difference may be presented in the form

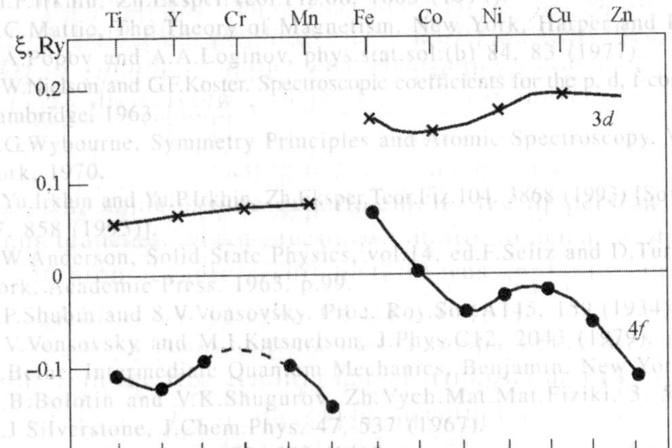

Fig. 2.28. Differences of correlation energies ξ for transitions $4f^{n+1}5d6s \rightarrow 4f^{n}5d6s$ in free rare earth atoms and for transitions $3d^{n+1} \rightarrow 3d^{n}$ between ground state multiplets of iron group atoms [133].

$$\left(-2\frac{b}{n}+\Delta b\right)F^{(2)}, \qquad \Delta b = b(f^n SL) - b(f^n S'L'), \tag{2.92}$$

where the coefficient $b(f^n SL)$ determines the contribution of the Slater integral $F^{(2)}$ to the SL-term energy, $2bF^{(2)}/n$ is the Coulomb energy per electron in the Hartree–Fock approximation, which is averaged over terms. In the rare earth series the value of Δb changes from 0 (f^0-configuration) to 113 (f^7-configuration), and $\Delta b(f^{n+7}) = \Delta b(f^n)$. Since the values of $F^{(2)} \simeq 0.05$eV are not too small, this contribution to the f-electron energy should be important.

In the relativistic calculations [134], an account of term structure was also performed. Final results for the effective one-electron energies differ weakly from those obtained in [133] (Table 2.5), but the changes in correlation corrections are appreciable. In particular, the correlation energy difference ξ^+, which is negative for most rare earths according to [133], becomes positive. One can interpret these data as increase of f–f correlation with increasing number of f-electrons.

Although the absolute calculated values of ε_{4f} and E_F are not quite reliable because of numerous approximations, general regularities in the rare-earth series seem to be qualitatively correct. This is confirmed by data on photoelectron spectra for the f-shell [136, 137].

A correct calculation should take into account also the change of wavefunctions in the final configuration f^{n-1}. The total relaxation in the final state ($f^n d^m \to f^{n-1} d^{m+1}$ transition, the hole in the 4f-shell is completely screened by additional electron in the 5d-state) was assumed in [133–134]. Such an assumption explains the decrease of 4f-electron binding energy in comparison with the Hartree–Fock approximation. An attempt of the calculation in absence of relaxation ($f^n d^m \to f^{n-1} d^m$ transitions) [134] agrees worse with experimental data.

Table 2.5. Values of $\Delta_- = E_F - \varepsilon(4f^n)$ and $\Delta_+ = \varepsilon(4f^{n+1}) - E_F$ for rare-earth metals.

$\Delta_\pm$ (eV)	γ-Ce	Pr	Nd	Pm	Sm	Eu	Gd	Tb	Dy	Ho	Er	Tm	Yb	Lu	La
Δ_-	2.1	4.1	5.1	-	5.6	1.5	8.7	4.6	5.6	6.0	6.0	6.2	0.7	-	-
Δ_-	1.9	3.8	5.1	-	5.5	1.9	8.3	3.3	4.7	5.5	5.4	5.4	1.1	7.5	-
Δ_-	1.9	3.3	4.5	4.5	4.5	-	7.0	1.9	3.4	4.8	4.8	5.2	-	8.5	-
Δ_+	3.3	2.6	2.1	-	0.5	-	3.3	2.1	1.8	1.5	1.5	1.3	-	-	-
Δ_+	3.1	2.0	1.3	-	0.2	-	3.2	2.1	1.4	1.6	1.6	1.0	-	-	4.7

Now we discuss briefly the results of band calculations for 5*f*-elements [138, 139, 57]. Band structure of actinides is to some extent similar to that of rare earths, but the strong spin-orbital coupling is accompanied with considerable delocalization of *f*-states. Thus the spectrum is appreciably disturbed owing to overlap and hybridization between *s,p*, 5*d*-states with 5*f*-bands. Concrete forms of the bandstructure and *f*-level positions in the self-consistent calculations depend strongly on the selection of the exchange-correlation potential. In the RAPW calculation of thorium by Keeton and Loucks [140] with the Slater potential ($\alpha = 1$) the position of the 5*f*-band (in the middle of the valence band) contradicted the de Haas–van Alphen data. However, the calculation [141] with the Gaspar–Kohn–Sham potential ($\alpha = 2/3$) yielded the correct 5*f*-band position (well above the Fermi level). Thus the band structure of Th is similar to that of *d*-metals. Despite the absence of *f*-electrons, the shape of the *d*-band is appreciably influenced by *d–f* hybridization.

According to [138], the 5*f*-bandwidth decreases with increasing atomic number from uranium (0.4 Ry) to plutonium (0.3 Ry). The electron spectrum of *U* near E_F contains a *s–d* conduction band which is strongly hybridized with a rather wide 5*f*-band. The 5*f*-bands in Pu are considerably more flat for $\alpha = 1$ than for $\alpha = 2/3$. This indicates that Pu lies on the boundary of the *f*-electron localization.

In the self-consistent calculation [142], the *f*-band turns out to be narrower than in [138]. The density of states of Pu contains sharp peaks with the height about 150 Ry^{-1}. The total value of $N(E_F)$ is 123.6 states/Ry (the *f*-contribution is about 50 states/Ry), which corresponds to the linear term in electronic specific heat with $\gamma =$ 21 mJ/mol · K^2. The experimental value makes up about 50 mJ/mol · K^2, so that the calculation seems to be qualitatively satisfactory, but correlation enhancement should be taken into account.

Relativistic calculations of the band structure in heavy actinides (Am, Cm, Bc) were also performed by Freeman and Koelling [138]. As compared to light actinides, the width of the 7*s*-band increases, 5*f*-bands become rather narrow and flat, the *sd*–5*f* hybridization being suppressed. Thus a strong localization of 5*f*-electrons takes place. As well as for rare earths, usual band description cannot be used in such a situation. An important role of correlations is confirmed by the high sensitivity of the spectrum to the number of *f*-electrons. The calculation for Am in the configuration $5f^7 7s^2$ (instead of the correct one, $5f^6 6d^1 7s^2$) yields a shift of 5*f*-states by 0.5 Ry. The delocalization of 5*f*-electrons in heavy actinides under pressure was investigated in the band calculations [143].

2.7 FERMI SURFACE

One of most powerful tools to verify the results of band calculations is investigation of the Fermi surface (FS). The structure of FS determines a number of the electronic properties of metals, in particular their anisotropy. The shape of FS depends on the geometry of Brillouin zones, crystal potential and interelectron correlations. The volume under FS is fixed by the Luttinger theorem: it has the same value as for the non-interacting electron system. It should be noted that this theorem is valid only provided that the Fermi-liquid picture (Landau theory) holds, and may be violated in system where strong Hubbard correlations result in splitting of one-electron bands (e.g., in systems which exhibit correlation-driven metal–insulator transitions [25]), so that the statistics of one-particle excitations changes.

In band calculations, FS is obtained as the constant energy surface which separates occupied and empty states at $T = 0$. If FS lies far from the boundaries of the Brillouin zones (e.g., at small band filling), it differs weakly from that of free electrons and has nearly spherical shape. At approaching the zone boundaries, the influence of crystal potential becomes stronger and anisotropy of FS increases. Especial strong singularities of FS occur at crossing a boundary. Since a Brillouin zone contains two electron states per atom (with account of the spin quantum number), in most cases one has to use several zones to place all the conduction electrons. Due to the anisotropy of the lattice and, consequently, of the electrons spectrum, zones with higher numbers start to be filled up when the lower zones are only partially filled, so that FS crosses several zones.

In transition *d*-metals, the valence electrons include *d*-electrons. An important feature of *d*-states is that they are highly anisotropic and poorly described by the free electron approximation even far from the Brillouin zone boundaries. Therefore the shape of the Fermi surface is very complicated and one has to use a large number of zones to describe it. Usually one considers Fermi surfaces of transition metals in the reduced zone picture.

2.7.1 Methods of Fermi surface investigation and de Haas-van Alphen effect

Experimental information on FS may be obtained from investigation of anisotropic electron characteristics of a metal. Most widely used methods of constructing FS which are based on magnetic field effects in electron spectrum. To first place, here belong so-called oscillation effects – de Haas–van Alphen (dHvA) effect (oscillations of magnetic susceptibility) and Shubnikov–de Haas effect (oscillations of conductivity), and also a number of magnetoacoustical and resonance effects [10].

Microscopical basis for the oscillations effects is the Landau quantization in a magnetic field. It is well known that at inclusion of the field H_z the orbital motion of a free electron becomes a quantum in the xy-plane. The quantization condition is written in terms of the area of the circumference with the radius $(k_x^2 + k_y^2)^{1/2}$:

$$S_\nu = \pi(k_x^2 + k_y^2) = \frac{2\pi eH}{\hbar c}\left(\nu + \frac{1}{2}\right), \tag{2.93}$$

where $\nu = 0, 1, \ldots$ are the numbers of Landau levels. The corresponding condition for energy reads

$$E_\nu \equiv E(k_x, k_y, k_z) - E(k_z) = \hbar\omega_c\left(\nu + \frac{1}{2}\right), \tag{2.94}$$

with

$$E(k_z) = \frac{\hbar^2 k_z^2}{2m}, \quad \omega_c = \frac{eH}{mc}. \tag{2.95}$$

being the kinetic energy of an electron moving in the z-direction and the classical cyclotron frequency which describes rotation of an electron in the xy-plane. Therefore the quantization in the magnetic field results in the occurrence of an oscillatory contribution to the electron energy. Comparing (2.93) and (2.94) we may write down

$$S_\nu = \frac{2\pi m_c}{\hbar^2} E_\nu, \tag{2.96}$$

where

$$m_c = \frac{\hbar^2}{2\pi}\frac{\partial S_\nu}{\partial E_\nu}. \tag{2.97}$$

the expression (2.97) being valid for an arbitrary dispersion law of conduction electrons [10–12]. Thus, at considering the electron motion in magnetic field the free-electron mass m is replaced by the cyclotron mass m_c. Unlike the band effective mass, which is defined at a point of the **k**-space, m_c is defined for a closed curve coresponding to an electron trajectory on the Fermi surface in the magnetic field.

The difference of energies for two neighbour Landau levels reads

$$E_{\nu+1} - E_\nu = \hbar\omega_c = \frac{2\pi eH}{\hbar c}\left(\frac{\partial S}{\partial E}\right)^{-1}. \tag{2.98}$$

Since the numbers ν are large and the energy differences are small in comparison with energies themselves, we may put

$$\frac{\partial}{\partial E}S(E_\nu) = \frac{S(E_{\nu+1}) - S(E_\nu)}{E_{\nu+1} - E_\nu}. \tag{2.99}$$

As follows from (2.98), (2.99), the areas of orbits in the **k**-space differ by the constant value

$$\Delta S = \hbar\omega_c \frac{2\pi eH}{\hbar c} \tag{2.100}$$

or

$$S(E_\nu, k_z) = (\nu + \lambda)\Delta S, \tag{2.101}$$

where λ does not depend on ν (typically, λ = 1/2).

Quantization of electron motion in *xy*-plane leads to the situation in which the quasi-continuous distribution of electron states is influenced by a discrete dependence on Landau level numbers. Therefore the resulting electron density of states $N(E)$ in the external magnetic field will exhibit an oscillating behaviour. To demonstrate the occurrence of oscillations, we suppose that the ν-th Landau level crosses E_F at some H for a given k_z, i.e.

$$E(k_z) + \hbar\omega_c(\nu + \lambda) = E_F. \tag{2.102}$$

At slightly decreasing H, ω_c increases and $N(E)$ decreases. However, with further decreasing field the Fermi level is crossed by the $(\nu + 1)$-th Landau level and $N(E)$ takes again the maximum value at $H = H'$. Subtracting the equalities (2.93) for two adjacent Landau levels one can see that the oscillating behaviour of $N(E)$ (and of related physical properties) as a function of the inverse magnetic field has the period

$$\Delta\left(\frac{1}{H}\right) = \frac{1}{H} - \frac{1}{H'} = \frac{2\pi e}{\hbar c}[S_\nu(E_F)]^{-1}. \tag{2.103}$$

To complete the consideration we have to sum over k_z which was up to now fixed. Since the energy E depends on k_z, for most k_z the condition (2.102) will be satisfied at different values of H and the summary oscillating contribution will be absent. However, this condition will approximately hold for a finite segment of k_z values provided that the energy varies very weakly in the k_z direction, i.e.

$$\frac{\partial E_{\mathbf{k}}}{\partial k_z} = 0, \tag{2.104}$$

or, which is equivalent,

$$\frac{\partial S_\nu(k_z)}{\partial k_z} = 0. \tag{2.105}$$

Thus the Landau quantization results in an oscillating behaviour of the electron density of states $N(E)$, this quantity having a singularity if E equals to the energy of a stationary (extremal) orbit determined by (2.105). The value of $N(E_F)$ which determines the most electron properties of a metal should be singular at

$$S_{\text{extr}}(E_F) = (\nu + \lambda)\frac{2\pi eH}{\hbar c}. \tag{2.106}$$

This permits to determine the extremal cross section areas by measuring the oscillation period.

At $T = 0$, near each energy E_0 which satisfies the quantization condition, $N(E) \sim |E - E_0|^{-1/2}$ (as well as in the case of one-dimensional Van Hove singularity). In real situations, the oscillation effects are smeared due to thermal excitations at the Fermi surface over the energy interval of order T or due to impurity scattering. The oscillations are observable at

$$k_B T < \hbar\omega_c \sim \frac{e\hbar}{mc}H = 2\mu_B H. \tag{2.107}$$

Usually one has to use the magnetic fields of order 10 kG at temperatures of a few K.

To obtain the oscillating contributions to physical quantities one has to perform the integration over k_z. In the case of free electrons the result for the magnetization at $T = 0$ reads [144]

$$M = \frac{eE_F}{4\pi^2\hbar c}(m\mu_B H)^{1/2}\sum_p \frac{(-1)^p}{p^{3/2}}\sin\left(\frac{\pi p E_F}{\mu_B H} - \frac{\pi}{4}\right). \tag{2.108}$$

The oscillation amplitude may be represented in the form

$$\bar{M} \sim \frac{1}{2\pi^3}\left(\frac{\mu_B H}{E_F}\right)^{1/2}\mu_B \tag{2.109}$$

and, for small H, is large in comparison with the smooth part of the magnetization (of the order of H). The diamagnetic susceptibility $\chi = dM/dH$ may be of order of unity, so that its oscillations are very strong.

The corresponding results for an arbitrary dispersion law are analyzed in detail in the monograph [10]. The oscillation period is obtained by the replacement

$$\pi\frac{E_F}{\mu_B} \to \frac{\hbar c}{e}S_{\text{extr}}(E_F)$$

in (2.108). Thus, investigations of magnetic susceptibility oscillations (de Haas–van Alphen effect) allow one to determine the areas of extremal cross sections of the Fermi surface. The temperature dependence of the magnetization oscillation amplitude enables one to determine the cyclotron mass:

$$\bar{M} \sim T \exp(-2\pi^2 m_c k_B T / e\hbar H). \tag{2.110}$$

In ferromagnetic metals, oscillation effects take place on the background of large spontaneous magnetization M_s. In particular, cross sections of the Fermi surface, which correspond to both spin projections, should be observed. For iron, the oscillation amplitude is by approximately nine orders smaller than M_s [145], but is still several orders larger than that which can be measured by the most sensitive modern technique.

2.7.2 Experimental and theoretical results on the Fermi surfaces

Investigation of the Fermi surface in TM is a considerably more difficult problem in comparison with simple metals. The reasons for this are as follows. High orbital degeneracy of *d*-electrons (unlike *s*,*p*-electrons) results in a large number of sheets of FS. Some of these sheets correspond to large cyclotron masses, which requires using strong magnetic fields in de Haas–van Alphen measurements. For magnetically ordered metals (Fe, Co, Ni, Mn, Cr) the picture is complicated by the presence of spin splitting of conduction band states. Besides that, there exist technical problems at preparing the samples which possess sufficient purity to observe oscillations. As a result, the situation in the fermiology of TM is far from total clearness – there are uncertainties in experimental results for a number of elements and ambiguities in theoretical models. Several *d*-metals (Sc, Hf, Mn, Tc, Ru) and many rare-earths and actinides have not been investigated. Data on cross sections with large m_c are often absent.

From the theoretical point of view, there exist problems connected with the applicability of one-electron picture in systems where electron correlations play an important role. Even within one- electron band calculations the results are rather sensitive to the choose of the crystal potential, so that the shape of FS may exhibit essential (in particular, topological) changes. This leads to difficulties in the interpretation of experimental data (e.g., in identification of different FS sheets and their details). The accuracy of band calculations does not as a rule exceed 0.5 eV. In a number of case, one has to fit theoretical results to experimental data by modifying some parameters, e.g., by shifting the Fermi level position or changing values of the gaps between different branches of electron spectrum.

However, some important results are now well established. First of all, existence of FS sheets with heavy electron effective masses, which are characteristic for TM and their compounds, should be mentioned. (Especially large values were found for the so-called heavy-fermion rare-earth and actinide systems which are discussed in detail in Chapter 6.) Thus FS investigations yield direct proof of delocalization of *d*-electrons. In particular, these data demonstrate the itinerant character of strong magnetism in iron group metals.

General information about the Fermi surfaces of TM (and, for comparison, of Li, Na and Ca) may be found in the Table 2.6 which contains data on classifications of sheets, stationary cross sections S and corresponding effective masses m_c. Since there exist a large number of different data for S and m_c/m, the results presented should not be considered as unambiguous. As a rule, we write down later data. In many cases, we restrict ourselves to writing down the interval of corresponding values. Most data on S are taken from the handbook [146], which contains the information up to 1981 and the detailed bibliography. More recent data are included provided that they seem to be important. Unlike [146], we give all the values of S in units of Å^{-2}. The transition from oscillation frequencies in Tesla (or periods in T^{-1}) may be made according to the formula

$$S\,(\text{Å}^{-2}) = 9.55 \cdot 10^{-5}\, F\,(\text{T}).$$

The values of m_c/m, which are also included in the Table 2.6, are of especial interest for transition metals.

Before discussing the FS characteristics of metals of each group (and, separately, of ferromagnetic TM Fe, Co, and Ni) we make some preliminary notes. As well as in the case of simple metals, correspondence between Fermi surfaces of TM with the same lattice structure may be established. In the rigid band approximation, the modification of FS reduces to a shift of the Fermi energy. This circumstance was used often (especially in the early works) for constructing FS at lack of theoretical calculations [11]. For example, the FS of Rh was obtained by Coleridge from that of Ni. The FS of bcc Fe was used to determine FS of a number of bcc TM (e.g., Cr, Mo, W). The type of FS is fairly good retained within a given column of the periodic system. However, with increasing the atomic number Z, the role of spin-orbital interaction increases. The latter may change qualitatively the form and topology of FS: because of the degeneracy lift for some energy bands, additional gaps in the spectrum occur and some open orbits may be transformed into closed ones. Deviations of FS in TM from that for free electrons become

Table 2.6. Characteristics of Fermi surfaces of some simple and transition metals, $e_N(i)$ and $h_N(i)$ stand for electron and hole orbits at the point N in the Brillouin zone with the number i, jg - jungle gym, mc - multiconnected. S are areas of cross sections; in some cases the corresponding k_F values are given in brackets. A_0, k_0 are the cross section and Fermi quasimomentum for free electrons.

Metal	Orbits theor.	S, Å^{-2} exp.	(k_F, Å^{-1}) theor.	m_c/m exp.	m_c/m theor	Ref.
Li	$e_\Gamma(1)$ (110) spher. (100) (111)	$0.976A_0$ $0.993A_0$ $1.011A_0$	$(1.06k_0)$ $(0.98k_0)$ $(0.99k_0)$		1.48 1.65 1.82	calc.147 – 149 dHvA 150
Na	$e_\Gamma(1)$ spher.	A_0	A_0	1	1	calc.147 – 149 dHvA 11
Ca	h(1) $e_L(2)$					calc.151,152 dHvA 153
Sc	$h_\Gamma(3,4)$ $e_{\Gamma k}(4)$, $h_{MK}(4)$					calc.154
Y		0.03 – 2.18				calc.155,156 dHvA 156,157
La	$h_{\Gamma A}(5)$, $h_\Gamma A(6)$ $e_{mc}(7)$ $e_{AH}(7)$, $e_{AH}(8)$					calc.158
Ti	$h_L(3,4)$,$e_\Gamma(5,6)$ or $h_\Gamma(3,4)$, $h_A(3)$, $e_H(5,6)$	0.10 – 0.72	0.18 – 0.64	0.65 – 1.95	1.6–2.8	calc.159 – 161 dHvA 162
Zr	$h_\Gamma(3,4)$ $e_H(5,6)$	0.22 – 1.34 2.29 – 3.36	2.0 –2.5		0.8, 1.2	calc.163 dHvA 164,165
V	$h_\Gamma(2)$, $h_{jg}(3)$, $h_N(3)$, closed	0.24 – 0.72	0.32 – 0.64			calc.165 – 167 dHvA 168 –171
Nb	$h_\Gamma(2)$, $h_{jg,N}(3)$ $h_{jg,\Delta}$	0.117 – 2.68	0.138 – 1.862	0.57 – 1.92	1.12 – 1.60	calc.172,173 dHvA 171,174
Ta	$h_\Gamma(2)$, $h_{jg,N}(3)$ $h_{jg,\Delta}$	0.292 – 2.00	0.279 – 2.00	0.84 – 1.66	1.09, 1.35	calc.172,173 dHvA 175
Cr	$h_H(3)$, $e_\Gamma(4)$ $h_{\Gamma H}(5)$	(0.109 – 0.226)	(0.17 – 0.27)			calc.176 – 181 dHvA 182,183
Mo	$h_H(3)$ $h_N(3)$,$e_\Gamma(4)$, $e_{\Gamma H}(5)$	0.06 – 2.25	0.05– 2.48	0.30	0.4 –1.5	calc.82,184 dHvA 145, 185,186
W	$h_H(3)$ $h_N(3)$ $e_\Gamma(4)$	0.014 – $0.43(2\pi/a)^2$	1 – 1.45 0.06– 0.23	0.9	0.25 – 1	calc.177 dHvA 185,187

Table 2.6. Continued

γ-Mn						calc 188,189
Re	h(5,6,7) $h_L(7)$ $e_{\Gamma A}(8)$ e(9)	0.48-1.18 4.57-6.11 6.6	0.42-1.6 4-6.7			calc.190 dHvA 191
Fe	$h_{H\uparrow}$ $h_{N\uparrow}$ $e_{\Gamma\downarrow}$	0.041- 4.16	0.037 4.16	0.36	0.71	calc.192-194 145 dHvA 195, 196
Co	$e_{\Gamma\uparrow}$, $e_{\Gamma\uparrow}$ $e_{ML\downarrow}$ $e_{L\downarrow}$	0.007- 1.42 0.025- 4.13	0.01- 0.16	0.1	0.08 0.2	calc.197-200 dHvA 201-203
Ni	e(5), e(6) h(3), h(4)	0.04- 0.75	0.03- 8.03	2.9, 3.7	1-1.9	calc.204-206, dHvA 207-209
Cu	neck, belly	$0.036A_0$ $0.98A_0$	$0.035A_0$ $0.98A_0$	0.41	0.46	calc.210 dHvA 211
Ru	e_Γ, e_{mc}, h_Γ, h_{LM}					calc.11 dHvA 212, 213
Os	h_{LM} (7) e_Γ (9) e_Γ (10) $h_{LM,\ mc}$	(0.057- 0.119)	0.015-0.030 1.71-2.19 1.33-1.64 0.73-1.18		0.12-0.2 1.2 1.1-1.5	dHvA 214
Rh	$e_{\Gamma 1}$ $e_{\Gamma 2}$ h_{x1} h_{x2} h_L	1.92-2.5 4.04-5.47 0.16-0.27 0.32-0.46 0.025	2.03-2.46 4.30 0.15-0.25 0.46 0.022	1.43-2.27 3.04-4.35 0.35-0.54 0.91-1.33 0.11	 1.2 0.11	calc.215 dHvA 216-219
Ir	$e_{\Gamma 1}$, $e_{\Gamma 2}$, h_{x1}, h_{x2} h_L	similar to Rh	similar to Rh	1.34-3.02 0.21-1.05		calc.215, 222 dHvA 216-221
Pd	e_Γ(s) h_x(d) h_{mc}(d) h_L	2.25-2.58 0.10-0.72 0.25, 1.32 0.02-0.03	2.30-2.59 0.10-0.60 0.05-0.08 0.28 0.24-0.98	1.34-1.47 0.41-0.7 1.37, 2.18 9.1, 10.53	2.2 0.63-1.91 2.40-8.1 0.84, 1.21	calc.215 dHvA 222-225
Pt	e_Γ(s) h_{open} (d) h_{closed} (d)	2.41-3.00 0.25-1.22 0.004	2.48-3.09 0.25-0.78 0.15	1.42-2.05 0.91-6.23 0.30	2.06-3.16 1.53-3.62 0.36	calc.215 dHvA 226,227
UPt_3	a-axis b-axis	0.057-0.63 0.041-0.21	0.058-0.63 0.044-0.27	1.6-5.3 1.3-4.2	25-90 13-50	dHvA 228

more appreciable with increasing Z in a given period, which agrees with the general statement about increasing the d-electron localization.

The role of d-electrons in the FS formation may be characterized by the value of m_c/m. One can see that this quantity increases in each d-period from left to right. In the beginning of d-periods (up to the Fe column) usually $m_c^{\text{theor}}/m < 1$ (except for Ti where $m_c^{\text{theor}}/m = 1.95$), but maximum values of m_c^{theor}/m are considerably larger. The latter fact is apparently connected with effective mass enhancement (electron–phonon, spin fluctuations, etc.) which is not taken into account in band calculations. At the same time, a satisfactory agreement between theory and experiment takes place for cross section areas. Starting from the Co column, large values $m_c^{\text{theor}}/m = 3.02–10.53$ occur, but the experimental values turn out to be smaller (e.g, for Ni, $m_c^{\text{theor}}/m \simeq 8$, $m_c^{\text{exp}}/m \simeq 1.9$; the largest values belong to palladium). This is probably due to the fact that the heaviest electron masses have not as yet been found. Unfortunately, incompleteness of experimental data does not permit to trace a regularity of m_c/m values in the periodic table.

Li, Na, Ca

To pick out peculiar features of TM, it is instructive to discuss first alkaline metals which are satisfactorily described by the free electron model (FEM). In particular, FEM yields correctly all the FS cross sections and effective masses of sodium (both experimental and theoretical). However, for lithium there exist some deviations, especially in m_c values. This is explained by a strong localization of the valence electron in Li near the nucleus, which is due to weakness of nuclear charge screening by the $1s^2$ shell (a similar effect takes place for d-electrons in TM since their mutual screening is also small). With increasing Z, disagreement with FEM increases and is appreciable for caesium.

The band spectrum and Fermi surface of calcium, which stands before d-metals in the periodic system, already differ qualitatively from predictions of FEM. Not all predicted stationary cross sections are experimentally observed. This may be explained by the presence of the above-discussed DOS peak near the Fermi level. Very strong are the deviations from FEM near the band top. Probably, they are connected with the hybridization with d-states which lie somewhat higher.

Sc, Y, La

Unfortunately, experimental data on *d*-elements of first column are poor because of absence of pure samples. Calculated FS of Y and Sc are similar and have multiconnected form with open orbits. They are strongly disturbed in comparison with FEM. Results of various calculations differ considerably, and their verification is difficult because of lack of experimental information.

Ti, Zr, Hf

The results of various band calculations demonstrate considerably different results. Thus unambiguous theoretical models of FS are absent for both Ti and Zr. In the work on the dHvA effect in Zr [164], an attempt of interpretation in the nearly-free electron model was performed. In the papers [159–161], experimental and theoretical cross section and m_c/m values for Ti are presented. The calculated FS is very sensitive to small variations of E_F position. The maximum value of m_c^{exp}/m is 2.8. The FS of Hf is not investigated, but theory predicts its similarity to FS of Zr.

V, Nb, Ta

The first attempt to construct the Fermi surfaces of these bcc metals was made in the rigid band model by using the calculation [192] for the bcc Fe. DHvA measurements in Nb and Ta [170, 171] yielded two oscillation periods which were related to holes at the point *N*. Other periods were attributed to 'jungle gym' features. A detailed comparison with theory for Nb and Ta was carried out in [172]. A satisfactory agreement (about 5%) was obtained for the cross sections areas of the types 2 and 3. The holes in the second zone at the point Γ were observed in dHvA experiments [171]. Maximum values of m_c/m is 1.8 which agrees with the data on specific heat.

Cr, Mo, W

The Fermi surface of chromium is strongly influenced by antiferromagnetism. The spin-density waves are connected with definite pieces of FS [11]. One of recent calculation was performed by Kulikov and Kulatov [181]. Unlike previous papers, appreciable deviations from the sphere were obtained. It was found that the electron masses of the ball orbits exceed considerably those of the ellipsoid orbits.

To understand main features of Mo and W, one has to remember that the Fermi level lies at a minimum of the density of states. Thus, the role of d-states is not so strong as in other TM. Probably, this is the reason for the orbits with large m_c/m being absent in both experiment and theory. For all the four FS sheets the values $m_c/m < 1$ are characteristic. The first calculations of FS in Cr, Mo, W used the band calculation for Fe [192]. More recent calculations [176, 177] supported in general the rigid band model. However, because of the strong spin-orbital interaction in W, the hole surface h_H in the third zone and electron surface $e_{\Gamma H}$ in the fifth zone vanish or strongly diminish. The LAPW calculation [178] confirms these old results.

Mn, Tc, Re

α-manganese, which has a complicated cubic structure and is antiferromagnetic at low temperatures, is poorly investigated, as well as radioactive technetium. Band calculations of bcc γ-Mn were made in [188, 189]. At the same time, there exist detailed results on rhenium, the theory [190] being in good agreement with experiment. Some uncertainty takes place for open orbits $e(8)$ (cylinder or torus?). Besides $m_c/m \cong 2.8$, there are very light masses $m_c/m \cong 0.07$.

Fe, Co, Ni

Extensive investigations of FS in Fe and Ni were performed by using dHvA and halvanomagnetic effects. It turned out that the dHvA data may be obtained in a usual way, although the oscillations are periodic in $1/B = 1/(H + 4\pi M)$ rather than in $1/H$. However, full picture of the Fermi surface is up to now absent.

Experimental data for Fe and Ni are discussed the review [145]. The main conclusion is the corroboration of the itinerant electron ferromagnetism picture: the sum of volumes, corresponding to spin up and down surfaces, agrees with the total number of conduction electrons, and the difference with the magnetization. It should be, however, noted that this conclusion is based not only on experimental results, but uses also some model notions. Besides that, the picture of spin-split bands is complicated by the spin-orbital interaction. The latter results in the degeneracy lift at points of accidental crossing of different spin and orbital subbands.

A detailed comparison of the dHvA results for Ni with the theoretical calculations is performed in [145]. The theoretical and experimental pictures demonstrate similarities. The s–d hybridization turns out to be important for understanding the experimental situation. The quasi-ellipsoidal pocket X_5 has an unusual form which is connected with the anisotropic interaction with the neighbour X_2 band. The largest experimental mass $m_c = 1.8m$ seem to correspond to the $X_{2\downarrow}$ pocket. The band calculation yields for the large belly surface in the [001] direction $m^{theor}/m \cong 8$ (from the electronic specific heat, $m^*/m \cong 10$). Unlike Pd and Pt, such large values were not observed in dHvA experiments.

The situation with FS of iron is considerably less clear. In particular, no theoretical calculations explain its shape in detail, and there exist several theoretical model to describe this [145].

Calculation of FS of cobalt was performed in [197]. In subsequent papers [198–200] detailed theoretical information on stationary cross sections and effective masses was obtained. Experimentally, a few of dHvA oscillation frequencies were found [201–203]. Two of them, α and β, correspond to cyclotron masses of $0.08m$ and $0.2m$.

Ru, Os

DHvA data for ruthenium [212] was interpreted in terms of the rigid band model by using the band structure of rhenium (see also [213]). For osmium, two groups of dHvA frequencies were found which differ by two orders of magnitude [214]. The largest ratio m_c/m is 1.5. The dependence m_c^{exp}/m on the area S turns out to be linear which corresponds to the quadratic dispersion law $E(\mathbf{k})$.

Rh, Ir, Pd, Pt

Despite the difference in the d-electron number, it is convenient to combine the d^7 and d^8 metals since they have the same fcc lattice and possess similar properties. Early experiments on the dHvA effect for Rh and Ir [216, 217] and Pd and Pt [223, 224] were interpreted in the rigid band model with the use of results for Cu and Ni. The model of the Fermi surface obtained is shown in Fig.2.29. The calculation in the relativistic APW method was performed in [215]. Comparison of theoretical and dHvA experimental values demonstrates an appreciable enhancement of m_c in Pt and Pd for light electron and holes. However, dHvA frequencies corresponding to heavy holes at open sheets (which yield about 80% of $N(E_F)$) with $m_c^{theor}/m = 9.1$ and 6.23 were not observed in early experimental works. Only in the paper [225], the values $m_c/m = 5.7$–8.1 were found for these holes in Pd. These values are still somewhat lower in comparison with the theoretical ones [215] and are by about 15% smaller than those required to explain data on electronic specific heat. At the same time, data on cross section areas [225] agree well with the theoretical predictions.

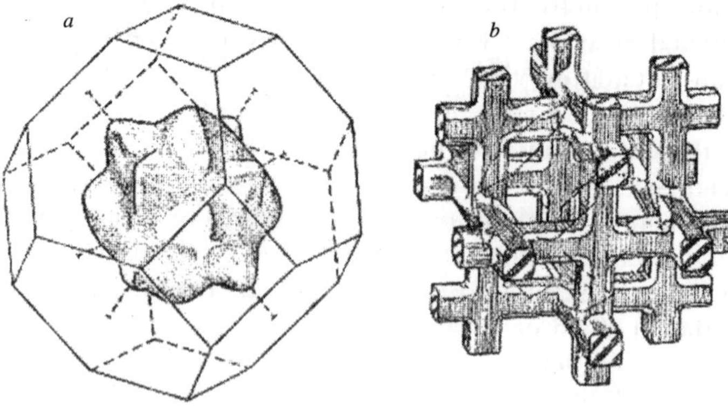

Fig. 2.29. The Fermi surface of Pd (a) electron surface at the point Γ (b) of multiconnected hole tubes [11].

RARE EARTHS AND ACTINIDES

Fermi surfaces of 4*f*- and 5*f*-metals were calculated in a number of works (see [11]). On the other hand, direct experimental data are not numerous. DHvA measurements in ytterbium [229] demonstrated the presence of hyperboloid pieces of FS. DHvA investigations of gadolinium [230] yielded the estimation of spin splitting Δ=0.57 eV.

Provided that 4*f*-states (that lie as a rule well below the conduction band) are not taken into account, FS of all the rare earth metals should be similar for the same crystal lattice. A FS model for the hcp structure which describes Gd and Tb is presented in Fig.2.30. The Fermi surface lies within third and fourth zones and is multiconnected. The shape of Fermi surface is in general similar to that of yttrium. At passing to heavy rare earths, changes occur which are connected with the degeneracy lift at some Brillouin zone points owing to the relativistic interactions. According to [231], these interactions result in that the 'arms' at the point M vanish, and the 'arms' at the point L touch, so that the similarity with FS of yttrium increases. Thus, one may expect that, at increasing atomic number, the Fermi surfaces of heavy hcp rare earths change their shape between those of Gd and Y. A number of attempts were made to use information on FS in rare earths for considering concrete physical effects. This explains the complicated non-collinear magnetic structures (Sect.4.7).

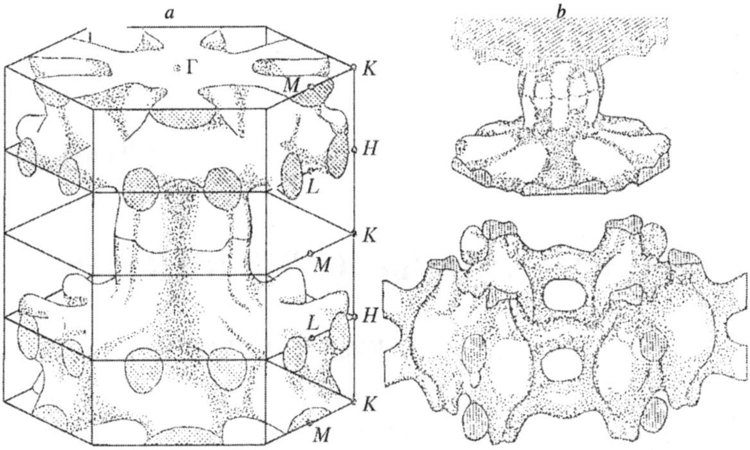

Fig. 2.30. (a) Theoretical model of the Fermi surface of gadolinium by Freeman, Dimmock and Watson (b) The Fermi surface of Tb in the one-band scheme: hole surface in the third zone and the electron surface in the fourth zone [11].

Actinides (5*f*-elements) have been investigated still less than rare earths. Unlike 4*f*-electrons, 5*f*-electrons, at least in light actinides, are delocalized and take part in the electron spectrum near the Fermi level. First FS calculations within the relativistic APW method for Th and Ac were performed in [232]. DHvA data for thorium, obtained in [233], were used to construct an empirical FS model. The latter contains a nearly spherical sheet at the centre of the zone and six closed sheets at the points *X* along the (100) directions. This model turned out to contradict the band calculation [140]. Some contradictions were eliminated in the subsequent calculation [234].

Chapter 3

THERMODYNAMIC PROPERTIES

3.1 COHESIVE ENERGY AND RELATED PROPERTIES

Condensation of atoms from the gaseous states into liquid and solid ones occurs due to various interatomic forces. The main contribution to binding energy comes from valence electrons which have large orbit radii, so that the wavefunctions at different lattice sites overlap appreciably. For metals, the valence electrons become itinerant and propagate freely in a crystal, the energy gain being determined by their interaction with positively charged ions.

As discussed in Chapter 1, *d*-electrons (and sometimes *f*-electrons) in transition group metals exhibit the behaviour which is intermediate between the localized and itinerant one. In particular, the collapse of *d*- and *f*-orbits in the beginning of periods takes place, which is accompanied by formation of high-energy states in the potential well (Fig.1.2). For *d*-elements the energy is close to that of valence *sp*-electrons (although the radius of *d*-shell is much smaller than the radii of *sp*-orbits). For 4*f*-electrons the well is much deeper, so that they may contribute appreciably to the binding energy only at the beginning of period, e.g., for cerium.

From the thermodynamical point of view, the total cohesive energy may be experimentally determined from the Born–Haber cycle. Thus main physical characteristics of the binding energy are the melting and boiling temperatures, T_m and T_b, and heats of fusion and sublimation, ΔH_f and ΔH_s. Besides that, the stability of the crystal is characterized by the density (or atomic volume V_a), ionization potentials, work function, Debye temperature, elastic moduli, activation energy for diffusion, specific heat etc. The corresponding values may be found in reviews and standard handbooks, e.g., [235–239]. Some of these quantities for *d*-, 4*f*- and 5*f*-metals are presented in Tables 3.1–3.3 and shown in Figs.3.1–3.8. As follows from comparison of the values of T_m for *d*-metals with those for preceding and following *sp*-elements, *d*-electrons yield an appreciable contribution to the binding

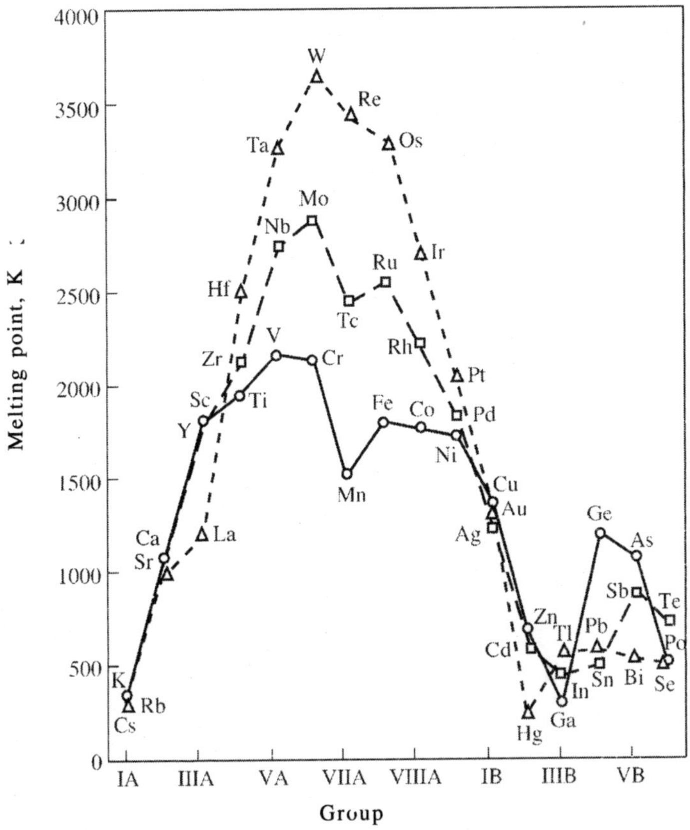

Fig. 3.1. Melting points of the IV, V and VI period elements of the periodic table [235].

energy. At the same time, the entropy change at melting ΔS_m in *d*-metals differs weakly from that in simple metals. In the periodic table, values of ΔS_m vary in a wide region (from 0.47 kcal/mol K for white phosphorus to 7.13 kcal/mol·K for Si). However, these regions become narrow for element groups with a given lattice: ΔS_m (in kcal/mol·K) equals to 1.76 ± 0.34 for bcc crystals, 2.28 ± 0.23 for fcc crystals, 2.33 ± 0.23 for hcp crystals and 6.50 ± 0.23 for the diamond structure [235].

The change of the thermodynamic and lattice parameters in the *d*-series enables one to investigate the dependence of the contribution of *d*-electrons on their number *n*. The strongest influence of *d*-electrons may be seen in the beginning of *d*-periods. The atomic volume decreases sharply (almost by two times) at passing from Ca to Sc, which is connected with the collapse of *d*-orbits. With further increasing *n*, V_a continues to decrease, but increases in Cu and Zn

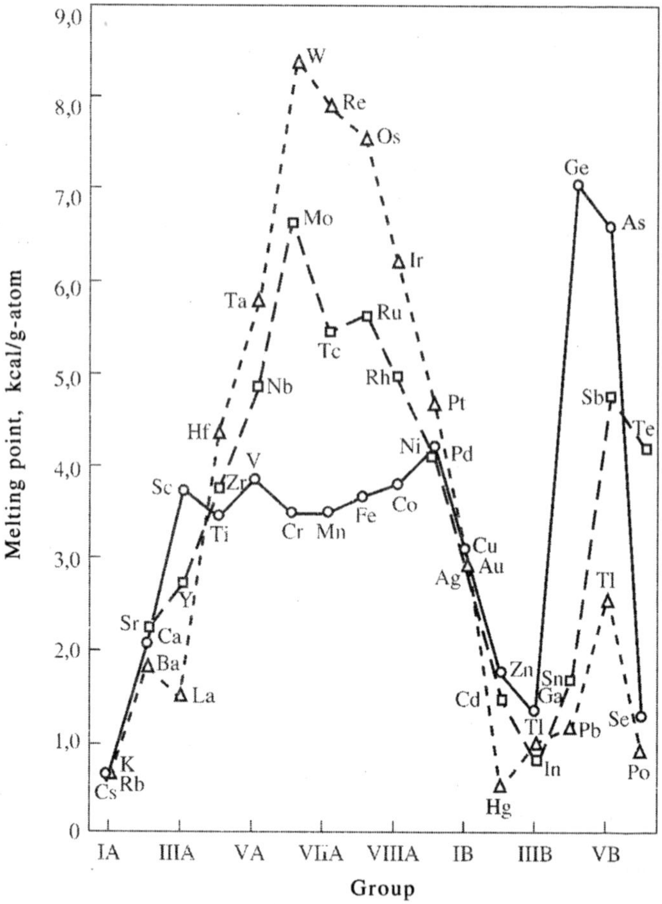

Fig. 3.2. Heat of fusion of the IV, V and VI period elements of the periodic table [235]. Open points are estimated values.

where the d-shell is completely filled. The values of the thermodynamic characteristics T_m, T_b, ΔH_f and ΔH_s, and elastic moduli of d-metals also exceed corresponding values for simple metals and demonstrate a smooth n-dependence. However, unlike V_a, these quantities have maxima in the middle of 3d-period (between V and Mn for different characteristics; a similar situation takes place for 4d and 5d-periods). We shall see in following Chapters that purely electronic characteristics (electronic specific heat coefficient γ and paramagnetic susceptibility χ) have a quite different (oscillating) n-dependence.

The simplest explanation of such a behaviour of the thermodynamic properties is the existence of several different factors which influence the total energy. In particular, with increasing number of

Table 3.1. Atomic volume V_a, Debye temperatureΘ_D, melting point T_m, boiling point T_b, linear coefficient of the thermal expansion α, self, self-diffusion activation energy Q, heat of fusion ΔH_f, cohesive energy ΔH^0, Young's modulus Y, shear modulus μ, Poisson's ratio σ, bulk modulus B, Leibfried, modified Leibfried and Bragg numbers, L, L' and B, for $3d$, $4d$ and $5d$ transition metals and neighbour elemets. For some metals, estimated values are presented [235, 238]

3d	Ca	Sc	Ti	V	Cr	Mn	Fe	Co	Ni	Cu	Zn
4d	Sr	Y	Zr	Nb	Mo	Tc	Ru	Rh	Pd	Ag	Cd
5d	Ba	La	Hf	Ta	W	Re	Os	Ir	Pt	Au	Hg
	d^0	d^1s^2	d^2s^2	d^3s^2	d^5s	d^5s^2	d^6s^2	d^7s^2	d^8s^2	$d^{10}s$	d^{10}
	d^0	d^1s^2	d^2s^2	d^4s	d^5s	d^5s^2	d^7s	d^8s	d^{10}	$d^{10}s$	d^{10}
	d^0	d^1s^2	d^2s^2	d^3s^2	d^4s^2	d^5s^2	d^6s^2	d^7s^2	d^9s	$d^{10}s$	d^{10}
V_a, cm³/mol	25.9	15.0	10.6	8.5	7.23	7.39	7.1	6.6	6.59	7.09	9.17
	33.9	19.9	14.0	10.8	9.39	8.63	8.18	8.29	8.88	10.3	13.0
	38.1	22.5	13.4	10.8	9.55	8.86	8.44	8.52	9.09	10.2	14.1
Θ_D, K	234	470	426	326	598	418	457	452	427	342	316
	147	268	289	241	459	351	600	480	283	228	352
	110	142	256	247	388	429	500	425	234	165	75
T_m, K	1112	1812	1941	2178	2148	1512	1808	1765	1726	1356	692
	1045	1775	2133	2741	2888	2443	2553	2233	1825	1234	594
	998	1193	2495	3271	3653	3433	3300	2716	2042	1336	234
T_b, K	1765	3537	3586	3582	2918	2368	3160	3229	3055	2811	1175
	1645	3670	4650	4813	5785	5300	4325	3960	3200	2468	1038
	1910	3713	4575	5760	6000	6035	5300	4820	4100	3240	630
$\alpha \cdot 10^6$, K^{-1}	22.4	10.0	8.35	8.3	8.4	22.6	11.7	12.4	12.7	16.7	29.7
	20	12.0	5.78	7.07	4.98	8.06	9.36	8.40	11.5	19.9	30.6
	18.8	10.4	6.01	6.55	4.59	6.63	4.7	6.63	8.95	14.1	61
$Q \cdot 10^3$, kcal/mol			48.8	91.5	73.2		64.0	61.9	67.0	48.9	23.9
		61.1	52.0	98.0	96.9				63.5	45.8	
		40.8	43.7	100	120				66.8	39.4	
ΔH_f, kcal/mol	2.07	3.70	3.42	3.83	3.47	3.50	3.67	3.70	4.21	3.12	1.76
	2.19	2.73	3.74	4.82	6.66	5.42	5.67	4.96	4.10	2.78	1.48
	1.83	1.48	4.39	5.76	8.42	7.86	7.56	6.22	4.70	2.95	0.55
ΔH^0 kcal/mol	42.1	80.1	112	122	94.5	66.9	99.4	102	102	80.8	31.0
	39.3	97.6	146	174	157	152	154	133	89.9	68.3	26.8
	42.8	102	145	187	200	186	187	159	135	87.6	15.4
$Y \cdot 10^{-6}$, kg/cm³	0.20	0.81	1.08	1.34	2.48	2.02	2.14	2.10	1.97	1.26	0.94
	0.14	0.66	0.94	1.07	3.34	3.76	4.20	3.70	1.26	0.82	0.63
	0.13	0.39	1.40	1.85	4.05	4.7	5.50	5.38	1.74	0.74	0.28
$\mu \cdot 10^{-6}$, kg/cm²	0.07	0.32	0.40	0.47	1.19	0.78	0.83	0.78	0.76	0.46	0.38
	0.05	0.26	0.35	0.38	1.18	1.45	1.63	1.50	0.52	0.29	0.25
	0.05	0.15	0.54	0.70	1.56	1.82	2.14	2.14	0.62	0.28	0.10
σ	0.31	0.27	0.34	0.36	0.21	0.24	0.28	0.33	0.30	0.34	0.29
	0.3	0.2	0.3	0.35	0.30	0.29	0.29	0.27	0.37	0.37	0.30
	0.28	0.29	3.30	0.35	0.28	0.29	0.28	0.26	0.38	0.42	0.36
$B \cdot 10^{-6}$, kg/cm²	0.15	0.58	1.07	1.65	1.94	0.61	1.72	1.95	1.90	1.33	0.61
	0.12	0.37	0.85	1.74	2.78	3.03	3.27	2.76	1.84	1.03	0.48
	0.10	0.25	1.11	2.04	3.30	3.79	4.26	3.62	2.84	1.77	0.29
L	4.80	3.20	3.42	4.66	2.12	2.24	2.60	2.87	2.90	3.51	1.69
	4.90	2.88	3.69	5.62	2.21	1.65	1.62	1.52	3.34	3.49	-
	4.44	2.95	2.91	3.67	2.08	1.80	1.55	1.26	3.06	3.94	3.51
L'	4.28	2.85	3.04	4.15	1.88	2.00	2.32	3.35	3.39	4.10	1.97
	4.36	2.56	3.29	5.00	1.97	1.93	1.90	2.78	3.90	4.07	-
	3.96	2.63	2.59	3.27	1.85	2.11	1.81	1.47	3.57	4.61	4.10
B	4.21	3.08	2.84	3.86	1.61	2.44	2.49	2.84	3.84	3.81	2.03
	4.84	2.09	3.06	4.66	2.40	1.73	1.70	1.59	3.54	3.70	-
	3.84	1.73	2.42	3.05	2.26	1.95	1.67	1.36	3.32	4.11	3.81

Table 3.2. Atomic volume V_a (cm^3/mol), Debye temperature Θ_D(K), melting point T_m(K), boiling point T_b (K), linear coefficient of thermal expansion α (10^{-6} K^{-1}), heat of fusion ΔH_f (kcal/mol), cohesive energy ΔH^0 (kcal/mol), Young's modulus Y (10^6 kg/cm^2), shear modulus μ (10^6 kg/cm^2), Poisson's ratio σ, bulk modulus B (10^6 kg/cm^2) for rare earth metals [235].

	La	Ce(γ)	Pr	Nd	Pm	Sm	Eu
	f^0d	f^1d	f^3	f^4	f^5	f^6	f^7
V_a	22.54	17.03	20.82	20.59	20.33	19.95	28.98
Θ_D	142	146	85	159	158	116	127
T_m	1193	1070	1208	1297	1308	1345	1099
T_b	3713	3972	3616	2956	2730	2140	1971
ΔH_f	1.48	1.24	1.65	1.71	1.94	2.06	2.20
ΔH^0	101.9	97.9	85.8	75.9	64	50.2	42.9
α	10.4	8.5	6.79	9.98	9.0	10.8	33.1
Y	0.387	0.306	0.332	0.387	0.43	0.348	0.155
μ	0.152	0.122	0.138	0.148	0.17	0.129	0.060
σ	0.288	0.248	0.305	0.306	0.278	0.352	0.286
B	0.248	0.244	0.312	0.333	0.360	0.300	0.150
	Gd	Tb	Dy	Ho	Er	Tm	Yb
	f^7d	f^8d	f^{10}	f^{11}	f^{12}	f^{13}	f^{14}
V_a	19.94	19.26	18.99	18.75	18.46	18.13	24.87
Θ_D	170	150	172	144	134	127	118
T_m	1585	1629	1680	1734	1770	1818	1097
T_b	3540	3810	3011	3228	3000	2266	1970
ΔH_f	2.44	2.46	2.49	3.38	2.62	4.22	1.83
ΔH^0	82.7	89.9	66.9	70.5	70.7	58.3	40.3
α	8.28	10.3	10.0	10.7	12.3	13.3	24.96
Y	0.573	0.586	0.644	0.684	0.748	0.77	0.182
μ	0.227	0.233	0.259	0.272	0.302	3.31	0.071
σ	0.259	0.261	0.243	0.255	0.238	0.235	0.284
B	0.391	0.407	0.392	0.404	0.419	0.405	0.135

d-electrons, *d*-levels become lower, so that the participation of *d*-states in the cohesion decreases. Besides that, the Coulomb repulsion among *d*-electrons also increases with n.

In the simple Friedel model the electron energy (2.84) has a minimum at $n = 5$, i.e. for the half-filled *d*-shell. This result is explained by that the energy of *d*-electrons decreases for the lower half of the *d*-band and increases for the upper one. Such a picture is in agreement with the situation of 3*d*, 4*d* and 5*d* periods except for an appreciable deviation for manganese.

Since the *d*-band width W_d is inversely proportional to some power

Table 3.3. Atomic volume V_a (cm^3/mol), Debye temperature Θ_D (K), melting point T_m (K), boiling point T_b (K), linear coefficient of thermal expansion α (10^{-6} K^{-1}), heat of fusion ΔH_f (kcal/mol), cohesive energy ΔH^0 (kcal/mol), Young's modulus Y (10^6 kg/cm^2), shear modulus μ (10^6 kg/cm^2), Poisson's ratio σ, bulk modulus B (10^6 kg/cm^2) for 5f-elements [235].

	Ac	Th	Pa	U	Np	Pu	Am	Cm	Bk	Cf
	f^0d^1	f^0d^2	f^2d^1	f^3d^1	f^5d^0	f^6d^0	f^7d^0	f^7d^1	f^8d^1	$f^{10}d^0$
V_a	22.56	19.79	15.03	13.16	13.11	12.06	17.78	12.8		
Θ_D	124	170	159	200	121	171				
T_m	1323	2024	1698	1404	910	913	1473	913		
T_b	3200	4500	4680	3950	4150	3727				
ΔH_f	3.03	3.56	2.99	2.47	1.60	0.68				
ΔH^0	104	136.7	132	125	113	91.8				
α	14.9	11.2	7.3	12.6	27.5	55	7.5			
Y	0.35	0.76	1.02	1.90	1.02	0.98				
μ	0.138	0.284	0.398	0.75	0.406	0.446				
σ	0.269	0.285	0.282	0.245	0.255	0.15				
B	0.25	0.553	0.078	1.007	0.694	0.546				

of the interatomic distance r_0, a dependence with a maximum for $n = 5$ should take place for the compressibility which is determined in terms of dW_d/dr_0.

An attempt to explain the anomaly of E_{coh} in Mn with account of the correlation effect were performed by Oles [240]. He started from the earlier paper [241] where the cohesive energy was estimated as the energy difference between the atomic limit and condensed state. We illustrate the idea of this explanation in the simplest case of s-band where the estimation gives

$$E_{\text{coh}} = \frac{n}{2}\left(1 - \frac{n}{2}\right) W - \frac{n^2}{4} U, \tag{3.1}$$

where U is the Hubbard parameter, $n < 1$. The expression (3.1) has a maximum at band filling

$$n = \frac{W}{W + U} < 1$$

Thus, a maximum is shifted from the middle of the band because of the Coulomb interaction. Since the the picture should be symmetric for the second half of the band (electron-hole symmetry), a minimum at $n = 1$ occurs.

Unlike d-metals, the most thermal and elastic characteristics of rare earths (Table 3.2, Fig.3.6) have only a weak (nearly linear) de-

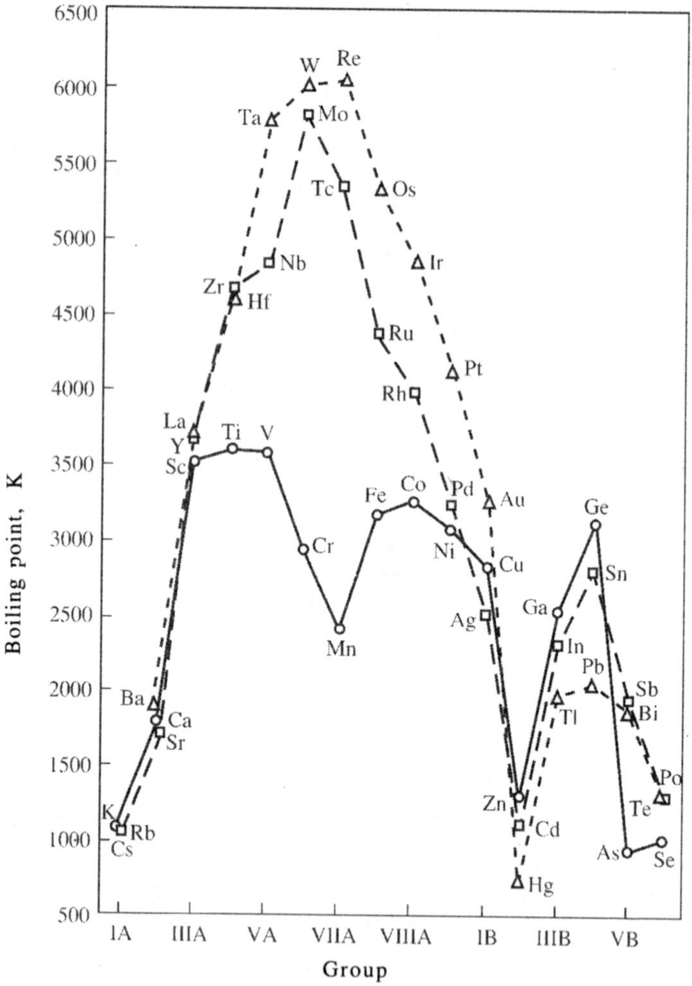

Fig. 3.3. Boiling points of the IV, V and VI period elements of the periodic table [235].

pendence on the number of f-electrons n_f. This agrees with the statement about a weak participation of $4f$-electrons in the chemical bond. At the same time, T_b and ΔH_0^0 (Fig.3.4) demonstrate a rather pronounced n_f-dependence. As discussed in [235], this fact may be connected with the difference in electron configurations of the solid and gaseous phases for some $4f$-elements. It is interesting that this dependence is reminiscent of the curves for the $4f$- electron energy which are obtained from spectral characteristics and band calculations [133–135] (see Sect.2.6). One can see that the strongest change

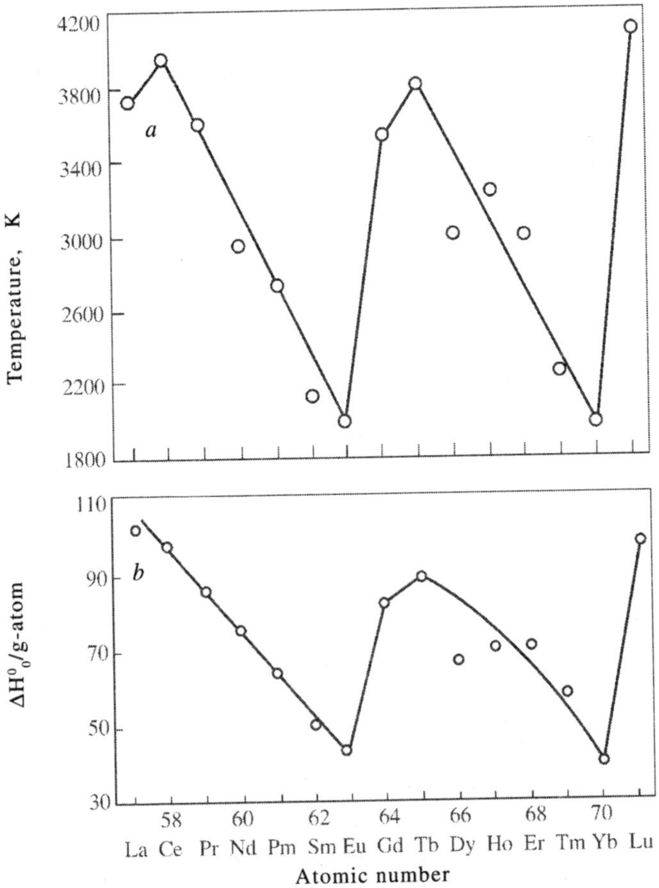

Fig. 3.4. (a) Boiling points of rare earth metals (b) Cohesive energy of rare earth metals. Open points are estimated values [235].

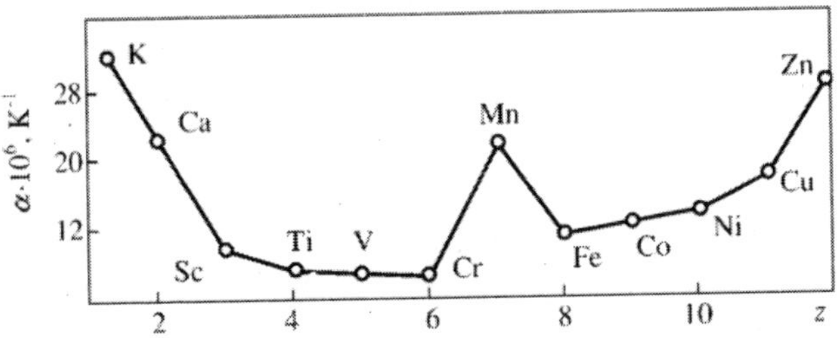

Fig. 3.5. Linear coefficient of thermal expansion of 3*d*-metals.

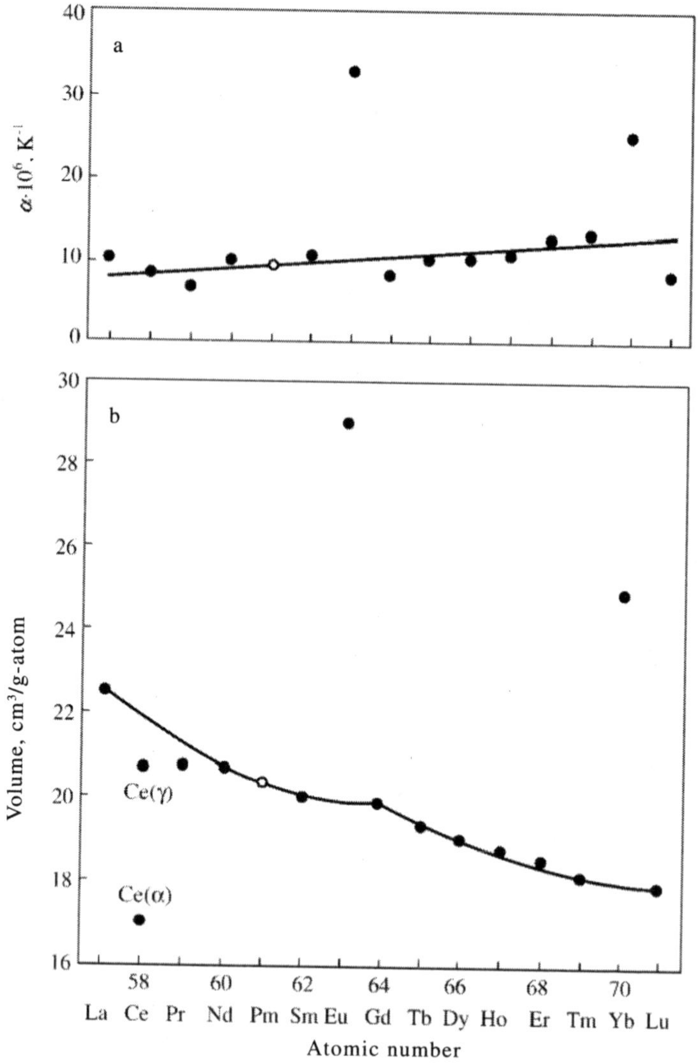

Fig. 3.6. (a) Linear coefficient of thermal expansion of rare earth metals (b) Atomic volume of rare earth metals. Open points are estimated values [235].

in T_b and ΔH_0^0 takes place at passing from Eu and Gd (unfortunately, there is only a rough estimation of T_b for Gd). In this sense, the situation is d and $4f$-periods is similar and may be related to high stability of the half-filled shells with the configurations d^5 and f^7.

The situation in actinides (where $5f$-electrons have an interme-

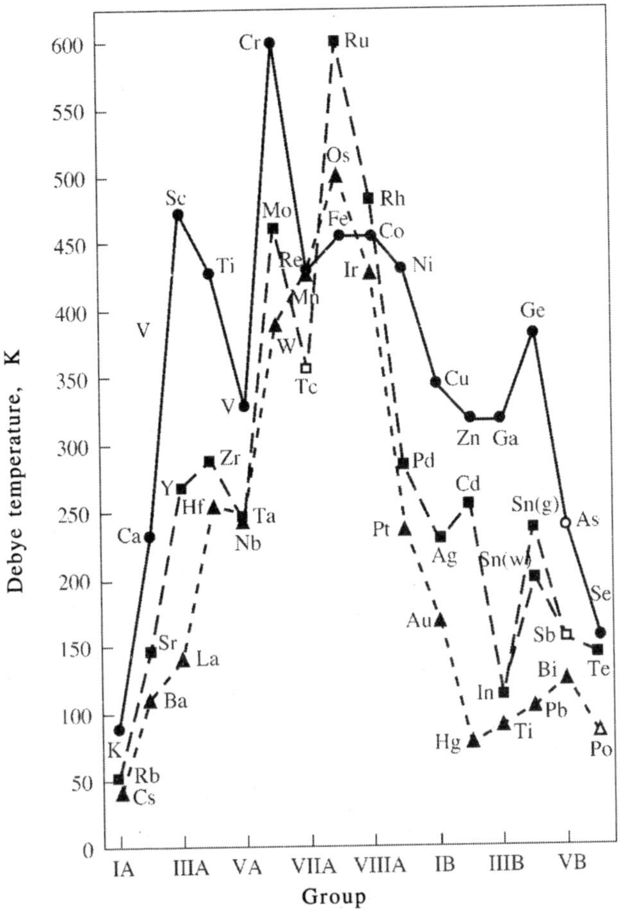

Fig. 3.7. Debye temperature at $T = 0$ as determined from specific heat data for the fourth, fifth and sixth periods of the periodic table [235]. Open points are estimated values.

diate degree of localization as compared to d and $4f$ electrons) is not clear in detail. According to Table 3.3, the atomic volume decreases considerably from Ac to Pu (there is an appreciable anomaly for Am). However, other thermodynamic and lattice characteristics do not exhibit a regular dependence on n_f. An accurate consideration should take into account the role of $6d$-electrons. The number of these for free atoms changes non-monotonously from 1 in Ac to zero in Pu with a maximum value of 2 in Th. However, the situation in metallic phase changes considerably since $5f$-states in light actinides where $5f$-electrons are strongly delocalized and hybridized with

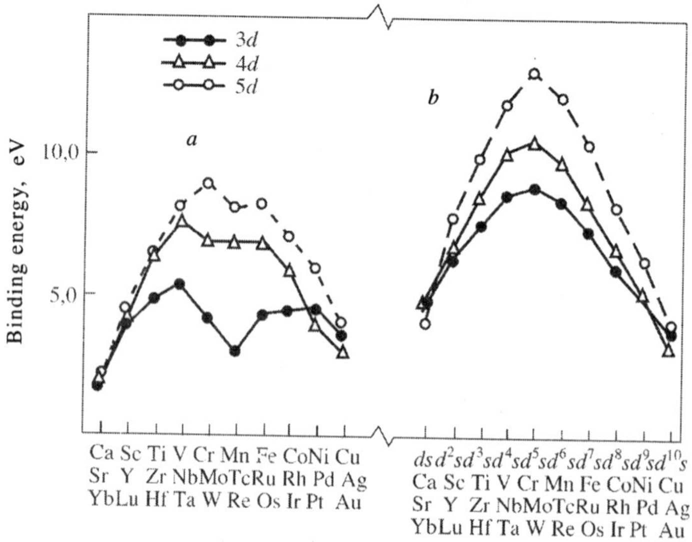

Fig. 3.8. (a) The measured cohesive energies of the *d*-metal series. (b) The corresponding calculated valence bond energies [236].

d-electrons. Experimental data for heavy actinides where 5*f*-electrons are well localized are insufficient to discuss regularities in thermodynamic properties.

Tables 3.1–3.3 demonstrate existence of some correlations between different thermodynamic and lattice characteristics. A number of theoretical and empirical relations, which describe these correlations, were established.

A correlation between the Debye temperature and T_m was obtained by Lindemann (1910). It is based on the idea that melting takes place when the atomic thermal displacement amplitude reaches some critical value $u = x_m d$ with d being the lattice constant. The corresponding calculation may be performed within the Debye model for lattice vibrations [6]. The displacement u of an ion ν may be written in Fourier representation

$$u_{\nu\alpha}(t) = \sum_{\mathbf{q}} u_{\mathbf{q}\alpha} \exp(i\mathbf{q}\mathbf{R}_\nu - i\omega_{\mathbf{q}\alpha}t), \tag{3.2}$$

($\mathbf{q}$ is the wavevector, α is the polarization). Using the virial theorem we obtain the expression for the vibration energy

$$\mathcal{E} = \sum_{\nu\alpha} M\,|\dot{u}_{\nu\alpha}|^2, \tag{3.3}$$

where M is the ion mass. On substituting (3.2) into (3.3) we obtain

$$\mathcal{E} = \sum_{\mathbf{q}\alpha} M \,|\, u_{\mathbf{q}\alpha} \,|^2 \, \omega_{\mathbf{q}\alpha}^2 \equiv \sum_{\mathbf{q}\alpha} E_{\mathbf{q}\alpha}. \tag{3.4}$$

Application of the Bose statistics yields for the averaged square of vibration amplitude

$$\overline{|\, u_{\mathbf{q}\alpha} \,|^2} = \bar{E}_{\mathbf{q}\alpha}(M\omega_{\mathbf{q}\alpha}^2) = \hbar\left(N_{\mathbf{q}\alpha} + \frac{1}{2} \right) \Big/ (M\omega_{\mathbf{q}\alpha}). \tag{3.5}$$

Performing the integration at high temperatures where $N_{\mathbf{q}} = T/\omega_{\mathbf{q}}$ in the Debye approximation ($\omega_{\mathbf{q}} = sq$, $q < q_{\mathrm{qD}}$) we get for the average square of atomic displacement

$$\overline{u^2} = 3\overline{|\, u_{\nu\alpha} \,|^2} = 3\sum_{\mathbf{q}} \overline{|\, u_{\mathbf{q}\alpha} \,|^2} = \frac{9k_{\mathrm{B}}T}{M\omega_{\mathrm{D}}^2} = \frac{9\hbar^2 T}{k_{\mathrm{B}} M \theta_{\mathrm{D}}^2}. \tag{3.6}$$

Then the melting temperature is estimated as

$$T_{\mathrm{m}} = x_{\mathrm{m}} M d^2 k_{\mathrm{B}} \theta_{\mathrm{D}}^2 / (9\hbar^2). \tag{3.7}$$

For most solids the parameter x_{m} equals 0.2–0.25. This universality supports strongly the adequacy of the Lindemann formula.

An empirical relation between the melting temperature T_{m} and cohesive energy reads [235]

$$T_{\mathrm{m}} = \frac{0.08 E_{\mathrm{coh}}}{3R \ln 2.045}, \tag{3.8}$$

where R is the gas constant. The relation between the thermal expansion coefficient α and T_{m} has the form [235]

$$\alpha T_{\mathrm{m}} = K. \tag{3.9}$$

For fcc, bcc and hcp crystals K = 0.0197–0.0051, and for crystals with the diamond structure K = 0.0039–0.0018. This correlation may be justified by that both a and T_{m} are connected with an anharmonic contribution to lattice vibrations.

From this point of view, it is instructive to reconsider the Lindemann correlation. Write down the expansion of the potential energy of a lattice with account of anharmonic terms

$$V(u) = V_0 = \frac{1}{2} a u^2 - b u^3 - c u^4 \tag{3.10}$$

(we confine ourselves for simplicity to the one-dimensional case). Then the average displacement reads

$$\bar{u} = \frac{3b}{a^2} k_{\mathrm{B}} T. \tag{3.11}$$

Assuming that melting takes place at $\bar{u} = u_{\mathrm{m}} = y_{\mathrm{m}} d$ we obtain the relation

$$k_B T_m = \frac{a^2}{3b} u_m . \tag{3.12}$$

Using the relation

$$\alpha = \frac{1}{d}\frac{d\bar{u}}{dT} = \frac{3b}{da^2} k_B \tag{3.13}$$

we derive the result (3.9) with $K = \bar{u}/d = y_m$, so that K is just the relative average displacement. An experimental verification of this statement is of interest.

In a similar way, the heat of fusion should be connected with the volume change at melting ΔV_m. However, systematic data for ΔS_m in transition metals are absent.

There exists also the relation between the melting temperature, the shear modulus $\mu = C_{11} - C_{12}$ and the atomic volume

$$rT_m/(\mu V_a) = \mathcal{L} \tag{3.14}$$

where the constant $\mathcal{L}$ is the Leibfried number, and a similar relation for the heat of fusion

$$\Delta H_f/(\mu V_a) = \mathcal{B}, \tag{3.15}$$

where $\mathcal{B}$ is the Bragg number. Since the entropy of fusion $\Delta H_f/T_m$ is approximately equal to R (the Richards rule [235]), we have $\mathcal{L} \cong \mathcal{B}$. However, the empirical values differ appreciably: $\mathcal{L} = 0.042$, $\mathcal{B} = 0.034$. One uses also the modified Leibfried number $\mathcal{L}' = K\mathcal{L}$ which takes into account the lattice structure ($K = 1.76$ for the bcc lattice and 2.29 for fcc and hcp lattices). The value of $\mathcal{L}'$ changes from 0.019 (Cr) to 0.041 (V) in the 3*d*-period and from 0.02 (Mo) to 0.05 (Nb) in the 4*d*-period. One can see from the Table 3.1 that the deviations of $\mathcal{L}$, $\mathcal{L}'$ and $\mathcal{B}$ from their average values have a rather regular character, so that correlation with electronic characteristics may be established (e.g., with specific heat, see Sect.3.3.2).

Although thermodynamic and lattice properties of metals are often considered in a phenomenological way, there exist also a large number of microscopical calculations within the band theory. The results on the cohesive energies, atomic volumes and bulk moduli for 3*d* and 4*d* row metals are presented in [24], and the atomic volumes and bulk moduli for 4*d* and 5*d* metals are given in [56].

Fig.3.8 shows, besides the measured cohesive energy E_{coh} of *d*-metals, the calculated valence bond energy

$$E_{vb} = E_{coh} + \Delta E_{at},$$

where ΔE_{at} is the preparation energy required to take the atom to a state corresponding to the non-magnetic ground-state configuration of the metal. This quantity is calculated as

$$\Delta E_{at} = E_p - E_{sp}^{LDA}.$$

Here E_p is the preparation energy required to take the atom to the magnetic ground state configuration, which may be obtained from experimental data [34]; the spin-polarization energy is the LSDA equivalent of the first Hund's rule energy. It is ΔE_{at} which behaves irregularly in the d-rows, and E_{vb} turns out to vary smoothly (approximately as a parabolic function).

Analysis of cohesive energy of all the $3d$- and $4d$ transition metals, and calculations of equilibrium lattice constant and bulk modulus of two representative elements, Cu and Ti, were performed in [242] within a renormalized-atom method (where the atomic valence wavefunction are truncated at the Wigner–Seitz radius and multiplied by a renormalization constant). Cohesive energy of a transition metal was presented as a sum of five terms: (i) the atomic preparation energy required to excite a free atom from its ground state to the d_s^{n-1} configuration appropriate to the crystal, $\Delta E_{at}(ds) - E_{cr}(d_s^{n-1})$ (ii) the difference in total Hartree–Fock energy between the free d_s^{n-1} atom and the renormalized atom (iii) the difference between the average energy of a free-electron band containing one electron and that of the renormalized atom s-level (iv) the change in one-electron energies which results from the broadening of the renormalized atom d-level into the d-band (v) the change in one-electron energy due to the hybridization between the conduction and d-bands. These contributions are shown in Fig.3.9. The problem of large cohesive energy in Cu, which exceeds by more than two times that of K, is discussed. This fact may be explained by the d-electron contribution owing to the s–d hybridization.

In the paper [243], results of calculation of the Debye temperature θ_D, Grueneisen constant and elastic moduli are presented for 14 cubic metals including TM. These results are obtained in the APW LDF approach with the use of Morse potential parametrization. The theoretical expression

$$k_B\theta_D = (48\pi^5)^{1/6}\hbar s(r_0 B\rho/M)^{1/2}, \qquad s = (B/\rho)^{1/2} \tag{3.16}$$

(r_0 is is the rigid-lattice equilibrium separation between ions, ρ is the density, B is the bulk modulus, s is the sound velocity) strongly overestimates the value of θ_D. Therefore the expressions (3.16) should be modified by introducing a scaling factor:

$$s = 0.617(B/\rho)^{1/2}, \qquad \theta_D = 41.63(r_0 B/M)^{1/2}, \tag{3.17}$$

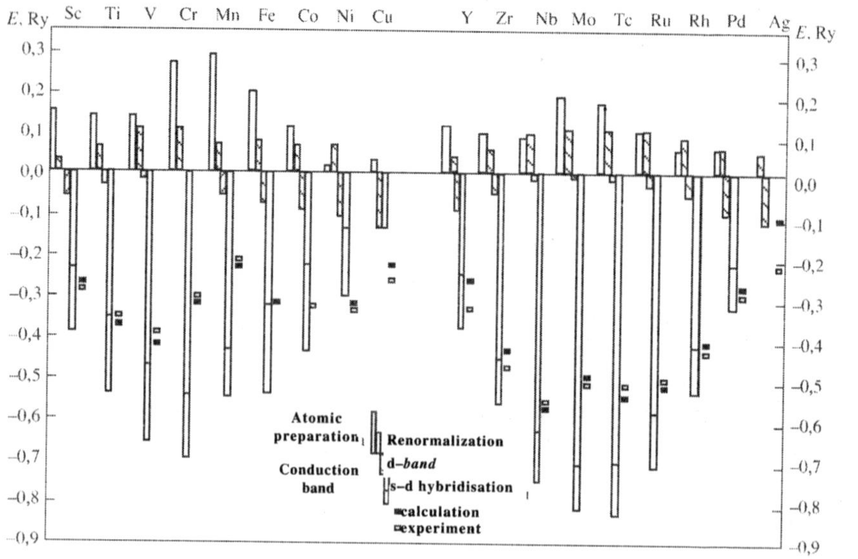

Fig. 3.9. Components of the cohesive energy for the 3d and 4d transition metals. The experimental value is denoted by the open box, and the calculated one by the filled box [242].

with r_0 in atomic units, B in kbars. The equilibrium value of r_0 is obtained by minimization of the free energy (including temperature-dependent vibrating-lattice contributions to entropy). The results for the temperature coefficient of linear expansion

$$\alpha(T) = d \ln r_0 / dT \tag{3.18}$$

are shown in Fig.3.10. The theory describes well the difference between the 'soft' alkaline metals and 'hard' transition metals. The calculated Debye temperatures determine the position of the 'knee', and the Grueneisen constants γ do the high-temperature amplitude of the $\alpha(T)$ curves. The theoretical values of γ turn out to be to high for Li and Sr and too low for Al, V, Cu and Nb. Authors of [243] conclude that the LDF approach may be used for description not only of ground states characteristics, but also of thermal properties.

Ab initio calculation of work function for 37 metals (including 3*d*, 4*d* and 5*d* series) was recently performed in the LMTO–ASA method [244]. The results agree within 15% with experimental data.

3.2 CRYSTAL STRUCTURE

Crystal structures of elements in the periodic system at $T = 0$ under normal pressure are shown in Fig.3.11.

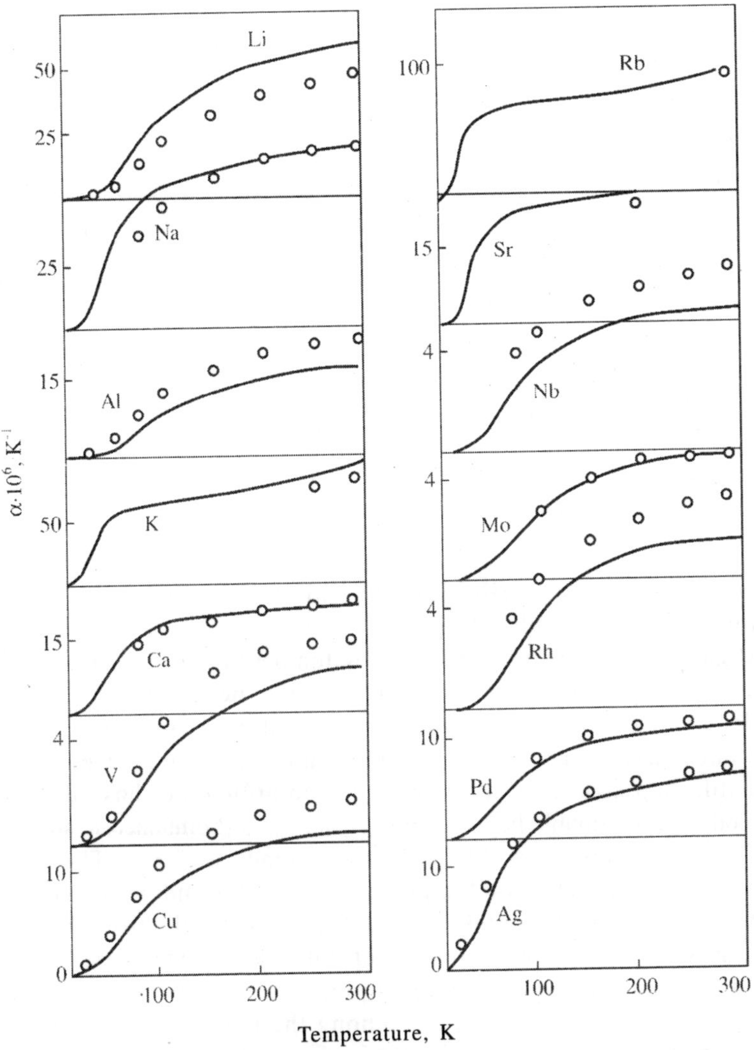

Fig. 3.10. The calculated coefficients of thermal expansion for non-magnetic cubic metals together with experimental points [243].

A number of elements exhibit phase transformations with a change of crystal structure (Table 3.4). In particular, for *d*-metals the closely packed fcc and hcp structures, which are stable in the ground state, transform often at high temperatures into a less dense bcc structure. This phenomenon may be explained by the softer phonon spectrum of the bcc lattice. Thus, with increasing temperature the free energy for the bcc structure decreases more rapidly owing to a large

0,0–0,8		1	2	3,5	5	6	7	8,5	9,5	10	n_d
Li hcp	Be hcp										
No hcp	Mg hcp										
K bcc	Ca fcc	Sc hcp	Ti hcp	V bcc	Cr bcc	Mn bcc	Fe bcc	Co hcp	Ni fcc	**Cu** fcc	Zn hcp
Rb bcc	Sr fcc	Y hcp	Zr hcp	Nb bcc	Mo bcc	**Tc** hcp	Ru hcp	**Rh** fcc	Pd fcc	Ag fcc	Cd hcp
Cs bcc	Ba bcc	Lu hcp	Hf hcp	Ta bcc	**W** bcc	Re hcp	Os hcp	Ir fcc	Pt fcc	Au fcc	**Hg** fcc
Fr bcc	Ra bcc										

Fig. 3.11. Crystal structures of elemental metals at low temperatures, n_d being the number of d-electrons in d-series [245].

negative entropy contribution and this structure becomes more favourable than the closely packed ones.

Most rare earth metals have at low temperatures the hcp structure and transform with increasing T into the bcc one [16, 246]. A tendency to decreasing the temperature interval for the bcc phase existence takes place with increasing atomic number. Cerium has a double hcp structure at $T = 0$ and a complicated phase diagram, praseodymium a double hcp structure, samarium a rhombohedric structure, and europium possesses the bcc lattice already at $T = 0$. There exist also data on existence of fcc phases for Pr, Nd and Yb [246, 247].

The light actinide metals with delocalized $5f$-electrons have the structures which are similar to those of d-metals, but the phase diagrams are more complicated. For example, plutonium has the largest variety of crystals modifications among the periodic table elements. The structures of heavy actinides (starting from americium) with localized $5f$-electrons are reminiscent of those of lanthanides.

Consider some approaches to the problem of determining crystal structures in metals from the point of view of electronic structure.

The simplest empirical correlation was established by Hume-Rothery [248] who found that crystal structures of stable phases for a wide class of metallic alloys were determined by the mean number of valence electrons per atom, n. A typical example is given by the Cu–Zn system [6]. From $n = 1$ (pure copper) to $n = 1.38$ the alloy has the fcc lattice (α-phase) At larger n there occurs the region of coexistence of α and β(bcc) phases, and starting from $n = 1.48$ the pure β-phase is stable. The γ-phase with a complicated cubic structure exists at

Table 3.4. Polymorphic transformations in Ca, Sr, d-metals, rare earths and actinides [238,139]. The temperature intervals (K) for stability of crystal structures under normal pressure are given in parentheses.

Ca	fcc (0–737), hcp (737–1123)
Sc	hcp (0–1223), bcc (1223–1811)
Ti	hcp (0–1158), bcc (1158–1938)
Mn	compl.cubic A12 (0–1000), compl.cubic A13 (1000–1365), fcc with tetragonal distortions (1000–1365), bcc (1405–1517)
Fe	bcc (α, 0–1183), fcc (γ, 1183–1163), bcc (δ, 1663, 1912)
Co	hcp (0–700), fcc (700–1763)
Sr	fcc (0–506), hcp (506–813), bcc (813–1163)
Y	hcp (0–1763), bcc (1763–1773)
Zr	hcp (0–1135), bcc (1135–2128)
La	hcp (0–583), fcc (583–1137), bcc (1137–1193)
Hf	hcp (0–2050), bcc (2050–2222)
Ce	fcc (α, 0–116), dhcp (β, 116–263), fcc (γ, 263–1003), bcc (δ, 1003–1068)
Pr	dhcp (0–1071), bcc (1071–1208)
Nd	dhcp (0–1141), bcc (1141–1297)
Pm	dhcp (0–1163), bcc (1163–1315)
Sm	rhomboedr.Sm-type (0–1190), fcc (1190–1345)
Eu	bcc (0–1099)
Gd	hcp (0–1535), bcc (1535–1585)
Tb	hcp (0–1560), bcc (1560–1633)
Dy	hcp (0–1657), bcc (1657–1682)
Ho	hcp (0–1701), bcc (1701–1743)
Er	hcp (0–1795)
Tm	hcp (0–1818)
Yb	hcp (0–1065), bcc (1065–1097)
Lu	hcp (0–1929)
Th	fcc (0–1400), bcc (1400–1750)
Pa	bct (0–1170), bcc (1170–1575)
U	orthorhomb.(0–662), tetr.(662–672), bcc (772–1132)
Np	orthorhomb.(0–278), tetr.(278–577), bcc (577–637)
Pu	monoclynic (α, 0–122), monoclynic (β, 122–206), orthorhomb. (206–310), fcc (310–458), bct (458–480), bcc (480–641)
Am	dhcp (0–1079), fcc (1079–1176)
Cm	dhcp, fcc (0–1340)
Bk	dhcp, fcc (0–986)
Cf	dhcp, fcc (α, 0–600), fcc (β, 600–725), (γ, >725)

1.58 $n < 1.66$. Finally, the hcp ε-phase becomes stable at $n = 1.75$.

A geometrical interpretation of this regularity may be obtained by considering the position of the Fermi surface in the reciprocal space with respect to the Brillouin zone boundaries [249]. Appearance of an energy gap in the electron spectrum near the Fermi level, which takes place near these boundaries, is energetically favourable. Indeed, it results in lowering of the electron system total energy since the energy of occupied states becomes lower. In the one-dimensional case, this effect results in distortion of a lattice with the formation of an insulating state (Peierls instability). Calculating the volume of the sphere, which is inscribed into the first Brillouin zone, we obtain the electron numbers $n = 1.36$ for the fcc lattice, 1.48 for the bcc lattice and 1.69 for the ideal hcp lattice in a satisfactory agreement with above experimental data.

Thus the Hume-Rothery rule relates the stability of a crystal structure to the touching of the Fermi surface with the Brillouin zone boundary where electron spectrum is maximally disturbed. Unfortunately, justification of this rule for simple metals within the qualitative band theory in the nearly-free electron approximation (pseudopotential theory) meets with difficulties: the singularity in the total energy turns out to be very weak in the three-dimensional situation (see discussion in [55]). However, one may expect that the influence of Brillouin zone boundary becomes appreciably stronger if density of states peaks are present near the Fermi surface.

For transition d-metals, the Hume-Rothery rule usually does not hold (at the same time, it may be applicable for rare earth alloys [16]). A different concept of correlation between crystal structure and n was put forward by Engel and Brewer (see [250, 251]). According to their theory, the type of lattice is determined by the number of s,p electrons. So, Na, Mg and Al, which possess one, two and three sp-electrons respectively, have bcc, hcp and fcc lattices. An attempt was made to generalize this concept to d-metals. In particular, the configuration d^5s (the same as for isolated atoms) is attributed to metals of the middle of d-periods (Cr, Mo, W) with the bcc structure. Formation of the same structure in V and its analogues is explained by high statistical weight of the d^4s-configuration (unlike isolated atoms). The hcp structure in the metals of III and IV groups corresponds to dominating role of configurations dsp and d^2sp respectively. The difference between bcc structure of iron (configuration d^7s) and hcp metals Ru and Os is related to strong s–d hybridization in the former.

At the same time, the applicability of the Brewer–Engel picture is rather restricted, so that it was criticized [251, 252]. Especially unsatisfactory is this theory for fcc metals. In particular, it yields incorrect predictions for alkaline earths Ca and Sr (s^2-configuration) and noble metals Cu, Ag, Au which have one *s*-electron. Generally, the supposition that *d*-electrons take part in chemical bonding but do not influence the lattice symmetry does not seem to be physically reasonable.

Of course, the direct calculation of the total energy from the real band electron spectrum is more justified for predicting stable structures, although this requires picking out small energy differences for various structures (of order of 10^{-2}–10^{-1} eV) on the background of large binding energy (of order of 10 eV). As a rule, such differences are smaller than real accuracy of band calculations. In particular, many-electron effects should give appreciable contributions and be taken into account in a quantitative theory of crystal structures. The problem of structural transitions at finite temperatures is still more complicated since the density functional approach becomes, generally speaking, inapplicable, and collective excitations of the lattice and electron subsystems should contribute to the entropy term. However, standard band calculations were rather successfully applied in the theory of crystal structures.

Stability of the bcc phase in the middle of the *d*-periods is connected with double peak structure of the density of states for this lattice. Maximal energy gain is achieved at filling of the lower peak. As follows from band calculations [24, 78], this peak is filled in vanadium at $n_d \simeq 4$ (which is by unity larger than in free V atom). For fcc and hcp lattices, the density of states has a more complicated many-peak structure, so that they should be stable at the beginning or end of *d*-periods.

From the atomic level point of view, one may consider in cubic lattices the states which have e_g and t_{2g} origin. The e_g-states lie lower than t_{2g} ones in the fcc lattice and higher in the bcc lattice. Since the e_g-band contains four electrons and the t_{2g}-band six electrons, the fcc structure should be favourable up to $n_d = 4$, and the bcc structure at n_d between 4 and 6. This agrees to some extent with the situation in the 3*d*-period (the bcc structure between Ti and Fe). However, Sc and Ti ($n_d < 4$) have the hcp (rather than fcc) structure, the stability of which is not described in terms of t_{2g}- and e_g-states.

Pettifor [253] calculated the energies of three basic structures (fcc, bcc and hcp) the 3*d*-metal series within the model pseudopotential approach with two *d*-resonance parameters. The stability of the crystal structures as a function of the number of *d*-electrons $n_d = n-2$ is described by the following sequence

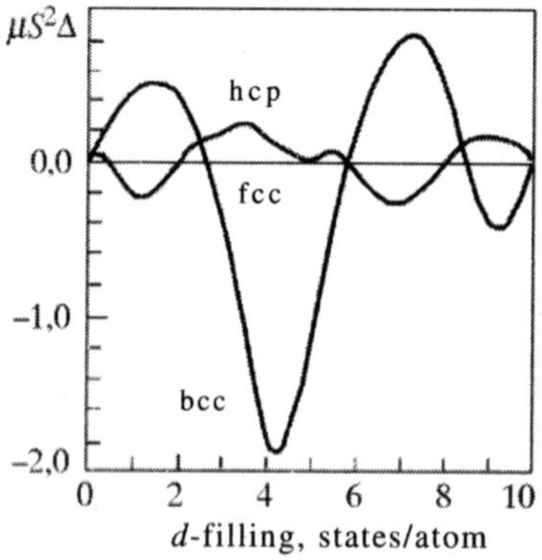

Fig. 3.12. Structural energy differences $\mu S^2\Delta$ ($\Delta = E_{bcc(hcp)} - E_{fcc}$, μ is the d-band mass, S is the atomic Wigner–Seitz radius), obtained from canonical d-bands, as a function of the canonical d occupation [245].

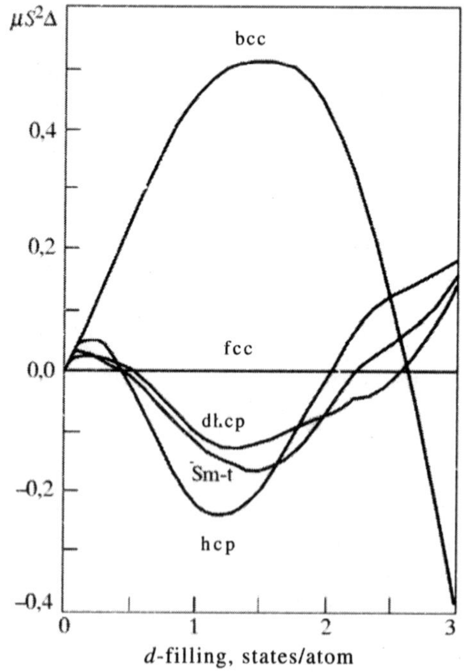

Fig. 3.13. Structural energy differences obtained from canonical d-bands as functions of the canonical d-occupation in the n_d-range, corresponding to the lanthanide crystal-structure sequence [245].

n_d	1−1.5	1.5−2	2−4.5	4.5−6.5	6.5−7.9
Structure	fcc	hcp	bcc	hcp	fcc

These results just correspond to the experimental situation in the 3d-row.

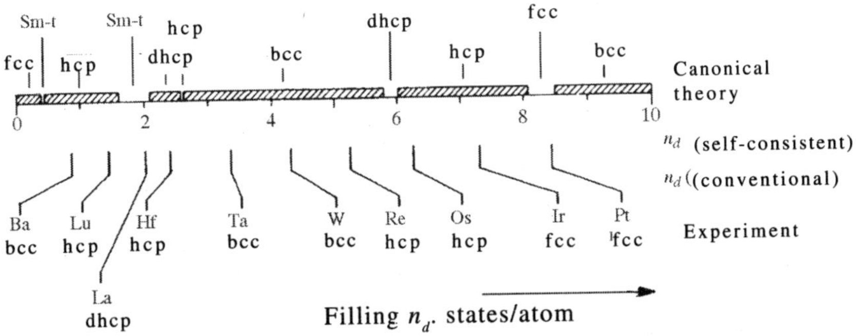

Fig. 3.14. Diagram of the most stable close-packed crystal structure as a function of the *d*-occupation number. Two estimates of the actual *d*-occupation numbers for the *d*-metals together with the experimental crystal structures are given below [245].

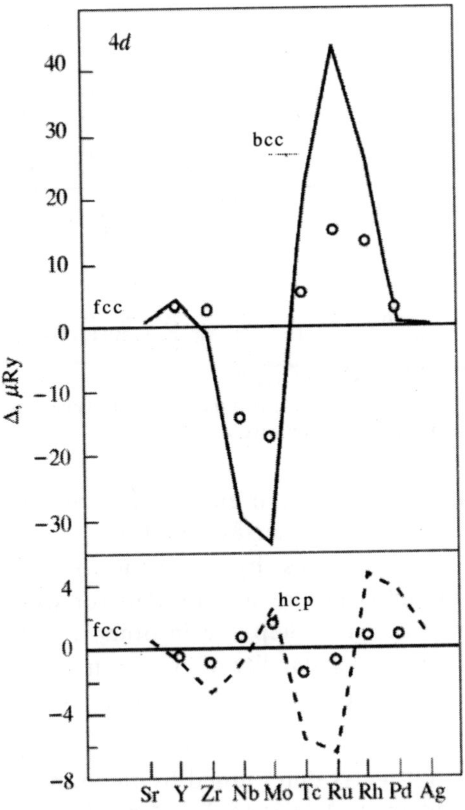

Fig. 3.15. Structural energy differences for 3*d*, 4*d* and 5*d* rows calculated at the experimentally observed equilibrium volume and plotted as functions of the *d*-occupation numbers [245].

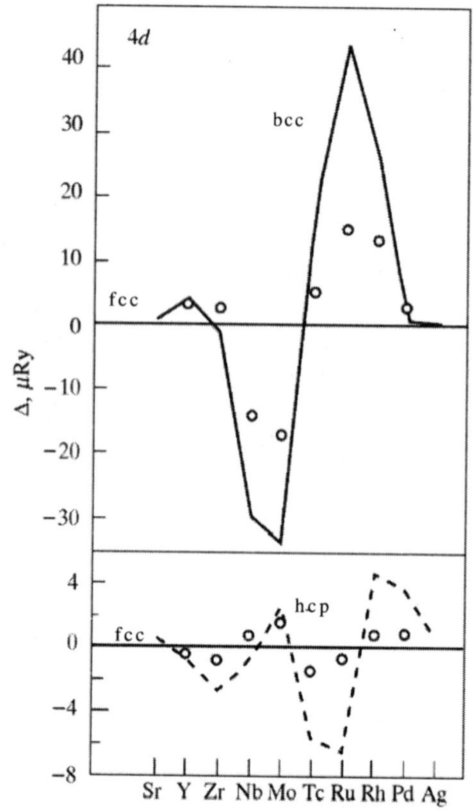

Fig. 3.16. Calculated bcc-fcc and hcp-fcc structural energy differences (solid and broken lines) for the 4*d*-metals [245] compared with the enthalpy differences derived from phase-diagram studies (open circles).

Skriver [245] evaluated the energy differences for various lattices in the 3*d*, 4*d* and 5*d* series and for rare earths (Figs. 3.12–3.16), and also for actinides, by using the force theorem approach in the LMTO method. The results by Pettifor were confirmed, and general sequence hcp–bcc–hcp–fcc in other *d*-periods was also reproduced. Investigations of stability of crystal structures in the lanthanide series yielded the sequence

$$\text{hcp} - \text{Sm-type} - \text{double hcp} - \text{fcc} - \text{fcc}'$$

The type of the lattice is determined by the number of 5*d*-electrons which is changed from 2 (La) to 1.5 (Lu) (Fig.3.14).

The one-electron band calculations [245] turn out to agree qualitatively with experimental data and give correctly the crystal structure of 35 metals from 42 investigated (in particular, the theory fails for Au,

Mn, Fe and Co). As to quantitative agreement (values of the phase diagram parameters, e.g. transition temperatures and pressures), the problem is considerably more difficult, so that satisfactory results are obtained for simple metals only. The calculated energy differences for *d*-metals [245] turn out to be by 3–5 times greater than enthalpy differences obtained from study of binary phase diagrams (Fig.3.16).

Structural and cohesive energies of 5*d*-elements were calculated in [254] within the linear augmented-Slater-type-orbital method. Recent calculation of total energies for 3*d*, 4*d* and 5*d* metals [255] was performed by scalar (semi-relativistic) RAPW method (unlike the calculations by Skriver [245], the frozen potential approximation was not used). The conclusions agree on the whole with the results for lattice parameters and bulk moduli [24] and crystal structure stability [245], relativistic corrections being not too important. However, the experimental structure for Au was obtained in this calculation.

Total energy calculations of crystal structure stability, equilibrium volume and bulk moduli for light actinides Th, Pa and U within the full-potential approach [256] yielded good agreement with experimental data.

3.3 SPECIFIC HEAT

In the previous sections of this Chapter we have treated energy characteristics of a crystal which yield a general description of the electron and ion systems. Here we consider the specific heat, which enables one to obtain a more detailed information about electron and lattice spectrum.

3.3.1 Lattice specific heat

Investigation of specific heat played an important role in the development of quantum solid state theory. We remind briefly of principal results. Specific heat at constant volume is defined in terms of the average energy

$$C_V = \left(\frac{\partial\bar{\mathcal{E}}}{\partial T}\right)_V . \qquad (3.19)$$

In the classical picture, each degree of freedom of a particle contributes $k_B T/2$ to the thermal energy. Thus the energy of a three-dimensional crystal which contains N ions and n electrons is given by

$$\bar{\mathcal{E}} = 3Nk_BT + \frac{3}{2}nk_BT \tag{3.20}$$

(the average energy of ions, which are considered as oscillators, is doubled owing to potential energy). Thus, the lattice specific heat of a monoatomic crystal should be equal to $3R$ ($R = k_BN_A$ = 8.3 J/mol · K is the gas constant, N_A is the Avogadro number). This result corresponds to the Dulong–Petit law which was established for a wide number of solids. However, experimentally this law holds as a rule in a not too wide region near the room temperature. With decreasing temperature, C_V decreases and tends, in agreement with the general Nernst theorem, to zero at $T \to 0$. On the other hand, C_V increases considerably at high T.

The quantum theory of lattice contribution to specific heat was developed by Einstein (1906) and Debye (1910). Einstein considered the crystal as a system of independent oscillators with the constant frequency ω_E. Then application of the Bose statistics yields

$$\bar{\mathcal{E}} = 3N\hbar\omega_E\left[N_B(\omega_E) + \frac{1}{2}\right], \tag{3.21}$$

with

$$N_B(\omega) = \left(\exp\frac{\hbar\omega}{k_BT} - 1\right)^{-1}.$$

Then

$$C_V = 3R\left(\frac{\hbar\omega_E}{k_BT}\right)^2 \exp\left(-\frac{\hbar\omega_E}{k_BT}\right)\Big/\left[\exp\left(-\frac{\hbar\omega_E}{k_BT}\right) - 1\right]^2. \tag{3.22}$$

Thus specific heat equals to $3R$ at high temperatures and is exponentially small at low T. Experimental data demonstrate a more slow decreasing of specific heat. This is due to that the Einstein model is well applicable only for optical branches of lattice oscillations (phonons), frequencies of which depend weakly on the wavevector. The calculation of acoustic branch contribution with the model dispersion law

$$\omega_\mathbf{q} = sq \quad (q < q_D), \qquad \omega_D = sq_D = k_B\theta_D/\hbar \tag{3.23}$$

was performed by Debye. The result reads

$$C_V = 3R\left(\frac{T}{\theta_D}\right)^3 3\int_0^{\theta_D/T} \frac{e^z z^4}{(e^z - 1)^2}dz. \tag{3.24}$$

At high T we come again to the Dulong–Petit law. At low $T \ll \theta_D$ specific

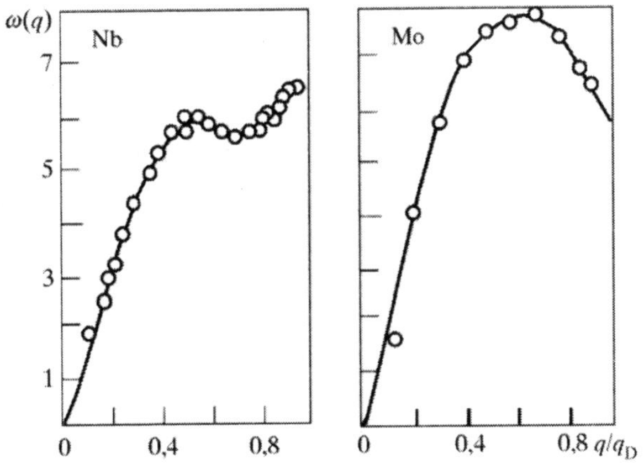

Fig. 3.17. Phonon dispersion curves in Nb and Mo [257].

heat obeys the Debye T^3-law

$$C_V = 12\frac{\pi^4}{5}R\left(\frac{T}{\theta_D}\right)^3. \tag{3.25}$$

A more accurate calculation of the phonon contribution to specific heat in a wide temperature region requires using a realistic dispersion law of phonons which may be very complicated. As an example, experimental data for acoustic branches in Nb and Mo are shown in Fig.3.17. One can see that the dependence $\omega(\mathbf{q})$ in Mo turns out to be non-monotonous. Anomalies in $\omega(\mathbf{q})$ curves are connected with strong electron–phonon interaction in transition metals.

Data on specific heat of simple and transition metals in a wide temperature region are presented, and regularities in $C(T)$ behaviour for different columns of the periodic table are discussed in the handbook [239]. The dependences $C(T)$ within a column turn out to be similar and differ mainly by singularities owing to structural and magnetic transitions. As an example, Fig.3.18 shows specific heat of vanadium where such phase transitions are absent, so that the dependence $C(T)$ has a simple form. One can see that the Dulong-Petit law holds in a not too wide temperature region and is violated at both low and high T. More complicated behaviour takes place in Zr (Fig.3.19) where the structural α–β transformation is present, and in heavy rare earths (Fig.3.20) which have a complicated magnetic phase diagram.

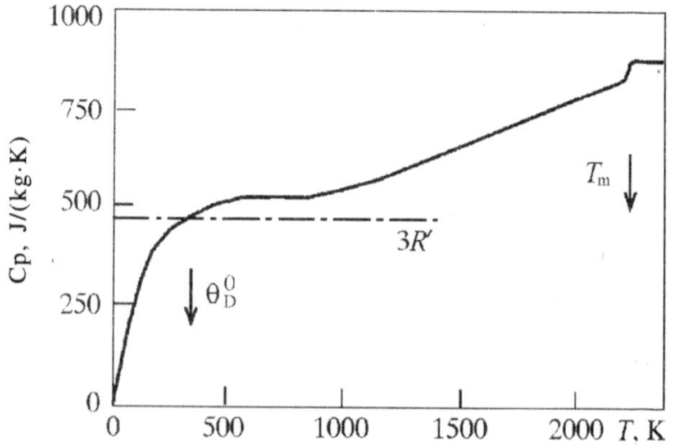

Fig. 3.18. Experimental specific heat of vanadium. The dash-dotted line correspond to the Dulong-Petit value ($R' = R/M$, M is the atomic weight) [239].

The jump of specific heat at the melting point, $\Delta C_{\mathrm{m}} = C(T > T_{\mathrm{m}}) - C(T<T_{\mathrm{m}})$, is as a rule positive for transition metals which exhibit phase transition (including light rare earths, Fe, Co, Ni) and negative for metals which do not (e.g., for Nb, Ta, although a small positive ΔC_{m} is observed in vanadium). It should be noted that the situation is different for simple metals: $\Delta C_{\mathrm{m}} > 0$ for Zn, Cd, Al, Ga, In, Pb despite the absence of structural phase transitions.

In the liquid phase, specific heat of d-metals may reach the values of $(4–9)R$, which are considerably larger than those for simple metals (see, e.g., the discussion in [258]).

The increase of C in comparison with $3R$, observed above the room temperature in most substances, is usually attributed to anharmonicity effects which result in occurrence of T-linear terms in the high-temperature lattice specific heat. We illustrate this phenomenon in the simplest one-dimensional model with the potential (3.10) (a more detailed discussion is given in [4]). Write down the partition function of an ion as

$$Z(T) = \int \exp\left(\left[-\frac{p^2}{2M} + V(u)\right] \Big/ k_{\mathrm{B}}T\right) dpdu. \tag{3.26}$$

Integrating over momentum p and expanding in the anharmonicity parameters up to b^2 and c we derive

$$Z(T) = (2\pi k_B TM)^{1/2} \int \exp\left(-\frac{au^2}{2k_B T}\right) \times$$
$$\times \left[1 + \frac{bu^3}{k_B T} + \frac{1}{2}\left(\frac{bu^3}{k_B T}\right)^2 + \frac{cu^4}{k_B T}\right] du. \tag{3.27}$$

Integration over u yields

$$Z(T) = 2\pi\left(\frac{M}{a}\right)^{1/2} k_B T\left[1 - \left(\frac{3c}{a^2} - \frac{15b^2}{a^3}\right)k_B T\right]. \tag{3.28}$$

Then the specific heat is given by

$$C(T) = -3N_A \frac{\partial}{\partial T}\left(T^2 \frac{\partial \ln Z(T)}{\partial T}\right) = 3R\left[1 + \frac{6}{a}\left(\frac{5b^2}{2a^2} - \frac{c}{a}\right)k_B T\right]. \tag{3.29}$$

Since $a/b \sim b/c \sim d$ and the average displacement is given by (3.11), the T-linear corrections owing to triple and quartic anharmonisms are of the same order of magnitude and may be estimated as $(\bar{u}/d)3R$. Therefore the anharmonicity may hardly explain the increase of specific heat at high temperatures in d-metals, which may be of order 100%, so that this problem needs further investigations.

3.3.2 Electronic specific heat

Besides the violation the Dulong–Petit law at low temperatures, the second difficulty of the classical theory of specific heat concerned the electronic contribution to specific heat. According to (3.20), this contribution should be equal to $3R/2$ for a metal with one electron per atom. However, such large contributions were never found in experiment. Only a small T-linear electron term in specific heat is observed practically at any temperatures.

This contradiction was resolved only in twentieth years of our century after formulation of the Fermi–Dirac statistics for electrons. The corresponding distribution function reads

$$f(E) = \left(\exp\frac{E - \zeta}{k_B T} + 1\right)^{-1}, \tag{3.30}$$

where $\zeta = \zeta(T)$ is the chemical potential, $\zeta(0)$ being equal to the maximum energy of occupied states – the Fermi energy E_F. The latter quantity is determined, according to the Pauli principle, by the number of electrons:

$$n = \int_{-\infty}^{\infty} f(E)N(E)dE \tag{3.31}$$

(in this Section the density of states $N(E)$ is determined for both spin projections). In metals, E_F/k_B is large and makes about 10^4–10^5 K. This explains the qualitatively small value of electronic specific heat. Indeed, only electrons in a narrow layer with a width about $\pi k_B T$ near the Fermi energy may change their energy and take part in thermal excitations, and most electrons are 'frozen'. The number of the 'thermal' electrons is estimated as

$$n^*/n \sim k_B T N(E_F) \sim k_B T/E_F, \tag{3.32}$$

so that

$$\delta \bar{\mathcal{E}}_e(T) \sim \frac{3}{2} n^* k_B T \sim k_B^2 T^2 N(E_F), \tag{3.33}$$

and

$$C_e \sim k_B^2 N(E_F) T.$$

A more accurate calculation (see below (3.40)) yields

$$C_e = \gamma T, \qquad \gamma = \frac{\pi^2}{3} k_B^2 N(E_F). \tag{3.34}$$

Thus electronic specific heat should be linear in a wide temperature interval $0<T<E/k_B$ (practically, up to temperatures which exceed the melting points). The electronic contribution may be picked out at low T where the lattice contribution rapidly decreases. To this end one writes down

$$C_V/T = \gamma + \alpha T^2 \tag{3.35}$$

and extrapolates the experimental dependence C_V/T vs. T^2 to zero temperature.

Table 3.5. Theoretical (obtained from band calculations) and experimental values of the coefficient of the linear specific heat in *d*-metals [78].

3d 4d 5d	Sc Y La	Ti Zr Nb	Y Nb Ta	Cr Mo W	Mn Tc Re	Fe Ru Os	Co Rh Ir	Ni Pd Pt	Cu Ag Au
γ_{theor} mJ/mol·K^2	5.25 5.46 –	1.40 1.37 1.15	431 3.44 2.96	1.67 1.40 0.99	6.49 – 1.91	2.63 2.00 1.65	2.08 3.24 2.20	4.13 5.58 5.18	0.70 0.62 0.69
γ_{exp} mJ/mol·K^2	10.7 10.2 10.1	3.35 2.80 2.16	9.26 7.79 5.90	1.40 2.0 1.3	9.20 – 2.3	4.98 3.3 2.4	4.73 4.9 3.1	7.02 9.42 6.8	0.69 0.65 0.73

According to (3.34), the value of γ is determined by the most important characteristic of electron system in a metal – the density of states at the Fermi level. Since transition metals are characterized by large values of $N(E_F)$, their γ's (Table 3.5) are considerably greater than for simple metals. Especially large values are observed in d-compounds with $N(E)$ peaks at the Fermi level. Here belong the superconductors with the A15 structure (for example, $\gamma = 33$ mJ/mol·K^2 for V_3Ga, see also Fig.6.1).

The experimental values of γ may be compared with those obtained from band calculation results,

$$\gamma\,(\mathrm{mJ/(mol\cdot K^2)}) = 0.1734 \cdot N(E_F)\,(\mathrm{states/Ry}).$$

As one can see from Table 3.5, a distinct correlation between these values takes place. As well as $N(E_F)$, γ_{exp} demonstrates a 'toothed' behaviour as a function of the d-electron number – large values for the configurations d^n with odd n and small values for even n (the ratio $\gamma_{exp}/\gamma_{theor}$ does not reveal such a behaviour). In the middle of d-series, $\gamma_{exp}/\gamma_{theor}$ is somewhat smaller than in the beginning or in the end.

The theoretical values of γ turn out to be as a rule smaller than experimental ones. This difference is larger for transition elements. So, for Sc, Ti, V, Y, Zr, Nb, Hf, Ta one has $\gamma_{exp}/\gamma_{theor} \sim 2$ and for neighbour simple metals Ca, Ba, Cu, Ag, Au this ratio is close to unity. Thus the deviation may be partially attributed to correlation effects which are more important in transition metals. In this connection we mention the recent first-principle band calculation beyond the density functional method [259] where a considerable change in $N(E_F)$ values is obtained, so that a better agreement with experimental data on γ takes place.

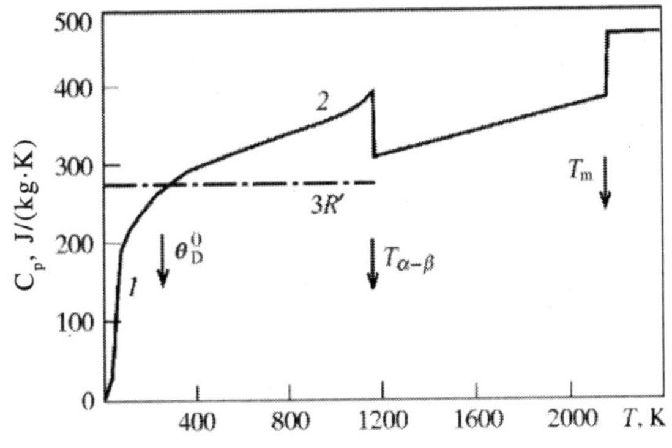

Fig. 3.19. Experimental specific heat of zirconium [239].

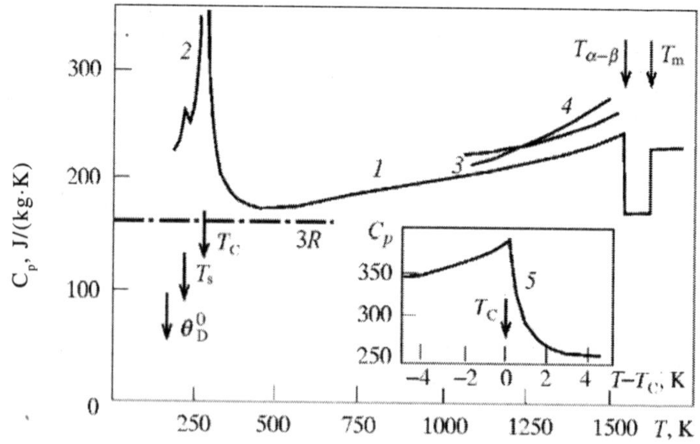

Fig. 3.20. Experimental specific heat of gadolinium. The inset shows the behaviour near the Curie point. The dependence near the melting point T_m is typical for all the heavy rare-earth metals [239].

Besides that, one has to bear in mind the inaccuracy in band calculations. Since the density of states in TM has a sharp energy dependence, small errors in the Fermi level position may influence strongly the value of $N(E_F)$. The relation $\gamma_{exp} < \gamma_{theor}$ for chromium is probably connected with the influence of antiferromagnetic ordering which is not taken into account in band calculations, but may disturb considerably electron structure near the Fermi level (formation of the antiferromagnetic gap).

The enhancement of γ_{exp} in comparison with γ_{theor} may be explained by effects of interaction of conduction electrons with various elementary excitations in a crystal. Most frequently one discusses the electron-phonon interaction which results in occurrence of a factor $1 + \lambda$ in the electron effective mass, and, consequently, in γ. The theoretical calculation of λ is a very difficult problem since this requires a detailed information on characteristics of electron and phonon subsystems.

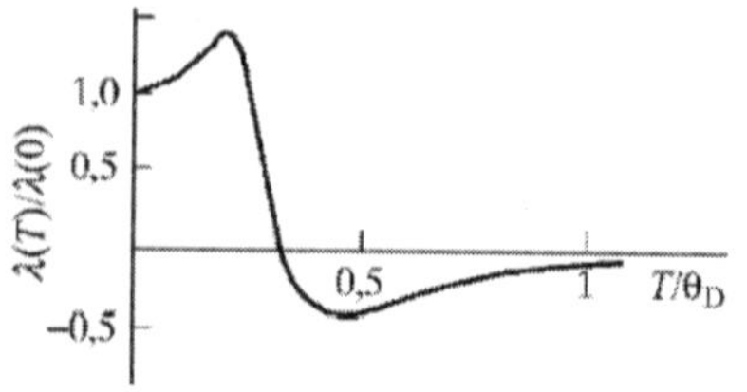

Fig. 3.21. Temperatu : dependence of the electron–phonon enhancement coefficient ($C_e(T) = \gamma_0[1+\lambda(T)]T$) in the Debye model [262].

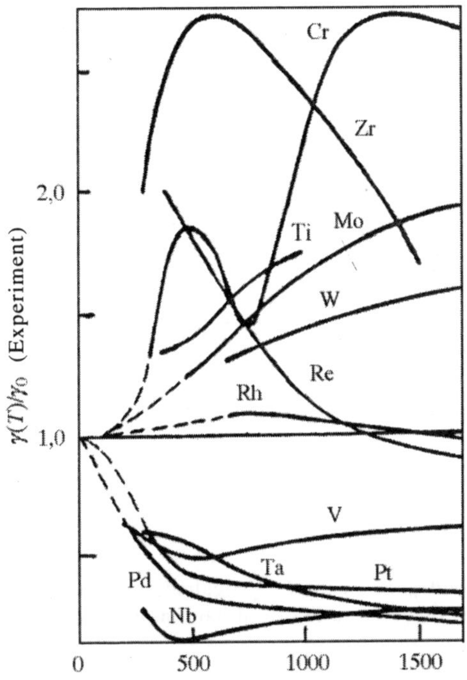

Fig. 3.22. Experimental temperature dependence $\gamma(T)/\gamma_0$ [262].

In transition metals the electron–phonon interaction is strong due to a strong dependence of total energy on the lattice parameter. The large value of λ leads also to superconductivity with rather high T_c [257]. It should be noted that λ should exhibit an appreciable temperature dependence. Calculations of this dependence were performed both in the Einstein and Debye models by Grimvall [260] and Osuzu [261] respectively. Results of [261] are shown in Fig.3.21. One can see that $\lambda(T)$ decreases rapidly at $T > 0.2\theta_D$.

Another important mechanism of specific heat enhancement is the interaction with spin fluctuations. This mechanism, which is especially important for weakly magnetic and nearly magnetic transition metals and their compounds (e.g., γ = 56 mJ/mol·K^2 for the nearly antiferromagnetic system $TiBe_2$ [507]), is discussed in Sect.4.4 and Appendix G.

An interesting example is metallic praseodymium where the interaction with crystalline electric field excitations is supposed to provide the enhancement $m^*/m \simeq 4$ [263]. Such effects of spin and charge fluctuations in anomalous rare-earth and actinide systems are considered in Sect.6.3.

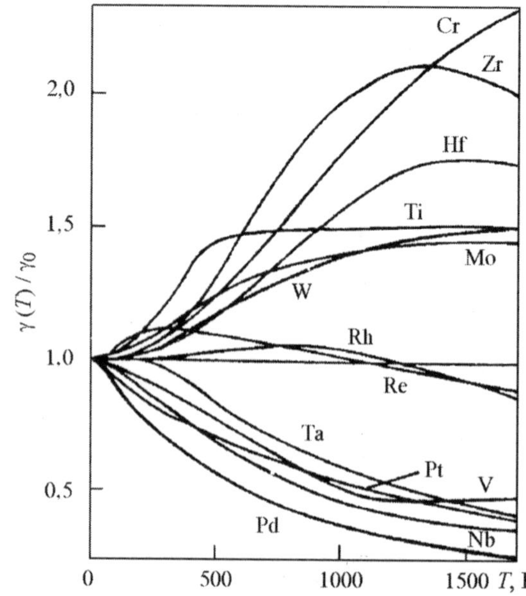

Fig. 3.23. Temperature dependence $\gamma(T)/\gamma_0$, calculated from the empirical density of states. The latter is obtained in the rigid-band model from the experimental γ_0 for alloys [262].

The temperature dependence of electronic specific heat in transition metals turns out to be strong. Corresponding experimental results are shown in Fig.3.22. It should be noted that picking out the electronic contribution on the background of the lattice one was performed in [262] within the Debye approximation. The latter is not quite satisafactory at high T where C_e makes up about 10% of the total specific heat.

The most simple reason for the dependence $\gamma(T)$ is smearing of the Fermi function with temperature. To consider this effect we consider the expressions for the number of particles (3.31) and for average energy

$$\bar{\mathcal{E}}_e = \int_{-\infty}^{\infty} Ef(E)N(E)dE. \tag{3.36}$$

Further we use the expansion of the Sommerfeld integrals

$$\int_{-\infty}^{\infty} f(E)F(E)dE = \int_{-\infty}^{\zeta} F(E)dE + \\ +\frac{\pi^2}{6}(k_B T)^2 F'(\zeta) + \frac{7\pi^4}{360}(k_B T)^4 F'''(\zeta) + \ldots, \tag{3.37}$$

which is obtained with the use of integrating by part. Then we derive

$$\zeta(T) = E_F - \frac{\pi^2}{6}(k_B T)^2 N'(E_F)/N(E_F) + \dots, \tag{3.38}$$

$$\bar{\mathcal{E}}_e = \int_{-\infty}^{E_F} N(E)dE + \gamma(T)T, \tag{3.39}$$

where

$$\gamma(T) = \frac{\pi^2}{3} k_B N(E_F)\left\{1 + \frac{1}{2}(\pi k_B T)^2\left[\frac{7}{5}\frac{N''(E_F)}{N(E_F)} - \left(\frac{N'(E_F)}{N(E_F)}\right)^2\right]\right\}. \tag{3.40}$$

The temperature corrections, although being formally small, are especially important in the case of a strong dependence $N(E)$, which is typical for transition metals. The first term in the square brackets means that finite-T specific heat is determined by the density of states, averaged over the energy interval of order of $\pi k_B T$ near E_F. In particular, at $N'(E_F) = 0$, $N''(E_F) > 0$ (minimum of the function $N(E)$) γ increases with T, and at $N''(E) < 0$ decreases with T. This result agrees qualitatively with the band calculations results and experimental data on the sign of the T-dependence (Fig.3.22): for odd d^n configurations (V, Ta, Nb, Pt, Pd) the great values of γ, which correspond to a $N(E)$ maximum, decrease with T, and for even configurations (Zr, Ti, Cr, Mo, W) the small values of γ correspond to a $N(E)$ minimum and increase with T. The second term in square brackets of

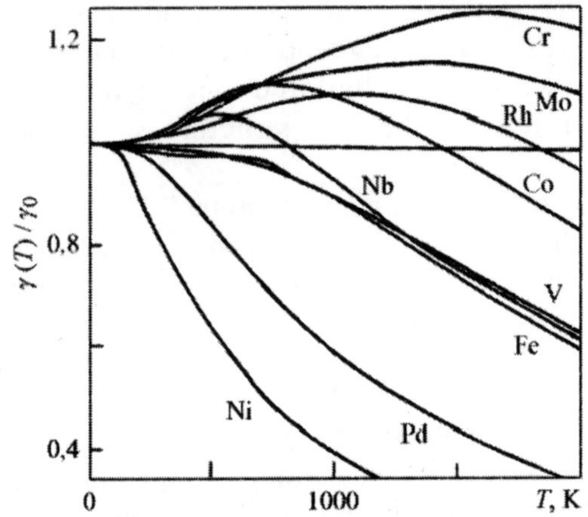

Fig. 3.24. Temperature dependence $\gamma(T)/\gamma_0$, obtained from the calculated density of states [262].

(3.40) occurs because of the temperature dependence of the chemical potential (3.38).

Shimizu [262] carried out numerical calculations to compare the experimental dependence $\gamma(T)$ with those following from the first-principle (Fig.3.23) and empirical (Fig.3.24) densities of states. Although a qualitative agreement takes place, the theoretical dependences turn out to be weaker and shifted towards high temperatures.

Provided that the function $N(E)$ possess a singular behaviour near E_F (narrow peaks), the expansion (3.37) does not hold and the analytical calculation is more complicated. The calculation of the dependence $\gamma(T)$ in the presence of a DOS peak was performed in [264]. The simplest model of $N(E)$ with triangle peaks at $E = E_i$ was used,

$$\delta N(E) = \sum_i \delta N_i(E), \tag{3.41}$$

$$\delta N(E) = \begin{cases} A_i(E - E_i), & E < E_i, \\ B_i(E_i - E), & E > E_i. \end{cases}$$

Unlike the expansion (3.40), the relative temperature variation of

$$\gamma(T) = \frac{1}{T^2}(I_2 - I_1^2/I_0), \tag{3.42}$$

with

$$I_n = \int_{-\infty}^{\infty} (E - E_F)^n N(E)\left(-\frac{\partial f(E)}{\partial E}\right) dE. \tag{3.43}$$

may be large. Using the position of the peaks E_i and the jumps of $N'(E)$ at these points as parameters, one can restore from the experimental dependence of γ the density of states and compare this with the calculated one. Such an analysis was performed in [264] for the A15 structure superconductors Nb_3Sn and V_3Si.

Besides the one-electron mechanism, the dependence $\gamma(T)$ is determined by the above-discussed temperature dependences of electron-phonon enhancement and correlation mechanisms which are expected to be rather strong for d-bands. The relative role of these effects in transition metals is still not investigated in detail.

3.3.3 Specific heat of magnetic metals

In magnetically ordered crystals, contributions to entropy and specific heat occur which are due to destruction of magnetic ordering with increasing temperature. Experimental separation of the magnetic entropy may be performed with the use of magnetocaloric effect (change of temperature at adiabatic magnetizing by the external magnetic field) [265].

In the model of localized spins with the value S the total magnetic entropy change is given by

$$\mathcal{S}_{\mathrm{mag}}(T \to \infty) = \int_0^\infty \frac{C_{\mathrm{mag}}(T)}{T} dT = R \ln(2S+1). \tag{3.44}$$

The magnetic specific heat C_{mag} is especially important at the magnetic phase transition points $T_{\mathrm{M}} = T_{\mathrm{C}}(T_{\mathrm{N}})$ where it has a singularity. In the simplest mean-field approximation C_{mag} is exponentially small at

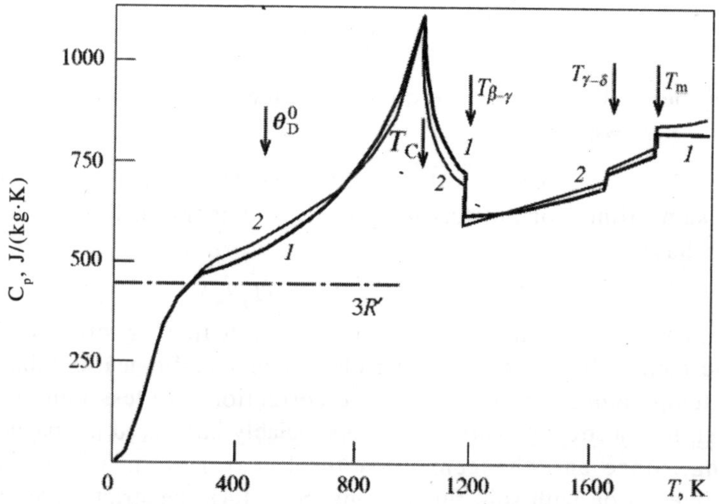

Fig. 3.25. Experimental specific heat of iron [239].

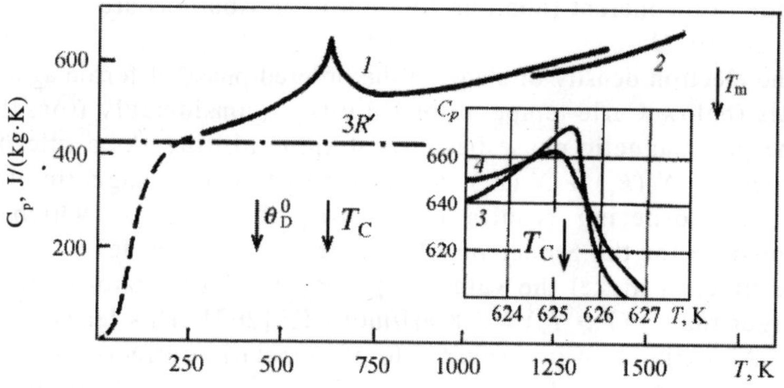

Fig. 3.26. Experimental specific heat of nickel according to the data of various authors [239].

low temperatures, exhibits a jump at crossing T_M and vanishes in the paramagnetic region, so that expression (3.44) holds at arbitrary $T > T_M$. In fact, the presence of short-range order, which is neglected in the mean-field theory, results in that C_{mag} is finite above T_M due to interaction with spin fluctuations. Besides that, fluctuations result in a weak power-law divergence of C_{mag} at T_M. The experimental temperature dependences of specific heat in ferromagnetic transition metals are shown in Figs.3.25, 3.26.

At low temperatures (well below T_M) there exist contributions owing to spin-wave (magnon) excitations (Appendix E). The corresponding average energy is given by

$$\bar{\mathcal{E}}_{sw}(T) = \sum_{\mathbf{q}} \hbar\omega_{\mathbf{q}} N_{\mathbf{q}}, \tag{3.45}$$

with $\omega_{\mathbf{q}}$ being the magnon frequency. Since in ferromagnets $\omega_{\mathbf{q}} \sim q^2$ at $q \to 0$, we have

$$\bar{\mathcal{E}}_{sw}(T) \sim T^{5/2}/T_C^{3/2}, \qquad C_{sw} \sim (T/T_C)^{3/2}. \tag{3.46}$$

At the same time, for antiferromagnets the dispersion law is linear and we have

$$\bar{\mathcal{E}}_{sw}(T) \sim T^4/T_N^3, \qquad C_{sw} \sim (T/T_N)^3. \tag{3.47}$$

The spin-wave contributions are present both in magnetic insulators and metals. In metals the linear electronic specific heat dominates at low temperatures and the spin-wave corrections are less important. However, the electronic contribution is appreciably influenced by magnetic ordering. Since an enhancement of the linear γT-term takes place due to interaction with spin fluctuations (Sect.4.4), the strict separation of magnetic and electronic contributions is, generally speaking, impossible. Some peculiar terms in specific heat of conducting ferromagnets which are due to incoherent (non-quasiparticle) contributions are discussed in Appendix G.

The electron density of states in the ordered phase of ferromagnetic metals (below Curie temperature T_C) differs considerably from that in the paramagnetic phase because of spin splitting. As a rule, total $N(E_F) = N_\uparrow(E_F) + N_\downarrow(E_F)$ is smaller in the ferromagnetic state since the ordering results in a shift of the peak, which is responsible for the Stoner instability, from the Fermi level. At the same time, in nickel the value of $\gamma\,(T<T_C) = 7.0$ mJ/mol·K^2 [266] is larger than $\gamma\,(T > T_C) = 5.8$ mJ/mol · K^2 [267]. This leads to the conclusion about an important role of correlation effects.

Consider the simplest theory of magnetic specific heat. In the mean-field approximation the magnetic (exchange) energy is expressed in terms of magnetization

$$\mathcal{E}_{\text{mag}} = -IM^2, \tag{3.48}$$

where I is an exchange parameter. Then the magnetic specific heat reads

$$C_{\text{mag}} = -2IM\frac{dM}{dT}. \tag{3.49}$$

In metals one has to take into account, besides the exchange energy, the kinetic energy of conduction electrons which depends on magnetization through the exchange splitting. The total energy in the Stoner model (see Sect.4.3) may be transformed as

$$\begin{aligned}\bar{\mathcal{E}} &= \sum_{\mathbf{k}\sigma} E_{\mathbf{k}} n_{\mathbf{k}\sigma} + In_{\uparrow}n_{\downarrow} = \sum_{\mathbf{k}\sigma} E_{\mathbf{k}\sigma} n_{\mathbf{k}\sigma} - In_{\uparrow}n_{\downarrow} = \\ &= \sum_{\sigma}\int Ef(E)N_{\sigma}(E)dE - I\left(\frac{n^2}{4} - M^2\right),\end{aligned} \tag{3.50}$$

where

$$n_{\sigma} = \frac{n}{2} + \sigma M, \qquad E_{\mathbf{k}\sigma} = E_{\mathbf{k}} - \sigma IM,$$

$$N_{\sigma}(E) = \frac{1}{2}N(E + \sigma IM). \tag{3.51}$$

Differentiating with respect to T one obtains

$$C(T) = C_{\uparrow}(T) + C_{\downarrow}(T) + \frac{\pi^2}{6}\left[\frac{N'_{\uparrow}(E_{\text{F}})}{N_{\uparrow}(E_{\text{F}})} - \frac{N'_{\downarrow}(E_{\text{F}})}{N_{\downarrow}(E_{\text{F}})}\right]T^2 I\frac{dM}{dT}, \tag{3.52}$$

where $C_{\sigma}(T)$ is the usual electronic specific heat, corresponding to the density of states $N_{\sigma}(E)$. One can see that the term of the type (3.49) is canceled by the corresponding contribution from the kinetic energy. The magnetic specific heat is determined by the second term in (3.52). Only electrons in the energy layer of order of T near the Fermi level contribute to C_{mag} so that it contains a small factor of order of $(k_{\text{B}}T/E_{\text{F}})^2$ and depends strongly on the DOS shape. The jump of specific heat at the Curie point is given by [26]

$$\Delta C = \frac{2}{3}(\pi T_{\text{C}})^2\left[N'(E_{\text{F}})\frac{\partial\zeta}{\partial M} + I^2 MN''(E_{\text{F}})\right]_{T=T_{\text{C}}} \tag{3.53}$$

The situation may change somewhat in the presence of DOS peaks. Consideration of specific heat of Ni with account of the realistic DOS structure is performed in [268]. As well as in the Heisenberg magnets, spin fluctuations result in a considerable modifications of C_{mag} behaviour at high temperatures [26, 268].

Chapter 4

MAGNETIC PROPERTIES

Strong magnetism is one of the most important peculiarities of transition d- and f-elements. Iron group metals (Fe, Co and Ni) exhibit ferromagnetic ordering and large spontaneous magnetization, chromium and manganese are antiferromagnetic–magnetic moments are ordered, but the summary magnetization is zero. Ferro- and antiferromagnetism are characteristic for most rare earth metals. Many alloys and compounds of transition metals also possess strong magnetic properties and have wide technical applications.

The basic magnetic characteristics of strongly magnetic substances are as follows

- the value of magnetic moment in the ground state
- behaviour of spontaneous magnetization with temperature and magnetic field; values of the ordering temperature T_C (T_N) and magnetic anisotropy constants; the saturation field H_a
- the type of temperature dependence of magnetic susceptibility (Pauli or Curie–Weiss) and its anisotropy; presence of local moments above T_C; values of the effective moment (Curie constant) and paramagnetic Curie temperature.

All these characteristics are described by some microscopical parameters (value of atomic magnetic moment in the localized model of magnetism or the exchange splitting in the itinerant electron model, exchange parameters, spin-orbital interaction constant). The main purpose of the theory is comparison of these parameters with observable physical properties. At present, alternative and in some respects contradicting explanations of transition metal magnetic properties exist, and the problem of their strong magnetism is far from the final solution.

4.1 EXCHANGE INTERACTIONS AND THE HEISENBERG MODEL FOR LOCALIZED SPINS

The history of theoretical investigation of ferromagnetism was rather dramatic. First attempts to explain this phenomenon gave rise to development of the theory of exchange interactions within the Dirac–Heisenberg model for localized spins. Although further theoretical developments demonstrated inapplicability of this model to metallic magnets, we discuss this model to first place.

Usual relativistic magnetic interactions between localized moments at different lattice sites corresponds to the energy

$$\mu_B^2/a^3 \sim 10^{-17}\,\mathrm{erg/cm^3} \sim 10^{-1}\,\mathrm{K},$$

so that magnetic ordering should be destroyed by thermal motion at very low temperatures. However, transition metals of iron group and a number of their compounds possess high values of ordering temperature (Curie point T_C) of order of 10^2–10^3 K. Such large interactions between magnetic moments are not explained by classical electrodynamics and require a quantum mechanical treatment.

In the absence of spin-orbital interaction the wave function of a two-electron system has the form of the product of spin and coordinate functions

$$\Psi(\mathbf{r}_1 s_1, \mathbf{r}_2 s_2) = \Psi(\mathbf{r}_1\mathbf{r}_2)\chi(s_1 s_2), \tag{4.1}$$

and, according to the Pauli principle, should be antisymmetric with respect to permutation of electrons. For symmetric (antisymmetric) spin function, which corresponds to the total spin $\mathbf{S} = \mathbf{s}_1 + \mathbf{s}_2$ of unity (zero), the trial coordinate wavefunctions may be written in the form

$$\Psi(\mathbf{r}_1\mathbf{r}_2) = \frac{1}{\sqrt{2}}[\psi_1(\mathbf{r}_1)\psi_2(\mathbf{r}_2) \mp \psi_1(\mathbf{r}_2)\psi_2(\mathbf{r}_1)] \tag{4.2}$$

respectively. Calculating the first-order perturbation theory correction to the total energy from the electrostatic interaction we obtain

$$\mathcal{E} = \int \Psi^*(\mathbf{r}_1\mathbf{r}_2)\frac{e^2}{|\mathbf{r}_1 - \mathbf{r}_2|}\Psi(\mathbf{r}_1\mathbf{r}_2)d\mathbf{r}_1 d\mathbf{r}_2 = Q \mp \mathcal{J}, \tag{4.3}$$

where we have introduced Coulomb integral

$$Q = \int \rho_1(\mathbf{r}_1)\frac{1}{|\mathbf{r}_1 - \mathbf{r}_2|}\rho_2(\mathbf{r}_2)d\mathbf{r}_1 d\mathbf{r}_2, \qquad \rho(\mathbf{r}) = -e\psi^*(\mathbf{r})\psi(\mathbf{r}) \tag{4.4}$$

and the exchange integral, which differs from (4.4) by permutation (exchange) of electrons

$$J = \int \psi_1^*(\mathbf{r}_1)\psi_2^*(\mathbf{r}_2)\frac{e^2}{|\mathbf{r}_1 - \mathbf{r}_2|}\psi_1(\mathbf{r}_2)\psi_2(\mathbf{r}_1)d\mathbf{r}_1 d\mathbf{r}_2. \tag{4.5}$$

Unlike the relativistic magnetic interaction, the exchange interaction has the electrostatic nature and is considerably larger. At the same time, exchange interaction between electrons at different lattice sites is strongly reduced by the squared factor of overlap of the corresponding wavefunctions. We may estimate

$$J \sim (10^{-2} - 10^{-3})Q \sim 10^{-3}\ \mathrm{eV} \sim 10\ \mathrm{K}.$$

The expression (4.3) may be rewritten in the form of the spin Hamiltonian

$$\mathsf{H} = Q - J\left(\frac{1}{2} + 2\mathbf{s}_1\mathbf{s}_2\right) = \begin{cases} Q - J, & S_{\mathrm{tot}} = 1, \\ Q + J, & S_{\mathrm{tot}} = 0, \end{cases} \tag{4.6}$$

where

$$2\mathbf{s}_1\mathbf{s}_2 = S_{\mathrm{tot}}^2 - 2s^2 = S_{\mathrm{tot}}(S_{\mathrm{tot}} + 1) - \frac{3}{2}. \tag{4.7}$$

Besides the ‘potential’ exchange interaction (4.5), there exists another exchange mechanism which is due to kinetic energy of electrons. In the simplest situation of s-band, we obtain (Appendix D)

$$\mathsf{H}' = -\frac{2t^2}{U}\left(2\mathbf{S}_1\mathbf{S}_2 - \frac{1}{2}\right), \tag{4.8}$$

with t being the transfer integral and U the intra-atomic Coulomb repulsion. This mechanism results in an antiferromagnetic interaction since the gain in the kinetic energy is achieved at antiparallel orientation of electron spins. As demonstrate numerical calculations (see, e.g., [265]), at realistic interatomic distances this contribution as a rule prevails over the ferromagnetic potential exchange. Thus the localized model does not explain ferromagnetism of iron group metals.

The derivation of Heisenberg model for orbitally degenerate atomic shells and arbitrary values of spin S is presented in Appendix D. It turns out that in this case the exchange Hamiltonian contains the interaction of not only spin, but also orbital moments. Exchange mechanisms in rare earths owing to indirect interaction via conduction electrons is discussed in Appendix K. In most cases, real exchange mechanisms are complicated, so that we have to consider the Heisenberg Hamiltonian as an effective one.

Consider the system of interacting magnetic moments which is described by the Heisenberg Hamiltonian in the magnetic field H

$$\mathsf{H} = -g\mu_B H \sum_i S_i^z - \sum_{ij} J_{ij} \mathbf{S}_i \mathbf{S}_j \tag{4.9}$$

with $J_{ij} > 0$ for nearest neighbours. Direct calculation of the magnetization $M = g\mu_B \langle S^z \rangle$ (hereafter we put $g\mu_B = 1$) in the model (4.9) is prevented by the non-linearity of the interspin exchange interaction in spin operators. To simplify the problem, we can linearize the Hamiltonian by introducing the effective mean field H^* which is expressed in terms of magnetization

$$\mathsf{H} = -\tilde{H} S^z = -(H + H^*) S^z, \qquad S^z = \sum_i S_i^z, \tag{4.10}$$

where

$$H^* = \lambda \langle S^z \rangle, \qquad \lambda = 2\sum_i J_{ij} = 2J_0, \qquad J_0 = J_{\mathbf{q}=0} \tag{4.11}$$

(in the nearest-neighbour approximation $J_0 = zJ$ with z being the number of the neighbour spins). In the case of spin $S = 1/2$ we have

$$\langle S^z \rangle = \frac{1}{2}(n_\uparrow - n_\downarrow), \qquad n_\uparrow + n_\downarrow = 1. \tag{4.12}$$

Using the Gibbs distribution

$$n_\sigma = \exp\frac{\sigma \tilde{H}}{T} \Bigg/ \left[\exp\frac{\tilde{H}}{T} + \exp\left(-\frac{\tilde{H}}{T} \right) \right], \tag{4.13}$$

we obtain the self-consistent mean-field equation for magnetization

$$\langle S^z \rangle = \frac{1}{2} \tanh \frac{H + \lambda \langle S^z \rangle}{2T}. \tag{4.14}$$

The result (4.14) is easily generalized to the case of arbitrary S

$$\langle S^z \rangle = S B_S \left(\frac{S\tilde{H}}{T} \right), \tag{4.15}$$

where

$$B_S(x) = \frac{d}{dx} \ln \sum_{m=-S}^{S} e^{-mx/S} = \left(1 + \frac{1}{2S} \right) \coth \left(1 + \frac{1}{2S} \right) x - \frac{1}{2S} \coth \frac{x}{2S}. \tag{4.16}$$

is the Brillouin function. In particular cases

$$B_{1/2}(x) = \tanh x, \qquad B_\infty(x) = \coth x - \frac{1}{x}. \tag{4.17}$$

The equation (4.15) has non-trivial solutions in the zero magnetic field provided that $T < T_C$, the Curie temperature being given by

$$T_C = S(S+1)\lambda = \frac{2}{3}S(S+1)J_0. \tag{4.18}$$

Using the expansion

$$\coth x = \frac{1}{x} + \frac{x}{3} - \frac{x^3}{45} + \ldots,$$

we derive near the Curie point

$$\langle S^z \rangle = \sqrt{\frac{5}{3}} \frac{S(S+1)}{(S^2+S+1/2)} \sqrt{1 - \frac{T}{T_C}}. \tag{4.19}$$

At low $T \ll T_C$ equation (4.15) may be solved by iterations and yields an exponential behaviour of the magnetization

$$\langle S^z \rangle = S[1 - S^{-1}\exp(-2T_C/T)]. \tag{4.20}$$

A more accurate description of the low-temperature behaviour is obtained in the spin-wave theory (Appendix E).

Consider the case of high temperatures $\mu_B H \ll T$. Performing expansion of (4.15) in H we obtain

$$\langle S^z \rangle = \frac{S(S+1)H}{3T[1 - \lambda S(S+1)/(3T)]}.$$

Thus we obtain the Curie–Weiss law for the paramagnetic susceptibility $\chi = M/H$

$$\chi(T) = \frac{C}{T-\theta}. \tag{4.21}$$

The Curie constant C is expressed in terms of the effective atomic moment,

$$C = \frac{1}{3}\mu_{\text{eff}}^2, \qquad \mu_{\text{eff}} = g\mu_B\sqrt{S(S+1)}. \tag{4.22}$$

The paramagnetic Curie temperature turn out to coincide in the simple model under consideration with T_C:

$$\theta = \lambda C = \frac{2}{3}S(S+1)J_0. \tag{4.23}$$

The case of antiferromagnetic exchange interaction $J < 0$ is considered in a similar way. In such a case, neighbour spins are ordered in an antiparallel way and the equation for the sublattice magnetization has the same form (4.15) with $J \to -J$. The magnetic susceptibility obeys the Curie–Weiss law with a negative value of θ. In the nearest-neighbour approximation, the Neel temperature T_N, where magnetic ordering disappears, equals to $|\theta|$.

4.2 MAGNETIC SUSCEPTIBILITY OF PARAMAGNETIC TRANSITION METALS

Now we return to discussion of situation in transition metals. Most *d*-elements (3*d*-metals Sc, Ti, V and all the 4*d* and 5*d*-metals) do not possess magnetic ordering. However, their magnetic properties differ considerably from those of simple (e.g. alkaline) metals which demonstrate a weak Pauli paramagnetism with $\chi \sim 10^{-4}$ emu/mol. The values of paramagnetic susceptibility of *d*-metals, which are presented in the Tables 4.1, 4.2 are by an order of magnitude larger.

There exist some regularities in the behaviour of χ in the transition metal rows. For metals of one group in the periodic system, the susceptibility decreases with increasing atomic number. The change of χ in a given period turns out to be non-monotonous: as a rule, metals with even number of *d*-electrons have lower susceptibility than those with odd number (Fig.4.1, Tables 4.1,4.2). Thus a correlation with electronic specific heat takes place. The temperature dependences of χ (Figs.4.2–4.6) are appreciably stronger than in simple metals. The sign of $d\chi/dT$ changes as a rule at passing to neighbour element. So, for metals of IV group (Ti, Zr and Hf), which

Table 4.1. Magnetic susceptibility (10^{-6} emu/mol) of paramagnetic d-metals with a cubic lattice. To exclude the influence of magnetic impurities, the signs of $d\chi/dT$ are given at not too low temperatures. References may be found in the review [270].

	V	Nb	Mo	Tc	Rh	Pd	Lu	Ta	W	Ir	Pt
	d^3s^2	d^3s^2	d^5s^1	d^5s^2	d^8s^1	d^{10}	d^1s^2	d^3s^2	d^4s^2	d^7s^2	d^9s^1
χ	300	212	89.2	270	107	550	336	162	53.3	24.1	192
$d\chi/dT$	-	-	+	-	+	-	-	-	+	+	-

Table 4.2. Longitudinal ($H \parallel c$) and transverse ($H \perp c$) magnetic susceptibility (10^{-6} emu/mol) of paramagnetic *d*-metals with a hcp lattice. The susceptibility of a polycrystal is given by $\chi = (\chi_\parallel + 2\chi_\perp)/3$. Two sets of data correspond to results of various authors (see [270]).

	Sc	Y	Ti	Zr	Hf	Re	Ru	Os
	d^1s^2	d^1s^2	d^2s^2	d^2s^2	d^2s^2	d^4s^2	d^7s^1	d^6s^2
$\chi_\parallel$	294, 281	174, 270	169	147, 151	95	68.3	35.2	5.4
$\chi_\perp$	232, 298	220, 445	145	86, 100	63	73.0	44.2	12.6
$d\chi/dT$	-	+	+	+	+	+	+	+

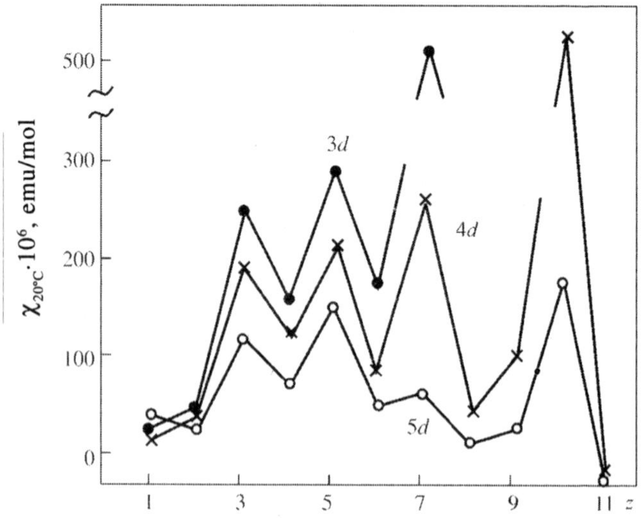

Fig. 4.1. Paramagnetic susceptibility at 293K in 3*d*, 4*d* and 5*d* rows as a function of valence electron number, $z = n_s + n_d$ [269].

have a small susceptibility, χ increases with T, and for metals of the V group (V, Nb, Ta) $d\chi/dT < 0$. The presented data for $d\chi/dT$ correspond to not too low temperatures, since at low T the behaviour of χ may be masked by the Curie–Weiss contribution from magnetic impurities. However, for Pd the intrinsic susceptibility has a maximum at 80 K, and for Y at 300 K.

Now we discuss various contributions to magnetic susceptibility of transition metals. To calculate the paramagnetic susceptibility we write down the magnetization of band electrons in a magnetic field H

$$M = \mu_B \int N(E)[f(E-\mu_B H) - f(E+\mu_B H)]dE \tag{4.24}$$

(the density of states $N(E)$ is put for one spin projection). Expanding (4.24) in H we derive

$$\chi = \frac{M}{H} = -2\mu_B^2 \int N(E)\frac{\partial f(E)}{\partial E}dE = 2\mu_B^2\left[N(\zeta) + \frac{\pi^2}{6}N''(\zeta)(k_B T)^2\right]. \tag{4.25}$$

Performing expansion in $\zeta - E_F$ and taking into account the temperature dependence of the chemical potential (3.38) we obtain

$$\chi(T) = 2\mu_B^2 N(E_F)\left\{1 + \frac{1}{6}(\pi k_B T)^2\left[\frac{N''(E_F)}{N(E_F)} - \left(\frac{N'(E_F)}{N(E_F)}\right)^2\right]\right\}. \tag{4.26}$$

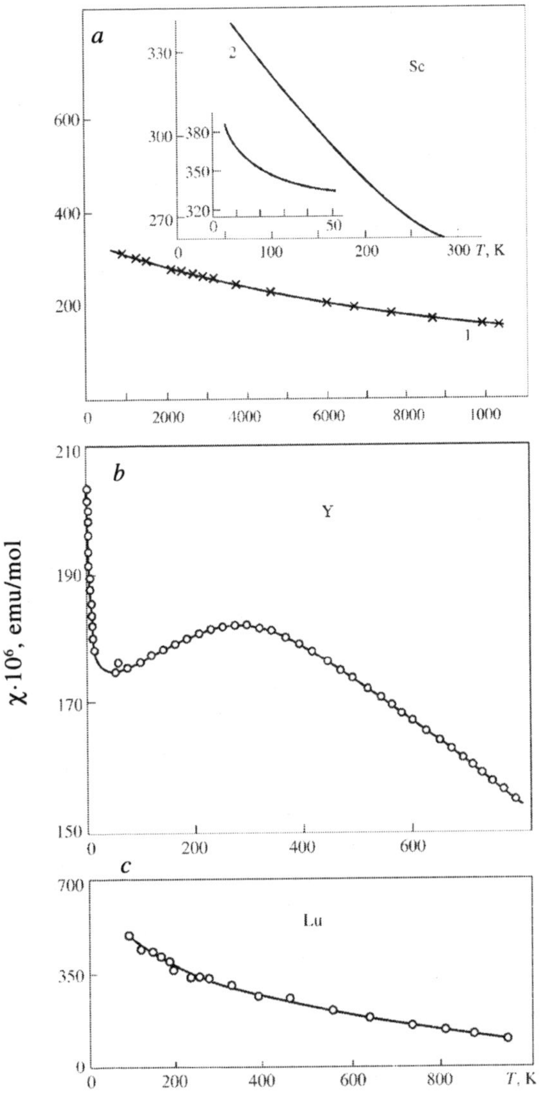

Fig. 4.2. Temperature dependence of magnetic susceptibility for transition metals of the third group [270]: (a) scandium (b) yttrium (c) lutetium.

Thus spin susceptibility, as well as specific heat, of conduction electrons is determined by the density of states at the Fermi level and is to leading order approximation temperature independent (Pauli paramagnetism). Such a behaviour, which is drastically different from the Curie–Weiss law for localized electrons, is explained by the Pauli principle: only a small part (of order of $k_B T/E_F$) electrons contribute

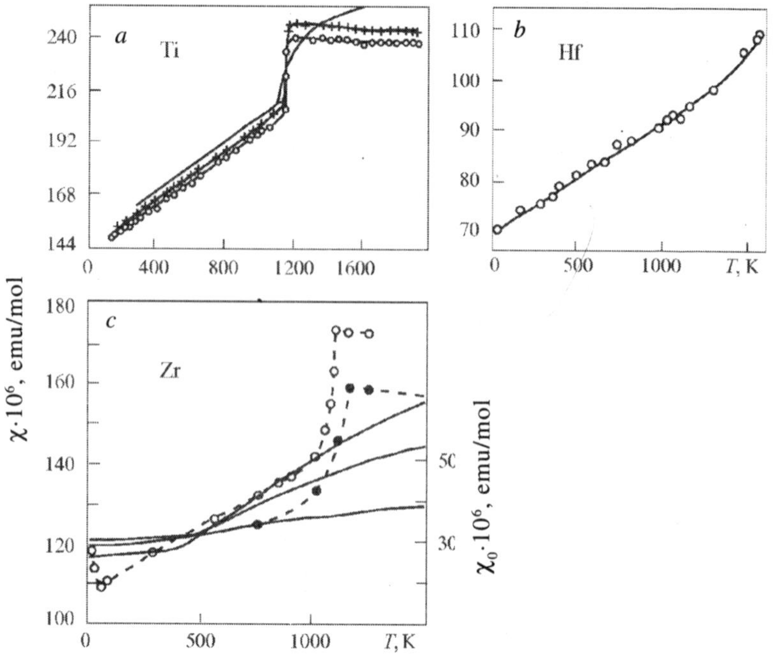

Fig. 4.3. Temperature dependence of magnetic susceptibility for transition metals of the fourth group according to data of various authors [270]: (a) titanium (b) hafnium (c) zirconium.

to χ. The dependence $\chi(T)$, which is obtained after account of next-order terms in T/E_F, turns out to be weak.

One can see that the sign of $d\chi/dT$ should be determined by the shape of density of states. Thus large values and strong T-dependence of χ in d-metals may be related to the presence of narrow d-bands near the Fermi level.

Due to presence of partially filled degenerate bands in d-metals, orbital contributions to susceptibility turn out to be important, which are analogous to the Van Vleck contribution for localized magnetic ions. The total paramagnetic susceptibility for a degenerate band may be represented in the form [271, 270]

$$\chi = \chi_S + \chi_L + \chi_{SL}, \tag{4.27}$$

with

$$\chi_S = 2\mu_B^2 \sum_{m,m'\mathbf{k}} \frac{f(E_{m\mathbf{k}}) - f(E_{m'\mathbf{k}})}{E_{m'\mathbf{k}} - E_{m\mathbf{k}}} |\langle m\mathbf{k} | 2\mathbf{S} | m'\mathbf{k}\rangle|^2; \tag{4.28}$$

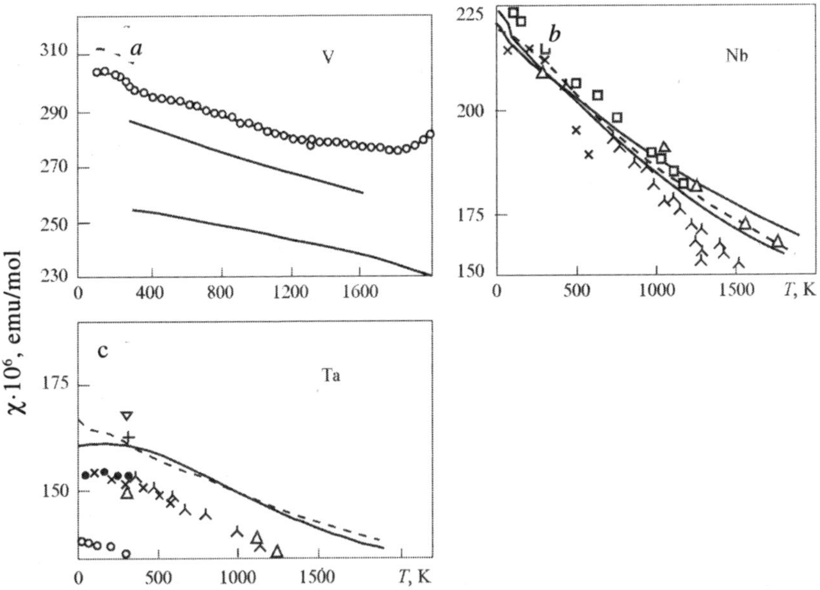

Fig. 4.4. Temperature dependence of magnetic susceptibility for transition metals of the fifth group [270]: (a) vanadium (b) niobium (c) tantalum.

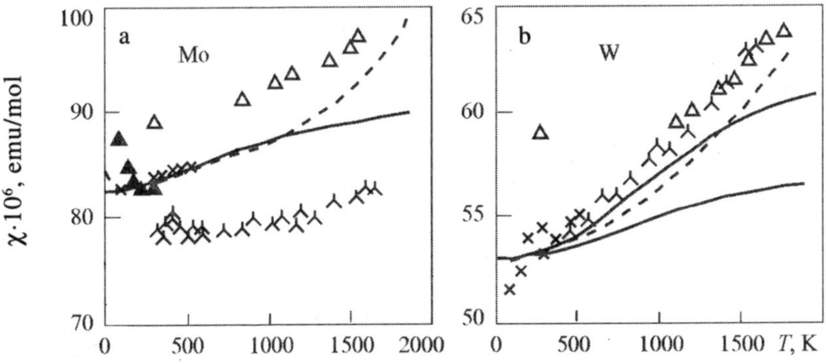

Fig. 4.5. Temperature dependence of magnetic susceptibility for transition metals of the sixth group [270]: (a) molybdenum (b) tungsten.

$$\chi_L = 2\mu_B^2 \sum_{m,m'\mathbf{k}} \frac{f(E_{m\mathbf{k}}) - f(E_{m'\mathbf{k}})}{E_{m'\mathbf{k}} - E_{m\mathbf{k}}} |\langle m\mathbf{k} | \mathbf{L} | m'\mathbf{k}\rangle|^2; \tag{4.29}$$

$$\chi_{SL} = 2\mu_B^2 \sum_{m,m'\mathbf{k}} \frac{f(E_{m\mathbf{k}}) - f(E_{m'\mathbf{k}})}{E_{m'\mathbf{k}} - E_{m\mathbf{k}}} \langle m\mathbf{k} | L | m'\mathbf{k} |\rangle\langle m\mathbf{k} | 2S | m'\mathbf{k}\rangle \tag{4.30}$$

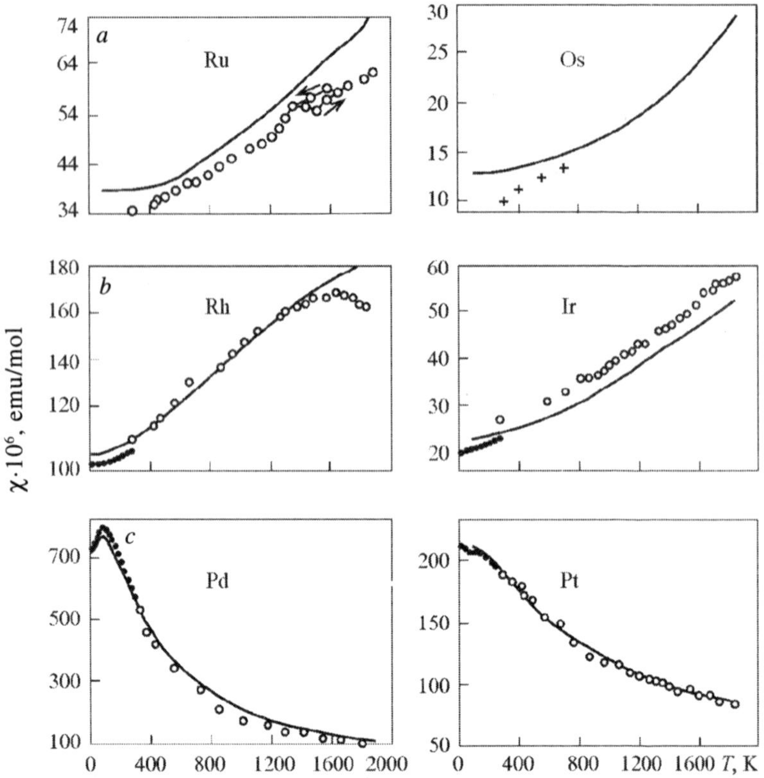

Fig. 4.6. Temperature dependence of magnetic susceptibility for paramagnetic transition metals of the eighth group [270]: (a) ruthenium and osmium (b) rhodium and iridium (c) palladium and platinum.

where m is the band index. A simple estimation of the orbital contribution in terms of the number of d-electrons n has the form

$$\chi_L \sim 2\mu_B^2 \frac{n(10-n)}{5W}, \tag{4.31}$$

where W is a characteristic energy difference between occupied and empty states, which equals approximately to the d-bandwidth. Separation of spin and orbital contributions may be performed by combined studying the dependences of $\chi(T)$ (with account of Stoner enhancement) and electronic specific heat $C_e(T)$, the orbital contribution being proposed to be weakly temperature-dependent [270]. Spin susceptibility may be also picked out by measuring the Knight shift.

According to the review [270], calculated orbital and spin susceptibilities of d-metals are of the same order of magnitude, the orbital susceptibility of d-electrons exceeding the spin susceptibility of s-electrons. The spin-orbit contribution is negligible, except for $5d$-metals. Band calculations of the orbital and spin-orbital susceptibility in bcc transition metals are presented in [272].

Data on orbital contributions are also obtained by measuring the gyromagnetic (magnetomechanical) ratio with the use of the Einstein – de Haas effect [273]

$$g' = \frac{M_L + M_S}{M_L + \frac{1}{2} M_S}. \tag{4.32}$$

This quantity differs somewhat from the g-factor

$$g = \frac{M_L + M_S}{\frac{1}{2} M_S}, \tag{4.33}$$

which is measured in magnetic resonance experiments (it is taken into account in (4.33) that the orbital mechanical momentum is not conserved in the lattice). Thus the factors g and g' are connected by

$$\frac{1}{g} + \frac{1}{g'} = 1, \qquad g' = \frac{g}{g-1}. \tag{4.34}$$

In the case of small orbital magnetization,

$$g = 2\left(1 + \frac{M_L}{M_S}\right), \qquad g' = 2\left(1 - \frac{M_L}{M_S}\right). \tag{4.35}$$

Investigations of the gyromagnetic ratio in paramagnetic metals are much more difficult than for ferromagnets since one has to measure very small rotation angles. The results of the experiments [274] for g' are as follows: V – 1.18; Nb – 1.05; Ta – 1.02; Pd – 1.77; Pt – 1.62. Strong deviations of values from 2 show the presence of appreciable orbital contributions. The measurements for iron group metals and ferromagnetic alloys yield $g-2 \sim 2-g' \sim (5-10)\%$ [273].

Finally, the diamagnetic contribution to χ in the one-band approximation is given by [275]

$$\chi_{\mathrm{dia}} = \frac{e^2}{6c} \sum_{\mathbf{k}} \frac{\partial f(E_{\mathbf{k}})}{\partial E_{\mathbf{k}}} \left[\frac{\partial^2 E_{\mathbf{k}}}{\partial k_x^2} \frac{\partial^2 E_{\mathbf{k}}}{\partial k_y^2} - \left(\frac{\partial^2 E_{\mathbf{k}}}{\partial k_x \partial k_y} \right)^2 \right] \tag{4.36}$$

(the magnetic field is directed along the z-axis). In the effective mass approximation where $E_{\mathbf{k}} = k^2/2m^*$ we obtain

$$\chi_{\text{dia}} = -\frac{2}{3}\mu_B^2 N(E_F)(m/m^*)^2. \qquad (4.37)$$

where c is the light velocity. Unlike the paramagnetic contribution, the diamagnetic susceptibility is inversely proportional to electron effective mass m^* since the influence of magnetic field on the orbital motion is directly proportional to electron velocity; for free electrons $\chi_{\text{dia}} = -\chi_{\text{para}}/3$. For the effective mass of d-electrons is large, the diamagnetic contribution may be as a rule neglected.

Investigations of single crystals of paramagnetic d-metals with the hcp structure demonstrated a considerable anisotropy of χ (Table 4.2). In Y, Re, Ru, Os, $\Delta\chi = \chi_{\parallel} - \chi_{\perp} < 0$, a typical difference being tens of percents (however, for osmium, $\Delta\chi/\chi \simeq 3$). For Ti, Zr and Hf, $\Delta\chi > 0$ and the anisotropy is also large. As a rule (with the exception of Os), the sign of $\Delta\chi$ is the same for the elements of one row, $\Delta\chi$ being positive for the atomic configuration d^n with even n and negative for odd n. This may indicate a partial adequacy of the atomic picture.

A reliable theory of susceptibility anisotropy is still absent. Anisotropic contributions may have orbital, spin-orbital and diamagnetic origin. The estimations of anisotropy of the orbital contribution to χ in the band theory for hexagonal metals ($\Delta\chi \sim \sqrt{8/3} - c/a = 1.633 - c/a$) [276] yielded the values $\Delta\chi/\chi \sim 0.02$ which are too small to explain experimental data. Recently [277] an attempt to calculate Dc for Sc and Y within relativistic calculations has been performed. Due to relativistic interactions, anisotropic part of spin susceptibility was obtained with $\chi_{\parallel} > \chi_{\perp}$, which disagrees with experimental data. According to [277], the agreement with experimental data should be restored after allowance for Van Vleck-type contributions. These conclusions agree on the whole with results of earlier calculations [278, 279].

4.3 ITINERANT ELECTRON FERROMAGNETISM AND THE STONER THEORY

Magnetic characteristics (values of paramagnetic and ferromagnetic Curie temperatures and of corresponding magnetic moments) for some ferromagnetic d- and f-metals and their compounds are presented in the Table 4.3. The correlation between the values of Curie temperature and the ratio of paramagnetic and ferromagnetic moments p_C/p_s is demonstrated by the Rhodes–Wohlfarth curve (Fig.4.7). In the Heisenberg model with spin S, $p_C = p_s = 2S$, so that p_C/p_s should

Table 4.3. Values of the ground state moment p_S determined from the saturation magnetization ($M_0 = p_S\mu_B$), ferromagnetic and paramagnetic Curie temperature T_C and Θ, and paramagnetic moment p_C, determined from the Curie constant ($C = \mu_{\text{eff}}^2/3 = p_C(p_C + 2)\mu_B^2/3$) for some d- and f-metals, and their alloys and compounds. Rough estimates for non-ferromagnetic transition metals Pd and Pt, where the Curie-Weiss law holds approximately, are also presented.

	p_S	T_C, K	p_C	Θ, K	Ref.
Fe	2.22	1044	2.3	1101	
Co	1.71	1388	2.3	1411	
Ni	0.62	627	0.9	649	
$CrBr_3$	3.0	37	3.0	37	280
CoS_2	0.84	116	1.00		
CrO_2	4.0	400			
Pd	–	–	0.6	–200	
Pd + 1% Fe	0.084	50	0.72		280
Pt	–	–	0.6	–1000	
$ZrZn_2$	0.12	21	0.66	35	280
Sc_3In	0.045	5.5	0.22	16	280
Ni_3Al	0.075	41	0.64		280
Co_2MnSi	5.10	1034	2.03	1044	281
Co_2MnSn	5.37	826	3.35	870	281
NiMnSb	4.2	728	2.06	910	282
PtMnSb	3.96	572	3.56	670	282
Pd_2MnSn	4.22	189	4.05	201	283
PtMnSn	3.5	330	4.2	350	282
Gd	7.13	289	7.05	316	
Tb	9.34	221	8.2	232	
EuO	6.80	69.4	6.8	76	280
$EuRh_3B_2$	0.56	46	3.5	–40	284
$CeRh_3B_2$	0.38	115	2.2	–373	284

be equal to unity. This condition is well satisfied for insulating localized-spin ferromagnets EuO and $CrBr_3$ and *f*-metals (some deviations in rare-earth metals may be explained by polarization of conduction electrons). One can see that for Fe and Ni p_C/p_s somewhat exceeds 1, and the values of p_C and p_s correspond to no values of S, so that magnetic moment turns out to be non-integer.

In metallic *d*-magnets, p_C/p_s is roughly inversely proportional to T_C. For intermetallic compounds with very small saturation moment M_s ($ZrZn_2$, Sc_3In, Ni_3Al), which are called weak itinerant ferromagnets, p_C/p_s is very large, so that the situation is drastically different from the localized magnetism picture. On the contrary, for Heusler alloys T_2MnZ, TMnZ we have $p_C/p_s < 1$. The latter case is discussed in Sect.4.4.

One can also see from Table 4.3 that the values of T_C and θ in some

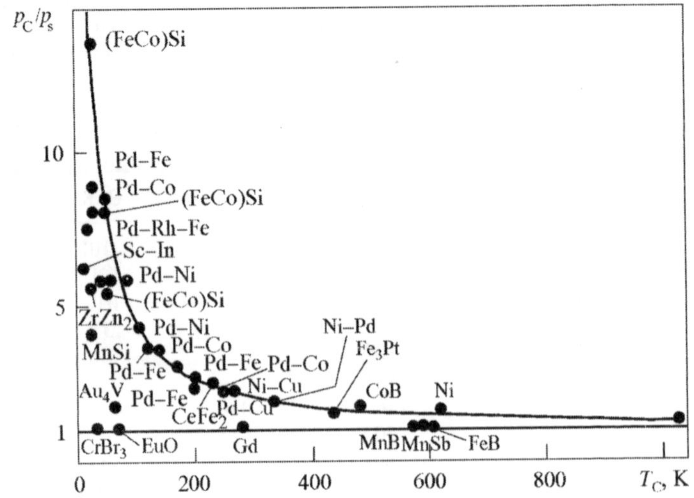

Fig. 4.7. The dependence of the ratio p_C/p_S on the Curie temperature (the Rhodes–Wohlfarth curve) [26].

systems are quite different. (Especially strange is the situation for the compound $CeRh_3B_2$; this belongs to anomalous systems where the Kondo effect plays apparently an important role, see Chapter 6.) Within the Heisenberg model, a not great difference only may be explained by next-nearest neighbour exchange interactions or fluctuation effects (Appendix E).

Thus the localized spin model is adequate for some insulator *d*- and *f*-compounds and for *f*-subsystem in rare earths, but does not describe magnetic properties of transition *d*-metals. In particular, the Heisenberg model does not explain non-integer values of magnetic moments and, in the case of magnetic ordering, the difference between values of the moments in the ground state and at high temperatures (as determined from the Curie constant).

To solve these problems, Stoner [285] put forward the model of itinerant magnetism based on the exchange interaction among band electrons, which was treated in the mean field approximation. The Fermi statistics of band electrons (contrary to atomic statistics of localized electrons in the Heisenberg model) leads to a radical change of magnetic properties. The magnetic ordering results in an increase of kinetic energy, but decreases the potential (exchange) energy. Thus the conditions for ferromagnetism to occur are rather restricted.

First calculations by Stoner [285] used the free-electron model and did not take into account correlation effects. The system of equations for the magnetization and number of electrons of an itinerant electron system in the molecular field $\tilde{H} = H + 2IM$ (I is the exchange parameter for itinerant electrons) has the form

$$M = \frac{1}{2}(n_\uparrow - n_\downarrow), \qquad n = n_\uparrow + n_\downarrow, \qquad n_\sigma = \int N(E) f\left(E - \frac{1}{2}\sigma\tilde{H}\right) dE. \quad (4.38)$$

This system is an analogue of the equations (4.13). For the free-electron density of states

$$N(E) = \frac{3}{4} n E^{1/2} / E_F^{3/2}$$

we have

$$n_{\uparrow,\downarrow} = \frac{3}{4}\left(\frac{T}{E_F}\right)^{3/2} F\left[\zeta \pm \frac{1}{2}(H + 2IM)/T)\right], \quad (4.39)$$

with

$$F(y) = \int_0^\infty \frac{z^{1/2}}{e^{z-y} + 1} dz.$$

At high temperatures we may expand the Fermi integrals to obtain

$$M = \frac{H + 2IM}{T} \frac{F'(\zeta)}{F(\zeta)} n. \quad (4.40)$$

Then the susceptibility reads

$$\chi = \frac{\chi_p}{1 - 2I\chi_p}, \quad (4.41)$$

with

$$\chi_p = \frac{n}{T} \frac{F'(\zeta)}{F(\zeta)}.$$

The condition of ferromagnetism (the Stoner criterion)

$$2I\chi_p > 1 \quad (4.42)$$

corresponds to divergence of the ground state susceptibility, and the Curie temperature is estimated as

$$2I\chi_p(T_C) = 1, \qquad T_C \sim (2I\chi_p(0) - 1)^{1/2} E_F. \quad (4.43)$$

Now we have to consider the quantity F'/F. If this would be weakly temperature dependent, the expression (4.41) would yield above T_C the Curie–Weiss law

$$\chi(T) = \frac{C}{T-\theta}, \qquad C = n\frac{F'}{F}, \qquad \theta = 2IC. \tag{4.44}$$

However, such a situation takes place only at very high temperatures $T > E_F$ which exceed in fact melting points. On the contrary, provided that the ratio T_C/E_F is small, i.e. $2I\chi(0)$ is close to unity, the susceptibility above T_C has the usual Pauli (square) temperature dependence. Thus, the Stoner theory does not explain in fact the Curie–Weiss law.

Solving the Stoner equations for the parabolic band (4.39) at $T = 0$ demonstrates that at

$$I > 2^{2/3} E_F/n \tag{4.45}$$

there exist solutions which describe the saturated ferromagnetic state with $M_0 = n/2$. At

$$2^{2/3} E_F/n > I > \frac{4}{3} E_F/n \tag{4.46}$$

we have non-saturated ferromagnetism with a partially filled spin down subband, and at smaller I magnetic ordering is absent. In the limit of small ground state magnetization we have the relation

$$\frac{2\sqrt{2}}{3\pi}\frac{M_0}{n} \sim \frac{T_C}{E_F}. \tag{4.47}$$

Thus M_0 is roughly proportional to the Curie point, in agreement with the Rhodes–Wohlfarth correlation (Fig.4.7).

The temperature dependence of magnetization at low T in the saturated state turns out to be exponential

$$M = \frac{n}{2}\left[1 - \frac{3\sqrt{\pi}}{4}\frac{T}{E_F}\exp\left(-\frac{E_F}{T}\left(\frac{In}{E_F} - 2^{2/3}\right)\right)\right]. \tag{4.48}$$

At the same time, corrections, which are proportional to T^2, occur in the non-saturated state owing to electron-hole spin-flip processes (Stoner excitations):

$$M \simeq M_0\left[1 - \frac{1}{2}\left(\frac{T}{T_C}\right)^2\right]. \tag{4.49}$$

Near the Curie point the magnetization has the square-root behaviour, as well as in the mean-field approximation in the Heisenberg model:

$$(M/M_0)^2 = 3(1 - T/T_C)\left[1 + \frac{2}{3}(2\pi)^{-1/2}(E_F/T_C)^{3/2}\right]. \tag{4.50}$$

Generally speaking, the results of the solution of the Stoner equations (4.38) turn out to be rather sensitive to the density of states shape. In particular, for the 'rectangle' band with constant density of states, considered in [286], only saturated ferromagnetic solutions are possible.

Consider the results of the Stoner theory for an arbitrary bandstructure. The criterion of ferromagnetism reads

$$IN(E_F) > 1. \tag{4.51}$$

Provided that $IN(E_F) - 1$ is small, the saturation magnetization equals [26]

$$M_0 = N(E_F)\left\{\frac{2[IN(E_F)-1)]}{[N'(E_F)/N(E_F)]^2 - \frac{1}{3}N''(E_F)/N(E_F)}\right\}^{1/2}. \tag{4.52}$$

According to (4.41), the value of T_C is

$$T_C^2 = \frac{6}{\pi^2}\frac{IN(E_F)-1}{IN(E_F)}\left\{\left[\frac{N'(E_F)}{N(E_F)}\right]^2 - \frac{N''(E_F)}{N(E_F)}\right\}^{-1}. \tag{4.53}$$

The susceptibility above T_C is presented as

$$\frac{1}{\chi(T)} = \frac{1}{\chi_p(T)} - \frac{1}{\chi_p(T_C)} = \frac{2I}{[IN(E_F)-1]^2}\left(\left(\frac{T}{T_C}\right)^2 - 1\right). \tag{4.54}$$

Of course, the expansion of the Fermi integrals (3.37), used at derivation of (4.52)–(4.54), holds only at $T_C \ll E_F$ and for a smooth (near the Fermi energy) density of states.

Modern numerical calculations with account of correlation effects (see, e.g., [287]) demonstrate that ferromagnetic ordering in the free electron system is possible at unrealistic large values of r_s (small densities). Thus, the occurrence of ferromagnetism is intimately related to the crystal lattice potential which modifies considerably the shape of electron density of states. In particular, small widths of d-bands result in a considerable localization of d-electrons and decrease of the kinetic energy, and consequently, favour magnetic ordering. First-principle band calculations [24] demonstrate that the values of the Stoner parameter I vary smoothly and not too strongly in the periodic table. At the same time, the values of $N(E_F)$ vary by orders of magnitude because of presence of density of states peaks for d-electrons. As a rule, such peaks are connected with the Van Hove electron spectrum singularities (Sect.2.4). In particular, for non-magnetic

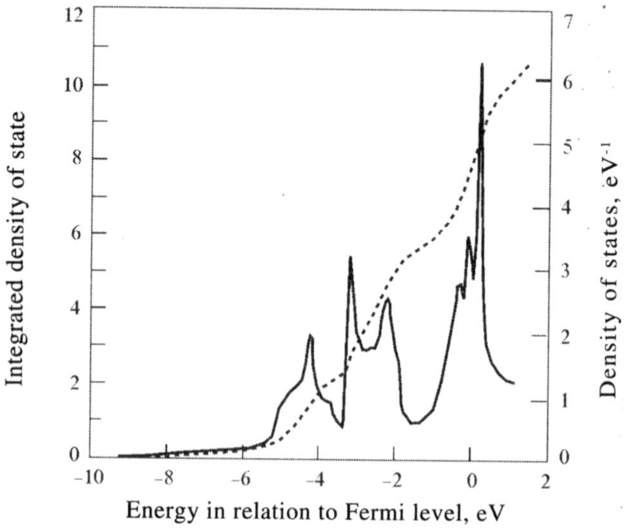

Fig. 4.8. Density of states (eV^{-1}, right-hand axis) and integrated density of states (dotted line, left-hand axis) of paramagnetic iron [24]. Energy is referred to the Fermi level.

iron (Fig.4.8) the peak occurs due to merging of the singularities for e_g-states along the *P*—*N* line [288]. The situation for Ni is discussed in [289]. For the weak itinerant ferromagnet $ZrZn_2$ the fulfilment of the Stoner criterion is connected with the presence of DOS peaks owing to C15 crystal structure [290].

Calculated values of the Stoner parameter turn out to describe satisfactorily ferromagnetism of transition metals in the ground state. On the other hand, the experimental data, which are discussed in Sect.4.2, demonstrate that the conclusions of the Stoner theory for finite temperatures turn out to be poor. First, this theory may yield only very weak dependences $\chi(T)$. At the same time, experimental data on transition metals demonstrate as a rule more strong dependences. One can see from Figs.4.2–4.6 that in the temperature interval under consideration $\chi(T)$ changes by tens of percents, whereas expression (4.41) yields the change of order of $(T/E_F)^2 \ll 1$. In a number of cases (Fe, Co, Ni, Pt, Pd), the Curie–Weiss law, i.e. the linear dependence $\chi(T)$ approximately holds. Second, experimental values of T_C for iron group metals (of order of 10^3 K) are considerably smaller than those predicted by (4.53).

To treat the latter problem in more detail, we consider the ratio of the ground state spin splitting $\Delta_0 = 2IM_0$ to the Curie temperature, $\delta = \Delta_0/T_C$. In the mean field Stoner theory this quantity turns out to be rather stable. This situation is reminiscent of the BCS theory of superconductivity where the ratio of the energy gap to the critical temperature is an universal constant, $2\Delta/T_c = 3.52$. Deviations from this value, which occur, e.g., in high-T_c superconductors, demonstrate inapplicability of the simple BCS theory.

At calculating δ for a ferromagnet with a smooth density of states we may use the expressions (4.52), (4.53) to obtain [291]

$$\delta = \frac{2\pi}{\sqrt{3}} \left\{ \frac{[N'(E_F)/N(E_F)]^2 - \frac{1}{3} N''(E_F)/N(E_F)}{[N'(E_F)/N(E_F)]^2 - N''(E_F)/N(E_F)} \right\}^{1/2} . \tag{4.55}$$

Restricting ourself to the case $N'' < 0$ we get

$$\frac{2\pi}{\sqrt{3}} \le \delta \le 2\pi. \tag{4.56}$$

In the case $N'' > 0$ (where ferromagnetism is a rule unstable since the Fermi level is near the density of states minimum) δ may become still smaller. Deviations of δ from the result (4.55) provide a measure of correlation (fluctuation) effects. The corresponding data according to various authors are presented in Table 4.4. The largest value of Δ are obtained by Gunnarsson [292]. The result of [293] is appreciably smaller because of an account of the magnetization dependence for the exchange-correlation energy.

One can see that the experimental values of δ exceed considerably theoretical ones and lie in the interval (4.56) for nickel only. However, the values of Δ_0 and T_C themselves do not agree with experiment in the latter case too. Besides that, angle-resolved

Table 4.4. Calculated and experimental values of the spin splitting Δ (eV), the Curie temperature T_C (K) and the ratio Δ/T_C for iron group metals according to data of various authors [291, 304].

	Fe	Co	Ni
$\Delta^{(1)}$ [292]	2	1.54	0.58
$\Delta^{(2)}$ [293]	1.41	1.11	0.38
Δ^{exp} [104]	1.5	1.1	0.3
$T_C^{(1)}$	5300	4000	2900
$T_C^{(2)}$	2500	2240	1790
T_C^{exp}	1040	1390	630
$\delta^{(1)}$	4.38	4.47	2.32
$\delta^{(2)}$	6.39	5.75	2.46
$\delta^{(exp)}$	16.7	9.16	5.5

photoemission and other spectral data (Sect.2.5) demonstrate an appreciable short-range order above T_C, which is not described by the mean-field theory.

The difficulties of the Stoner theory require new physical ideas. One of simple ways to obtain a more strong temperature dependence of paramagnetic susceptibility and lower the Curie point is changing the dependence $\chi_p(T)$. Evidently, presence of a T-linear term in the latter would give (after allowance for the Stoner enhancement) the Curie–Weiss law for $\chi(T)$. The dependence $\chi_p(T)$. for complicated electron densities of states was analyzed in [294]. The expansion (4.26) is quite correct for smooth functions $N(E)$. However, the situation may be changed drastically in the case of a sharp energy dependence, e.g., in the presence of a break owing to the van Hove singularity.

Consider the simplest model of $N(E)$ with a break

$$N(E)=\tilde{N}(E)+N'(E)(E-E_0), \tag{4.57}$$

$$N'(E)=\begin{cases} A_1, & E<E_0, \\ A_2, & E>E_0, \end{cases}$$

where $\tilde{N}(E)$ is a smooth function. Then exact integration in (4.25) gives

$$\chi_p(T)=\frac{1}{2}\left[N(\zeta)+(A_2-A_1)T\ln\left(1+\exp\left(-\frac{|E_0-\zeta|}{T}\right)\right)+O\left(\frac{T^2}{\zeta^2}\right)\right]. \tag{4.58}$$

The effect of singularity vanishes at $|E_0 - \zeta| \gg T$.

According to the band calculations for palladium, there exists the van Hove singularity with $|E_0 - E_F| = 300$ K. To describe correctly the experimental behaviour of $\chi(T)$ in Pd (in particular, the position of the maximum), it is necessary to shift the singularity to the Fermi level up to $|E_0 - E_F| = 100$ K.

Presence of DOS peaks at the Fermi level, which is, as discussed above, typical for both strong and weak itinerant ferromagnets, may influence appreciably the $\chi(T)$ behaviour and the T_C value even in the Stoner theory of ferromagnetism. In particular, the triangle peak model was used to describe properties of weak itinerant ferromagnets [280]. Under condition of the linear dependence $\chi^{-1}(T)$ owing to a DOS singularity the Curie temperature of a ferromagnetic metal is estimated as

$$T_C \sim (2I\chi_p(0)-1)E_F \tag{4.59}$$

and may be considerably smaller than (4.53).

4.4 SPIN-FLUCTUATION THEORIES

A more radical way to improve the results of the Stoner theory for temperature dependences of magnetic properties is going out beyond the one-electron mean-field theory. Formally, the Stoner model corresponds to the interaction Hamiltonian of an infinite range,

$$\mathsf{H}_{\text{int}} = In_{\uparrow}n_{\downarrow} = I\left(\frac{n^2}{4} - \langle S^z \rangle^2\right), \qquad n_\sigma = \sum_{\mathbf{k}} n_{\mathbf{k}\sigma}, \tag{4.60}$$

so that the mean-field approximation is exact. A more rich physics is described by models which include electron correlations. The simplest among them is the Hubbard model (G.1) with the on-site Coulomb repulsion U. The Stoner approximation for the electron spectrum in this model corresponds to the simplest Hartree–Fock decoupling and may be justified only in the small-U limit which is inconsistent with the ferromagnetism criterion. Therefore correlation (spin-fluctuation) effects, generally speaking, turn out to be important.

As well as for Heisenberg magnets, low-temperature thermodynamics of itinerant magnets is determined by spin-wave contributions (see, e.g., [295]). Spin-wave theory of metallic ferro- and antiferromagnets is considered in Appendix G. The simple spin-wave treatment enables one also to reproduce some of the results of spin-fluctuation theories.

Spin-fluctuation theories, which describe thermodynamics of itinerant magnets in a wide temperature region, were developed during last twenty years [26, 296–302]. They turned out to be especially successful for weak itinerant ferromagnets which are characterized by extremely small values of saturation magnetic moment M_0 and large ratios p_C/p_s (Table 4.3, Fig.4.7). These substances correspond to a near vicinity of the Stoner instability and possess strong spin fluctuations. However, unlike systems with local magnetic moments, the fluctuations are localized not in real, but in reciprocal space (near the wavevector $q = 0$).

Presence of the small parameter $2I\chi - 1$ yields a possibility of an analytical treatment of weak itinerant ferromagnetism. The simple RPA expression for magnetic susceptibility (G.12), which follows from the spin-wave theory, turns out to be thermodynamically inconsistent. In the theory of Moriya and Kawabata [296] the expression for paramagnetic susceptibility was taken in the form

$$\chi(T) = \frac{\chi_p(T)}{1 - 2I\chi_p(T) + \lambda(T)}, \tag{4.61}$$

where $\lambda(T)$ is proportional to the squared amplitude of spin fluctuations. Calculation of λ in a self-consistent way by using the expression for the free energy yields in the immediate vicinity of T_C

$$\lambda(T) \sim (T/E_F)^{4/3}.$$

Thus λ dominates over $2I\chi$ in the temperature dependence of the inverse susceptibility $\chi^{-1}(T)$. With increasing T the dependence $\chi^{-1}(T)$ becomes practically linear. It should be noted that the Curie–Weiss behaviour in the spin-fluctuation approach has a quite different origin in comparison with the Heisenberg model. The Curie–Weiss law holds approximately not only for weakly ferromagnetic, but also for almost ferromagnetic metals where

$$0 < 1 - 2I\chi_p(0) \ll 1,$$

e.g., for Pd, $HfZr_2$. The Curie constant is determined by the electron structure near the Fermi level only and, unlike the Heisenberg magnets, does not depend on the saturation magnetization.

At $0 < 2I\chi_p(0) - 1 \ll 1$ fluctuations result in a considerable decrease of the Curie temperature in comparison with the Stoner theory:

$$T_C \sim (2I\chi_p(0) - 1)^{3/4} E_F. \qquad (4.62)$$

Near T_C one has

$$M(T) \sim \left(T_C^{4/3} - T^{4/3}\right)^{1/2}. \qquad (4.63)$$

Corresponding results for a weak itinerant antiferromagnet have the form

$$T_N \sim \left(2I\chi_Q - 1\right)^{2/3}, \qquad M_Q(T) \sim \left(T_N^{3/2} - T^{3/2}\right)^{1/2}. \qquad (4.64)$$

Here χ_Q and M_Q are the non-uniform (staggered) susceptibility of the non-interacting system and the sublattice magnetization, which correspond to the wavevector of the antiferromagnetic structure **Q**, $2I\chi_Q - 1 \ll 1$. The electronic specific heat of weakly and almost ferromagnetic metals is substantially enhanced by spin fluctuations:

$$C(T) \sim T \ln(\min[|1 - IN(E_F)|, T/E_F]) \qquad (4.65)$$

(4.65) (see also Appendix G). In the case of a weak antiferromagnet, the logarithmic factor is absent, and one has for the spin-fluctuation contribtion [26]

$$\delta C(T) \sim |1 - 2I\chi_Q|^{1/2}\, T. \qquad (4.66)$$

Thus main difficulties of the Stoner theory are eliminated for weak itinerant magnets. A number of attempts to generalize the spin-fluctuation approach on strong itinerant ferromagnets (e.g., transition metals) were performed [26, 297, 300–302]. In particular, an interpolation theory within the path integral method is considered in [26]. An important shortcoming of such theories is that spin-fluctuation effects are taken into account for finite-temperature effects only, but the ground state is described by the Stoner theory. This approach is inapplicable in the case of strong correlations which is discussed in Sect.4.6.

Generally speaking, equations of spin-fluctuation theories are rather complicated and require numerical calculations. By these reasons, simple approaches, which provide analytical expressions, are of interest. A version of such approach may be based on introducing the effective temperature T_{eff} which is related to the effect of smearing the electron density of states by spin fluctuations [303]. This quantity may be substituted into Fermi distribution functions at calculating thermodynamical averages. Consider the local approximation for spin fluctuations [300–303]. For a fivefold degenerate d band the average square of fluctuation amplitude in the paramagnetic region is estimated as

$$v^2 = \frac{3UT}{10(1-2U\chi_0)}, \tag{4.67}$$

where U is the on-site Hubbard repulsion,

$$\chi_0 = -\frac{1}{2\pi}\int f(E)\,\mathrm{Im}\,[G(E)]^2\, dE \tag{4.68}$$

is the local susceptibility, the quantity

$$G(E) = \int \frac{N(E')}{E-E'-\Sigma(E)}dE' \tag{4.69}$$

being the site-diagonal Green's function, $\Sigma(E)$ the self-energy. Note that, unlike the Stoner criterion $2U\chi_p > 1$, the condition $2U\chi_0 > 1$ which corresponds to the formation of local magnetic moments (or to occurrence of the Hubbard splitting, which means a reconstruction of the ground state) is usually not satisfied in such approaches.

The effective smeared density of states is given by

$$\tilde{N}(E) = -\frac{1}{\pi}\int \frac{N(E')\,\mathrm{Im}\,\Sigma(E)}{[E-E'-\mathrm{Re}(E)]^2 + [\mathrm{Im}\,\Sigma(E)]^2}dE'. \tag{4.70}$$

To the lowest order in fluctuations

$$\mathrm{Im}\,\Sigma(E) = -\pi N(E) v^2, \tag{4.71}$$

so that the fluctuations result in the Lorentz smearing. At calculating the integrals which include the Fermi function, this smearing may be taken into account by the effective smearing the Fermi distribution

$$\frac{\partial f(E)}{\partial E} \to -\frac{1}{\pi}\int \frac{\operatorname{Im}\Sigma(E')}{[E'-E-\operatorname{Re}\Sigma(E')]^2+[\operatorname{Im}\Sigma(E')]^2}\frac{\partial f(E')}{\partial E'}dE'. \qquad (4.72)$$

Using near the Fermi energy the linear approximation

$$\operatorname{Re}\Sigma(E) = \Sigma' E, \qquad f(E) \simeq \frac{1}{2} - \frac{E}{4T}, \qquad (4.73)$$

we obtain the simple analytical result

$$\frac{\partial \tilde{f}(E)}{\partial E} = -\frac{\kappa^{-1}}{4T} = -\frac{1}{4T_{\text{eff}}}, \qquad (4.74)$$

$$\kappa = \left[\frac{2}{\pi}(1-\Sigma')^{-1}\arctan\frac{2T(1-\Sigma')}{\pi v^2 N(E_{\text{F}})}\right]^{-1}.$$

The effective temperature $T_{\text{eff}} = \kappa T$ defined by (4.74) may be considerably larger than T. Then the Curie temperature $T_{\text{C}} = T_{\text{S}}/\kappa$ is appreciably renormalized in comparison with the Stoner value T_{S}. The effect of lowering the Curie temperature becomes more appreciable in the presence of DOS singularities.

Another approach based on introducing a spin-fluctuation temperature was developed by Mohn and Wohlfarth [304]. Following to [305], they used the Ginzburg–Landau expansion for the inverse susceptibility

$$\frac{H}{M} = A + BM^2 + B(3\langle m_{\parallel}^2\rangle + 2\langle m_{\perp}^2\rangle), \qquad (4.75)$$

where, as well as in the Stoner theory,

$$A = -\frac{1}{2\chi(0)}\left(1-\frac{T^2}{T_{\text{S}}^2}\right), \qquad B = \frac{1}{2\chi(0)M_0^2}, \qquad (4.76)$$

and $\chi(0)$ is the enhanced susceptibility given by

$$\chi(0)^{-1} = N_{\uparrow}^{-1}(E_{\text{F}}) + N_{\downarrow}^{-1}(E_{\text{F}}) - \frac{I}{2}. \qquad (4.77)$$

The equilibrium magnetic moment is determined as

$$M_0^2 = |A/B|_{T=0},$$

$\langle m_{\parallel}^2\rangle$ and $\langle m_{\perp}^2\rangle$ are the mean square values of the longitudinal and transverse local fluctuating moments (the coefficients in (4.75) are determined by the condition of vanishing of the scalar products (**Mm**) in the free energy). A renormalization of the coefficients A and B owing to the last term in (4.75) was considered by expanding in powers of magnetization

$$B(3\langle m_\parallel^2\rangle + 2\langle m_\perp^2\rangle) = a_1 - a_2 M^2 + a_3 M^4. \tag{4.78}$$

The expression for a_1 may be expressed in terms of the spin-fluctuation susceptibility

$$a_1(T) = (1 - t_C^2)\chi_{sf}^{-1}(T), \qquad t_C \equiv \frac{T_C}{T_S}, \tag{4.79}$$

where T_C is the true value of the Curie temperature, renormalized by spin fluctuations. At $T = T_C$ we have

$$\chi_{sf}(T_C) = \frac{\langle m^2\rangle}{3T_C}, \tag{4.80}$$

$$\langle m^2\rangle = \frac{1}{3}(3\langle m_\parallel^2\rangle + 2\langle m_\perp^2\rangle).$$

The amplitude of the fluctuating moment at T_C is given by

$$\langle m^2\rangle_{T_C} = \frac{3}{5}M_0^2(1 - t_C^2), \tag{4.81}$$

so that

$$a_1(T_C) = 5T_C/M_0^2. \tag{4.82}$$

The equation for T_C, $A' = A + a_1 = 0$, yields

$$\left(\frac{T_C}{T_S}\right)^2 + \frac{T_C}{T_{sf}} - 1 = 0, \tag{4.83}$$

where

$$T_{sf} = M_0^2/[10\chi(0)]. \tag{4.84}$$

is a characteristic temperature describing the influence of fluctuations.

The expression (4.83) differs somewhat from the corresponding result of the more rigorous theory for weak itinerant ferromagnets by Lonzarich and Taillefer [305]

$$\left(\frac{T_C}{T_S}\right)^2 + \left(\frac{T_C}{T_{sf}}\right)^{4/3} - 1 = 0, \tag{4.85}$$

where the $T^{4/3}$-dependence is due to influence of a finite cutoff wavevector.

The paramagnetic susceptibility within the approach [304] is given by

$$2\chi(0)/\chi(T) = t_C^2(T/T_C)^2 + (1 - t_C^2)(T/T_C) - 1. \tag{4.86}$$

The curves $\chi(T)$ interpolate between a Curie–Weiss law (spin fluctuations

Table 4.5. Values of the Stoner and renormalized Curie points, T_S and T_C, spin-fluctuation temperature T_{sf}, and the fluctuating magnetic moment at T_C^{calc} for iron group metals; $t_C = T_C^{exp}/T_S$ [304].

	T_S	T_{sf}	T_C^{calc}	T_C^{exp}	$\langle m^2\rangle^{1/2}$	t_C
Fe	2560	1293	1068	1043	1.52	0.41
Co	2240	2439	1436	1388	0.93	0.62
Ni	1790	759	656	631	0,42	0.35

predominate, $t_C = 0$) and the Stoner square law (single-particle excitations play the main role). The Curie constant is determined entirely by the ferromagnetic susceptibility $\chi(0)$.

Determining T_S and $\chi(0)$ from band calculations, Mohn and Wohlfarth calculated the values of Curie temperature for iron group metals (Table 4.5) in a good agreement with experiment (however, a considerable uncertainty of the parameters used should be noted). Calculations for some Y–Co and Y–Fe compounds and borides were also performed. The results turn out to be close to the universal curve which shows the dependence of T_C/T_{sf} as a function of T_S/T_{sf} (Fig.4.9).

The results enable on e to estimate the role of fluctuation effects in various metals. The value of t_C may be a quantitative measure of these effects: according to [304], we have for $t_C > 0.5$ the Stoner systems and for $t_C < 0.5$ the fluctuation systems. Large values of

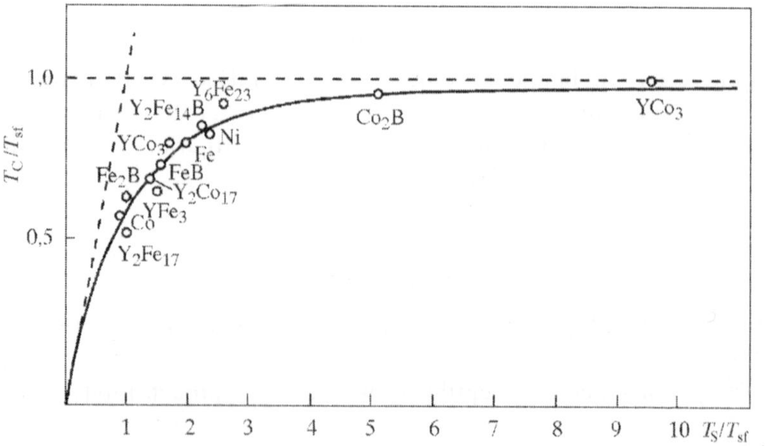

Fig. 4.9. Comparison between equation (4.86) and experimental Curie temperatures for iron group metals and intermetallic systems. The asymptotes corresponds to pure fluctuation behaviour and the tangent to pure Stoner behaviour.

M_0 suppress fluctuations owing to strong molecular field, and the high-field susceptiblity $\chi(0)$ works in the opposite direction by lowering T_{sf} and favouring magnetic moment fluctuations at low temperatures. Although the value of M_0 in iron exceeds by 15 times that in nickel, the value of $\chi(0)$ in Ni is smaller since it is a 'stronger' ferromagnet which is close to the saturation limit. Therefore fluctuation effects in both metals turn out to be comparable. At the same time, the large value of M_0 in cobalt is sufficient to suppress fluctuation effects, so that T_{sf} is even larger than T_S, and T_C is determined mainly by single-particle excitations. It should be noted that these conclusions differ from the results of [291], according to which fluctuations play an important role in Fe and Co, whereas Ni is satisfactorily described by the Stoner theory with account of DOS singularities.

4.5 ELECTRONIC STRUCTURE AND PROPERTIES OF HALF-METALLIC FERROMAGNETS

The limiting case of 'strong' ferromagnets with a large spin splitting is opposite to weak itinerant ferromagnets. In the old Stoner theory (Sect.4.3), a strongly ferromagnetic solution, where the spin splitting exceeds the Fermi energy and one spin subband is empty or completely filled, was considered. It was believed that such a situation (for the hole states) corresponds to ferromagnetic nickel. However, modern band calculations [24] within the LSDF approach disproved this assumption (the spin up density of states at the Fermi level turned out to be small but finite, Fig.2.25).

At the same time, band calculations did lead to discovery of real magnets, which are similar to strong Stoner ferromagnets. Calculation by de Groot et al of the band structure for the Heusler alloy NiMnSb [306], PtMnSb [306–308] with the C_{1b} (MgAgAs) structure demonstrated that the Fermi level for one of spin projections lies in the energy gap. Since these systems behave for one of spin projections as insulators, they were called 'half-metallic ferromagnets' (HMF). Later a similar picture was obtained for CoMnSb [309], ferrimagnet FeMnSb [310], antiferromagnet CrMnSb [308]. Band structure calculations for a large group of ferro- and antiferromagnetic Heusler alloys from another series T_2MnZ (T = Co, Ni, Cu, Pd) with the $L2_1$ structure demonstrated that a state, which is close to HMF ($N(E)$ is practically zero), takes place in systems Co_2MnZ with Z = Al, Sn [311] and Z = Ga, Si, Ge [312]. (In the $L2_1$ structure, all the four sublattices in the fcc lattice are filled by the atoms T, Mn and Z, whereas in the the C_{1b} structure some of the positions are empty, so that the symmetry lowers to tetrahedral one). Besides that, a half-metallic state was found in band calculations of CrO_2 (rutile structure) [313, 314], UNiSn (C_{1b} structure) [315, 316], Fe_3O_4 [317].

The situation of strongly different spin up and spin down states, which is realized in HMF, is of interest for the general theory of itinerant magnetism [318]. The scheme of the 'half-metallic' state formation in Heusler alloys may be described as follows [306, 308, 311, 316]. At neglecting hybridization of T- and Z-atom states, *d*-band of manganese for above structures is characterized by a wide energy gap between bonding and antibonding states. Due to strong intra-atomic (Hund) exchange for manganese ions, in the ferromagnetic state spin up and down subbands are moved apart considerably. One of spin subbands comes closely to ligand *p*-band, and therefore the corresponding gap is partially or completely smeared by *p–d* hybridization. The other subband retains the gap, and the Fermi level may find itself at the latter, The energy gap in other subband retains and may coincide under certain conditions with the Fermi level, which just yields the HMF state. For the C_{1b} structure we have the true gap, and for the $L2_1$ structure a deep pseudogap. This is connected with a considerable change in the character of the *p–d* hybridization (especially between *p* and t_{2g} states) in the absence of inversion centre, as it takes place for the C_{1b} structure. Thus the latter structure is more favorable for the HMF state.

According to [324], similar factors are responsible for the gap in the partial density of states for one of manganese position (Mn(I)) in the compound Mn_4N, the structure of which is obtained from the T_2MnZ structure by removing some atoms. A qualitatively similar mechanism, which is based on the strong Hund exchange and hybridization between *d*-states of chromium with *p*-states of oxygen, is considered in [313] for CrO_2. As discussed in [311], stability of the ferromagnetic state itself is a consequence of difference in the *p–d* hybridization for states with opposite spin projections. To describe such a situation, Kuebler et al [311] have introduced the term 'covalent magnetism' and stressed the distinction from the picture of the spectrum in the Stoner model where the densities of states with opposite spin projections differ by the constant spin splitting. The results of band structure calculations are shown in Figs.4.10–4.13.

The interest in half-metallic ferromagnets (HMF) was connected to the first place with their unique magnetooptical properties [307, 308], which are intimately related to the electronic structure near

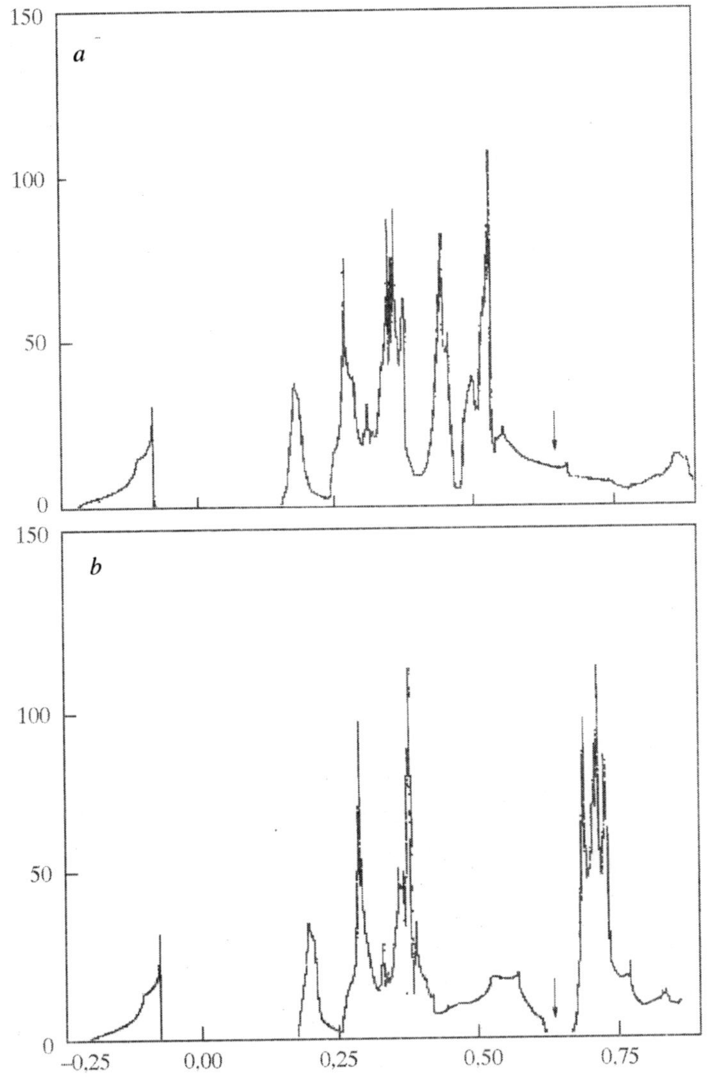

Fig. 4.10. Density of states (Ry^{-1}) in PtMnSb [307] (a) spin up (b) spin down.

the Fermi level (absence of states with one spin projection), which results in strong asymmetry of optical transitions (see the discussion in Sect. 5.7.3). In particular, large magnetooptical Kerr effect was observed in PtMnSb (for other compounds of this series the effect is smaller because of smallness of relativistic interactions for light atoms).

Besides that, the HMF are important in connection with the problem

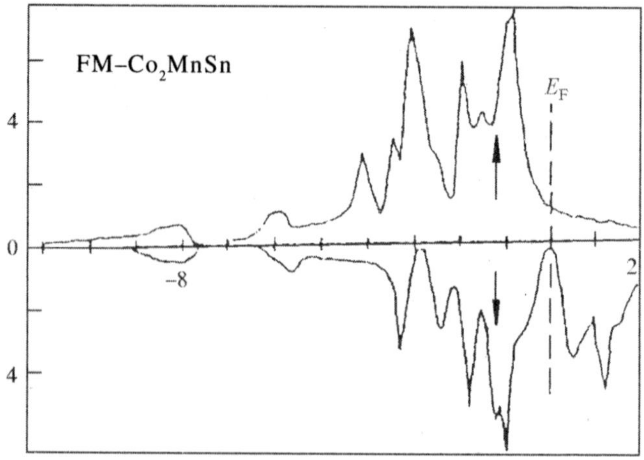

Fig. 4.11. Density of states for the Heusler alloy Co_2MnSn [311].

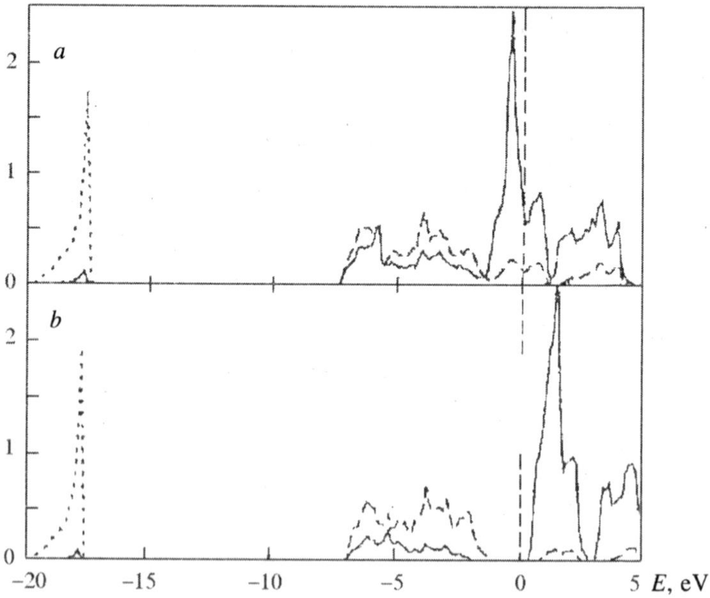

Fig. 4.12. Partial densities of states (1/eV atom, energy in eV) for CrO_2 [313]. The solid line corresponds to $3d$ states of Cr, dotted and broken lines to $2s$ and $2p$ states of oxygen (a) spin up (b) spin down.

of obtaining the large saturation magnetic momentum, since, evidently, their electron spectrum is favourable for maximum spin polarization (further increase of the spin splitting in the half-metallic state does not result in an increase of magnetic moment). In fact, the elec-

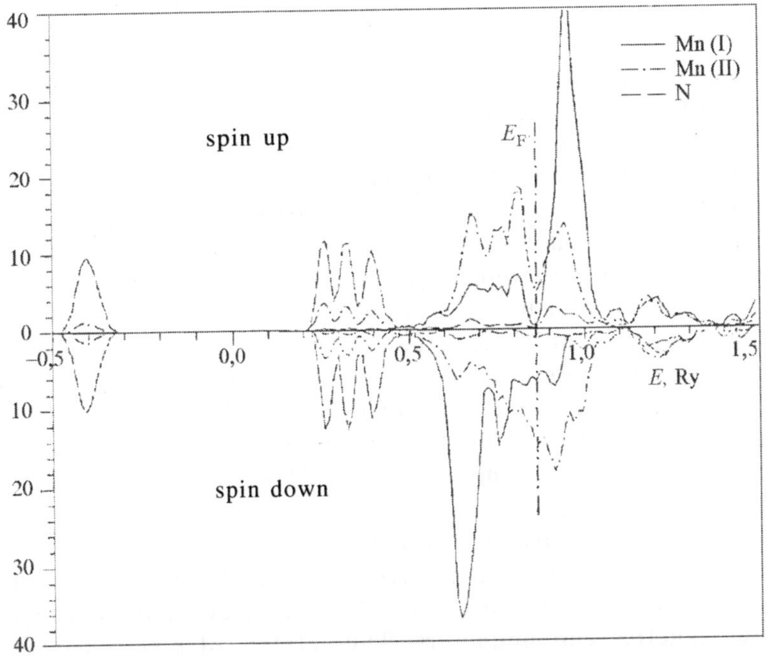

Fig. 4.13. Partial densities of states (1/Ry atom) for Mn_4N [324].

tronic structure, which is reminiscent of the half-metallic one (a deep minimum of DOS at E_F for $\sigma = \downarrow$) was found in the system of alloys Fe–Co [319] and in the compounds R_2Fe_{17}, $R_2Fe_{14}B$ with the record value of M_0 [320, 321]. Such a minimum is typical for systems with well-defined local magnetic moments and takes place also in pure iron (Fig.2.24). Well-pronounced minima for the states with $\sigma = \downarrow$ occur in the compounds RCo_5 (R = Y, Sm, Gd) [322]. A comparison of magnetic properties of a large series of alloys Y_nCo_m, Y_nFe_m with their electronic structure, calculated by the recursion method, was performed in [323]; the state, close to HMF, was obtained for YCo_5, YCo_7, Y_2Co_{17}. The small partial values of $N_\uparrow(E_F)$ were also found in Mn_4N, Fe_4N for the crystallographic Mn(I) position [324–326].

Recently, the interest in half-metallic ferromagnetism has greatly increased in connection with the discovery of giant magnetoresistivity in ferromagnetic manganites [688].

The magnetic properties of some half-metallic Heusler alloys are presented in the Table 4.3. These compounds possess large ground state moment and high Curie temperature. Strong ferromagnetism of Heusler alloys is mainly due to local moments of well-separated

Mn atoms. An interesting feature is that the Rhodes–Wohlfarth ratio p_C/p_s exceeds considerably unity. Moreover, the effective moment in the paramagnetic state, determined from the paramagnetic susceptibility, decreases appreciably with temperature.

This behaviour may be explained by that the change of electronic structure in half-metallic magnets at destruction of the magnetic ordering is especially strong – the gap in the electron spectrum vanishes. The temperature dependence of magnetic moment in the paramagnetic state may be due to short-range magnetic order (local densities of states are similar to those in the ferromagnetic state). From the many-electron point of view, the decrease of the local moment with increasing temperature is connected with the absence of corrections to ground state magnetization of the type (G.49). However, such corrections do occur at high temperatures.

From the theoretical point of view, HMF are also characterized, due to quasimomentum conservation law, by absence of decay of a spin wave into a pair electron–hole with opposite spins (Stoner excitations). Therefore spin waves are well defined in the whole Brillouin zone, similar to Heisenberg ferromagnets and degenerate ferromagnetic semiconductors. Thus effects of electron-magnon interactions (spin-polaron effects) are not masked, unlike usual itinerant ferromagnets, by the Stoner excitations and may be studied in a pure form. At present, experimental data on neutron scattering in the Heusler alloys Pd_2MnSn, Ni_2MnSn [327] and Cu_2MnAl [328] are available. Spin waves turn out to be well defined over the whole Brillouin zone, which makes up, according to [26], a criterion of validity of the localized moment model.

From the point of view of band theory, the smallness of magnon damping may be explained by that the partial Mn-atom spin-up density of states is small since the corresponding subband is almost empty [311]. One may expect that the damping will be still smaller provided that the Fermi level for one of spin projections lies in the energy gap. Thus a purposeful investigation of spin-wave damping at comparing results for various Heusler alloys from the T_2MnZ and TMnZ series is of great interest for verification of the theory.

As discussed above, the *p–d* hybridization plays an important role in the Heusler alloys, so that they should be described within a generalized Anderson-lattice model (Sect.6.7). However, this model is reduced in limiting cases to the Hubbard or *s–d* exchange models. Within the latter models, HMF are described as saturated ferromagnets where the spin splitting Δ exceeds the Fermi energy.

Consider the picture of density of states of a HMF within the

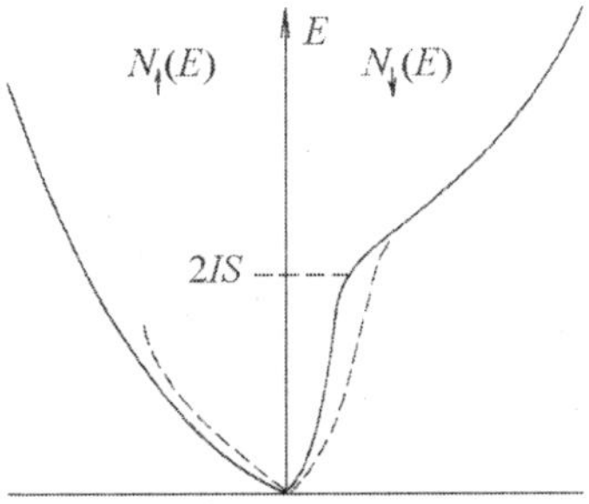

Fig. 4.14. Density of states in the *s–d* model in the case of empty conduction band ($I > 0$). At $T = 0$ (solid line) the spin-polaron tail of spin-down states reaches the band bottom. The broken line corresponds to finite temperatures [329].

s–d exchange model with account of correlation effects [329, 734]. Writing down the expansion of the Dyson equation (G.30) we obtain

$$N_\sigma(E) = -\frac{1}{\pi}\mathrm{Im}\sum_{\mathbf{k}} G_{\mathbf{k}\sigma}(E) =$$
$$= \sum_{\mathbf{k}}\delta(E - t_{\mathbf{k}\sigma}) - \sum_{\mathbf{k}}\delta'(E - t_{\mathbf{k}\sigma})\mathrm{Re}\Sigma_{\mathbf{k}\sigma}(E) - \frac{1}{\pi}\sum_{\mathbf{k}}\frac{\mathrm{Im}\Sigma_{\mathbf{k}\sigma}(E)}{(E - t_{\mathbf{k}\sigma})^2}. \tag{4.87}$$

The second term in the right-hand side of (4.87) describes renormalization of quasiparticle energies. The third term, which arises from the branch cut of the self-energy $\Sigma_{\mathbf{k}\sigma}(E)$, describes the incoherent (non-quasiparticle) contribution owing to scattering by magnons. One can see that this does not vanish in the energy region, corresponding to the 'alien' spin subband with the opposite projection $-\sigma$. The picture of the density of states for the empty conduction band is shown in Fig.4.14. The $T^{3/2}$-dependence of the magnon contribution to the residue (G.53), which follows from (G.34), i.e. of the effective electron mass in the lower spin subband, and an increase with temperature of the incoherent tail from the upper spin subband result in a strong temperature dependence of partial densities of states $N_\sigma(E)$, the corrections being of opposite sign. The corresponding behaviour of conduction electron spin polarization $P(T) \simeq \langle S^z\rangle$ (see (G.39)) is confirmed by experimental data on field emission from ferromagnetic semiconductors [330] and transport properties of half-metallic Heusler alloys [331].

The picture of $N(E)$ near the Fermi level in HMF (or degenerate semiconductor) turns out to be also non-trivial. If we neglected magnon frequencies in the denominators of (G.34), the partial density of incoherent states should occur by a jump above or below the Fermi energy for $I > 0$ and $I < 0$ respectively owing to Fermi distribution

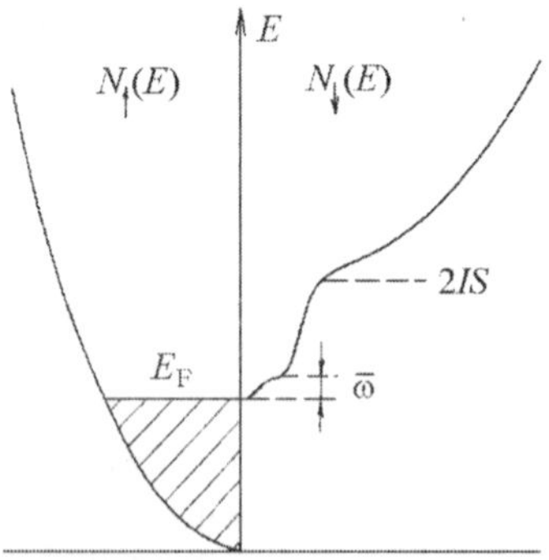

Fig. 4.15. Density of states in a half-metallic ferromagnet with $I > 0$. Non-quasiparticle states with $\sigma = \downarrow$ are absent below the Fermi level [329].

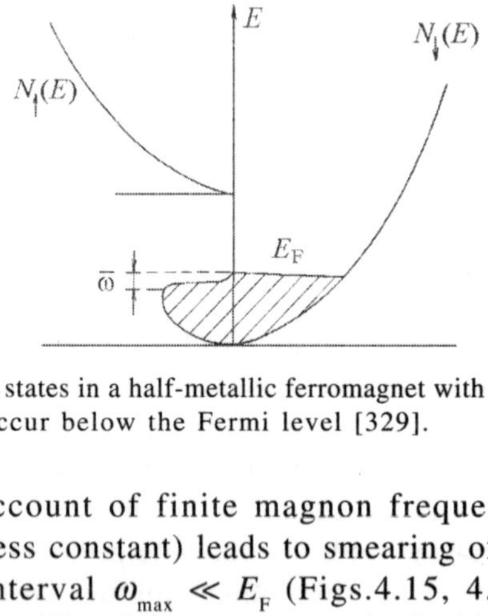

Fig. 4.16. Density of states in a half-metallic ferromagnet with $I < 0$. Non-quasiparticle states with $\sigma = \uparrow$ occur below the Fermi level [329].

functions. An account of finite magnon frequencies $\omega_{\mathbf{q}} = Dq^2$ (D is the spin stiffness constant) leads to smearing of these singularities on the energy interval $\omega_{max} \ll E_F$ (Figs.4.15, 4.16), $N_{-\alpha}(E_F)$ being equal to zero. For $|E - E_F| \ll \omega_{max}$ we obtain [329, 332, 734]

$$\frac{N_{-\alpha}(E)}{N_\alpha(E)} = \frac{1}{2S}\left|\frac{E - E_F}{\omega_{max}}\right|^{3/2} \theta(\alpha(E - E_F)), \quad \alpha = \operatorname{sign} I. \tag{4.88}$$

With increasing $|E - E_F|$, $N_{-\alpha}/N_\alpha$ tends to a constant value which is of order of I^2 within perturbation theory. In the strong coupling limit where $|I| \to \infty$ we have

$$\frac{N_{-\alpha}(E)}{N_\alpha(E)} = \frac{1}{2S}\theta(\alpha(E - E_F)), \quad |E - E_F| \gg \omega_{max}. \tag{4.89}$$

Similar calculations for a Hubbard ferromagnet with strong correlations and electron concentration $n > 1$ yield [333]

$$N_{\uparrow}(E)=\sum_{\mathbf{k}\sigma}f(t_{\mathbf{k}+\mathbf{q}})\delta(E-t_{\mathbf{k}+\mathbf{q}}+\omega_{\mathbf{q}})=\begin{cases}N_{\downarrow}(E), & E_{\mathrm{F}}-E\gg\omega_{\max},\\ 0, & E>E_{\mathrm{F}}\end{cases} \qquad (4.90)$$

(cf. (J.22)). This result has a simple physical meaning. Since the current carriers are spinless doubles (doubly occupied sites), the electrons with spins up and down may be picked up with an equal probability from the states below the Fermi level of doubles. On the other hand, according to the Pauli principle, only the spin down electrons may be added in the singly occupied states in the saturated ferromagnet.

Thus the spin polarization $P(E)$ in strong itinerant ferromagnets changes sharply near E_{F}. In the case of an antiferromagnetic s–d exchange (or in a Hubbard ferromagnet) there exist occupied non-quasiparticle states which may result in electron depolarization in photoemission experiments. At the same time, empty non-quasiparticle states may be observed by inverse photoemission. Since emission experiments have not too high energy resolution, appreciable deviations from 100% spin polarization should be observed. Data on photoemission in the HMF NiMnSb [334] have yielded the polarization about 50%. The above-discussed picture can be also experimentally verified by core level spectroscopy [734] and novel spectroscopic methods, e.g., spin-polarized scanning tunneling microscopy (STM, see [736]).

The incoherent contribution to the density of states should also result in peculiar contributions to thermodynamic and transport properties. The non-quasiparticle terms in electronic specific heat are discussed in Appendix G (see (G.65)–(G.67)). Since the impurity resistivity is determined by the dependence $N(E)$ near the Fermi level, it contains also non-quasiparticle contributions (Sect.5.3). Asymmetry of $N(E)$ should lead to an appreciable contribution to thermoelectric power.

Peculiarities of electronic structure of HMF should be clearly observed at investigation of nuclear magnetic relaxation [318, 731]. The T-linear Korringa contribution to the longitudinal nuclear relaxation rate, which dominates as a rule for ferromagnetic metals, is determined by the transverse spin susceptibility and given by the Moriya formula [335, 336]

$$\frac{1}{T_1}=-\frac{A^2T}{2\pi\omega_{\mathrm{n}}}\mathrm{Im}\sum_{\mathbf{q}}\langle\langle S^+_{\mathbf{q}}\,|\,S^-_{-\mathbf{q}}\rangle\rangle_{\omega}=\pi A^2TFN_{\uparrow}(E_{\mathrm{F}})N_{\downarrow}(E_{\mathrm{F}}), \qquad (4.91)$$

where ω_{n} is the MNR frequency, γ_{n} is the nuclear hyromagnetic ratio, A is the hyperfine interaction constant, F is the exchange enhancement factor. (A detailed discussion and comparison of different contributions to $1/T_1$, including orbital ones, is given in paper [335]). At the same time, the transverse relaxation rate contains a contribution from the longitudinal susceptibility:

$$\frac{1}{T_2}=\frac{1}{2T_1}+\frac{\pi}{4}A'^2TF'\left[N_\uparrow^2(E_F)+N_\downarrow^2(E_F)\right] \tag{4.92}$$

(generally speaking, A' and F' differ from A and F). Neglecting for simplicity the exchange splitting and spin dependence of hyperfine interaction matrix elements, we derive

$$\frac{1/T_2}{1/T_1}=\frac{[N_\uparrow(E_F)+N_\downarrow(E_F)]^2}{4N_\uparrow(E_F)N_\downarrow(E_F)}>1. \tag{4.93}$$

One can see that $1/T_1 = 1/T_2$ in paramagnetic metals, but the ratio (4.93) should considerably differ from unity in itinerant ferromagnets with an appreciable dependence $N(E)$. In fact, $1/T_2$ exceeds by several times $1/T_1$ in iron and nickel [336].

In HMF the contribution (4.91) vanishes, which is connected with absence of processes of magnon decay into Stoner excitations, whereas $1/T_2$, which is determined by electron transitions without spin flips, should have usual behaviour. To obtain the temperature dependence of $1/T_1$ in HMF one has to consider the contribution of two-magnon processes. Using (G.24) we obtain [337, 338]

$$\frac{1}{T_1^{(2)}}=\frac{A^2TS}{\pi\omega_n}\sum_{\mathbf{q}}\frac{\gamma_{\mathbf{q}}^{(2)}(\omega_n)}{\omega_{\mathbf{q}}^2}=\frac{12\pi^{1/2}}{S}\left(\frac{v_0}{16\pi^2}\right)^2\zeta\left(\frac{3}{2}\right)\frac{T^{5/2}}{D^{7/2}}(k_{F\uparrow}^2+k_{F\downarrow}^2). \tag{4.94}$$

Qualitatively, the dependence $T^{5/2}$ may be interpreted as the Korringa law with the value $N_\downarrow(E_F)$ being replaced with the 'thermal' value of the density of non-quasiparticle states, which is proportional to $T^{3/2}$. Experimentally, strong deviations from the linear Korringa law were observed at measuring $1/T_1$ in the HMF NiMnSb [339]. At not too low temperatures $T > 250$ K ($T_C = 750$ K) the dependence of the form

$$1/T_1(T)=aT+bT^{3.8}.$$

was obtained.

In the above-discussed ferrimagnet Mn_4N with $T_C = 750$ K the nuclear magnetic relaxation was investigated for the Mn(I) position (a narrow NMR line was obtained only for this position) [340]. At low temperatures $T < 77$ K the behaviour $1/T_1(T)$ is linear, and at higher temperatures a square law holds. It should be noted that the two-magnon contribution may modify considerably the temperature dependence of $1/T_1$ in 'usual' ferromagnets, a crossover from $T^{5/2}$ to $T^{3/2}$ dependence of the correction taking place with increasing temperature [731].

4.6 MAGNETISM OF HIGHLY-CORRELATED *d*-SYSTEMS

The most difficult case for standard approaches in the itinerant electron magnetism theory (band calculations, spin-fluctuation theories) is that of systems where strong interelectron correlations lead to a radical reconstruction of the electron spectrum – formation of the Hubbard's subbands. Examples of such systems are oxides and sulphides of transition metals with a large energy gap, e.g. MeO (Me = Ni, Co, Mn), NiS_2 [25], basis systems for copper-oxide high-T_c superconductors La_2CuO_4 and $YBa_2Cu_3O_6$. At low temperatures these compounds are antiferromagnetic, so that the gap may be described as the Slater gap owing to AFM ordering. However, the gap retains in the paramagnetic region and has therefore a Mott–Hubbard nature.

Standard band calculations within the density functional method [341] did not yield satisfactory explanation of insulating properties of the MeO compounds. The reasonable value of gap in MeO and CuO_2-planes may be obtained by using special calculation methods which take into account in a sense the Hubbard correlations (e.g., the self-interaction correction (SIC) approach [77, 342] which considers the attraction of an electron to the field of the hole formed, Sect.2.3). First-principle calculations [343] yield the reasonable values of the Hubbard parameter in TM oxides, U = 6–10 eV.

The Hubbard splitting is apparently present also in some metallic ferromagnets, e.g., the solid solution $Fe_{1-x}Co_xS_2$ which has the pyrite structure, CrO_2 [26]. For the latter compound, corresponding direct optical data are present [344]. The spontaneous spin splitting above the Curie point, which is observed in iron group metals, may be also interpreted as related to the Hubbard subbands.

To treat the problem of ferromagnetic ordering in narrow bands from experimental point of view, we discuss the system $Fe_{1-x}Co_xS_2$. According to [345], its electronic structure is rather simple: all the electrons, responsible both for conductivity and magnetism, belong to the same narrow e_g-band. CoS_2 is a ferromagnetic metal with strong correlations, and FeS_2 is a diamagnetic insulator. Recent band calculations of these compounds [346] demonstrate that CoS_2 has an almost HMF structure.

Experimental investigations of magnetic properties of $Fe_{1-x}Co_xS_2$ have been performed in [347]. The most salient feature is the onset of ferromagnetism at surprisingly small electron concentrations $n = x < 0.05$. The magnetic moment equals to $1\mu_B$ in the wide concentration region $0.15 < n < 0.95$, and magnetic state is non-saturated for $n < 0.15$. However, unlike usual weak itinerant ferromagnets, there is no indication of an exchange enhancement of the Pauli susceptibility above T_C, and the Curie–Weiss law works well at arbitrary electron concentrations, the Curie constant being proportional to n. Such a behaviour cannot be explained within one-electron Stoner-like approaches, e.g. the T-matrix approximation [348], and demonstrates an important role of local magnetic moments (LMM). The inconsistency of one-electron approach for highly-correlated systems is demonstrated in Appendix H.

The problem of the description of LMM is a crucial point of the strong itinerant ferromagnetism theory. In the spin-fluctuation theories [26], LMM are introduced essentially *ad hoc* (e.g., the static approximation in the path integral, which corresponds to replacement of the translation-invariant system by the disordered one with random magnetic fields). On the other hand, the many-electron (atomic) picture describes LMM naturally. The role of strong electron correlations in the formation of LMM can be qualitatively demonstrated in the following way. If the Hubbard intrasite repulsion U is large enough, the electron spectrum contains the Hubbard subbands of singly and doubly occupied states. At $n < 1$, the number of doubles is small and tends to zero at $U \to \infty$. Then the singly occupied states make up LMM, and the empty site (holes) are the current carriers.

The simplest way to describe the formation of Hubbard subbands is calculation of electron Green's function with the use of Hubbard atomic representation (Appendix H). The formation of the Hubbard subbands contradicts the Fermi-liquid picture, in particular the Luttinger theorem about the conservation of the volume under the Fermi surface: each from two Hubbard subbands, originating from a free-electron band, contains one electron state per spin. Thus the energy spectrum of itinerant electron systems with LMM, contrary to weak itinerant magnets, differs essentially from that of a normal Fermi liquid. In the simplest Hubbard-I approximation, magnetic ordering results in narrowing of spin subbands rather than in a constant spin splitting (see (H.10)). Therefore the condition of ferromagnetism in such systems should not coincide with the Stoner criterion which corresponds to instability of the non-magnetic Fermi liquid with respect to a small spin polarization. Below we discuss the problem of ferromagnetism in strongly correlated Hubbard systems.

Rigorous investigation of ferromagnetism in the Hubbard model with $U \to \infty$ was performed by Nagaoka [349]. He proved that the ground state for simple cubic and bcc lattices in the nearest-neighbour approximation with electron number $N_e = N + 1$ (N is the number of lattice sites) possesses maximum total spin, i.e. is saturated ferromagnetic. The same statement holds for the fcc and hcp lattices with the transfer integral $t < 0$, $N_e = N + 1$, or $t > 0$, $N_e = N - 1$. (For other sign combinations, the ground state is more complicated because of divergence of the density of states at the band edge.) The physical sense of the Nagaoka theorem is rather simple. For $N_e = N$, $U = \infty$ each sites is singly occupied and the motion of electrons is impossible, so that the energy of the system does not depend on spin configuration. At introducing of an excess electron or hole, its kinetic energy turns out to be minimum for the uniform ferromagnetic spin alignment since this does not prevent their motion. It should be, however, noted, that in fact the proof of the Nagaoka theorem uses non-trivial topological consideration. In particular, it does not work in the one-dimensional case where the dependence of the kinetic energy on spin configurations is absent because of absence of closed trajectories [349]. Evidently, the picture of saturated ferromagnetism is preserved at small finite concentrations of current carriers.

In the case of the half-filled band ($N_e = N$), $|t| \ll U$ the ground state is antiferromagnetic because of Anderson's kinetic exchange interaction (Appendix D). This interactions is due to the gain in the kinetic energy at virtual transitions of an electron to a neighbour site, which are possible provided that the electron at this site has an opposite spin directions. In systems with finite U and $N_e \neq N$, a competition between ferro- and antiferromagnetic ordering occurs (it should be noted that, as discussed in Appendix D, the kinetic antiferromagnetic interaction emerges even in the formal limit $U = \infty$ due to non-orthogonality corrections, see also Ref.[727]). As follows from the calculation of the spin-wave energy [349], ferromagnetism preserves provided that

$$|t|/U < \alpha n, \tag{4.95}$$

where the constant $\alpha \sim 1$ depends on the lattice structure. At the same time, antiferromagnetism is stable at $N_e = N$ only. It was supposed in early papers that a canted magnetic structure is formed in the intermediate region [350]. However, numerical calculations [351] demonstrated that in fact a phase separation takes place into insulating antiferromagnetic and metallic ferromagnetic regions, all the current carriers being localized in the latter ones. Such a phenomenon seems to be observed in some highly doped magnetic semiconductors [352].

The problem of electron and magnon spectrum of a Hubbard ferromagnet with the nearly half-filled band may be considered rigorously in the spin-wave temperature region (Appendix J). To obtain a simple interpolation scheme for arbitrary current carrier concentrations and temperatures, we use the simplest Hamiltonian of the Hubbard model with $U \to \infty$, $n < 1$ in the many-electron representation with inclusion of the external magnetic field H:

$$\mathsf{H} = \sum_{\mathbf{k}\sigma} \varepsilon_{\mathbf{k}} X_{-\mathbf{k}}(0\sigma) X_{\mathbf{k}}(\sigma 0) - \frac{H}{2} \sum_{i} [X_i(++) - X_i(--)], \qquad (4.96)$$

with $\varepsilon_{\mathbf{k}} = \to -t_{\mathbf{k}}$. This Hamiltonian describes the motion of current carriers (holes) on the background of local magnetic moments – singly occupied sites. Following to [353], we consider the dynamic magnetic susceptibility (H.14). Its calculation (Appendix H) yields

$$G_{\mathbf{q}}(\omega) = \left(2\langle S^z \rangle + \sum_{\mathbf{k}} \frac{(\varepsilon_{\mathbf{k-q}} - \varepsilon_{\mathbf{k}})(n_{\mathbf{k}\uparrow} - n_{\mathbf{k-q}\downarrow})}{\omega - H - E_{\mathbf{k}\uparrow} + E_{\mathbf{k-q}\downarrow}} \right) \times$$
$$\times \left(\omega - H - \sum_{\mathbf{k}} \frac{(\varepsilon_{\mathbf{k-q}} - \varepsilon_{\mathbf{k}})(\varepsilon_{\mathbf{k}} n_{\mathbf{k}\uparrow} - \varepsilon_{\mathbf{k-q}} n_{\mathbf{k-q}\downarrow})}{\omega - H - E_{\mathbf{k}\uparrow} + E_{\mathbf{k-q}\downarrow}} \right)^{-1}, \qquad (4.97)$$

where the Hubbard-I approximation energies $E_{\mathbf{k}\sigma}$ and the occupation numbers $n_{\mathbf{k}\sigma}$ are given by (H.12).

At calculating the static magnetic susceptibility χ, one has tc treat carefully the limits $H \to 0$, $\omega \to 0$, $q \to 0$ because of non-hergodicity of the ferromagnetic ground state. Indeed, simple putting $H = \omega = q = 0$, which was made in some papers [354–356], yielded only the Pauli susceptibility rather than the Curie–Weiss behavior, which is physically unreasonable. To avoid the loss of the Curie–Weiss contribution from local moments, we apply the approach, employed by Tyablikov [357] for the Heisenberg model (see Appendix E). Using the sum rules

$$n_\sigma = \frac{1-c}{2} + \sigma \langle S^z \rangle = \sum_{\mathbf{q}} \langle X_{-\mathbf{q}}(\sigma, -\sigma) X_{\mathbf{q}}(-\sigma, \sigma) \rangle$$

and the spectral representation (E.18) we derive the equation for magnetization

$$\langle S^z \rangle = \frac{1-c}{2} + \frac{1}{\pi} \int_{-\infty}^{\infty} N_{\mathrm{B}}(\omega) \, \mathrm{Im} \sum_{\mathbf{q}} G_{\mathbf{q}}(\omega) d\omega. \qquad (4.98)$$

For small concentration of holes $c \ll 1$ one obtains, in agreement with the consideration by Nagaoka [349], the saturated ferromagnetism with

$$\langle S^z\rangle = \frac{1-c}{2} - \sum_{\mathbf{p}} N_B(\omega_{\mathbf{p}}), \tag{4.99}$$

$$\omega_{\mathbf{p}} = \sum_{\mathbf{k}} (\varepsilon_{\mathbf{k-p}} - \varepsilon_{\mathbf{k}}) f(\varepsilon_{\mathbf{k}}).$$

The magnon spectrum in (4.99) coincides with the exact result to lowest order in the inverse nearest-neighbour number $1/z$ (Appendix J).

The equation (4.98) can be simplified under the condition $\langle S^z\rangle \ll 1$, which holds both in the paramagnetic region ($\langle S^z\rangle = \chi H$, $H \to 0$) and for $n \ll 1$ at arbitrary temperatures. Expansion of denominator and numerator of (4.9) in $\langle S^z\rangle$ and H has the form

$$G_{\mathbf{q}}(\omega) = \frac{\omega A_{\mathbf{q}\omega} + \langle S^z\rangle B_{\mathbf{q}\omega} + HC_{\mathbf{q}\omega}}{\omega - \langle S^z\rangle D_{\mathbf{q}\omega} - HP_{\mathbf{q}\omega}}. \tag{4.100}$$

Here the quantity

$$A_{\mathbf{q}\omega} = \sum_{\mathbf{k}} \frac{f_{\mathbf{k-q}} - f_{\mathbf{k}}}{\omega + E_{\mathbf{k-q}} - E_{\mathbf{k}}} \left(1 + \frac{2}{1+c} \sum_{\mathbf{k}} \frac{E_{\mathbf{k-q}} f_{\mathbf{k-q}} - E_{\mathbf{k}} f_{\mathbf{k}}}{\omega + E_{\mathbf{k-q}} - E_{\mathbf{k}}}\right)^{-1} \tag{4.101}$$

determines the value of the effective magnetic moment, and

$$D_{\mathbf{q}0} = -\sum_{\mathbf{k}} \frac{8E_{\mathbf{k}}}{(1+c)^2} \left(E_{\mathbf{k}} \frac{\partial f_{\mathbf{k}}}{\partial E_{\mathbf{k}}} - E_{\mathbf{k+q}} \frac{f_{\mathbf{k-q}} - f_{\mathbf{k}}}{\omega + E_{\mathbf{k-q}} - E_{\mathbf{k}}} \right) \times$$

$$\times \left(1 + \frac{2}{1+c} \sum_{\mathbf{k}} \frac{E_{\mathbf{k-q}} f_{\mathbf{k-q}} - E_{\mathbf{k}} f_{\mathbf{k}}}{\omega + E_{\mathbf{k-q}} - E_{\mathbf{k}}}\right)^{-1} \equiv J_0^{\text{eff}} - J_{\mathbf{q}}^{\text{eff}} \tag{4.102}$$

describes the effective exchange interaction owing to motion of current carriers. Roughly speaking, this differs from the RKKY interaction by the replacement of the *s–d(f)* exchange parameter to the transfer integral. Such a replacement is characteristic for the narrow-band limit.

The ground state magnetization for small electron concentration $n \ll 1$ is given by

$$S_0 = \frac{n}{2} + \frac{1}{\pi} \int_0^\infty \operatorname{Im} \sum_{\mathbf{q}} A_{\mathbf{q}\omega} \, d\omega, \tag{4.103}$$

The non-zero value of S_0 occurs because of the second term in brackets in (4.11) and is formally small in the parameter $1/z$ (in the simplest 'Debye model' for the electron spectrum $\varepsilon_\kappa = a + bk^2$, $k < k_D$, one obtains $S_0 = n/8$ [353]). Thus we have for small n the non-saturated ferromagnetic state, in agreement with the experimental data [347]. Calculation of the critical electron concentration for onset of ferromagnetism apparently requires more advanced approximations.

At $T > T_C$, putting in (4.100) $\langle S^z \rangle = \chi H$ one obtains the equation for paramagnetic susceptibility

$$\frac{1-c}{2}+\frac{1}{\pi}\int_0^\infty \coth\frac{\omega}{2T}\,\mathrm{Im}\sum_{\mathbf{q}} A_{\mathbf{q}\omega}\,d\omega = T\sum_{\mathbf{q}}\left(\frac{\chi B_{\mathbf{q}0}+C_{\mathbf{q}0}}{\chi D_{\mathbf{q}0}+P_{\mathbf{q}0}}+A_{\mathbf{q}0}\right). \tag{4.104}$$

The Curie temperature is determined by the condition $\chi(T_C) = \infty$. At small n we have (cf.(E.25))

$$T_C = \frac{S_0}{2}\left(\sum_{\mathbf{q}} D_{\mathbf{q}0}^{-1}\right)^{-1}. \tag{4.105}$$

Expanding the right-hand side of (4.104) in χ at $T_C \ll T \ll E_F$ one gets

$$\chi = -\frac{1}{4}\sum_{\mathbf{k}}\frac{\partial f(\varepsilon_{\mathbf{k}})}{\partial \varepsilon_{\mathbf{k}}}+\frac{C}{T-\theta}, \tag{4.106}$$

where the first term corresponds to the Pauli contribution and the second one to the Curie–Weiss contribution of LMM,

$$C = \frac{1}{2}S_0, \qquad \theta = \sum_{\mathbf{q}} D_{\mathbf{q}0} > T_C \tag{4.107}$$

being the Curie constant and the paramagnetic Curie temperature. To the lowest order approximation in n,

$$\theta \simeq T_C \simeq C\frac{v_0}{2\pi^2}m^* k_F \varepsilon_{\max}^2, \tag{4.108}$$

with

$$k_F = (3\pi^2 n)^{1/3}, \qquad \frac{1}{m^*} = \left.\frac{\partial^2 \varepsilon_{\mathbf{k}}}{\partial k^2}\right|_{\varepsilon_{\mathbf{k}}=\varepsilon_{\max}}.$$

Thus the many-electron approach provides a simple derivation of the Curie–Weiss law. This approach may be generalized to the narrow-band s–d model [353].

The above concepts may also be useful for the general theory of metallic magnetism. Obviously, the assumption about strong (in comparison with the total bandwidth) interelectron repulsion is not valid for transition d-metals. However, it may hold for some electron groups near the Fermi level. This idea was used in [353] to treat ferromagnetism of iron group metals. The narrow-band Hubbard model was applied to describe the group of 'magnetic' states which form the narrow density of states peak owing to the 'giant' Van Hove singularities (Sect.2.4). Correlations for these states are to be strong because of small peak width $\Gamma \simeq 0.1$eV. The rest of s, p, d-electrons form broad bands and are weakly hybridized with 'magnetic' electrons of the peak. The peak states were assumed to be responsible for LMM formation and other magnetic properties in Fe and Ni. Such a model enables one to explain simply the low (as compared to the Fermi energy) values of the Curie temperature which is, as follows from the above consideration, of order of Γ.

In the ferromagnetic phase the splitting of spin up and down peaks is $\Delta \simeq 1$–2 eV $\gg \Gamma$ and structures of both peaks are similar. Since the lower peak is completely filled, the situation for 'magnetic' electrons turns out to be close to the saturated ferromagnetism, i.e. to the half-metallic state in the usual Hubbard model with large U. In such a situation one may expect strong (even in comparison with the Heusler alloys) non-quasiparticle effects. These may be important for the explanation of experimental data on spin polarization, obtained by photo- [358] and thermionic [359] emission, which contradict drastically to band calculation results for Fe and Ni.

4.7 MAGNETISM OF RARE EARTHS AND ACTINIDES

Since well-localized $4f$-electrons retain their magnetic moment in a crystal, rare earth metals exhibit strong magnetism. All the 'light' rare earths (RE) from Ce to Eu have at low temperatures complicated antiferromagnetic structures. All the 'heavy' RE from Gd to Tm (except for Yb which is a Pauli paramagnet) are ferromagnetic at low temperatures. With increasing temperature, the ferromagnetic ordering is as a rule changed by antiferromagnetic spiral structures, except for Gd which passes immediately into paramagnetic state. The period of spiral structures turns out to be appreciable temperature dependent, the spiral angle decreasing with lowering T.

Unlike d-ions, $4f$-ions are as a rule rather weakly influenced by the crystal field (Sect.1.3). Therefore magnetic moments of rare earth metals should be close to those of the free ion R. However, there are a number of important exceptions, especially among light rare earth elements. Consider magnetic structures of RE metals [16, 17, 246].

Cerium exhibits AFM ordering in the β (double hcp) phase with $T_N = 12.5$ K. The ordered moment $0.62\mu_B$ is small in comparison with that for Ce^{3+} ion ($2.51\mu_B$). Probably this is connected with strong crystal field influence on the not too stable f-shell.

Magnetism of praseodymium is determined to a large measure by existence of low-lying excited states of Pr^{3+} ion. Praseodymium exhibits AFM ordering below 25 K with ordered moment about $1\mu_B$.

Neodymium has a complicated AFM structure. The moments at hexagonal sites of the double hcp lattice order at T_N = 19.2 K, their value being modulated along the [1010] axis in the basal plane with the amplitude of $2.3\mu_B$. The moments at cubic sites (in neighbour planes) order at 7.8 K, the modulation amplitude being $1.8\mu_B$. These moment values are considerably smaller than that of free ion ($3.2\mu_B$) owing to crystal field.

Preliminary neutron investigations of promethium [360] demonstrate below 98 K a possible existence of ferromagnetic domains with ordered moment of $0.24\mu_B$. These data require a verification.

The crystal structure of samarium contains sites with cubic and hexagonal symmetry. Magnetic moments in the hexagonal layers order in a collinear AFM structure at 106 K, and in cubic layers at 13.8 K. The ordered moments are extremely small (about $0.1\mu_B$).

Europium, which has the bcc structure, possesses below 90 K a helcoidal AFM structure with the [100] axis and ordered moment of $5.9\mu_B$.

Magnetic structures of heavy rare earths are shown in Fig.4.17. Gadolinium is ferromagnetic with T_C = 293 K (see also Table 4.3). Below 232 K, there occurs a finite angle of the ferromagnetic spiral, which reaches the maximium value of 75° at 180 K and decreases up to 32° at low temperatures.

Terbium and dysprosium are ferromagnetic below 221 and 85 K respectively. At higher T, they have a simple helicoidal structure, the spiral angle being also appreciably temperature dependent.

Holmium has below 20 K a ferromagnetic spiral structure with

$$\mu_i^x = \mu_\perp \cos \mathbf{QR}_i, \qquad \mu_i^y = \mu_\perp \sin \mathbf{QR}_i, \qquad \mu_i^z = \mu_\parallel \tag{4.109}$$

and at higher T the helicoidal structure.

In erbium, the ferromagnetic spiral state (T < 20 K) changes into a complicated spiral structure

$$\mu_i^x = \mu_\perp \cos \mathbf{QR}_i, \qquad \mu_i^y = \mu_\perp \sin \mathbf{QR}_i, \qquad \mu_i^z = \mu_\parallel \cos \mathbf{QR}_i. \tag{4.110}$$

In the interval 52 < T < 84 K, a static longitudinal spin wave state occurs with

$$\mu_i^x = \mu_i^y = 0, \qquad \mu_i^z = \mu \cos \mathbf{QR}_i. \tag{4.111}$$

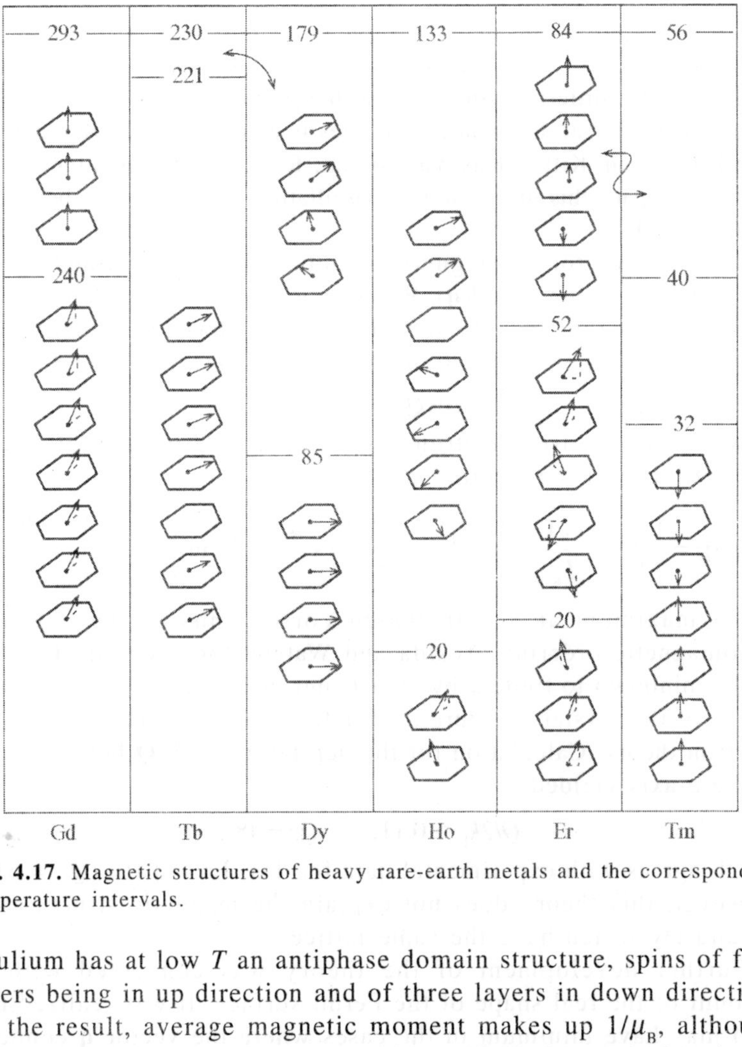

Fig. 4.17. Magnetic structures of heavy rare-earth metals and the corresponding temperature intervals.

Thulium has at low T an antiphase domain structure, spins of four layers being in up direction and of three layers in down direction. As the result, average magnetic moment makes up $1/\mu_B$, although the moment at each site is $7\mu_B$. Above 32 K the static longitudinal spin wave state is realized.

Ytterbium is diamagnetic for $T < 270$ K and paramagnetic for $T > 270$ K. Paramagnetism is connected with presence of an "intrinsic" fracture of trivalent ions Yb^{3+}.

Lutetium trivalent ions have a completely filled 4f-shell and do not possess magnetic moment. Therefore Lu is a Pauli paramagnet.

Theory of magnetic properties of rare earths was extensively developed during 60s–70s [15, 16, 265]. Here we discuss briefly some points with especial attention to connections between magnetism and electronic structure.

Although, unlike d-metals, magnetic moments in $4f$-metals have an atomic nature, electronic structure of conduction band influences also magnetic ordering since the exchange interaction between well-separated f-spins at different lattice sites is mediated by current carriers (the Ruderman–Kittel–Kasuya–Yosida, RKKY, indirect interaction). In the real space this interaction has oscillating and long-range behaviour (Appendix K).

The wavevector of magnetic structure **Q** is determined by the maximum of the function $J(\mathbf{q})$. Phenomenologically, formation of spiral magnetic structures may be described in the Heisenberg model with interactions between nearest and next-nearest layers J_1 and J_2 where

$$\cos Qc = -J_1/4J_2. \tag{4.112}$$

However, in the free-electron approximation, $J(\mathbf{q})$, as well as the susceptibility of the electron gas,

$$\chi(\mathbf{q}) = \frac{1}{2}\sum_{\mathbf{k}} \frac{f(E_{\mathbf{k}}) - f(E_{\mathbf{k+q}})}{E_{\mathbf{k+q}} - E_{\mathbf{k}}} = \frac{m^* k_{\mathrm{F}}}{8\pi^2}\left(1 + \frac{4k_{\mathrm{F}}^2 - q^2}{4k_{\mathrm{F}}q}\ln\left|\frac{q+2k_{\mathrm{F}}}{q-2k_{\mathrm{F}}}\right|\right), \tag{4.113}$$

has a maximum at $q = 0$, which corresponds to the collinear ferromagnetic ordering. Yosida and Watabe [361] obtained a shift of the minimum to finite q by an account of reciprocal lattice vectors **g** (i.e. of the crystal structure), other factors being taken in the free-electron theory. Calculation for the hcp lattice with **Q** being parallel to the c-axis yielded

$$Q/2k_{\mathrm{F}} \simeq 0.11, \qquad Qc = 48^\circ,$$

which agrees with experimental data for Tm, Er and Ho ($Qc = 51°$). However, this theory does not explain the magnetic structures of Tb and Dy which have the same lattice.

Further development of the theory was connected with an account of the real shape of the Fermi surface. It was realized that $J(\mathbf{q})$ may have minimum in the cases where the vector **q** connects large congruent pieces of the Fermi surface (the 'nesting' condition). Such a situation takes place in the itinerant antiferromagnet Cr. It was supposed in [362] that in $4f$-metals **Q** corresponds to the vector of the 'arms' near the point L in the Brillouin zone of rare earths (Sect.2.7). Change of this vector in the rare earth series explains satisfactorily the experimental tendency in the spiral angle. In particular, absence of the 'arms' in gadolinium corresponds to ferromagnetic ordering.

Besides the topology and shape of the Fermi surface, an important role belongs to energy gaps near the superzone boundaries, which are determined by the magnetic structure. The largest energy gain takes place provided that the gaps occur near the Fermi surface (the situation which is reminiscent of the Hume-Rothery rule in the theory of crystal structures, sect.3.3). The influence of the superzone boundary on the spiral period was studied by Elliott and Wedgwood [363]. They demonstrated that the value of Q, obtained in [361], decreases with increasing the energy gap. This fact explains readily the observed T-dependence of the spiral angle with temperature, since the sublattice magnetization decreases with T.

An alternative mechanism which influences the spiral angle was proposed by de Gennes [364]. He took into account the finite value of the mean free path l for conduction electrons which mediate the exchange interaction among $4f$-moments. The damping of electron states, which is due to scattering, weakens the long-range antiferromagnetic components of the oscillating RKKY interaction and favours the ferromagnetic ordering. Therefore Q should decrease with decreasing l. Since the value of localized spin decreases from Gd to Tb, the magnetic resistivity changes from $10\mu\Omega$. cm in Tm to $120\mu\Omega \cdot$ cm in Gd, which corresponds to l of 40 and 5 Å. Thus the strong magnetic scattering may explain the absence of antiferromagnetism in gadolinium. The increase of l with lowering T should give the tendency which is opposite to the experimental data and to influence of superzone boundaries. Combined effect of both mechanisms on the spiral angle was considered by Miwa [365].

Fedro and Arai [366] investigated the influence of higher order Kondo-like logarithmic corrections (see (G.73)) to the electron spectrum on the spiral angle. They obtained the strong effect for the negative sign of the s–f exchange parameter I, as well as in the Kondo problem (Chapter 6). However, this assumption is unrealistic for elemental rare earths where $I > 0$. Besides that, an account of spin dynamics results in a considerable smearing of the logarithmic Kondo divergences [367].

Phenomenological treatment of formation of commensurate and incommensurate magnetic structures in rare earth and of their temperature evolution was performed by Dzyaloshinsky [368].

Last time, inhomogeneities of spiral structures were investigated. In particular, so called 'bunching' and 'spin sleep' phenomena were discovered (see, e.g., [369]). Methods of Moessbauer and μSR spectroscopy are extensively applied to this problem [370].

The magnetic properties of actinide series metals are determined by localization of $5f$-electrons. As discussed in Sect 1.1, its degree increases with increasing atomic number. Light actinides (Th, Pa, U) exhibit usual Pauli paramagnetism. These metals are satisfactorily described by the model where itinerant $5f$-electrons form wide bands so that the Stoner criterium is not satisfied. Neptunium and plutonium have apparently, similar to Pt and Pd, 'nearly ferromagnetic' properties. Their magnetic susceptibility has a non-monotonous temperature dependence at low T, and appreciable spin-fluctuation T^2-contributions to resistivity and enhancement of specific heat are observed.

The magnetic susceptibility of americium, although being somewhat larger, depends weakly on temperature. However, its magnetic properties may be already described within the model of localized $5f$-electrons since the Am^{3+} ion (configuration $5f^6$) possesses, as well as Sm^{3+} ion, the ground term with zero value of J, the first excited state with $J = 1$ lying higher by about 500 K. Thus americium should exhibit van Vleck paramagnetism.

In Cm, Bc and Cf (other heavy actinides are not investigated) the Curie–Weiss law holds at high temperatures, the effective magnetic moments being in agreement for those of corresponding three-valent ions [371]. Thus the Russel–Saunders coupling scheme turns out to be a good approximation. Neutron scattering data demonstrate that the double-hcp α-curium is ordered antiferromagnetically below 50 K. At the same time, the metastable fcc β-curium possesses below 205 K a finite saturation moment about $0.4\mu_B$ and probably has a ferrimagnetic order. α-Bc is antiferromagnetic with $T_N = 22$–34 K according to data of various authors. The data on magnetism of α-Cf are contradictory. In the work [372], AFM ordering below about 60 K was found which was changed by ferro- or ferrimagnetic ordering below 32 K. Only one transition into ferromagnetic (or ferrimagnetic) state with $\mu_0 = 6.1\mu_B$ was observed at 51 K in [373]. No magnetic ordering was found in the cubic β Bc and β Cf.

Actinide intermetallic and insulating compounds demonstrate a rich variety of magnetic properties. According the so-called Hill criterion (see [371]), the f–f overlap ceases and formation of magnetic moment takes place at sufficiently large distances between the actinide ions exceeding $d_c = 3.25$–3.50 Å (Fig.4.18).

Quantitative theory of magnetic structure of actinides is now developed insufficiently. Large spin-orbit coupling results in strong orbital contributions to magnetic properties. For example, the relativistic spin-polarized self-consistent band calculation of δ-Pu [374] yielded the following values of magnetic momenta (in μ_B): $M_S = 5.5$, $M_L = -3.4$, $M_J = M_S + M_L = 2.1$ whereas the experimental value of M_J makes up about $1\mu_B$. There is an open question about the accuracy of the atomic sphere approximation (ASA). The effect of the non-sphericity of the potential should be appreciable for orbital contributions.

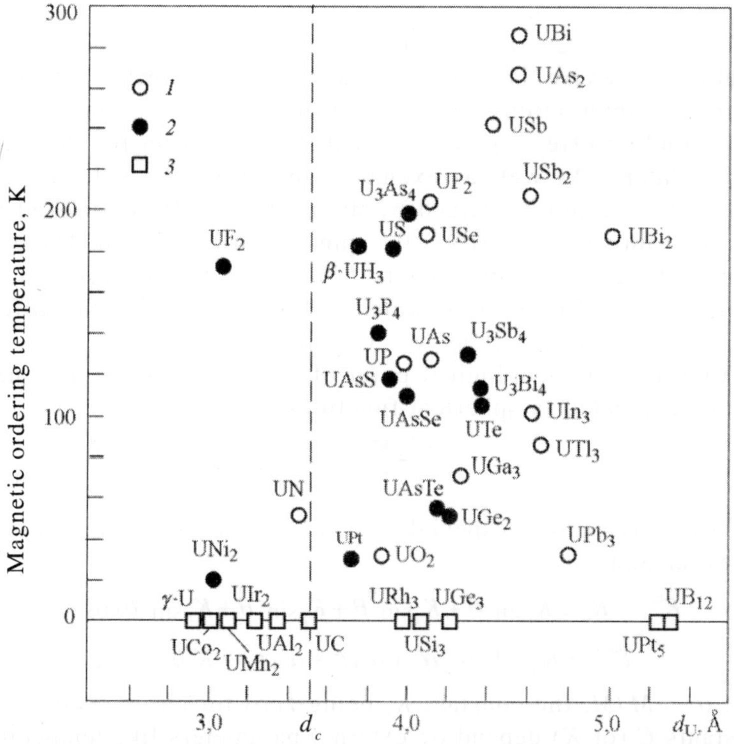

Fig. 4.18. The Hill diagram for uranium compounds which demonstrates dependence of type and temperature of magnetic ordering on the distance between uranium ions; 1 – antiferromagnetic ordering, 2 – ferromagnetic ordering, 3 - temperature-independent paramagnetism [371].

Band calculation of hcp Bk [375] yielded the value $M_L = 2.36$, which is comparable with the spin magnetic momentum (but is still smaller than the atomic value for the f^8 configuration). This demonstrates considerable narrowing of $5f$-band in heavy actinides.

4.8 MAGNETIC ANISOTROPY

In zero magnetic field a ferromagnetic sample becomes divided in regions (domains) which are magnetized along the easy magnetization axes of the crystal, so that the total magnetic moment is zero. Existence of the easy and hard directions is just determined by crystalline magnetic anisotropy (MA) which is characterized by the dependence of the total energy for magnetically saturated states on the magnetization direction.

Investigation of MA is of interest for two reasons. First, MA determines main technical properties of ferromagnetic materials – size and shape of magnetic domains, magnetization process in the external field, coercive force, etc. Second, MA reflects weak, but important relativistic interactions among electrons in a crystal and study of corresponding effects gives a valuabie information for the microscopic theory. As well as exchange integrals, the values of MA parameters are not explained by dipole forces. It is believed now that the main source of MA is the spin-orbital interaction. The orders of magnitude of MA parameters turn out to be different for iron group metals and rare earths, which is directly related to the different role of various interactions for corresponding ions.

From a phenomenological point of view, the anisotropy energy may be expanded in spherical functions.

$$\mathcal{E}_{\mathrm{a}} = \sum_{\lambda\mu} C_{\lambda\mu} Y_{\lambda\mu}. \tag{4.114}$$

One uses also often (especially for hexagonal and cubic crystals) the expansion

$$\mathcal{E}_{\mathrm{a}}^{\mathrm{hex}} = K_0 + K_1 \sin^2\theta + K_2 \sin^4\theta + K_3 \sin^6\theta + K_4 \sin^6\theta \cos 6\phi,$$

$$\mathcal{E}_{\mathrm{a}}^{\mathrm{cub}} = K_0 + K_2(\alpha_x^2\alpha_y^2 + \alpha_y^2\alpha_z^2 + \alpha_z^2\alpha_x^2) + K_2\alpha_x^2\alpha_y^2\alpha_z^2, \tag{4.115}$$

with $\alpha_i = M_i/M$, the constant K_1 being zero for cubic crystals. The constants C (or K) depend on external parameters like temperature, pressure etc.

Values of MA constants for iron group metals and gadolinium are given in Table 4.6. Gd possesses, unlike other rare earths, zero atomic orbital momentum and is similar in this respect to cobalt. Therefore, MA of Gd is rather small, the corresponding experimental data being contradictory and the constant K_1 demonstrating a non-monotonous temperature dependence with the sign change [381]. Other rare earths

Table 4.6. Magnetic anisotropy constants (10^5/cm^3) and anisotropy field (Oe) for iron group metals and gadolinium. Data of [265] with some corrections are used.

	Fe (bcc)	Ni (fcc)	Co (hcp)	Gd (hcp)
K_1 (293K)	–	–	43	–
K_1 (4.2K)	–	–	77	–8.5
K_2 (293K)	4.8	–0.49	12	–
K_2 (4.2K)	6	–12	10	25
K_3 (293K)	2	0.4	–	–
K_3 (4.2K)	–	6		
K_a (293K)	560	205	9500	–

Table 4.7. Magnetic anisotropy constants, 10^8 erg/cm^3 at low temperatures [15, 381], and anisotropy of paramagnetic Curie temperature $\Delta\Theta = \Theta_\parallel - \Theta_\perp$, K [39] for heavy rare earth metals. The corresponding theoretical estimates are presented for the crystal field and exchange mechanism.

	Tb	Dy	Ho	Er	Tm
K_1^{exp}	–5.5	–5	–2.2	–	–
K_2^{exp}	–0.45	0.54	–1.7	–	–
K_1^{cf}	–5.5	–5.05	–1.98	1.97	5.5
K_1^{exch}	–5.5	–4.6	–1.43	1.1	1.85
$\Delta\Theta^{exp}$	44	48	15	–29	–58
$\Delta\Theta^{cf}$	44	38	15	–16	–44
$\Delta\Theta^{exch}$	48	40	12	–25	–24

have MA constants which are by one–two orders of magnitude larger (Table 4.7).

Theoretical calculation of the signs and values of the MA constants is a complicated problem which is far from complete solution. First-principle calculations of MA in iron group metals turn out to be highly sensitive to their electronic structure, so that reliable results are now absent.

Especially difficult is the explanation of the MA temperature dependences. In the phenomenological approach, the MA constants are proportional to the corresponding powers of magnetization. In particular, performing the simplest averaging procedure we obtain the Akulov–Zener law [265]

$$K_n(T) = K_n(0)\left[\frac{M(T)}{M(0)}\right]^{n(2n+1)}, \tag{4.116}$$

so that

$$K_1 \sim M^3, \qquad K_2 \sim M^{10}. \tag{4.117}$$

A more accurate treatment of the dependences $K_i(T)$ in the spin-wave theory of a Heisenberg ferromagnet was performed in papers [376–378]. These results describe well the behaviour of MA constants in insulators (e.g., yttrium garnets). However, experimental T-dependences in ferromagnetic d-metals turn out to be more complicated. The behaviour $K_2(T)$ in iron and nickel corresponds to higher powers of magnetization than given by (4.117). The dependence $K_3(T)$ in iron is non-monotonous. A change in $K_1(T)$ sign occurs in cobalt near 500 K [265] (note that the hcp–fcc transition takes place at 700 K).

4.8.1 Quenching of orbital momenta by periodic lattice potential and magnetic anisotropy of *d*-metals

As discussed above, mechanisms of magnetic anisotropy are intimately related to spin-orbital interactions and orbital magnetic contributions: in the absence of SOI electron spins do not 'feel' their orientation. In contrast with rare earths, a very strong crystal field in the $3d$-magnets destroys the atomic structure of the SL-terms of the d^n configurations and particularly the multiplet structure of the total angular momentum J. Therefore, we have to consider the orbital momenta of single electrons. Experimental results demonstrate that they are almost completely quenched ($\langle l \rangle$ is close to zero for the ground state in not too strong magnetic fields). In the case of the atomic levels this occurs when the ground state is a singlet (i.e. spherically symmetric). For d-electrons in a crystal, this can take place, e.g., in the presence of a crystal field of sufficiently low point symmetry, which splits off a sublevel $m = 0$ from other states (cf. Sect.1.3). However, real ferromagnetic metals are characterized by high symmetry and the quenching of l is clearly due to a different mechanism associated with the periodic crystal potential.

The idea of this quenching mechanism is as follows [379, 39]. The degeneracy of atomic levels, which is retained by the local crystal potential, becomes lifted at spreading them into the energy bands in the $\mathbf{k}$-space. Indeed, the dispersion law of two bands $E_1(\mathbf{k})$ and $E_2(\mathbf{k})$, corresponding to atomic levels E_1 and E_2 is, generally speaking, different. The accidental degeneracy takes place only for those $\mathbf{k}$ where the bands are crossing. The electron orbital momentum operator $\mathbf{l} = [\mathbf{rk}]$ is diagonal in quasimomentum $\mathbf{k}$. Then all the diagonal matrix elements l_{ii} between crystal functions $\psi^i_{\mathbf{k}}$ vanish because of their symmetry (or antisymmetry) in the sign of projection m, whereas the off-diagonal matrix elements l_{ij}, although non-zero, correspond to different energies $E_1(\mathbf{k})$ and $E_2(\mathbf{k})$ for a fixed $\mathbf{k}$. This means that the contribution of the latter matrix elements to $\langle l \rangle$ in the presence of a perturbation with matrix elements h_{ij} is of order of $l/\Delta_{ij}(\mathbf{k})$ ($\Delta_{ij}(\mathbf{k}) = E_i(\mathbf{k}) - E_j(\mathbf{k})$) and, generally speaking, small.

Consider the simple example of the triple representation t_{2g} for a cubic crystal. It contains three functions ψ^i which are linear combinations of the functions with specific values of m:

$$\psi^{1,2} = \frac{1}{\sqrt{2}}(\phi_1 \pm \phi_{-1}), \qquad \psi^3 = \frac{1}{\sqrt{2}}(\phi_2 - \phi_{-2}). \tag{4.118}$$

As is clear from the symmetry properties, we have $l = 0$ for all i.

On the other hand, e.g., $\langle\psi^1|l^z|\psi^2\rangle = 1$. With account of the point crystal potential only we have the degenerate local level situation and the existence of $l^z_{ij} \neq 0$ means that in fact the orbital momenta are not quenched. Indeed, in a magnetic field H_z a regrouping of the functions takes place and the degeneracy is lifted by the values of the Zeeman energy $g\mu_B mH_z$. Thus effective orbital momenta with the value $l = 1$ occur. However, for the crystal functions of the band type the degeneracy of ψ^i_k is lifted for almost all values of $\mathbf{k}$. This means that, as long as $g\mu_B mH_z \ll \Delta(\mathbf{k})$, strong regrouping of the functions in a field H_z is energetically unfavourable and the value of orbital momentum unquenched by the field is proportional to a small quantity $g\mu_B mH_z/\Delta(\mathbf{k})$.

Intrinsic l-dependent mechanisms also mix the states $\psi^i_{\mathbf{k}}$ and thus unquench l. Especially important from this point of view is the spin-orbit coupling (Appendix L) since this connects the vectors $\mathbf{s}$ and $\mathbf{l}$. This yields a small unquenched momentum $\bar{l} \sim \lambda/\Delta$, which becomes oriented in the crystal. Again, the spin momentum is oriented due to spin orbit coupling, so that the magnetic anisotropy energy $\mathcal{E}_a$ is proportional to λ^2 for uniaxial crystals (and to λ^4 for cubic crystals). In fact, $\mathcal{E}_a$ corresponds to the spin-orbital coupling energy for the unquenched momentum $\bar{l}$:

$$\mathcal{E}_a \simeq \tilde{\mathcal{E}}_{so} = \lambda(\tilde{\mathbf{l}}\mathbf{s}) \sim \lambda^2/\Delta. \tag{4.119}$$

For $\lambda \sim 10^{-14}$ erg and $\Delta \sim E_F \sim 10^{-11}$ erg we have $\mathcal{E}_a \sim 10^{-17}$ erg $\sim 10^{-1}$ K.

The magnetization process is then as follows (Fig.4.19). In zero field the vectors $\mathbf{s}$ and $\tilde{\mathbf{l}}$ are oriented along the easy axis z. The application of field H_x in a hard direction tilts the spin momentum from the z-axis as the ratio $H_x/\tilde{\mathcal{E}}_{so}$ increases. (For simplicity, we do not take into account the contribution of the l-component, which is

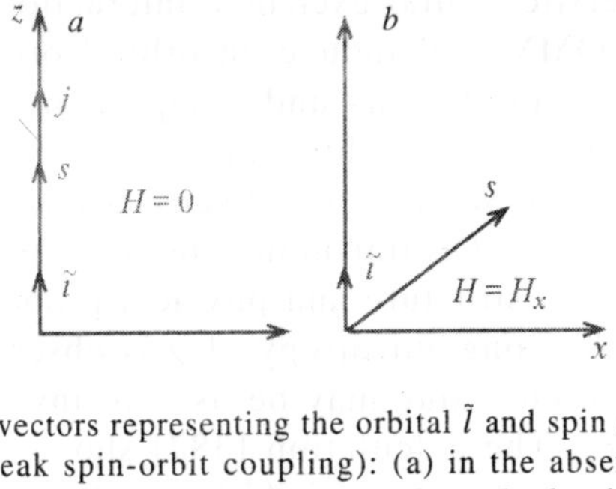

Fig. 4.19. Positions of vectors representing the orbital $\tilde{l}$ and spin s angular momenta in $3d$-ferromagnets (weak spin-orbit coupling): (a) in the absence of an external field (b) in the presence of the field H directed along the hard axis.

possible provided that corresponding splitting is large). The direction of $\tilde{\mathbf{l}}$ remains practically unchanged, since this is fixed by strong CF ($\varepsilon_{cr} \gg \tilde{\varepsilon}_{so}$). At $H = H_a \sim \tilde{\varepsilon}_{so}/\mu_B$ the coupling between the vectors $\tilde{\mathbf{l}}$ and $\mathbf{s}$ is broken and the magnetization is saturated, its value being smaller by the quantity $\tilde{l}$ than along the easy direction. The quantity H_a is called the anisotropy field and is an important characteristic of the magnetic hardness of a material.

Concrete calculations of magnetic anisotropy of transition metals with account of realistic band structure are discussed in papers [380, 381]. The degeneracy points in the **k**-space, where $\Delta(\mathbf{k}) = 0$, turn out to play an important role, quantitative results being strongly influenced by band structure details. The unusual orientation dependence of the magnetic anisotropy energy in nickel was attributed in [382] to existence of very small pockets of the Fermi surface, which undergo dramatic changes in size as a function of magnetization [145]. However, such pockets were not found from dHvA or galvanomagnetic effects data. Recent LMTO-ASA calculations of anisotropy energy in iron group metals [383] have yielded the values, which are small in comparison with experimental data.

The orbital contributions increase noticeably for surface and impurity states, and in layered systems, since here the quenching mechanism owing to periodic lattice potential becomes less effective. So, an appreciable enhancement of orbital magnetic momenta (OMM) at surfaces of Fe, Co, Ni was obtained in the band calculation [384]. Further investigations of these problems within many-electron models are required.

The unquenching of OMM due to Coulomb interaction, discussed above in Sect.1.3, should be once more mentioned. The latter tends to satisfy the Hund rules. Up to now, this mechanism was practically misregarded (see, however, the paper [385] where such a possibility was considered within a spin-density functional approach). On the other hand, intersite orbital exchange interaction (Appendix D) can also unquench OMM and induce an orbital ordering. Corresponding quantitative calculations and comparison of the electrostatic mechanisms with the spin-orbital one would be very interesting.

To conclude this Section, we discuss some examples of OMM effects in *d*-systems. Orbital momenta seem to play an important role in the electron structure and physical properties of high-T_c superconductors. A strong anisotropy of χ is observed above T_c [386]. Its sign and absolute value may be used to investigate the electron structure near E_F. The calculation [387] shows that the sign of $\Delta\chi$ corresponds to the contribution of off-diagonal matrix elements between the states x^2-y^2 at E_F and the xy-states, which lie below by about 1eV. Recent investigations of polarized X-ray spectra of copper ions [388] confirm that the bands of the type x^2-y^2 with a small admixture of z-type bands are present at the Fermi surface. Thus unquenching of OMM is due to transitions between the states $2^{-1/2}(|2\rangle \pm |-2\rangle)$.

A number of d-compounds (e.g., with perovskite, spinel and rutile structure) contain Jahn–Teller ions which possess an orbitally degenerate ground state. Strong anisotropy of magnetization and g-factors takes place in such systems (e.g., for the ions Cu^{2+} $g_{\parallel} = 2.4$, $g_{\perp} = 2.08$ [274]). A gain in the total energy is achieved by the degeneracy lift at symmetry lowering which is due to lattice distortion. A 'ferro' – or 'antiferromagnetic' orbital ordering may accompany the cooperative Jahn–Teller effect. Description of such phenomena and their interpretation is usually given within an one-electron approach [42]. Apparently, in a number of cases the interaction of ion L-momenta with crystal field should be considered in the many-electron scheme with account of spin-orbital coupling. However, such a consideration seems to be up to now not performed.

4.8.2 Magnetic anisotropy of rare earths

Now we consider the situation in rare earths [39]. For $4f$-electrons the energy of crystal field $\mathcal{E}_{cr}$ is small in comparison with not only the electrostatic interaction, but also with the spin-orbital energy, so that the the total atomic quantum numbers S,L,J are retained also in a crystal (Sect.1.3). Consequently, the magnetization process in RE magnets occurs in a different way as compared to the case of d-magnets.

In the case of easy-axis anisotropy, at $H = 0$ the vector $\mathbf{J} = \mathbf{L} + \mathbf{S}$ lies along the z-axis governed by the minimum of $\mathcal{E}_{cr}$ for the angular momentum $\mathbf{L}$. In a field H_x the coupling between $\mathbf{L}$ and $\mathbf{S}$ is retained (because $\mathcal{E}_{cr} \ll \mathcal{E}_{so}$) and they rotate as a whole, approaching the x-direction. The angle of rotation is determined by the ratio $g\mu_B JH/\mathcal{E}_{cr}$. Although the energy $\mathcal{E}_{cr} \sim 10^{-2}$–$10^{-3}$ eV is the smallest among the energies of other interactions for $4f$-electrons, it is considerably higher (by two or three orders of magnitude) than the effective energy $\tilde{\mathcal{E}}_{so}$ responsible for the magnetic anisotropy of $3d$-metals with quenched orbital momenta. This accounts for the giant anisotropy exhibited by rare-earth magnets.

The existence of unquenched OMM results in another important consequence in the theory of magnetic properties of rare earths. As demonstrated in Appendix K, the s–f exchange interaction leads, in the second order of perturbation theory, not only to an indirect

f–f exchange of the Heisenberg (or de Gennes) type, but also results to the exchange interactions, which are determined by orbital momenta (see (K.11)). Such interactions (in contrast to spin exchange) readily become anisotropic after allowance for the anisotropy of the crystal. The contributions of the anisotropic exchange to the magnetic anisotropy energy were calculated in [389]. The total result for the hexagonal lattice with the parameters c and a reads

$$\mathcal{E}_{cr} = (K_1^{cr} + K_1^{exch})\cos^2\theta + \ldots,$$

$$K_1^{cr} = \alpha_J J\left(J - \frac{1}{2}\right)\frac{Z^{eff}e^2}{a}\frac{\langle r_f^2\rangle}{a^2}1{,}2\left(c/a - \sqrt{8/3}\right),$$

$$K_1^{exch} \sim (g-1)D_1 J^2 I_{sf}^2 N(E_F). \qquad (4.120)$$

Here α_J is the Stevens parameter, Z^{eff} is the effective ion charge, $\langle r_f^2\rangle$ is the averaged square of the f-shell radius, D_1 is defined in (K.10). The expressions (4.120) yield estimates of the order of magnitude of K_1^{cr} and K_1^{exch}. Since for heavy rare earths $\alpha_J \sim 10^{-2} - 10^{-3}$, we get $K_1^{cr} \sim 10^7$–10^8 erg/cm^3. Then, $D_1 \sim 10^{-2}$, so that $K_1^{exch} \sim 10^6$–10^7 erg/cm^3.

Thus the magnetic anisotropy of rare earths is one or two order of magnitude higher than that of the most strongly anisotropic hexagonal d-magnets. This difference is a consequence of the fact that, for RE, magnetic anisotropy is determined by the electrostatic interactions of the crystal field or an anisotropic exchange type and not by the weak spin-orbital interaction (as for d-magnets).

The MA constant K_1 and the anisotropy of the paramagnetic Curie temperature $\Delta\theta = \theta_{\parallel} - \theta_{\perp}$ of heavy of rare earths are given in Table 4.7. The latter quantity is found by extrapolating the Curie-Weiss law $\chi^{-1} = C^{-1}(T - \theta)$ for corresponding crystal directions up to the temperatures where $\chi^{-1} = 0$. For clarity, a comparison with the experimental resuts is made separately for the crystal field and exchange mechanisms. The largest difference between two theoretical values is found in the case of Tm, which is therefore of the greatest interest from the point of view of the magnetic anisotropy.

Although a complete comparison with experiment is difficult because of absence of precise data, the crystal field contribution is probably dominant and the anisotropic exchange contributes only 10–20%. A reliable experimental determination of the latter would be of fundamental interest in the theory of exchange interactions. In contrast with the single-ion crystal-field mechanism, the anisotropic exchange gives rise to a two-ion anisotropy so that it can be separated on the

Table 4.8. Values of orbital momentum and the type of magnetic anisotropy for rare earth ions.

R^{3+}	Ce	Pr	Nd	Pm	Sm	Eu	Gd	Tb	Dy	Ho	Er	Tm	Yb	
	f^1	f^2	f^3	f^4	f^5	f^6	f^7	f^8	f^9	f^{10}	f^{11}	f^{12}	f^{13}	
L	F 3	H 5	I 6	I 6	H 5	F 3	S 0	F 3	H 5	I 6	I 6	H 5	F 3	

basis of the alloy composition dependence. Therefore experimental investigations of magnetic anisotropy of RE alloys may be useful. It is worthwhile also to mention the methods for detecting aniso-ropic exchange, which are based on hyperfine interactions [390].

Now we discuss the sign of magnetic anisotropy. Here the theory provides precise predictions. The signs of both α_J and D_1 are reversed on transition from f^3 (f^{10})) to f^4 (f^{11}) configurations in the first (second) halves of the RE series and also on transition from the first to the second half. If we bear in mind that Pm has not yet been investigated, and Eu and Yb have cubic lattices, the agreement with the experimental results is quite satisfactory.

This mathematical result has a clear physical meaning. The magnetic anisotropy is related to the magnitude and orientation of the orbital components of the total orbital momenta in an electric crystal field. One can see from the Table 4.8 that, besides the trivial electron-hole symmetry in the values of L between the first and second halves of the series, there is also a symmetry within each half associated

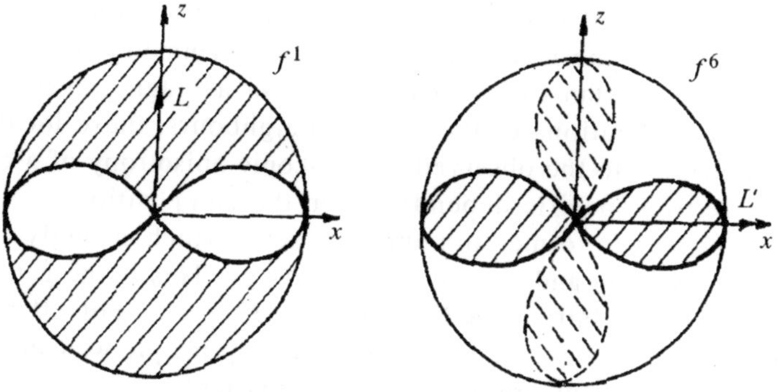

Fig. 4.20. Anisotropic distribution of the f-electron density expected for the f^1 and f^6 configurations with $L = 3$. The solid curve in the right part represents an orbital hole which appears when f^1 is replaced by f^6. In the presence of magnetic anisotropy, the hole and its vector L' are rotated by 90° (dashed curve) relative to the electron [39].

with the occupancy of orbital quantum states. For example, f^1- and f^6-states have the same value $L = 3$ and it might seem that the anisotropy should also be the same. However, we should take into account that $L(f^1)$ is the angular momentum of one electron, whereas $L(f^6)$ is the orbital momentum of a hole in the spherical f^7 configuration characterized by $L = 0$. Thus the anisotropy of the electric charge will be opposite for the f^1 and f^6-configurations (Fig.4.20).

The same reversals of the sign of the first anisotropy constant in the RE series are as a rule observed not only for pure RE metals, but also for their alloys and compounds. Consider the situation in the practically important intermetallic systems RCo_5. The theory predicts the orientational plane-axis transition for the compounds with R = Ce, Pr, Nd, Tb, Dy, Ho. Indeed, in the case of these elements, in the ground state the angular momentum **J** is oriented in the basal plane of the hexagonal lattice. The Co ions have an easy c-axis, but because of the exchange interaction with the R ions they are oriented so that their magnetic moments are aligned in a plane at low temperatures. As a sample is heated, the magnetization of the R sublattice decreases faster than that of the Co sublattice. Thus the gain in the anisotropy energy of the Co sublattice, which is characterized by a larger magnetization, becomes dominating, the plane-axis transition takes place at some temperature [391]. Such transitions are observed experimentally [392].

The possibility of occurrence of strong local anisotropy effects in RE alloys because of lowering of the local symmetry of the environment for a given RE ion is also of interest. An effect of this kind was considered in [393] with application to the system $RCo_x Ni_{5-x}$. It was found that, besides the usual anisotropy constant corresponding to the macroscopic (average) symmetry of the alloy there are also constants of a local type. After averaging over various possible configurations of the nearest neighbours, these local constants contribute to the observed macroscopic anisotropy of a crystal giving rise to a specific dependence on the composition.

It is worthwhile to mention another factor which strongly affects both the magnitude and sign of the anisotropy. This is the geometric factor of the hcp lattice $\sqrt{8/3} - c/a = 1.633 - c/a$. For all the RE metals, $c/a = 1.58 - 1.61 < 1.63$. However, in principle we can change the sign of the anisotropy by changing this factor.

Chapter 5

TRANSPORT PROPERTIES

Besides practical significance, electronic transport phenomena in metals are rather important from the theoretical point of view. They have played a great role in formulation of principal statements of modern quantum solid state physics. In particular, the classical theory does not explain the zero value of electrical resistivity at $T = 0$ which should be determined by cross sections of electron scattering by ions. In quantum theory, electrons in a periodical crystal are described by Bloch states with a definite wavevector **k** and carry current without a loss of energy. The scattering of electrons is determined by various mechanisms in the electron-lattice system which disturb the periodicity (impurities, thermal vibrations etc.). It also turns out that, as well as for electronic specific heat and magnetic properties, the degenerate statistics of current carriers and the existence of the Fermi surface are crucial for transport phenomena.

The theoretical description of transport phenomena includes a number of parameters: concentration n and charge sign of current carriers, effective mass m^*, electron velocity at the Fermi level v_F, parameters of various interactions: electron–lattice, electron–impurity, electron-spin, electron–electron. These interactions influence not only energy spectrum, but also the mean free path l and the relaxation time τ. The simplest expressions for the conductivity and the Hall coefficient

$$\sigma = \frac{e^2 n\tau}{m^*}, \qquad R = -\frac{1}{nec}. \tag{5.1}$$

enable one to determine n and τ (the effective mass may be determined from de Haas-van Alphen effect data).

In real substances the situation is more complicated than that described by (5.1). This is connected with the presence of several groups of current carriers, combined influence of different scattering mechanisms etc. The main problem of the microscopic theory in such a situation is separating these factors and determining their relative contributions. This problem is rather important for transition metals which demonstrate a great variety of transport phenomena. In this Chapter we pay especial attention to peculiarities which occur in comparison with simple metals. Let us list some of these peculiarities.

1. Existence of current carriers with strongly different characteristics (s- and d-electrons). Although $4f$-electrons usually do not take part in conductivity, they may influence the current carriers due to s–f hybridization, especially in some rare-earth compounds. Besides that, current carriers with opposite spins have different characteristics in ferromagnetic metals.
2. Presence of internal partially filled d-shells (narrow bands) may be an subsidiary scattering source (due to the s–d transitions).
3. Occurrence of additional scattering mechanisms by d- and f-shell magnetic moments.
4. Anomalous (spontaneous) transport phenomena (e.g., the Hall, Faraday, Kerr and $\Delta\rho/\rho$ effects), which are connected with the magnetization of $d(f)$-ions rather than with external magnetic field.
5. Anomalous behaviour of transport coefficients near magnetic ordering points.
6. Large correlation effects and possibility of electron localization in narrow d-bands.

Thus, main peculiarities of transition d-metals are connected with the intermediate degree of d-electron localization. This fact leads to d-electrons playing simultaneously two roles: of current carriers and of scattering centres. Below we analyze a number of scattering mechanisms which are specific for transition metals.

5.1 GENERAL CLASSIFICATION OF TRANSPORT PHENOMENA

The densities of charge and heat current **j** and **W**, produced by the external electric field $\mathbf{E} = -\text{grad}\ \phi$ and temperature gradient, are determined by the equations

$$j_\alpha = \sigma_{\alpha\beta} E_\beta + \lambda_{\alpha\beta} \text{grad}_\beta T; \tag{5.2}$$

$$W_\alpha = q_\alpha - \phi j_\alpha = \nu_{\alpha\beta} E_\beta - \kappa_{\alpha\beta} \text{grad}_\beta T, \tag{5.3}$$

where the electric and thermal conductivities σ and κ and the coefficients λ and ν are tensor quantities, which determine transport properties of the crystal (summation over repeated indices is assumed). The term $\phi\mathbf{j}$, which yields transfer of electron potential energy in the electric field, is subtracted in (5.3) from the total energy current $\mathbf{q}$.

At considering transport phenomena, it is important to take into account the Onsager symmetry principle. Within the framework of linear theory (small deviations of the system from the equilibrium state) the coefficients in the system

$$\frac{\partial X_i}{\partial T} = \sum_j \gamma_{ij} \frac{\partial S}{\partial X_j} \tag{5.4}$$

(X_i are quantities determining the state of the system, S is the density of entropy) are symmetric in indices i, j:

$$\gamma_{ij} = \gamma_{ji}. \tag{5.5}$$

Equations (5.2), (5.3) are a particular case of the system (5.4) with

$$\frac{\partial X_1}{\partial T} = \mathbf{j}, \qquad \frac{\partial X_2}{\partial T} = \mathbf{W},$$

$$\frac{\partial S}{\partial X_1} = \frac{\mathbf{E}}{T}, \qquad \frac{\partial S}{\partial X_2} = -\frac{\mathrm{grad}T}{T^2}. \tag{5.6}$$

The relations (5.6) are readily obtained from the expression

$$\frac{dS}{dT} = \int \frac{\mathrm{div}\mathbf{q}}{T} d\mathbf{r} = \int \frac{\mathbf{Ej}}{T} d\mathbf{r} - \int \frac{1}{T^2} \mathbf{W} \mathrm{grad} T d\mathbf{r}. \tag{5.7}$$

In the presence of the magnetic field $\mathbf{H}$ the transport coefficients are functions of $\mathbf{H}$. Since the equations of motion in non-magnetic crystals are invariant under simultaneous replacement $\mathbf{H} \to -\mathbf{H}$, $t \to -t$, the Onsager relations read

$$\sigma_{\alpha\beta}(\mathbf{H}) = \sigma_{\beta\alpha}(-\mathbf{H}), \quad \kappa_{\alpha\beta}(\mathbf{H}) = \kappa_{\beta\alpha}(-\mathbf{H}), \quad \nu_{\alpha\beta}(\mathbf{H}) = T\lambda_{\beta\alpha}(-\mathbf{H}). \tag{5.8}$$

The transport coefficients may be expanded in $\mathbf{H}$, the transport phenomena being subdivided into even and odd ones with respect to the magnetic field. Practically, it is sufficient to retain linear and quadratic terms only to obtain from (5.1), (5.2)

$$\begin{aligned} E_\alpha &= (\rho_{\alpha\beta} + R_{\alpha\beta\gamma} H_\gamma + r_{\alpha\beta\gamma\delta} H_\gamma H_\delta) j_\beta + \\ &+ (\alpha_{\alpha\beta} + Q_{\alpha\beta\gamma} H_\gamma + \Delta\alpha_{\alpha\beta\gamma\delta} H_\gamma H_\delta) \mathrm{grad}_\beta T; \end{aligned} \tag{5.9}$$

$$\begin{aligned} W_\alpha &= (\alpha_{\alpha\beta} + Q_{\alpha\beta\gamma} H_\gamma + \Delta\alpha_{\alpha\beta\gamma\delta} H_\gamma H_\delta) j_\beta + \\ &+ (-\kappa_{\alpha\beta} + \mathcal{L}_{\alpha\beta\gamma} H_\gamma + \mathcal{B}_{\alpha\beta\gamma\delta} H_\gamma H_\delta) \mathrm{grad}_\beta T, \end{aligned} \tag{5.10}$$

Here ρ is the electrical resistivity tensor, R is the Hall coefficient, r determines the magnetoresistivity ($\Delta\rho/\rho$-effect), α is the thermoelectric power coefficient, the Nernst–Ettingshausen coefficient Q determines the change of thermoelectric power in the magnetic field, and the coefficients $\Delta\alpha$ describe the longitudinal and transverse magnetothermoelectric effect. In the equation for the thermal current (5.10), κ is the thermal conductivity tensor, $\mathcal{L}$ determines the occurrence of transverse temperature gradient in magnetic field (the Righi–Leduc effect), $\mathcal{B}$ describes thermal magnetoresistivity; other coefficients coincide with those in (5.9) due to the Onsager relations.

Now we discuss magnetically ordered crystals. In ferromagnets we have to take into account, besides the external magnetic field, magnetization **M**. In the cases of antiferro- or ferrimagnetic ordering we have to consider the magnetization of each magnetic sublattice. In the simplest two-sublattice case it is convenient to use the variables $\mathbf{M} = \mathbf{M}_1 + \mathbf{M}_2$ and $\mathbf{L} = \mathbf{M}_1 - \mathbf{M}_2$. Expanding transport coefficients in **H**, **M**, **L** we write down by analogy with (5.9), (5.10)

$$E_\alpha(\mathbf{H},\mathbf{M},\mathbf{L}) = E_\alpha(0) + \left(\sum_i R^i_{\alpha\beta\gamma} Y^i_\gamma + \sum_{ij} r^{ij}_{\alpha\beta\gamma\delta} Y^i_\gamma Y^j_\delta\right) j_\beta + \\ + \left(\sum_i N^i_{\alpha\beta\gamma} Y^i_\gamma + \sum_{ij} \Delta\alpha^{ij}_{\alpha\beta\gamma\delta} Y^i_\gamma Y^j_\delta\right) \mathrm{grad}_\beta T,$$

$$W_\alpha(\mathbf{H},\mathbf{M},\mathbf{L}) = W_\alpha(0) + \left(\sum_i N^i_{\alpha\beta\gamma} Y^i_\gamma + \sum_{ij} \Delta\alpha^{ij}_{\alpha\beta\gamma\delta} Y^i_\gamma Y^j_\delta\right) T j_\beta + \\ + \left(\sum_i \mathcal{L}^i_{\alpha\beta\gamma} Y^i_\gamma + \sum_{ij} \mathcal{B}^{ij}_{\alpha\beta\gamma\delta} Y^i_\gamma Y^j_\delta\right) \mathrm{grad}_\beta T, \tag{5.11}$$

where the indices $i, j = H, M, L$ and $\mathbf{Y}^i = \mathbf{H}, \mathbf{M}, \mathbf{L}$. A concrete form of the tensor coefficients in (5.11) may be found for each crystal structure [394].

One can see that, besides usual transport coefficients, there exist the Hall and $\Delta\rho/\rho$ effects owing to the spontaneous magnetization. Such effects are called spontaneous or extraordinary (anomalous). From the phenomenological point of view, the spontaneous effects are similar to the corresponding effect in non-magnetic crystals. However, in fact they possess essential peculiarities. As a rule, the corresponding coefficients are large in comparison with those for 'normal' effects. So, the spontaneous Hall coefficient R^M in metals depends strongly on temperature, its absolute value at high temperatures exceeding by several orders that of R^H. Thus the expressions (5.11) may not in fact be treated as just a formal expansion, and the spontaneous effects may not be reduced to normal ones, e.g., by introducing an effective field.

Thus we meet with the problem of constructing a microscopic theory of spontaneous transport effects on the basis of new mechanisms of interaction of current carriers with the lattice ions in magnetic crystals. Evidently, one of such mechanisms is the exchange interaction. Surprisingly, spin-orbital interaction (Appendix L) plays also an important role in a number of effects.

Of interest are also terms in (5.11) which contain the vector **L**. In particular, the $R^L L$-term results in the even Hall effect and the $r^{HL}\,HL$-term in the odd $\Delta\rho/\rho$-effect. Consider the corresponding experimental data. At investigating the Hall effect in the ferrimagnet Mn_5Ge_2 [395] it was found that the Hall coefficient R changes its sign at the compensation temperature where the magnetizations of sublattices are opposite. This phenomenon was explained in paper [396]. According to (5.11),

$$R = \frac{E_y}{j_x M_z} = R^M + R^L \frac{L_z}{M_z}, \tag{5.12}$$

One can see that the second term in (5.12) diverges at the compensation point ($M_z = 0$), L_z changing its sign, which yields an effective sign change in R_s. Thus existence of the compensation point enables one to separate the ferromagnetic and antiferromagnetic Hall effects. The odd $\Delta\rho/\rho$-effect in Mn_5Ge_2 was picked out in the paper [397].

The situation in true antiferromagnets where magnetic sublattices are equivalent is different. The terms, which are linear in L, should vanish in crystals where the sublattices are connected with crystal symmetry transformations. However, these terms may occur in crystals where the period of magnetic structure coincides with the crystallographic period and magnetic sublattices transform one into another under the antisymmetry operations (i.e. usual symmetry operations combined with time reversion). Since these operations change the sign of magnetic moment, the **L**-linear terms turn out to be not forbidden by symmetry requirements. According to [398], such a situation takes place in the hematite Fe_2O_3 where the even Hall effect was observed [399]. In this case one may also expect the odd magnetoresistivity

$$\left(\frac{\Delta\rho}{\rho}\right)_{\text{odd}} = \frac{\rho(\mathbf{H}) - \rho(-\mathbf{H})}{2\rho(0)} \propto \mathbf{H}[\mathbf{L}(\mathbf{H}) - \mathbf{L}(-\mathbf{H})], \tag{5.13}$$

which was also found in hematite [400]. It should be noted that the microscopic theory of **L**-linear effects is up to now absent.

Recently, the linear term in the magnetoresistivity was observed in the normal phase of the high-T_c system $YBa_2Cu_3O_y$ [401], the value of this term strongly increasing in the narrow region of oxygen concentration. 6.88< y <6.95. The authors relate the effect with existence of dynamically correlated antiferromagnetic regions in these sample.

5.2 CALCULATION OF TRANSPORT COEFFICIENTS

To evaluate transport coefficients (e.g., resistivity) we need to know the scattering probability which is determined by the values and **k**-dependences of interaction matrix elements. The simplest method to solve this problem is consideration of the transport equation for the electron distribution function in a crystal in the presence of external fields. In the stationary regime, the evolution of this function is described by the equation

$$\frac{df}{dt}=\left(\frac{\partial f}{\partial t}\right)_{\text{field}}+\left(\frac{\partial f}{\partial t}\right)_{\text{collis}}=0. \tag{5.14}$$

so that the effect of acceleration by external electric field E_x is balanced by collisions for some non-zero electron velocity v_x along the field direction. To linear approximation in the field we may write down

$$f=f_0+f_1, \tag{5.15}$$

$$\left(\frac{\partial f}{\partial t}\right)_{\text{field}}=\left(\frac{\partial f_0}{\partial t}\right)_{\text{field}}=\frac{\partial f_0}{\partial E_{\mathbf{k}}}=-eE_x\frac{\partial f_0 v_x}{\partial E_{\mathbf{k}}}\frac{\partial E_{\mathbf{k}}}{\partial \mathbf{k}}\frac{d\mathbf{k}}{dt},$$

where f_0 is the equilibrium Fermi distribution and f_1 is the linear correction. Introducing the scattering probability $W_{\mathbf{kk}'}$ we represent the equation (5.14) in the form

$$\sum_{\mathbf{k}'}(W_{\mathbf{k}'\mathbf{k}}f_{1\mathbf{k}}-W_{\mathbf{kk}'}f_{1\mathbf{k}'})=-eE_x v_x\frac{\partial f_{0\mathbf{k}}}{\partial E_{\mathbf{k}}}. \tag{5.16}$$

Thus we have obtained the integral equation for the function $f_{1\mathbf{k}}$. After its solution, the electric current and conductivity are calculated as

$$j_x=-e\int v_x f_1 d\mathbf{k},\qquad \sigma_{xx}=j_x/E_x. \tag{5.17}$$

The quantity W may be calculated for each scattering mechanism (impurities, phonons, spin inhomogeneities, etc.). In the case of the independent mechanisms

$$\rho(T)=\rho_i+\rho_{\text{ph}}+\rho_{\text{mag}}+\ldots \tag{5.18}$$

The additivity of various mechanism contributions is called the Matthiessen rule. Generally speaking, deviations from this rule occur which are connected with interference of different scattering processes.

An alternative method for calculating the transport relaxation time is the use of the Kubo formula for conductivity [402]

$$\sigma_{xx} = \frac{1}{2T} \int_{-\infty}^{\infty} \langle j_x(t) j_x \rangle dt. \tag{5.19}$$

where

$$\mathbf{j} = -e \sum_{\mathbf{k}\sigma} \mathbf{v}_{\mathbf{k}\sigma} c^{\dagger}_{\mathbf{k}\sigma} c_{\mathbf{k}\sigma}, \qquad \mathbf{v}_{\mathbf{k}\sigma} = \frac{\partial E_{\mathbf{k}\sigma}}{\partial \mathbf{k}} \tag{5.20}$$

is the current operator. Representing the total Hamiltonian in the form $\mathsf{H} = \mathsf{H}_0 + \mathsf{H}'$, the correlator in (5.19) may be expanded in the perturbation H' [403, 404]. In the second order we obtain for the electrical resistivity [403]

$$\rho_{xx} = \sigma_{xx}^{-1} = \frac{k_B T}{\langle j_x^2 \rangle} \int_0^{\infty} \langle [j_x, \mathsf{H}'(t)][\mathsf{H}', j_x] \rangle dt, \tag{5.21}$$

where $\mathsf{H}'(t)$ is calculated with the Hamiltonian H_0. Provided that the perturbation Hamiltonian has the form

$$\mathsf{H}' = \sum_{\mathbf{k}\mathbf{k}'\sigma\sigma'} \hat{W}^{\sigma\sigma'}_{\mathbf{k}\mathbf{k}'} c^{\dagger}_{\mathbf{k}\sigma} c_{\mathbf{k}'\sigma'} \tag{5.22}$$

(in particular, for the phonon and magnon scattering) we obtain

$$\rho = \frac{k_B T}{\langle j_x^2 \rangle} e^2 \sum_{\mathbf{k}\mathbf{k}'\sigma\sigma'} (\mathbf{v}_{\mathbf{k}\sigma} - \mathbf{v}_{\mathbf{k}'\sigma'})^2 \int_0^{¥} \langle \hat{W}^{\sigma\sigma'}_{\mathbf{k}\mathbf{k}'}(t) \hat{W}^{\sigma'\sigma}_{\mathbf{k}'\mathbf{k}} \rangle \exp[i(E_{\mathbf{k}\sigma} - E_{\mathbf{k}'\sigma'})t] dt, \tag{5.23}$$

with

$$\langle j_x^2 \rangle = e^2 \sum_{\mathbf{k}\sigma} (v^x_{\mathbf{k}\sigma})^2 n_{\mathbf{k}\sigma}(1 - n_{\mathbf{k}\sigma}). \tag{5.24}$$

This approach is equivalent to solution of the transport equation by the variational method [7, 8].

The transport equation for elastic impurity scattering is considered in Appendix M. The result for the impurity resistivity in the lowest-order Born approximation has the form

$$\rho_i = \sigma_i^{-1} = \frac{e^2 n \tau_i}{m^*}, \qquad \tau_i = \frac{2\pi\hbar^4}{n_i \bar{\phi}^2 (2m^*)^{3/2} E_F^{1/2}}. \tag{5.25}$$

Here $\bar{\phi}$ is the average impurity potential, n_i is the impurity concentration. Thus the resistivity ρ_i is temperature-independent.

The case of the scattering by acoustical phonons may be considered with the use of either the transport equation (5.15) [1] or the formula (5.23) [7, 8]. The situation is most simple in the case of high $T > \theta_D$. Then the phonon frequency is small in comparison with temperature, so that the scattering is elastic and we may put for the phonon occupation numbers

$$N_{\mathbf{q}} = N_B(\omega_{\mathbf{q}}) \simeq k_B T/\omega_{\mathbf{q}}.$$

Then the resistivity is proportional to the number of scattering particles, as well as in the case of impurity scattering, and is linear in temperature:

$$\sigma_{ph} = \rho_{ph}^{-1} = \frac{e^2 n}{\pi^3 \hbar^3} \frac{M}{k_F} \left(\frac{k}{C} \frac{dE_{\mathbf{k}}}{d\mathbf{k}} \right)_{k_F} k_B \theta_D \frac{\theta_D}{T}, \tag{5.26}$$

where M is the ion mass and

$$C = \frac{\hbar^2}{2m} \int |\operatorname{grad} u_{\mathbf{k}}(\mathbf{r})|^2 d\mathbf{r}, \tag{5.27}$$

is the Bloch constant, the function u_k being determined by (2.1).

At $T \ll \theta_D$ the number of phonons decreases rapidly with lowering T. Since the phonons with small q yield main contribution, replacement $\omega_{\mathbf{q}}/T = x$ in the corresponding integral yields $q^2 N_{\mathbf{q}} dq \sim T^3$. Besides that, only a small part of the electron quasimomentum k, of order of q^2/k^2, is lost at scattering by long-wave phonons. This results in an extra factor of $\omega_{\mathbf{q}}^2/k^2 \sim T^2$. Thus the total scattering probability is proportional to T^5 (the Bloch law). An interpolation expression for the phonon contribution may be presented in the form [1]

$$\sigma_{ph} = \frac{e^2 n}{4\pi^3 \hbar^3} \frac{M}{k_F I(T)} \left(\frac{k}{C} \frac{dE_{\mathbf{k}}}{d\mathbf{k}} \right)_{k_F}^2 k_B \theta_D \left(\frac{\theta_D}{T} \right)^5, \tag{5.28}$$

where

$$I(T) = \int_0^{\theta_D/T} \frac{e^x x^5}{(e^x - 1)^2} dx. \tag{5.29}$$

At low temperatures

$$I(T) \simeq I(0) = 5 \int_0^{\infty} \frac{x^4}{e^x - 1} dx = 124.4.$$

The Bloch theory of resistivity describes well experimental data for simple metals in a wide temperature region. So, for Ag and Cu the agreement is within 5%.

5.3 RESISTIVITY

Examples of $\rho(T)$ dependences for transition metals are shown in

Figs.5.1–5.8 [239]. Some of the figures show also different contributions to resistivity (impurity, ρ_i, phonon, ρ_{ph}, magnetic, ρ_{mag}, Mott s–d scattering, ρ_{sd}, electron–electron, ρ_{ee}, etc.) which are discussed below.

The values of both ρ and $d\rho/dT$ are considerably larger than for simple metals. There exist two types of $\rho(T)$ behaviour – convex and concave one. The type is determined to some extent by the column number in the periodic table: the convex dependence is observed for atomic d^n-configurations with odd n (such a behaviour takes place also in most rare-earth metals). However, this regularity is not universal. In particular, the concave dependence is observed in Cr, Mo, W (d^4) and Ru, Os (d^6), but not in Ti, Zr, Hf (d^2). The convex behaviour reflects a saturation tendency which becomes more pronounced with increasing absolute values of ρ.

The connection of conductivity with the energy dependence of density of states may be obtained in the simplest approximation of the relaxation time τ [8] where

$$f_1 = -\tau \mathbf{E}\mathbf{v} \frac{\partial f_0}{\partial E_{\mathbf{k}}}. \tag{5.30}$$

so that

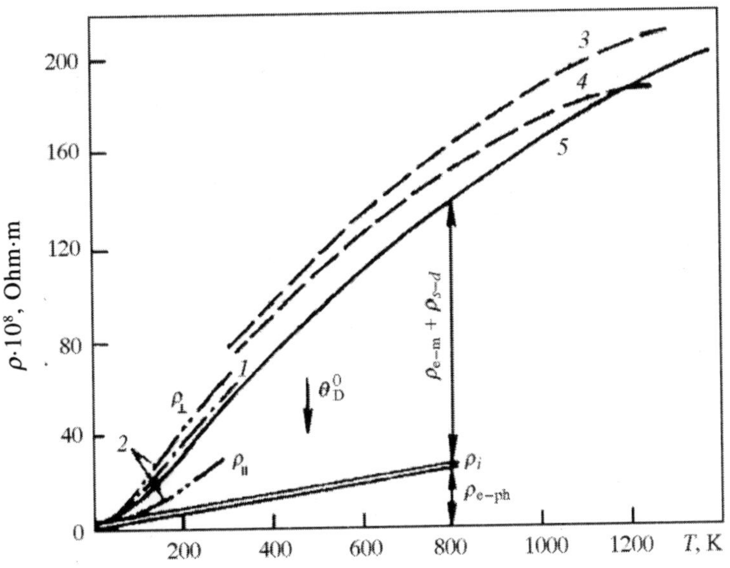

Fig. 5.1. The temperature dependence of electrical resistivity of scandium according to data of various authors [239].

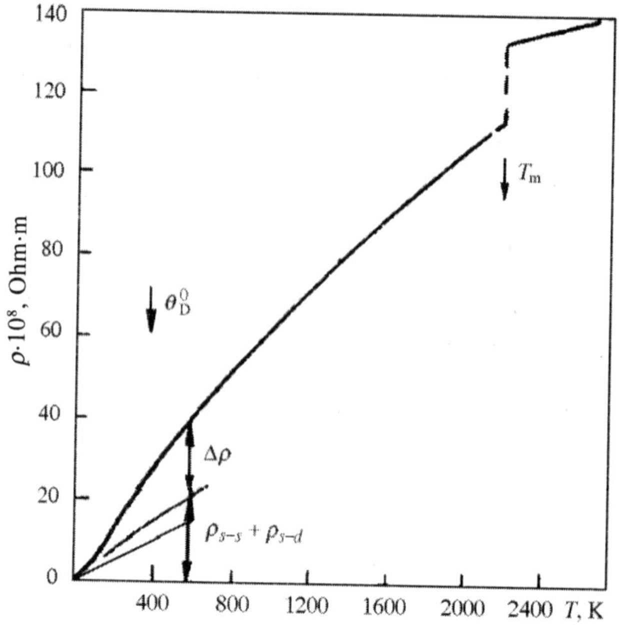

Fig. 5.2. The temperature dependence of electrical resistivity of vanadium [239].

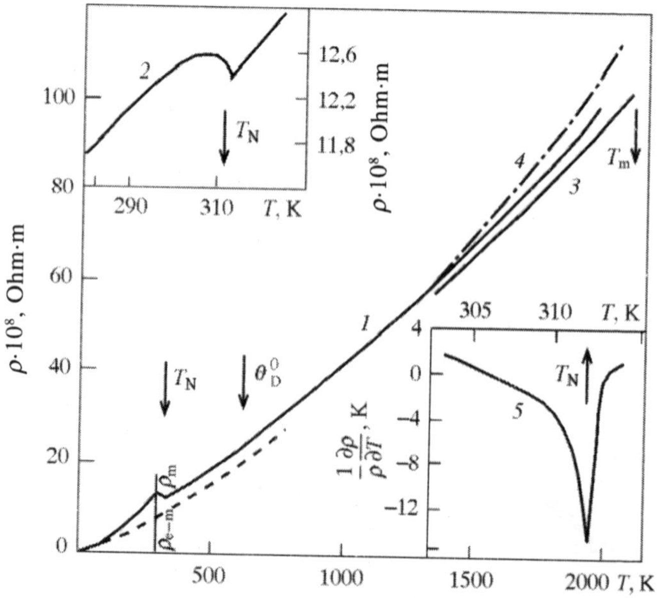

Fig. 5.3. The electrical resistivity cf chromium according to data of various authors (1,3,4); the calculated anomaly near the Neel point (2,5) [239].

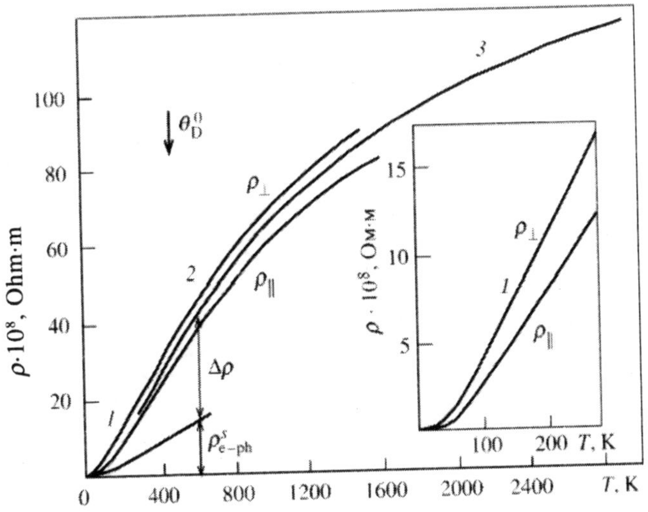

Fig. 5.4. The electrical resistivity of rhenium in the direction of the hexagonal axis and perpendicular to it, and for a polycrystal (3) [239].

$$\mathbf{j} = -\frac{1}{3}e^2\mathbf{E}\sum_{\mathbf{k}}\tau v^2\frac{\partial f_0}{\partial E_{\mathbf{k}}} = -\frac{e^2\mathbf{E}}{24\pi^3}\int d\varepsilon\frac{\partial f_0(\varepsilon)}{\partial\varepsilon}\int\frac{dS}{|\operatorname{grad}_{\mathbf{k}}\varepsilon|}\tau v^2 \equiv$$

$$\equiv \mathbf{E}\int\sigma(\varepsilon)\left(-\frac{\partial f_0(\varepsilon)}{\partial\varepsilon}\right)d\varepsilon. \qquad (5.31)$$

For the spherical Fermi surface

$$\sigma(\varepsilon) = -\frac{4e^2}{3m^*}\varepsilon N(\varepsilon)\tau(\varepsilon). \qquad (5.32)$$

To lowest-order approximation we have

$$\sigma = \frac{j_x}{E_x} = \sigma(E_F) = \frac{e^2 n\tau(E_F)}{m^*}. \qquad (5.33)$$

Using the expansion (3.37) we obtain

$$\sigma = \sigma(E_F) + \frac{\pi^2}{6}(k_B T)^2\left(\frac{\partial^2\sigma(\varepsilon)}{\partial\varepsilon^2}\right)_{\varepsilon=E_F}, \qquad (5.34)$$

The second term in (5.34) should be important at high temperatures.

The role of the $N(E)$ dependence in transport properties for an arbitrary electron system (including one with strong correlations) may be illustrated by simple consideration of impurity scattering [405]. To this end we expand the one-electron Green's function to

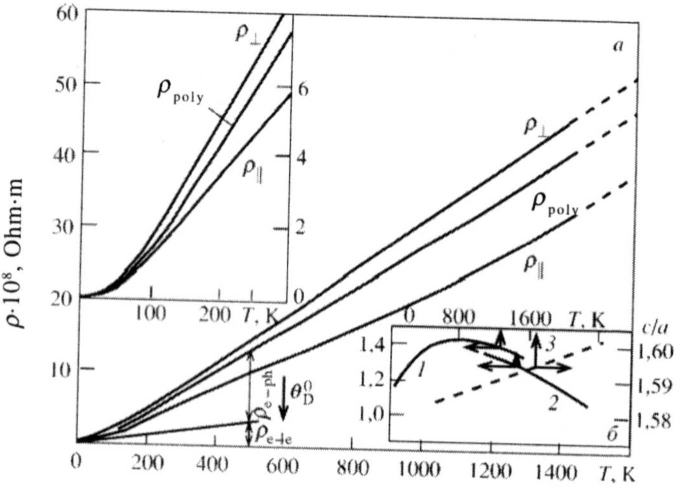

Fig. 5.5. (a) The electrical resistivity of ruthenium in the direction of the hexagonal axis and perpendicular to it, and for polycrystalline sample; (b) the anisotropy of resistivity (1) and diffusivity (2), and the ratio *c/a* (3) [239].

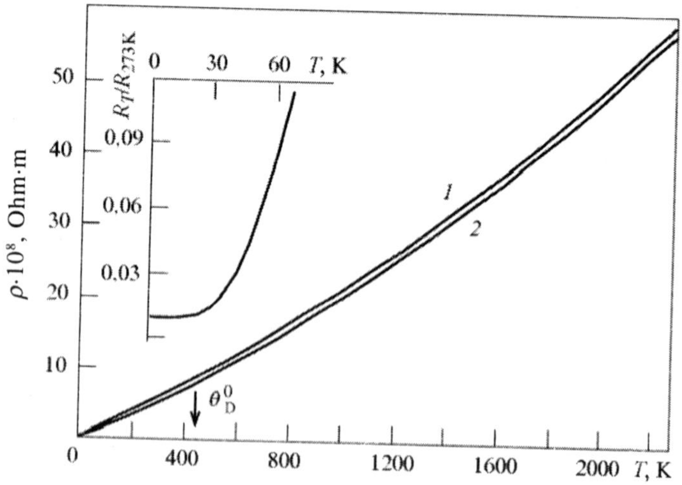

Fig. 5.6. The electrical resistivity of iridium (1); 2) the calculation based on the Mott s-d model [239].

second order in the impurity potential V

$$\langle\langle c_{\mathbf{k}} \mid c_{\mathbf{k}'}^{\dagger}\rangle\rangle_E = \delta_{\mathbf{kk}'} G_{\mathbf{k}}(E) + G_{\mathbf{k}}(E) T_{\mathbf{kk}'}(E) G_{\mathbf{k}'}(E),$$

$$T_{\mathbf{kk}'}(E) = V + V^2 \sum_{\mathbf{p}} G_{\mathbf{p}}(E) + \ldots, \tag{5.35}$$

where G are exact Green's functions for the ideal crystal. Then,

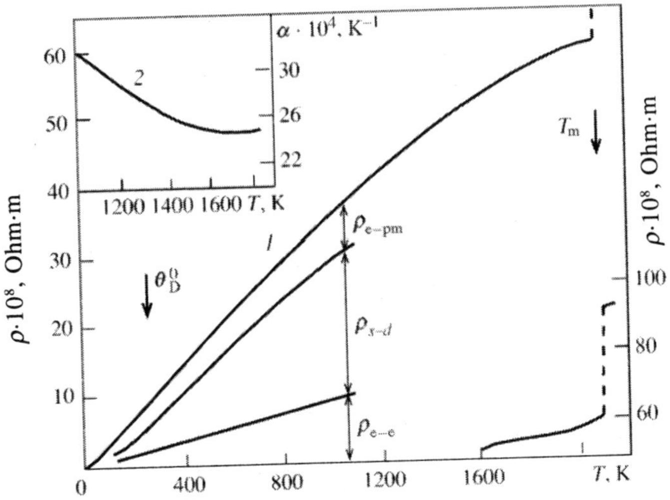

Fig. 5.7. The electrical resistivity of platinum (1); 2) the temperature coefficient of resistivity α [239].

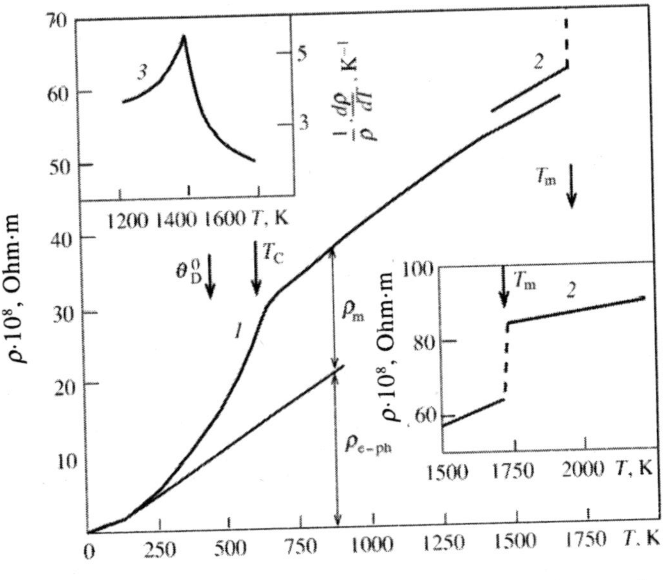

Fig. 5.8. The electrical resistivity of nickel (1); 2) the anomaly near the melting point; 3) the temperature coefficient of resistivity as a function of $t = (T - T_C)/T_C$ near the Curie point [239].

at neglecting vertex corrections, the transport relaxation time is determined from the imaginary part of T-matrix

$$\tau^{-1}(E) = -2V^2 \operatorname{Im} \sum_{\mathbf{p}} G_{\mathbf{p}}(E) = 2\pi V^2 N(E), \tag{5.36}$$

which yields the required connection with the density of states.

5.3.1 Electron–electron scattering

The low-temperature resistivity of most transition metals is satisfactorily described by the formula

$$\rho = \rho_0 + AT^2 + BT^5. \tag{5.37}$$

According to [406], the fitting yields negligible values of the coefficients at T, T^3 and T^4 terms (however, the T^4-term was observed in V and Ta [407]). The T^5-term is to be attributed to electron-phonon scattering, and the T^2-term, which dominates at $T < 10$ K, may be connected with various mechanisms. The simplest among them is the electron–electron scattering. The considerations of this mechanism were performed in a number of papers [408–411]. It was demonstrated that in the case of a single group of current carriers the scattering is possible provided that Umklapp processes are taken into account. These processes lead to an momentum transfer from electrons to lattice so that the conservation law

$$\mathbf{k}_1 + \mathbf{k}_2 - \mathbf{k}_1' - \mathbf{k}_2' = \mathbf{g}$$

holds at non-zero reciprocal lattice vectors g. The calculations in the case of the screened Coulomb interaction, which use solving the transport equation by the variational approach [7], yield the estimate for the electron–electron resistivity

$$\rho_{ee} \simeq \frac{\pi^2 z'}{32} \frac{e^2}{v_F E_F} G^2 \frac{g k_F}{\kappa^2} \left(\frac{k_B T}{E_F} \right)^2. \tag{5.38}$$

where κ is the inverse screening radius, z' is the nearest-neighbour number in the reciprocal lattice, G is an analogue of the atomic form factor (e.g., in X-ray scattering theory):

$$G \simeq \frac{1}{v_0} \int |u(\mathbf{r})|^2 e^{i\mathbf{g}\mathbf{r}} d\mathbf{r}, \tag{5.39}$$

with $u(\mathbf{r})$ being the Bloch modulation function, v_0 the lattice cell volume. The occurrence of the factor $(T/E_F)^2$ is connected with that the scattering is possible in a narrow layer near the Fermi level with the width of order of T only. The formula (5.38) may be presented in the form

$$\rho_{ee} = \left(\frac{e^2 n\tau_e}{m^*}\right)^{-1}, \qquad \tau_e = \frac{\hbar E_{kin}}{E_{coul}^2}\left[\frac{\pi^2 z'}{32} G^2 \frac{g k_F}{\kappa}\left(\frac{k_B T}{E_F}\right)^2\right]^{-1}, \tag{5.40}$$

with

$$E_{kin} = \frac{\hbar^2}{m^* a_0^2}, \qquad E_{coul} = \frac{e^2}{a_0}$$

being effective kinetic and potential energies of the electron gas, a_0 the lattice constant. Estimating numerical coefficients in (5.38) for a simple metal we obtain [7]

$$\rho_{ee} \sim 5 \cdot 10^{-3} G^2 \left(\frac{k_B T}{E_F}\right)^2 \ \Omega \cdot \text{cm}. \tag{5.41}$$

This quantity would yield an appreciable part of the room-temperature resistivity provided that $G \sim 1$. However, apparently $G \ll 1$ which is confirmed by absence of the T^2-term in simple metals even at low temperatures.

The contribution of electron–electron scattering with account of the **k**-dependence of the relaxation time was considered by Schroeder [411]. Then the effect is determined by the quantity

$$\mathbf{k}_1\tau(\mathbf{k}_1) + \mathbf{k}_2\tau(\mathbf{k}_2) - \mathbf{k}'_1\tau(\mathbf{k}'_1) - \mathbf{k}'_2\tau(\mathbf{kk}'_2), \tag{5.42}$$

which does not vanish at $\mathbf{g} = 0$.

The role of electron–electron mechanism may increase in the presence of several carrier groups where scattering is possible without Umklapp processes. For spherical Fermi sheets of s- and d-electrons with the effective masses m_s, m_d an analogue of the expression (5.38) reads

$$\rho_{ee} \simeq \frac{\pi^4}{16} \frac{e^2}{v_s E_F^s} \frac{k_F}{\kappa}\left(\frac{v_s - v_d}{v_d}\right)^2 \left(\frac{k_B T}{E_F^s}\right)^2, \tag{5.43}$$

so that the small factor G^2 is absent, and there occurs a large factor $(v_s/v_d)^2$.

The expression (5.43) may explain the value of the observed T^2-contribution in transition metals. The corresponding experimental situation by the beginning of 70s is described in the review [406]. Measurement of low-temperature resistivity of pure d-metals permitted to determine the coefficients A at the T^2-term and perform comparison with various theoretical models. An important question is the correlation between A and electron density of states. This question was discussed

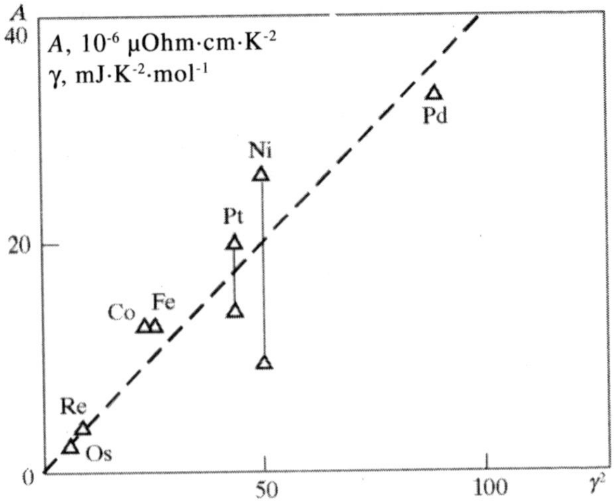

Fig. 5.9. Coefficient at the T^2-term A vs γ^2 for transition metal elements [410]

by Rice [410]. As follows from (5.43), ρ_{ee} should be proportional to squared d-electron mass, i.e. to squared coefficient at linear specific heat γ. Fig.5.9 displays the coefficient A as a function of γ^2 for some d-metals. An approximate relation between these quantities is

$$A\,(\mu\Omega\cdot\mathrm{cm/K^2}) \simeq 0.4\cdot10^{-6}\left[\gamma\,(\mathrm{mJ/mol\cdot K^2})\right]^2. \tag{5.44}$$

The correlation between A and γ is also evident from the Table 5.1. However, the value of the coefficient in (5.44) may vary considerably. For Zr, Ta, Hf, W this coefficient makes up (2–10)·10^{-6}. Possibly, this is connected with insufficient purity of the samples. For very pure samples of W and Re (ρ(273 K)/ρ(4.2 K) ~ 10^4–10^5) one has $A \sim 10^{-6}\gamma^2$.

Later investigations demonstrated that the scattering at the sample surface may be important in pure metals provided that the mean free path is large in comparison with the sample size (e.g., with diameter of a wire). The interference of electron–phonon and surface scattering results in occurrence of a T^2-like contribution [414, 8]. It was demonstrated in papers by Volkenshtein et al [412] that the value of A in pure W, Re, Os is determined by the dimensional effect, and the contribution of electron–electron scattering is smaller than $0.05\cdot10^{-12}\Omega\cdot$cm/$K^2$. In impure samples (e.g., in Ta and V) a considerable contribution may originate from scattering by thermal vibrations of impurity ions, according to the theory [415].

At the same time, in a number of other pure transition metals the coefficient A remains large even after excluding the dimensional

Table 5.1 Electrical resistivity ρ ($\mu\Omega\cdot$cm) at room temperature, the coefficient at the T^2 term A (10^{-6} $\mu\Omega\cdot$cm/K^2), the coefficient at linear specific heat γ (mJ/mol K^2), and the ratios A/γ^2 and ρ/γ in 3*d*, 4*d* and 5*d* rows [406, 412, 413]]

3d 4d 5d	Sc Y La	Ti Zr Nb	V Nb Ta	Cr Mo W	Mn Tc Re	Fe Ru Os	Co Rh Ir	Ni Pd Pt	Cu Ag Au
ρ	52	48	20	12	144	10.2	6	7.4	1.7
	67	43	15	5	20	7.5	5	10.8	1.6
	62	34	13	5	20	10.6	5.3	10.8	2.3
A	–	–	3.3[a]	–		–	–	–	–
	100[b]	80	32	2		2.7		33	–
	–	15	70[c]	1[c]	4.5[c]	0.2-0.5	0.9	13-19	–
γ	9	4.5	8.5	2	17	5	5.1	7	1
	8	2.8	7.8	2	–	2.95	4.65	9.57	
	–	2.16	6.3	0.84	2.3	2.35	3.14	6.41	
A/γ^2	–	–	0.045	–	–	–	–	–	–
	1.6	10.2	0.53	0.5	–	0.31	–	0.36	–
		3.2	1.78	1.41	0.95	0.36	0.09	0.38	–
ρ/γ	6	11	2.2	6	8.5	2	1.2	1.05	1
	8	15.5	1.9	2.75	–	2.6	1.1	1.13	–
	–	15.75	2.15	6.48	8.68	4.5	1.7	1.68	–

[a] ρ/ρ (4.2K) = 1400
[b] ρ/ρ (4.2K) = 700
[c] After excluding dimensional effect, A is very small

effect. Especially surprising is the situation for Mo where $A = 1.2 \cdot 10^{-12}\Omega\cdot$cm/K^2 exceeds by several tens the value for an analogous metal W. This strong difference may be hardly explained by the electron-electron scattering since the values of γ differ by 2.5 times only. Thus the question about origin of the T^2-contribution in paramagnetic transition metals remains open. Further investigation of this problem is prevented by absence of reliable experimental data on some *d*-metals, in particular for two first columns of periodic table. Data for Sc and Ti are absent, and the purity of investigated samples of Zr and Hf is insufficient. Data for Y [412] yielded a large value of $A \sim 10^{-10}\Omega\cdot$cm/K^2. Since the value of γ in Y is rather high, this result supports the idea of $A - \gamma^2$ correlation.

From the theoretical point of view, various interference mechanisms for different scattering processes may be considered. However, more simple interpretations of observed correlations between ρ and other characteristics seem to be not excluded.

As will be discussed in Sect.5.3.3, an alternative mechanism for

T^2-dependence of resistivity in ferromagnetic metals is scattering by spin waves. Separation of electron-electron and electron-magnon scattering is a rather difficult problem. Since the T^2-term is comparable in ferromagnetic and paramagnetic metals, main role is often attributed to the electron-electron mechanism (see, e.g., [406]). However, spin-density fluctuations, which occur at finite temperatures even in paramagnets, should result in scattering of current carriers due to exchange interaction. This scattering may be treated as enhancement of the exchange part of electron-electron scattering by spin fluctuations. Such an enhancement should be appreciable in metals like Pd. Thus it is difficult to discriminate electron-electron and spin-fluctuation contributions even in principle. Quantitative theoretical description of spin-fluctuation effects in elemental paramagnetic d-metals is now absent. Spin-fluctuation resistivity enhancement in weakly and almost magnetic metals (Sect.4.4) is considered in [416, 26]. In weakly and nearly ferromagnetic metals we have

$$\rho_{\text{mag}} \sim |1 - IN(E_{\text{F}})|^{-1/2}\, T^2. \tag{5.45}$$

In the antiferromagnetic case

$$\rho_{\text{mag}} \sim |1 - 2I\chi_{\text{Q}}|^{-1/2}\, T^2 \tag{5.46}$$

(cf. (4.65),(4.66)). At the boundary of the magnetic instability, where the prefactors in (5.45), (5.46) diverge, the type of the temperature dependence changes:

$$\rho_{\text{mag}} \sim \begin{cases} T^{5/3}, & \text{FM}, \\ T^{3/2}, & \text{AFM}. \end{cases} \tag{5.47}$$

These results are confirmed by data on resistivity of some d-compounds (see [26]). A Stoner-type renormalization of electron-electron scattering amplitude by spin-spin interactions and its role in resistivity of transition metals is considered in [417].

5.3.2 Mott *s–d* scattering mechanism

An independent mechanism for resistivity of transition metals is the *s–d* scattering considered by Mott [418]. This mechanism is based on the assumption that an appreciable part of scattering corresponds to transitions of main current carriers (s-electrons) to the unoccupied part of d-band, the states of which possess much smaller mobility. Such interband transitions may occur due to any scattering mechanism (impurities, phonons, spin excitations etc.), their probability being large because of high density of d-states near the Fermi level. At the same time, the inverse *d–s* transitions may be neglected at calculating conductivity since $N_s(E_{\text{F}}) \ll N_d(E_{\text{F}})$ (electron numbers in the subbands are restored mainly due to thermal relaxation). Thus we may write down

$$\sigma = \sigma_s + \sigma_d = \rho_s^{-1} + \rho_d^{-1},$$

with

$$\rho_s = \rho_{ss} + \rho_{sd}, \qquad \rho_d = \rho_{dd} + \rho_{ds} \simeq \rho_{dd}. \tag{5.48}$$

The Mott's model is widely used to explain concentration dependences of resistivity of transition metal alloys. In particular, the resistivity decreases at filling of d-shells of transition metal ions by electrons of another alloy component, which results in prohibition of s–d transitions.

The s–d scattering owing to phonons was also considered by Wilson [479, 2]. In the case $T \gg \theta_D$ he obtained the result

$$\rho_{sd}^{\mathrm{ph}} = \left(\frac{3}{4\pi}\right)^{1/3} \frac{m_s^{1/2} m_d}{e^2 n M a_0} \frac{(E_{\mathrm{F}}^s)^{3/2}}{k_{\mathrm{B}}\theta_{\mathrm{D}}^2} T \left(1 - \frac{\hbar s \Delta k_{sd}}{k_{\mathrm{B}}\theta_{\mathrm{D}}}\right), \tag{5.49}$$

where s is the sound velocity, $\Delta k_{sd} = k_{\mathrm{F}}^s - k_{\mathrm{F}}^d$ is the minimum scattering quasimomentum of a phonon, as determined by the momentum conservation law. Temperature corrections to (5.49) are obtained by the expansion of the Fermi distribution functions which yields

$$\rho_{sd}^{\mathrm{ph}} = aT\left[1 - b\left(\frac{k_{\mathrm{B}}T}{E_{\mathrm{F}}}\right)^2\right]. \tag{5.50}$$

The formulas (5.49), (5.50) may explain values and temperature dependence of transition metal resistivity at high T. On the other hand, at low T phonons with small quasimomenta play the dominant role, so that we should have

$$\rho_{sd}^{\mathrm{ph}} \sim \exp\left(-\frac{\hbar s \Delta k_{sd}}{k_{\mathrm{B}}T}\right), \tag{5.51}$$

(if one takes into account the overlap of s- and d-sheets of the Fermi surface, $\rho_{sd}^{\mathrm{ph}} \sim T^5$). However, such a strong decrease is not observed experimentally, which leads to difficulties of the theory. In particular, Wilson [2] claimed that s–d transitions do not play an important role in the resistivity of transition metals.

Generally, s–d transitions may take place for all scattering mechanisms and are especially important for elastic processes. At low temperatures the resistivity is determined mainly by impurity scattering. Due to the strong energy dependence of the density of states in the d-band, this contribution may exhibit a considerable temperature dependence according to (5.34). For the parabolic s- and d-bands one obtains [8]

$$\rho_{sd}^{i}(T) = \rho_{sd}^{i}(0)\left[1 - \frac{\pi^2}{6}\left(\frac{k_B T}{E_d - E_F}\right)^2\right]. \tag{5.52}$$

Unlike the electron–electron contribution ρ_{ee}, ρ_{sd} is proportional to the density of states of d-electrons at E_F and consequently to the first power of γ. This difference may be used to separate the s–d scattering contribution. One can see from Table 5.1 that the ratio of the T-linear resistivity term to γ in the beginning of periods (of order of 10) exceeds considerably that in the end of periods. Possibly this tendency is connected with different role of s–d transitions. This seems be important for Sc and Ti columns, decreases appreciably in the V column and further on (as well as for other properties, the Mn column makes up an exception). Such a behaviour agrees with that of the coefficient A at the T^2-term and reflects the tendency towards lowering and narrowing of d-bands to the end of periods. The narrowing may result in hampering s–d transitions because of energy and quasimomentum conservation laws.

5.3.3 Resistivity of magnetic metals

Existence of magnetic moments in transition elements results in additional factors which influence the behaviour of current carriers in external electric field. Firstly, thermal fluctuations in the system of magnetic moments provide a new scattering mechanism owing to s–d exchange interaction. Secondly, the electron spectrum of magnetic crystals depends appreciably on spontaneous magnetization (or sublattice magnetization in antiferromagnets) and, consequently, on temperature.

The first effect may be described by introducing an additional contribution to resistivity,

$$\rho_{tot} = \rho + \rho_{mag}. \tag{5.53}$$

The second effect cannot be described by the simple expression (5.53). In the simplest case where the influence of magnetic ordering is small we may perform the expansion in magnetization to obtain

$$\rho_{tot}(M) = \rho_{tot}(0) + aM^2. \tag{5.54}$$

Unlike (5.53), the sign of the second term in (5.54) needs not to be positive. The expansion (5.54) does not hold in the cases where the gap in the spectrum modifies strongly the states near the Fermi level. In such situations the occurrence of magnetic splitting may result in considerable anomalies of resistivity and other transport properties at the magnetic ordering point.

For the antiferromagnetic structure with the wavevector **Q**, the disturbance of electron spectrum by magnetic ordering is especially strong at **2k** = **Q**. Then we obtain from (G.70)

$$E_{\mathbf{k}}^{1,2} = t_{\mathbf{k}} \pm | I\bar{S} |. \tag{5.55}$$

This disturbance may influence strongly transport properties provided that the Fermi level coincides with the antiferromagnetic gap. The corresponding resistivity anomaly at the Neel point was discussed in [420]. The result in the mean-field approximation has the form

$$\rho(T) = aT + b[1 - \bar{m}^2(T)] + c\bar{m}^2(T)T, \tag{5.56}$$

where

$$\bar{m}(T) = \bar{S}/S \sim (1 - T/T_N)^{1/2},$$

is the relative sublattice magnetization. With lowering temperature, this contribution increases rapidly at passing T_N and may lead to a $\rho(T)$ maximum. Such a behaviour is observed in α-Mn (Fig.5.10), Cr (Fig.5.3) and rare-earth metals [265, 406]. An alternative explanation of the maximum is based on the critical scattering near the second-order magnetic transition point. Investigations of Dy and Ho in strong magnetic fields [421] demonstrated that the resistivity decreases sharply at field-induced transition into ferromagnetic state with disappearance of the spiral magnetic superstructure.

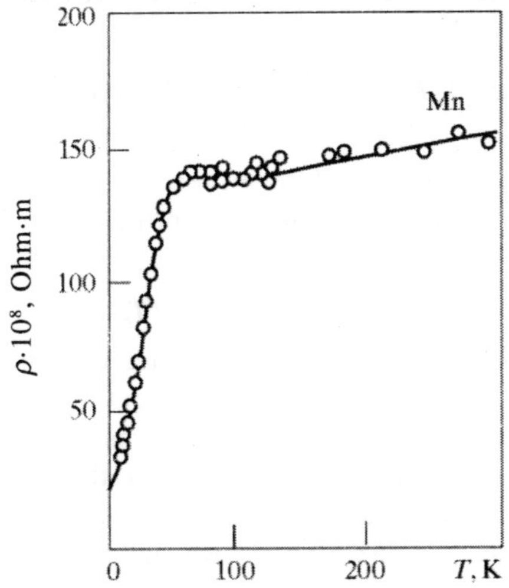

Fig. 5.10. Anomaly of the resistivity $\rho(T)$ (10^{-6} Ω·cm) at the Neel point (T_N= 96K) in α-Mn [265].

Consider exchange scattering of conduction electrons by spin disorder within the s–d exchange model. The result for the magnetic resistivity at high temperatures in the mean-field approximation is given by (M.84). For spin $S = 1/2$ it takes the form

$$\rho_{\mathrm{mag}} = \frac{9\pi}{2}\frac{m^*}{ne^2}\frac{I^2}{E_{\mathrm{F}}}\left(\frac{1}{4} - \overline{S}^2\right). \tag{5.57}$$

In a far paramagnetic region we have for arbitrary S

$$\rho_{\mathrm{mag}} = \frac{3\pi}{2}\frac{m^*}{ne^2}\frac{I^2}{E_{\mathrm{F}}}S(S+1). \tag{5.58}$$

The result of the type (M.84) was first obtained by Kasuya [422]. It explains rather well experimental data on the temperature dependence of resistivity of ferromagnetic metals near the Curie point. For rare earth metals, the expression (5.58) with the replacement

$$S(S+1) \to (g-1)^2 J(J+1)$$

describes satisfactory the change of high-temperature spin-disorder resistivity in the 4f-series [16].

Consider magnetic scattering at low temperatures. Passing to magnon operators with the use of the Holstein–Primakoff representation (E.1) we obtain from (5.23)

$$\rho = \frac{\pi k_{\mathrm{B}} T}{\langle j_x^2 \rangle} 2I^2 S e^2 \sum_{\mathbf{kq}} (v^x_{\mathbf{k}\uparrow} - v^x_{\mathbf{k+q}\downarrow})^2 N_{\mathbf{q}} n_{\mathbf{k}\uparrow}(1 - n_{\mathbf{k+q}\downarrow})\delta(E_{\mathbf{k}\uparrow} - E_{\mathbf{k+q}\downarrow} + \omega_{\mathbf{q}}). \tag{5.59}$$

Integrating over $\mathbf{k}$ we derive the result for resistivity

$$\rho = C_1 T^2 \int_{T_0/T}^{\infty} \frac{x}{\sinh x} dx + C_2 T_0 T \ln\coth\frac{T_0}{2T}, \tag{5.60}$$

where the constants C_i are determined by the electron spectrum, C_2 being non-zero only for a non-parabolic electron spectrum, the quantity

$$T_0 \sim T_{\mathrm{C}} q_0^2 \sim (I/E_{\mathrm{F}})^2 T_{\mathrm{C}} \tag{5.61}$$

coincides with the boundary of the Stoner continuum ω_- (Appendix G.1), $q_0 = 2|IS|/v_{\mathrm{F}}$ is the threshold vector for the one-magnon scattering processes. At very low temperatures $T < T_0$ the one-magnon resistivity (5.60) is exponentially small since the quasimomentum and energy conservation law cannot be satisfied at characteristic thermal magnon quasimomenta. At $T \gg T_0$ we have

$$\rho_0(T) \sim T^2 N_\uparrow(E_{\mathrm{F}}) N_\downarrow(E_{\mathrm{F}}) \tag{5.62}$$

(the second term in (5.60) yields small corrections of order T_0T ln T and is usually neglected). The same results follow from the solution of the transport equation (Appendix M.3). Thus spin-wave scattering in the wide temperature interval $T_0 < T < T_C$ results in the square temperature dependence of resistivity. The difference from the electron-phonon scattering (see (5.28)) is explained by the square dispersion law of magnons, so that their number is proportional to $T^{3/2}$ rather than T^3.

The T^2-dependence was obtained by Turov [425] and Kasuya [426], and further confirmed by many authors. However, at very low temperatures in ferromagnetic transition metals there exist contributions to resistivity, which are proportional to $T^{3/2}$ or T [265, 406]. Linear temperature corrections owing to relativistic interactions were found in [425]. However, such corrections are too small to explain the experimental data. An attempt was made in [288] to explain the $T^{3/2}$-term by the non-quasiparticle contributions to the impurity resistivity which occurs due to strong energy dependence of incoherent states near the Fermi level. Indeed, taking into account (5.36) we obtain the correction to conductivity

$$\delta\sigma(E) \sim -V^2 \int \left(-\frac{\partial f(E)}{\partial E} \right) \delta N(E) dE \sim -T^{3/2}.$$

It should be noted that the T^2-term (5.62) is absent in the case of a half-metallic ferromagnets (HMF, Sect.4.5) where the states with one spin projection only exist at the Fermi level and one-magnon scattering processes are forbidden in the whole spin-wave region. This seems to be confirmed by comparing experimental data on resistivity of Heusler alloys TMnSb (T = Ni, Co, Pt, Cu, Au) and PtMnSn [331]. The T^2-contribution from one-magnon processes to resistivity for half-metallic systems (T = Ni, Co, Pt) was really not picked out, whereas the dependences $\rho(T)$ for "usual" ferromagnets were considerably steeper (Fig.5.11). In the HMF situation, as well as for usual ferromagnets at $T < T_0$, the resistivity is determined by two-magnon scattering processes. This results in a weak temperature dependence of resistivity, $\rho(T) \sim T^{9/2}$ for $T < T_0$ and $\rho(T) \sim T^{7/2}$ for $T > T_0$ [427, 428, 735], which is due to vanishing of the electron-magnon scattering amplitude at zero magnon wavevector (Appendix G.1).

Consider the situation in rare-earth metals which are ferromagnetic at low temperatures. Because of strong anisotropy, the dispersion law of spin waves differs from that in d-metals. The magnon spectrum in rare earths contains a gap of order $T^* \sim 10$ K; in the absence of anisotropy in the basal plane we have the linear law $\omega_{\mathbf{q}} \sim q$.

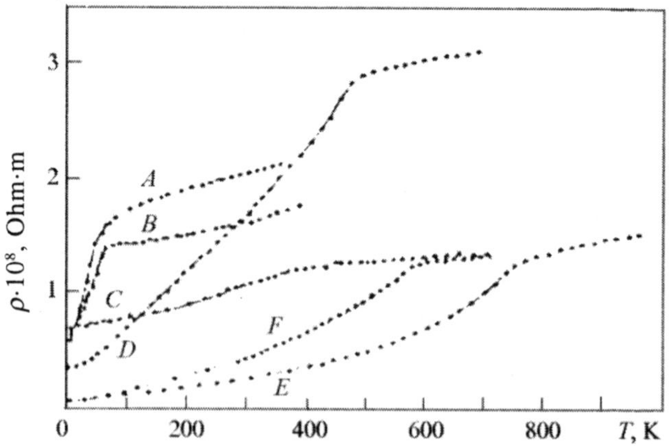

Fig. 5.11. Temperature dependences of resistivity in the Heusler alloys TMnSb (T = Cu (A), Au (B), Co (D), Ni (E), Pt (F)) and PtMnSn (C) [331].

The gap results in the occurrence of an exponential factor exp $(-T^*/T)$ in the magnetic resistivity. For linear dispersion we obtain dependence $\rho \sim T^4$ [429] instead of T^2 since each power of q yields at integration a factor of T/T_C (instead of $(T/T_C)^{1/2}$ at $\omega_{\mathbf{q}} \sim q^2$). The latter result was confirmed by the experimental dependence $\rho \sim T^{3.7}$ for gadolinium in the region 4–20 K [430].

Using the formula (5.23) with the Hamiltonian of the s–d model (G.2) in the spin-wave region, we obtain for the low-temperature magnetic resistivity of antiferromagnetic metals

$$\rho = \frac{\pi k_B T}{\langle j_x^2 \rangle} 2I^2 S e^2 \sum_{\mathbf{kq}} (v_{\mathbf{k}}^x - v_{\mathbf{k+q}}^x)^2 n_{\mathbf{k}} (1 - n_{\mathbf{k+q}}) [N_{\mathbf{q}} (u_{\mathbf{q}} + v_{\mathbf{q}})^2) \times$$
$$\times \delta(E_{\mathbf{k}} - E_{\mathbf{k+q}} + \omega_{\mathbf{q}}) + N_{\mathbf{q+Q}} (u_{\mathbf{q+Q}} - v_{\mathbf{q+Q}})^2) \delta(E_{\mathbf{k}} - E_{\mathbf{k+q}} + \omega_{\mathbf{q+Q}})]. \tag{5.63}$$

The resistivity at very low temperatures is determined by contributions of small q in (5.63), i.e. by transitions inside antiferromagnetic subbands (Appendix G.2). Due to the linear dispersion law of magnons, such transitions result, as well as electron–phonon scattering, in a T^5-dependence of resistivity [428]. (The earlier result $\rho \sim T^4$ [426] was erroneous since the coefficients of the Bogoliubov transformation (E.10), which have an essential q-dependence, were not taken into account.) Because of the singularity in uv-transformation coefficients, the contribution from the region of small $|\mathbf{q} - \mathbf{Q}|$ (i.e. of the intersubband contibutions) is, generally speaking, larger. However, as well as for ferromagnets, it is impossible to satisfy the quasimomentum conservation law at $\mathbf{q} \to \mathbf{Q}$ because of antiferromagnetic splitting, so that this contribution is in fact exponentially small at

$$T < T_0 = \omega(q_0) \sim (|IS|/E_F)T_N, \tag{5.64}$$

with $q_0 = 2|IS|/v_F$ being the threshold value of $|\mathbf{q} - \mathbf{Q}|$. (Note that the boundary temperature is not so small as for a ferromagnet, cf. (5.61).) At higher temperatures $T > T_0$ the intersubband contributions yield the T^2-behaviour of resistivity [433]. In the two-dimensional situation these contributions become T-linear which may explain the characteristic dependence $\rho(T)$ in high-T_c superconductors.

5.3.4 Resistivity of transition metal alloys

Investigations of transport phenomena in alloys as a function of concentration of various components give a possibility to obtain information on their electron structure. In particular, data on concentration dependence of residual resistivity $\rho(c)$ in transition metal alloys yields an important information on change of d-states.

In disordered alloys of simple metals with the same valence (metals with different valence do not form usually the continuous solid solution series), the Nordheim rule usually holds

$$\rho(c) = \rho_0 c(1-c), \tag{5.65}$$

$$\frac{1}{\rho_0}\left(\frac{d\rho(c)}{dc}\right)_{c=0,1} = \pm 1, \qquad \frac{1}{\rho_0}\left(\frac{d^2\rho(c)}{dc^2}\right)_{c=0,1} = -2.$$

In TM alloys the symmetry of $\rho(c)$ curve is violated (see Fig.5.12). An explanation of this violation can be obtained within the Mott s–d transition model. To demonstrate this we write down the conductivity of the alloy $A_{1-c}B_c$, with A being a transition metal and B a simple metal, in the form

$$\sigma = \sigma_s + \sigma_d \simeq \sigma_s = \rho_s^{-1}, \qquad \rho_s = \rho_{ss} + \rho_{sd}. \tag{5.66}$$

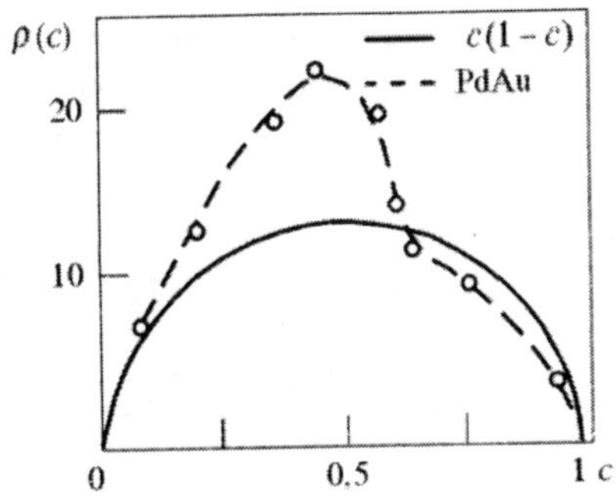

Fig. 5.12. The concentration dependence of resistivity ($\mu\Omega \cdot$ cm) in Pd–Au alloys.

Taking into account the relations

$$\rho_{ss} = \rho_s^0 c_B(1-c_B), \qquad \rho_{sd} = ac_B(1-c_B)N_d(E_F) \tag{5.67}$$

and the concentration dependence

$$N_d(E_F) \simeq c_B(1-c_B)N_A(E_F), \tag{5.68}$$

we obtain [434]

$$\rho \simeq \rho_s = \rho_s^0 c_B(1-c_B) + \tilde{a}c_B(1-c_B)^2, \tag{5.69}$$

which yields a deviation from the Nordheim rule.

Other factors which lead to violation of this rule may exist. So, for Cu–Ni alloys, d-electrons are at localized levels for small nickel concentrations and form band states for $c > 40\%$ only [435]. At the critical concentration, an abrupt change of the alloy properties takes place.

5.3.5 Two-current model of ferromagnetic metals

The two-current models considers two types of current carriers in an itinerant ferromagnet which have different spin projections. Unlike the Mott model, where the current of d-states is neglected, contributions of both types of states are comparable. Effects of strong spin polarization are especially important for d-bands, but may be appreciable for s-type current carriers too.

The phenomenological treatment of the model is rather simple [436]. The total current and, consequently, conductivity, are represented as a sum of contributions from majority and minority current carriers:

$$j = \sigma E, \qquad \sigma = \sigma_\uparrow + \sigma_\downarrow. \tag{5.70}$$

Then the resistivity takes the form

$$\rho = \sigma^{-1} = \left(\frac{1}{\rho_\uparrow} + \frac{1}{\rho_\downarrow}\right)^{-1} = \frac{\rho_\uparrow \rho_\downarrow}{\rho_\uparrow + \rho_\downarrow}, \tag{5.71}$$

which corresponds to a parallel junction. Further we have to take into account transitions between both the types of carriers. These are due to spin-flip scattering processes by spin inhomogeneities, spin waves, magnetic impurities etc. Then we have

$$\rho_\uparrow = \rho_\uparrow^0 + \rho_{\uparrow\downarrow}, \qquad \rho_\downarrow = \rho_\downarrow^0 + \rho_{\downarrow\uparrow} \tag{5.72}$$

(consecutive junction). On substituting (5.72) into (5.71) we obtain

$$\rho = \frac{\rho_\uparrow^0 \rho_\downarrow^0 + \rho_\downarrow^0 \rho_{\uparrow\downarrow} + \rho_\uparrow^0 \rho_{\downarrow\uparrow} + \rho_{\uparrow\downarrow}\rho_{\downarrow\uparrow}}{\rho_\uparrow^0 + \rho_\downarrow^0 + \rho_{\uparrow\downarrow} + \rho_{\downarrow\uparrow}}. \tag{5.73}$$

Such an approach may be applied to describe transport effects in ferromagnetic transition metals and their alloys. Unfortunately, reliable microscopic calculations of the quantities which enter (5.73) are hardly possible, and it is more convenient to determine them from experimental data. One uses often the data on the deviation from the Matthiessen rule.

To illustrate this approach we consider a ternary alloy $M_{1-x-y}A_xB_y$. We have for the impurity resistivity ($\rho_{\uparrow\downarrow} = \rho_{\downarrow\uparrow} = 0$)

$$\rho_{AB} = \rho_A + \rho_B + \Delta\rho, \tag{5.74}$$

where $\rho_{A,B}$ are the resistivities of the corresponding binary alloys,

$$\Delta\rho = (\alpha_A - \alpha_B)^2 \rho_A \rho_B [(1+\alpha_A)^2 \alpha_B \rho_A + (1+\alpha_B)^2 \alpha_A \rho_B]^{-1}, \tag{5.75}$$

with

$$\alpha_{A,B} = \rho_{A,B\downarrow} / \rho_{A,B\uparrow} \cdot$$

Varying the concentrations x and y one can experimentally determine the parameters $\alpha_{A,B}$.

At finite temperatures, additivity of the impurity residual resistivity ρ_i and T-dependent contributions for a binary alloy is also violated:

$$\rho_{\uparrow,\downarrow}(T) = \rho_{i\uparrow,\downarrow} + \rho_{l\uparrow,\downarrow}(T), \tag{5.76}$$

$$\rho(T) = \rho_i \left[1 + \left(\frac{\alpha - \mu}{1+\alpha}\right)^2\right] + \rho_l(T) + \left(\frac{\alpha - 1}{\alpha + 1}\right)^2 \rho_{\uparrow\downarrow}(T).$$

with

$$\alpha = \rho_{i\downarrow} / \rho_{i\uparrow}, \qquad \mu_l = \rho_{l\downarrow} / \rho_{l\uparrow},$$

$$\rho_i = \frac{\rho_{i\uparrow}\rho_{i\downarrow}}{\rho_{i\uparrow} + \rho_{i\downarrow}}, \qquad \rho_l = \frac{\rho_{l\uparrow}\rho_{l\downarrow}}{\rho_{l\uparrow} + \rho_{l\downarrow}}, \tag{5.77}$$

where l stands for the index of a concrete temperature-dependent scattering mechanism.

Using experimental data on ternary and binary alloys, parameters α_A, α_B, μ_l, ρ_i, $\rho_l(T)$ and $\rho_{\uparrow\downarrow}$ were determined for the ferromagnetic transition metals [436]. In a number of cases, the parameters α deviate strongly from unity. As an example, data on the alloy $NiCo_{1-x}Rh_x$ are shown in Fig.5.13. Due to strong non-linearity of the $\rho(x)$ dependence, the parameters $\rho_\uparrow$ and $\rho_\downarrow$ turn out to be considerably different. Note that the deviations from the linear dependence at small x cannot be explained by the Nordheim rule.

In a metal with transition metal impurities the scattering of conduction electrons into d-resonance impurity states near the Fermi level plays the dominant role. The behavior of the quantity α in the $3d$-

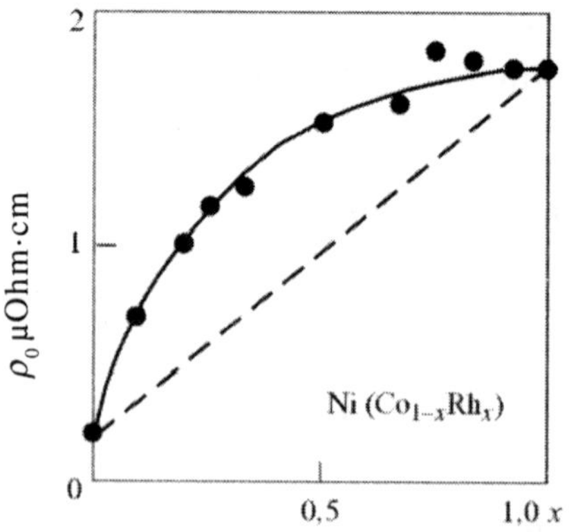

Fig. 5.13. Residual resistivity of NiCo$_{1-x}$ Rh$_x$ alloy. The large deviation from the 'Matthiessen rule (linear dependence) is accounted for strongly different values of $\alpha_{Co}$and α_{Rh}; the solid curve corresponds to their values of 13 and 0.3 respectively [436].

impurity series for Fe and Ni hosts is presented in the Table:

T	Ti	V	Cr	Mn	Fe	Co	Ni	Cu
α (*Ni*T)	1.9	0.52	0.3	10	15	20	–	3.4
α (*Fe*T)	0.38	0.12	0.22	0.13	–	2.1	5	–

As discussed in [436], qualitatively this behavior may be explained on the basis of the Friedel concept of the virtual bound *d*-state. The spin-up nickel *d*-band is practically filled, and only spin down states may screen the impurity charge perturbation. For early 3*d*-impurities (Cr, V, Ti) in Ni the spin-up impurity *d*-state is repelled above the spin-up *d*-band, so that the impurity density of states $g_{d\uparrow}(E_F)$ is rather large, the magnetic moment being opposite to the host one. At the same time, for strongly magnetic Co, Fe and Mn impurities only *s*-states are present at E_F for spin up. The value of $g_{d\downarrow}(E_F)$ is rather large for all the 3*d*-impurities in the Ni host. Thus α strongly increases at passing from the first half of 3*d*-series to the second one.

Magnetic moments of 4*d*-impurities are created mainly by host magnetization, the magnetization perturbation being strongly delocalized. The behavior of α for the first half 4*d*-series impurities in Ni is similar to that in 3*d*-series.

In pure iron the Fermi level is just below the spin-up *d*-band. Therefore in the Fe host the repulsive impurity potential pushes up the spin up *d*-level for Mn, Cr, V and Ti impurity series through the Fermi energy, the magnetic moment decreasing. At the same time, for Co and Ni impurities $g_{d\downarrow}(E_F)$ is small. For all the 3d-impurities in iron $g_{d\downarrow}(E_F)$ is rather small. The qualitative consideration of the problem may be performed also within the Anderson impurity model (see [717]). Quantitatively, the local density of states and other characteristics of the impurity electronic structure are obtained from band calculations. Such calculations were performed for a large number of impurities (including 3d- and 4d-impurities) in the Ni and in the Fe host [718–722]. One can see from the results that even for 3d-impurities in Ni the simple Anderson-model picture is in fact poor: the local impurity DOS $g_{d\sigma}(E)$ is strongly influenced by the host. Besides that, the Fe host, which has (unlike the Ni host) a large magnetization, determines in a great measure formation of magnetic moment even for later 3d-impurities. The 4d-impurity states are strongly hybridized with the valence host states so that the picture of narrow virtual bound impurity d-state is not applicable. This is especially obvious for Y, Zr and Nb impurities in Ni where $g_\sigma(E_F)$ for both σ are quite small and differ by a small exchange spin splitting only [719]. However, for Tc to Pd impurities rather sharp hybridization peaks are present. As compared to Mn, Fe and Co, the spin up peaks for Tc, Ru and Rh are broader and extent somewhat above E_F. The most consistent calculation of resistivity may be performed on the basis of the scattering phase shift analysis on the base of band structure calculations. Such calculations were carried out for d-impurities in Cu [723], the role of magnetic moments being neglected, and in Ni [724]. In the latter case the agreement with experiment was not quite satisfactory. A simplified estimation in the case of a ferromagnetic host with appreciable magnetization perturbations was performed in [717] by using the results of band structure calculations and the Friedel sum rule.

A general (with neglecting crystal-field effects) expression for resistivity per impurity in terms of the phase shifts $\eta_{l\sigma}$ has the form

$$\rho_\sigma = \rho_{u\sigma}\sum_l (l+1)\sin^2(\eta_{l\sigma} - \eta_{l+1,\sigma}), \tag{5.78}$$

where $\rho_{u\sigma}$ is the resistivity unitary limit for a given spin projection per scattering channel (see Sect.6.1). In the approximation of free conduction electrons we have

$$\rho_{u\sigma} = 2m^*/(\pi z_\sigma e g_\sigma).$$

where m^*, z_σ and g_σ are the effective mass, concentration and density of states of conduction electrons at the Fermi level for a given spin projection, e is the electronic charge. To estimate the phase shifts we use for each spin projection (spin-flip processes are neglected) the Friedel sum rule

$$\Delta n_\sigma = \frac{1}{\pi}\sum_l (2l+1)\eta_{l\sigma}. \tag{5.79}$$

The changes of electron numbers are determined by

$$\Delta n_{\uparrow,\downarrow} = \frac{1}{2}(\Delta Z \pm \Delta M),$$

where ΔZ is excess charge introduced by the impurity ion, i.e. the difference between impurity and host atomic numbers, ΔM is the total magnetization change induced by the impurity (in μ_B). Similar to the standard Friedel approach for non-magnetic hosts, only d-scattering ($l = 2$) contribution can be taken into account. This is generally speaking, sufficient for rough estimations. However, in a number of cases s,p-contributions play an important role. In particular, for the impurities in the Ni host, which do not destroy strong ferromagnetism ($\Delta Z = -\Delta M$), the density of spin-up d-states near the Fermi level is rather small, and their disturbance by impurity is practically absent [720]. A similar situation takes place for d-impurities from the beginning of d-series where d-states are almost empty and $\Delta Z + \Delta M = 10$. The sp-contributions are also important for non-magnetic sp-impurities which introduce a strong charge perturbation.

The values of ΔM and some data on the partial contributions $\Delta n_{l\sigma} = (2l + 1)\eta_{l\sigma}/\pi$ may be obtained from the band calculations results [720] for the Ni host and in [721] for the Fe host. First we discuss the case of the Ni host. To perform calculations we have to specify the values of $\rho_{u\sigma}$. For the Cu host one puts usually $\rho_u = \rho_{u\sigma}/2 = 3.8\ \mu\Omega\cdot$cm/at % [725]. As demonstrate band structure calculations [24], the value of total density of states with spin up for metallic nickel is two times larger, but the s-contribution is considerably smaller than for Cu: although the spin up d-subband is practically filled, its tail dominates at E_F, so that $g_{d\uparrow}/g_{s\uparrow}{\sim}10$. Still larger d-contributions occur for spin down where a large density of states peak is present at E_F and $g_{d\downarrow}/g_{s\downarrow} \sim 100$. Thus, although d-electrons are usually assumed to possess small mobility, the problem of their contribution to conductivity should be investigated.

For non-magnetic impurities Cu, Zn, Ga, Ge in Ni the partial values of Δn_σ may be estimated from results of Ref.720. The calculations according to (5.78) with $\rho_{u\downarrow} = 15\mu\Omega \cdot$ cm/at % yield satisfactory values of ρ. However, the estimation of $\rho_{\downarrow}$ with the same value of $\rho_{u\downarrow}$ yields very large values of α. To reduce α up to reasonable values we have to put $\rho_{u\uparrow}/\rho_{u\downarrow}\approx 5$.

Thus the estimations of Ref.[717] lead to the conclusion that *d*-electrons make an important contribution to conductivity, especially to the spin-down current. Indeed, the *s*-electron contribution cannot provide such a large ratio $\rho_{u\uparrow}/\rho_{u\downarrow}$; besides that, this is expected to yield an opposite tendency since $g_{s\uparrow}>g_{s}\downarrow$ [24]. Note that in the situation of several conduction electron groups the quantities $\rho_{u\sigma}$ should be considered as phenomenological fitting parameters. Similar statement about the important role of *d*-states in electronic transport was made by Kondorsky [502,503] on the basis of data on the anomalous Hall effect.

Now we consider briefly the case of the Fe host. Unlike nickel, iron is not a strong ferromagnet since *d*-states with both spin projections are present at E_F. Moreover, the Fermi level lies in a pseudogap for the spin down *d*-states. Therefore one may expect that spin up *d*-states should make a larger contribution to the transport properties (besides that, band calculations [24] give a strong spin polarization of *sp*-electrons at E_F). Indeed, for *Fe*Ni and *Fe*Co systems we have $\Delta M > 0$, but $\alpha > 1$, so that we may estimate $\rho_{u\uparrow}/\rho_{u\downarrow} = 0.2$.

At present, the two-current model was also applied to consider anisotropy of electrical resistivity in magnetic field, temperature dependence of the normal Hall effect, thermoelectric power, magnetoresistivity, transport effects in bulk samples [726] and multilayers (especially, giant $\Delta\rho/\rho$-effect [437]).

5.4 THERMOELECTRIC POWER

In the presence of a temperature gradient, the electric and heat currents are given by the linear relations (5.1) and (5.2). For $j = 0$ we obtain

$$\mathbf{E} = -\frac{\lambda}{\sigma}\mathrm{grad}T \equiv \alpha \mathrm{grad}T, \tag{5.80}$$

where α is the absolute differential thermoelectric power.

The coefficients in (5.1) and (5.2) are determined by the disturbance of the equilibrium distribution function by external fields. Provided that the **k**-dependent relaxation time τ may be introduced,

$$f_{1\mathbf{k}} = -\tau_{\mathbf{k}} v_{\mathbf{k}} \left[eE + (\varepsilon_{\mathbf{k}} - E_F)\frac{1}{T}\mathrm{grad}T \right]. \tag{5.81}$$

Using the expressions for electric and heat current, we obtain

$$\sigma = e^2 K_0, \qquad \lambda = v/T = eK_1/T,$$

where

$$K_n = -\frac{1}{3}\sum_{\mathbf{k}} v_{\mathbf{k}}^2 \tau_{\mathbf{k}} (\varepsilon_{\mathbf{k}} - E_F)^n \frac{\partial n_{\mathbf{k}}}{\partial \varepsilon_{\mathbf{k}}}. \tag{5.82}$$

Then we have

$$\alpha = -\frac{K_1}{eTK_0}. \tag{5.83}$$

After expanding (5.83) in T/E_F up to the second order we obtain

$$\alpha = -\frac{\pi}{3}\frac{k_B^2 T}{e}\left(\frac{\partial \ln \sigma(E)}{\partial E}\right)_{E=E_F}, \tag{5.84}$$

where $\sigma(E)$ is the conductivity as a function of the Fermi level position (see (5.31),(5.32)). Thus, the thermopower is expressed in terms of conductivity and its energy derivative.

The sign of α is determined by the sign of electric charge (or effective mass). In particular, it should be reversed when the Fermi level approaches the Brillouin zone boundaries (a becomes positive).

In the simplest approximation we may write down

$$\sigma(E) = e^2 n(E)\tau(E)/m^*, \tag{5.85}$$

where $n(E)$ is the number of electrons in the **k**-space under the surface with a given $E = E_F$, so that $dn(E)/dE$ equals to the density of states $N(E)$. Then we have

$$\alpha = -\frac{\pi^2}{3}\frac{k_B}{e}k_B T\left[\frac{N(E)}{n(E)} + \frac{1}{\tau(E)}\frac{\partial \tau(E)}{\partial E}\right]_{E=E_F}. \tag{5.86}$$

The expression (5.86) contains the concentration contribution which is determined by the number of electrons, and the relaxation contribution which depends on the function $\tau(E)$. The value of the first contribution is estimated as

$$\alpha \sim -\frac{\pi^2}{3}\frac{k_B}{ne}\frac{k_B T}{E_F} \approx -0.9\cdot 10^2 \frac{k_B T}{E_F}\frac{\mu\text{V}}{\text{K}}, \tag{5.87}$$

which agrees roughly with experimental data ($\alpha \sim 1\,\mu$V/K at $k_B T/E_F \sim 10^{-2}$). At the same time, in semiconductors α does not contain the small factor $k_B T/E_F$ and is considerably larger.

The dependences $\tau(E)$ are different for various scattering mechanisms. At high temperatures,

$$\tau(E) \sim E^{3/2}, \qquad n(E) \sim E^{3/2}, \qquad \sigma(E) \sim E^3.$$

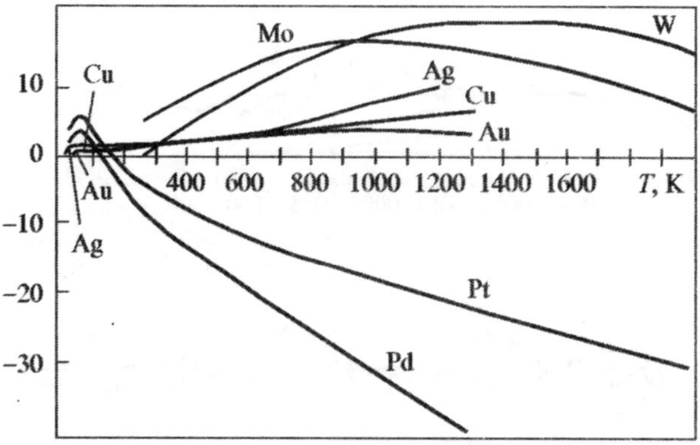

Fig. 5.14. The typical temperature dependences of thermoelectric power (μ V/K) in metals (a) noble metals and refractory *d*-metals in a wide temperature region (b) noble metals at intermediate temperatures (c) alkaline metals at low temperatures [8].

At low temperatures, where impurity scattering dominates and the electron mean free path is constant,

$$\tau(E) \sim E^{-1/2}, \qquad \sigma(E) \sim E.$$

Then (5.84) yields

$$\alpha = -\frac{\pi^2 k_B^2 T}{eE_F}\begin{cases} 1, & T > \theta_D, \\ 1/3, & T \ll \theta_D. \end{cases} \tag{5.88}$$

The relation (5.88) gives reasonable results for Na and K which are described by the free electron model, so that α and the Hall coefficient are negative. However, generally speaking, such simple dependences do not reproduce experimental data even for simple metals (see Fig.5.14). One can see that α may become positive and in most cases the dependences $\alpha(T)$ are non-monotonous. The high-temperature behaviour is usually attributed to the effect of phonon drag [8]. The corresponding contribution in the free electron model may be presented in the form

$$\alpha_{ph} = -\frac{C_{ph}}{3ne} = \frac{1}{3}C_{ph}R, \tag{5.89}$$

where C_{ph} is the lattice specific heat, R is the Hall coefficient.

The dependences $\alpha(T)$ in transition metals are still more complicated. Especially large values of α are observed in Pd and Pt. An important fact is a similarity in the $\alpha(T)$ behaviour within each column of the periodic table. The generalized dependences $\alpha(T)$ are shown in Fig. 5.15. A correlation between signs of α and the Hall coefficient is as a rule absent, which demonstrates inapplicability

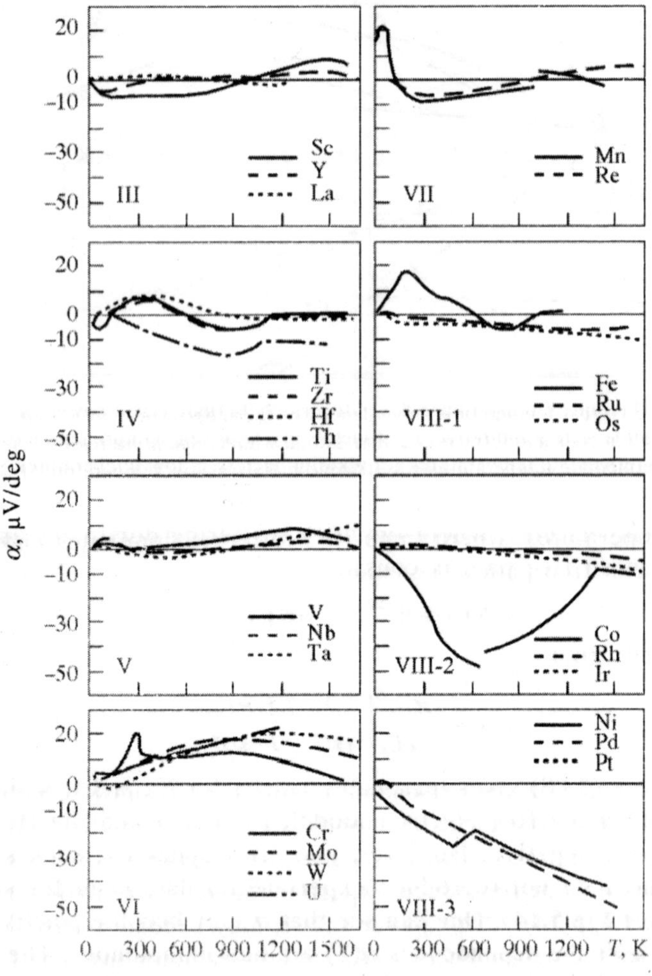

Fig. 5.15. A summary of thermoelectric power data for transition metals, U and Th [438]. Roman numerals indicate the periodic table column.

of simple theories to these quantities in transition metals. In some cases (e.g., for La near the α–β transformation and for Ti near 500K), a simultaneous sign change takes place in $R(T)$ and $\alpha(T)$. In other cases (e.g., for Sc and Hf) the sign inversion temperatures are considerably different. Sometimes the change of $\alpha(T)$ sign is not accompanied with that for $R(T)$ (however, one has to bear in mind that $R(T)$ is measured as a rule in a more narrow temperature interval). As well as other transport properties, thermoelectric power exhibits also anomalies at the magnetic ordering points (Fig.5.16).

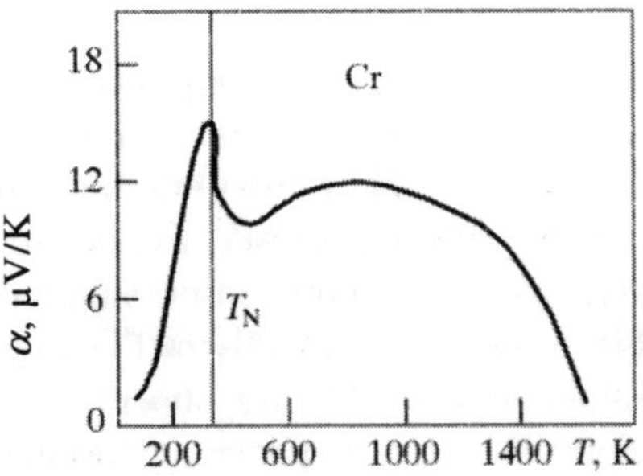

Fig. 5.16. The anomaly of thermoelectric power of chromium at the Neel point [438].

To explain complicated $\alpha(T)$ behavior one has apparently to take into account presence of several current carrier groups and scattering mechanisms. One of the possible approaches is using the Mott model of *s–d* transitions. In this model, main contribution to scattering is connected with transitions of *s*-electrons into the *d*-band. The corresponding relaxation time is proportional to the inverse density of *d*-states. Assuming that the relaxation contribution in (5.86) dominates we derive

$$\alpha(T) = -\frac{\pi^2}{3e} k_B^2 T \left(\frac{\partial \ln N_d(E)}{\partial E} \right)_{E=E_F} . \tag{5.90}$$

The expression (5.90) is used frequently to describe the data on the thermopower for metals in the end of periods (e.g., Pd and Pt) and for alloys like Cu–Ni, Pd–Ag. It reproduces satisfactorily the temperature and concentration dependences of α in the cases where a filling of transition metal *d*-band by electrons of second alloy component takes place near its top where the value of $dN_d(E)/dE$ is very large.

We may conclude that the investigation of the $\alpha(T)$ behavior permits to study the *d*-electron density of states. As discussed above, a sharp dependence $N(E)$ influences considerably a number of physical properties of transition metals. However, the anomalies in thermopower are expected to be especially sensitive to electron structure details due to presence of the factor $d\tau(E)/dE$ in (5.86). However, one has to exclude other factors which may influence $\alpha(T)$.

In the presence of density of states singularities (e.g., Van Hove ones), the standard scheme of calculating $\alpha(T)$ should be modified. This is due to that the expansion of the integrals which determine thermopower (see (5.83)) in T/E_F becomes impossible for peaks with the width of order k_BT. Direct integration was performed in paper [439] with application to palladium. A triangle model with a jump of $dN(E)/dE$ near E_F was used. Although an agreement with experimental data was obtained at high temperatures, the maximum at low temperatures remained not explained. This maximum may be due to the phonon drag effect [7] (however, the signs of R_0 and α are opposite in this temperature region). Freezing out of the *s–d* transitions and occurrence of a many-electron mechanism at low temperatures may also play a role in the $\alpha(T)$ dependence.

Non-monotonous dependences $\alpha(T)$ are observed also in a number of transition metal compounds, e.g., copper-oxide systems which provide a basis for high-T_c superconductors (see experimental data [440, 441] and theoretical considerations [405, 442]), Kondo lattices and heavy-fermion compounds (see discussion in Sect.6.4). In such systems, very narrow density of states peaks near the Fermi level have apparently a correlation (many-electron) origin and are not obtained in band calculations.

5.5 THE HALL EFFECT

Investigation of electric properties of metals in an external magnetic field lead to discovery of a number of interesting physical effects which provided a powerful tool for analyzing their electron structure. This was realized already in the beginning of modern solid state physics development. So, the physical basis of galvanomagnetic effects (the Hall effect and magnetoresistivity) is described correctly in the classical monograph [1]. However, concrete methods of the Fermi surface reconstruction with the use of such effects were developed only after the extensive theoretical work in 50–60s (see [10]).

Modern theory explains well most effects in normal metals. At the same time, the situation in transition metals is less satisfactory, which is connected with their complicated electronic structure and the presence of spontaneous magnetization (in ferromagnets).

The Hall effect in transition metals includes in fact two effects of different microscopic nature

1. the normal effect connected with the Lorentz force
2. (ii) the anomalous (spontaneous) Hall effect owing to spin-orbit coupling.

The corresponding Hall constants are defined by

$$E_y = R_0 j_x H_z + R_1 j_x M_z, \qquad R_1 = R^M \equiv 4\pi R_s. \tag{5.91}$$

Since $R_s \gg R_0$, the anomalous Hall effect dominates in ferromagnetic crystals. The Hall effect in magnetically ordered metals will be considered in Sect.5.7.1. Here we note that the spontaneous effect takes place in paramagnets too. Putting in (5.91) $M = \chi H$ we obtain

$$E_y/j_x = R_0^* H_z, \qquad R_0^* = R_0 + 4\pi\chi R_s. \tag{5.92}$$

Thus the spontaneous Hall effect may be picked out in the case of a strong dependence $\chi(T)$ and is appreciable for large χ, especially near the Curie point.

Consider the simple quasiclassical theory of the normal Hall effect. Writing down the phenomenological equations

$$j_x = \sigma_{xx}E_x + \sigma_{xy}E_y, \qquad j_y = \sigma_{yx}E_x + \sigma_{yy}E_y, \tag{5.93}$$

where off-diagonal components are determined by the magnetic field we obtain for cubic crystals ($\sigma_{xx} = \sigma_{yy} = \rho^{-1}$)

$$j_x = \left(\sigma_{xy} - \frac{\sigma_{xx}\sigma_{yy}}{\sigma_{yx}}\right)E_y, \qquad E_y \simeq -\rho^2\sigma_{yx}j_z. \tag{5.94}$$

The Hall coefficient is introduced by

$$R = \frac{E_y}{j_x H_z} = -\rho^2 \frac{\sigma_{yx}(H_z)}{H_z}. \tag{5.95}$$

In the presence of the field H_x and the Lorentz force

$$F_y = -\frac{e}{c}[\mathbf{vH}]_y = \frac{e}{c}v_x H_z.$$

the off-diagonal component σ_{yx} is connected with the diagonal one by the dimensionless parameter $\omega_c\tau$ (ω_c is the cyclotron frequency, τ is the relaxation time):

$$\sigma_{yx} = \omega_c\tau\sigma_{xx}, \tag{5.96}$$

so that the conductivity tensor in the field H_z has the form

$$\hat{\sigma} = \frac{e^2 n\tau/m^*}{1+(\omega_c\tau)^2}\begin{pmatrix} 1 & \omega_c\tau & 0 \\ \omega_c\tau & 1 & 0 \\ 0 & 0 & 1+(\omega_c\tau)^2 \end{pmatrix}. \tag{5.97}$$

It is interesting that according to (5.96) $\sigma_{yx} \ll \rho^{-1}$. Then we obtain from (5.95)

$$R = -\rho^2 \frac{e\tau\sigma_{xx}}{m^* c} = -\frac{1}{enc}. \tag{5.98}$$

Thus, in the simplest approximation the normal Hall effect coefficient is a constant which does not depend on the scattering mechanism. We shall see below that the situation changes drastically for the

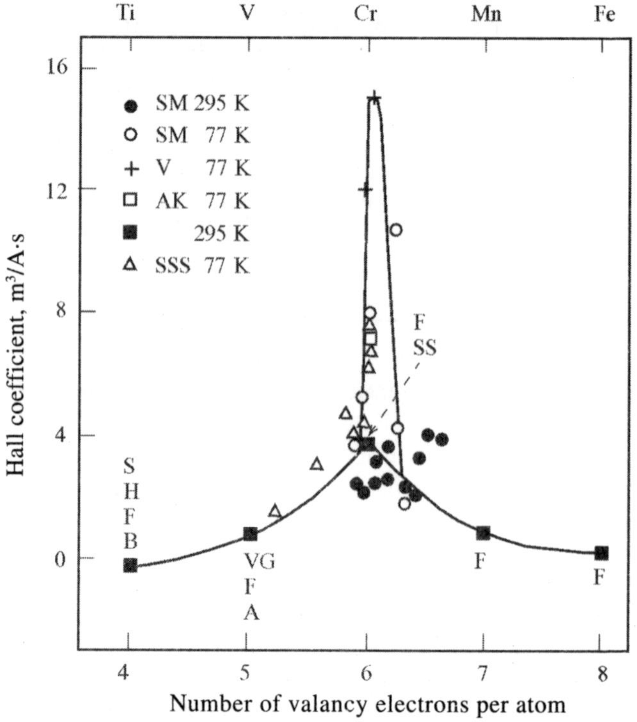

Fig. 5.17. The Hall coefficient as a function of the number of valence electrons for polycrystalline transition metals and alloys according to data of different authors [443].

spontaneous Hall effect. In particular, in the case of the phonon scattering (Appendix M.2) the expansion of σ in the scattering amplitude may start, in contrast with (5.96), from zero-order terms, so that $R_s \sim \rho^2$.

Now we consider the experimental situation. In simple (in particular, alkaline) metals the Hall effect is satisfactorily described by the formula (5.98) which may be used to determine the carrier concentration n. However, in polyvalent metals, where the Fermi surface crosses the boundaries of the first Brillouin zone and several current carrier group exist, there occur considerable deviations and the T-dependence rises [8]. Thus the Hall effect may be used in principle to determine characteristics of the electron structure and the number and mobility of current carriers in various regions of the Brillouin zone. The behaviour of the Hall coefficient in the d-series is shown in Fig.5.17.

Unlike $\Delta\rho/\rho$-effect (Sect.5.6), the Hall effect in non-ferromagnetic transition metals has essentially peculiar features. The normal Hall

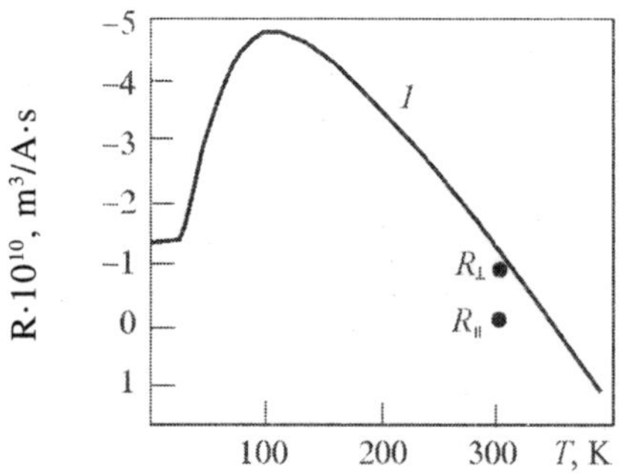

Fig. 5.18. The temperature dependence of the Hall coefficient in scandium according to data of different authors [239].

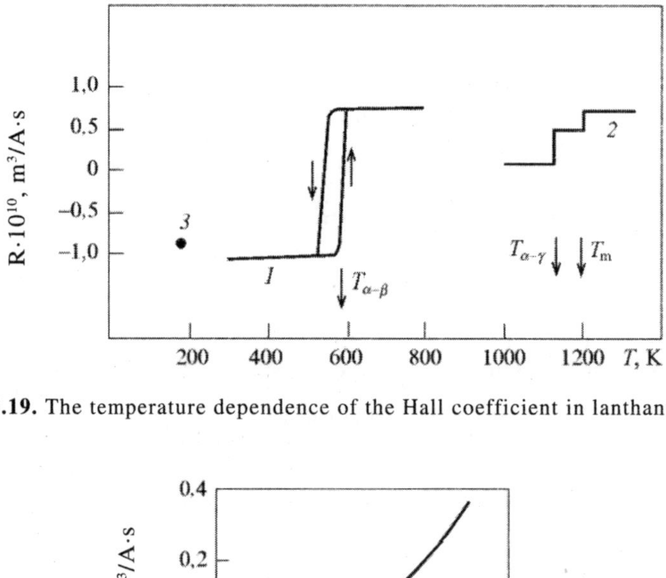

Fig. 5.19. The temperature dependence of the Hall coefficient in lanthanum [239].

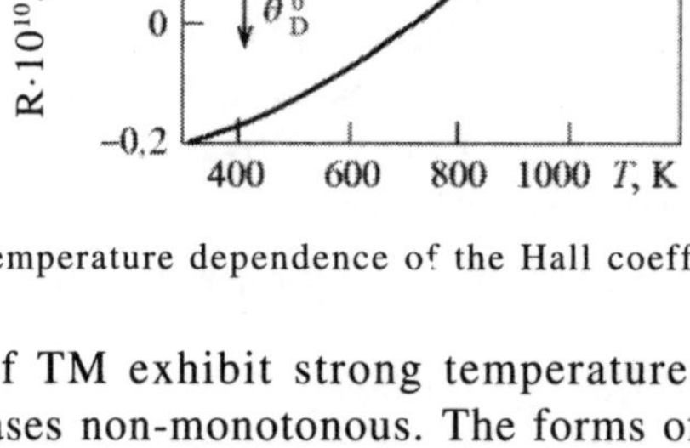

Fig. 5.20. The temperature dependence of the Hall coefficient in titanium [239].

coefficients of TM exhibit strong temperature dependences which are in some cases non-monotonous. The forms of the $R_0(T)$ behaviour are as a rule similar in the periodic table columns. These dependences are shown in Figs.5.18–5.28. A weak T-dependence is observed for

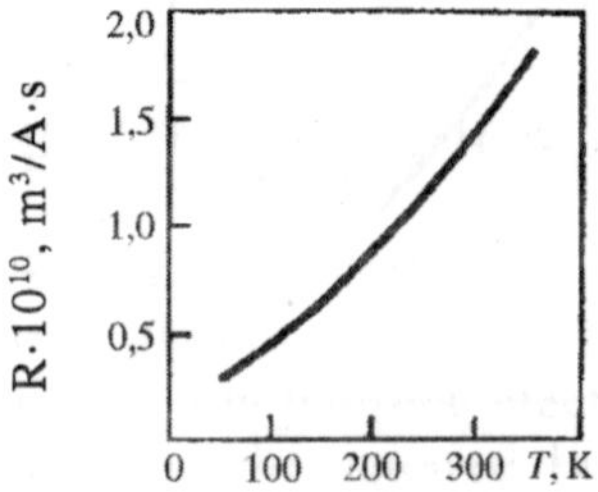

Fig. 5.21. The temperature dependence of the Hall coefficient in zirconium [239].

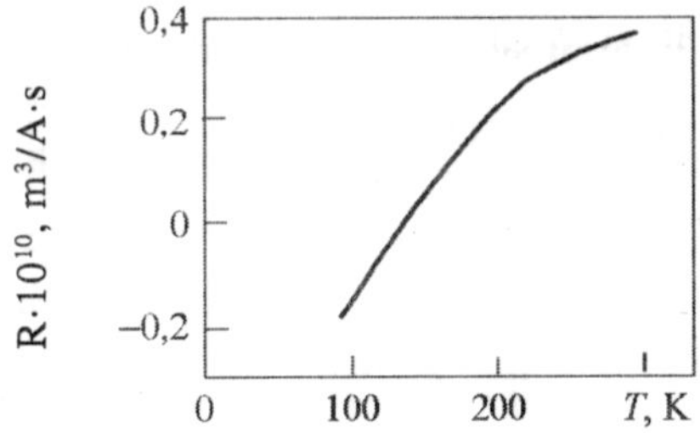

Fig. 5.22. The temperature dependence of the Hall coefficient in hafnium [239].

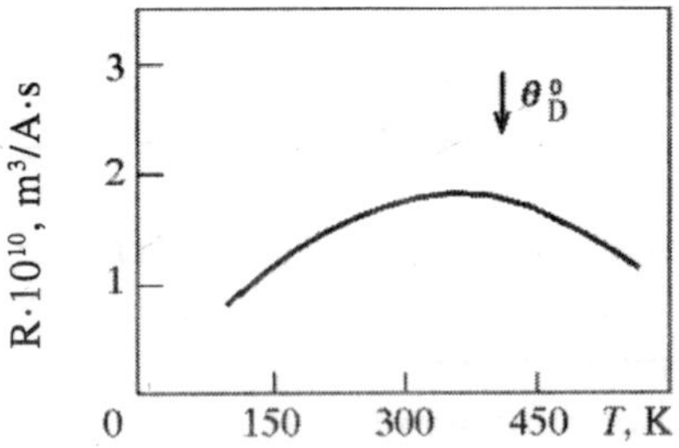

Fig. 5.23. The temperature dependence of the Hall coefficient in molybdenum [239].

Mo and W only. The local minimum of $R_0(T)$ dependence in Mn (Fig.5.29) is probably connected with antiferromagnetism of this metal.

In principle, the complicated behaviour $R_0(T)$ may be explained by the presence of several groups of current carriers. Using (5.97) we obtain in the case of two groups

$$R = \left(R^{(1)}\sigma_1^2 + R^{(2)}\sigma_2^2\right)/(\sigma_1 + \sigma_2)^2 = \rho^2\left(R^{(1)}\sigma_1^2 + R^{(2)}\sigma_2^2\right), \quad (5.99)$$

where

$$R^{(i)} = -\frac{1}{en_i c}, \qquad \sigma_i = \frac{e^2 n_i \tau_i}{m_i^*}.$$

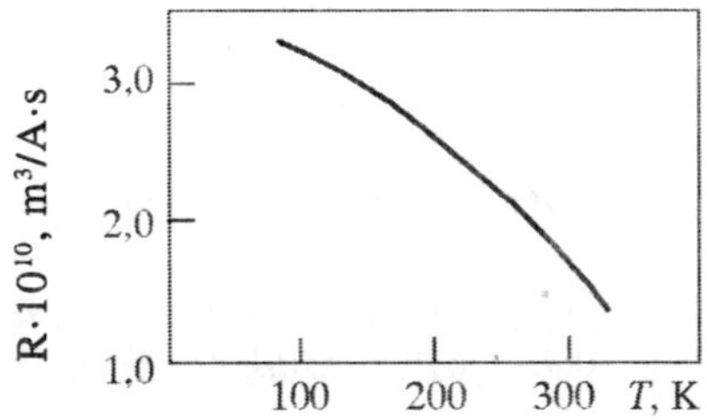

Fig. 5.24. The temperature dependence of the Hall coefficient in rhenium [239].

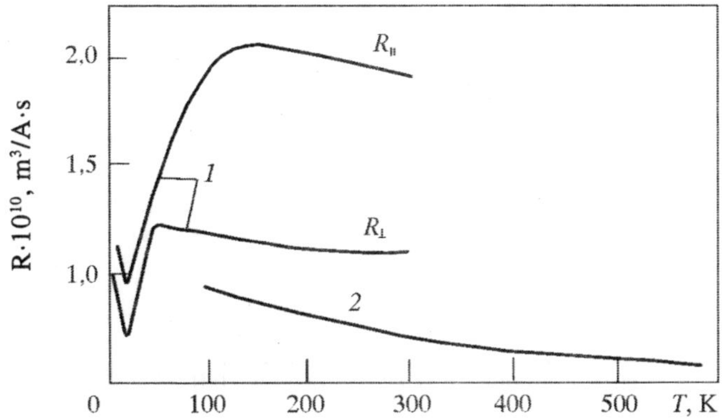

Fig. 5.25. The temperature dependence of the Hall coefficient in ruthenium according to data of different authors [239].

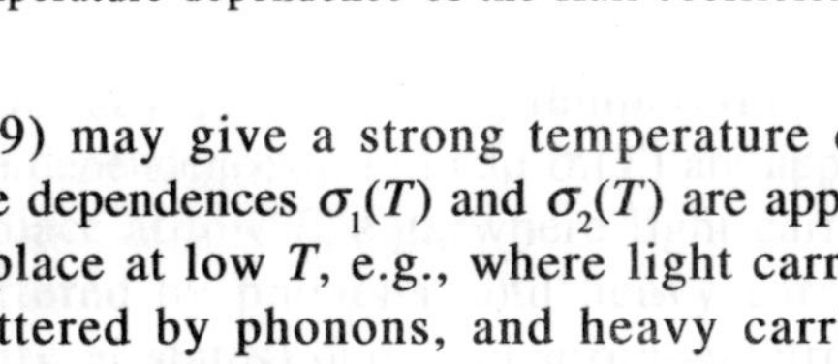

Fig. 5.26. The temperature dependence of the Hall coefficient in rhodium [239].

The result (5.99) may give a strong temperature dependence only provided that the dependences $\sigma_1(T)$ and $\sigma_2(T)$ are appreciably different. This may take place at low T, e.g., where light carriers (s-electrons) are mainly scattered by phonons, and heavy carriers (d-electrons with large density of states) due to electron–electron collisions. For example, this mechanism may be responsible for a maximum of $|R_0(T)|$ in Cu at low temperatures (Fig.5.30). Similar (but appreciably more pronounced maxima) are present in Pd and Pt. It is known that in all these three metals two groups of carriers with considerably different effective masses exist which correspond to neck and belly Fermi surfaces. Ziman [444] proposed that the increase of $|R_0(T)|$ in Cu

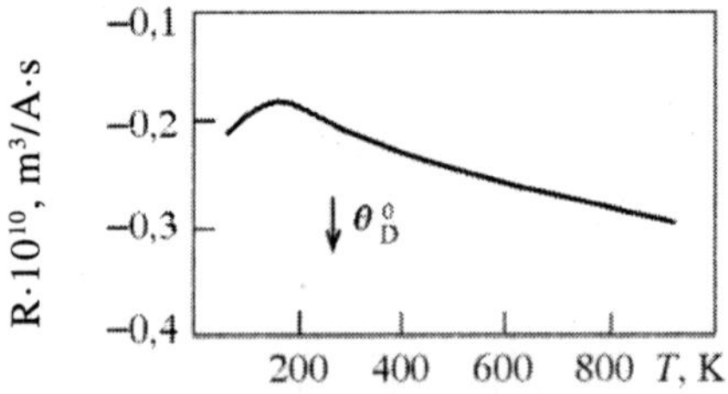

Fig. 5.27. The temperature dependence of the Hall coefficient in platinum [239].

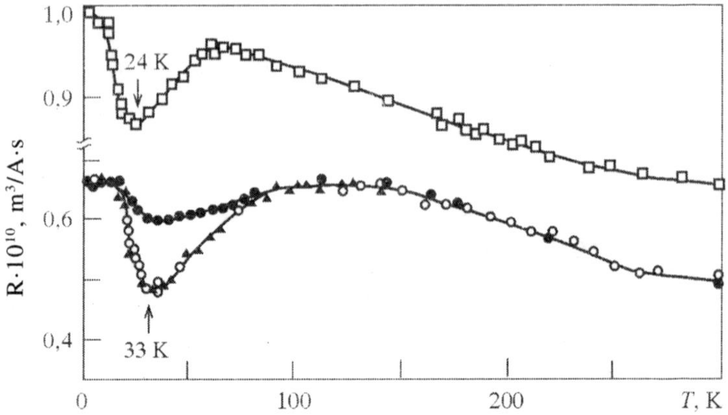

Fig. 5.28. The temperature dependences of the Hall coefficient in the field H = 13.5 kOe for Ta (open squares) and V (different measurement series, other symbols) [446].

below 100 K is connected with freezing out Umklapp processes, the lifetime of light-mass belly states becoming lengthened with respect to one on the neck region, $\tau_B/\tau_N \gg 1$ whereas $\tau_B \sim \tau_N$ at room temperatures. These arguments are confirmed by the calculations [445]. The occurrence of the $|R_0(T)|$ maximum may be connected with the transition to strong field regime $\omega_c\tau \gg 1$ and with strong anisotropy of τ. This hypothesis seems to be confirmed by investigations of single crystals (see discussion in [443]). At the same time, the explanation of the maximum in Pd and Pt at $T \sim$ 200K is more difficult.

At high $T \sim$ 100–1000K the electron–phonon mechanism dominates for all the carriers in paramagnetic metals. Therefore, the quantitative explanation of the strong $R_0(T)$ change (by several times in Sc, Ti, Zr, Hf, V, Re) is hardly possible on the basis of (5.99). The non-monotonous dependence $R_0(T)$ in V and Ta (Fig.5.28) is discussed in [446]. The authors claim that usual Umklapp processes do not explain the $R_0(T)$ minimum at T = 20–30 K. Since the temperatures

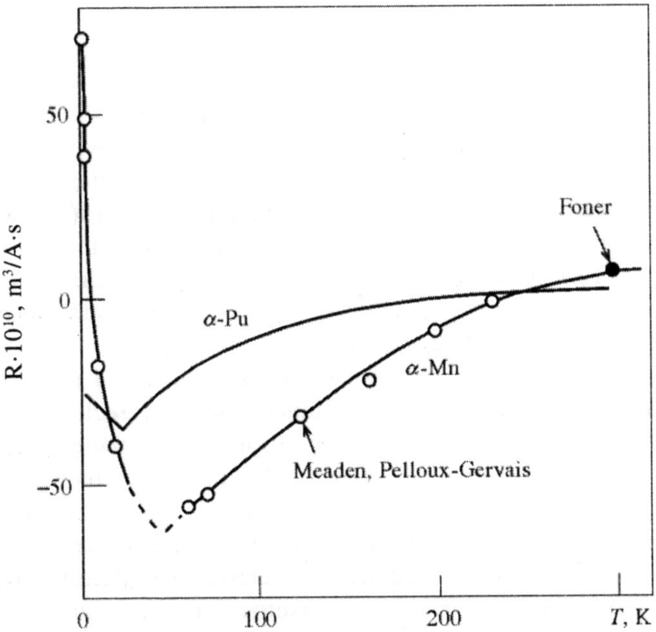

Fig. 5.29. Temperature dependence of the Hall coefficient of polycrystalline α -Mn and α-Pu [443].

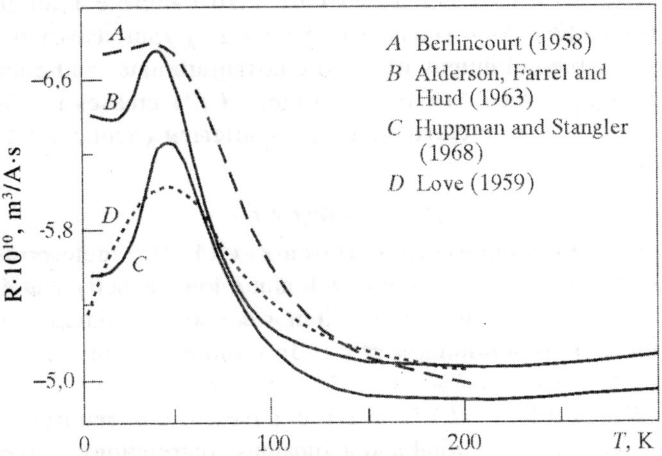

Fig. 5.30. The temperature dependence of the Hall coefficient of polycrystalline Cu [443].

of their freezing in these metals make up about 300 and 200K respectively for closed sheets, and for open sheets the Umkplapp processes do not freeze out up to $T = 0$. Only Umklapp processes between closed hole sheets $h_N(3)$ and the open hole surface in the ΓNH plane, which are localized in the region of minimum distance between the sheets, yield reasonable values of the freezing temperature T^*. Below T^* the anisotropy of scattering decreases which results in $R_0(T)$ increase. Adding impurities suppresses the anisotropy too and leads to vanishing of $R_0(T)$ minima in agreement with experimental data. This interpretation agrees with the theory [415] which considers the anisotropy of non-equilibrium distribution function.

Another mechanism of strong dependence $R(T)$ in paramagnetic metals may be the influence of the anomalous Hall effect. As follows from (5.92), such a dependence may be connected with the dependences $\chi(T)$ and $R_s(T)$. This possibility was noted in early papers, in particular, in connection with the anomalies at the Curie point. Kondorsky [447] used this idea for explaining the $R(T)$ behaviour in Zr and Re. Of course, reliable independent methods for determining R_s in non-ferromagnetic metals are absent. The expression (5.92) may give required values provided that $R_s/R_0 \sim 10^2$–10^4, i.e. R_s is nearly the same or somewhat larger in comparison with ferromagnetic metals. This assumption seems to be reasonable since the spin-orbital coupling has the same order of magnitude for all the transition metals.

The expression (5.92) may explain also concrete temperature dependences $R(T)$. As discussed in Sect.4.2, χ increases with T in the periodic table columns with even configurations d^n and decreases for odd configurations. At the same time, $R_s(T)$ increases in absolute value as $\rho^2(T)$ due to electron–phonon scattering (Sect.5.7.1). Then we obtain

$$R_0^* = R_0 + 4\pi\chi(T)bT^2, \tag{5.100}$$

The sign of the temperature correction in (5.100) is determined by the sign of R_s. In fact, the value of b is not known exactly and should be used as a fitting parameter. Then we may reproduce various experimental dependences $R(T)$. For example, one has to put $R_0 < 0$, $R_s > 0$ for Ti, $R_0 > 0$, $R_s > 0$ for Zr, $R_0 > 0$, $R_s < 0$ for Re with $R_s \sim \rho^2(T) \sim T^2$ in all the cases. Unfortunately, it is difficult to separate the normal and anomalous contributions. Apparently, the value of R_0 may be determined from low-temperature data for high-purity samples where $R_s \to 0$.

5.6 MAGNETORESISTIVITY

The resistivity change in magnetic field (magnetoresistivity, $\Delta\rho/\rho$ effect) corresponds to square terms in H, i.e. is an even effect. One distinguishes the longitudinal ($\mathbf{H} \parallel \mathbf{E}$) and transverse ($\mathbf{H} \perp \mathbf{E}$) effects. As well as the Hall effect, magnetoresistivity occurs due to distortion of electron path owing to external field or magnetization. Naively, one might assume that this distortion should diminish the velocity component $\mathbf{v}$ along the electric field and result in a decrease of current, i.e. an increase of resistivity. In fact, the situation is more complicated since one has to take into account the Hall field E_y which compensates the influence of magnetic field.

The change of resistivity in magnetic field may be described by the Frank law

$$\Delta\rho/\rho \equiv \frac{\rho(H)-\rho(0)}{\rho(0)} = \frac{BH^2}{1+CH^2}. \tag{5.101}$$

In weak fields ($CH^2 \ll 1$) we have the square-law increase, and in strong fields the resistivity is saturated,

$$\Delta\rho/\rho = B/C = \text{const.}$$

In some situations, the linear dependence $\rho(H)$ (the Kapitza law) is observed in strong fields (see [1,10]). Besides that, for Mo, Re, Pt, Fe and Pd the experimental dependences may be fitted as [448]

$$\Delta\rho/\rho \sim H^m, \quad m<2. \tag{5.102}$$

According to the Kohler rule,

$$\Delta\rho/\rho = f\left(\frac{H}{\rho}\right), \tag{5.103}$$

with f being an universal function. The Kohler plots for some simple and transition metals are shown in Fig.5.31. The Kohler rule holds in wide intervals of ρ and H almost in all cases.

Let us try to estimate the values of the coefficients B and C in (5.101) in the simple theory which considers the motion of an electron in a crystal in external electric and magnetic fields. We use the expansion in T/E_F and the relaxation time approximation. Then the longitudinal $\Delta\rho/\rho$ effect turns out to be absent to the second-order terms, and for the transverse effect one obtains

$$B = \frac{\pi^2}{3}\left(\frac{e\overline{l}}{(m^*)^2}\frac{k_B T}{\overline{v}^3}\right)^2, \qquad C = \left(\frac{e\overline{l}}{m^*\overline{v}}\right)^2, \tag{5.104}$$

where $\overline{l} = \overline{v}\tau$ is the electron mean free path, the averages are taken

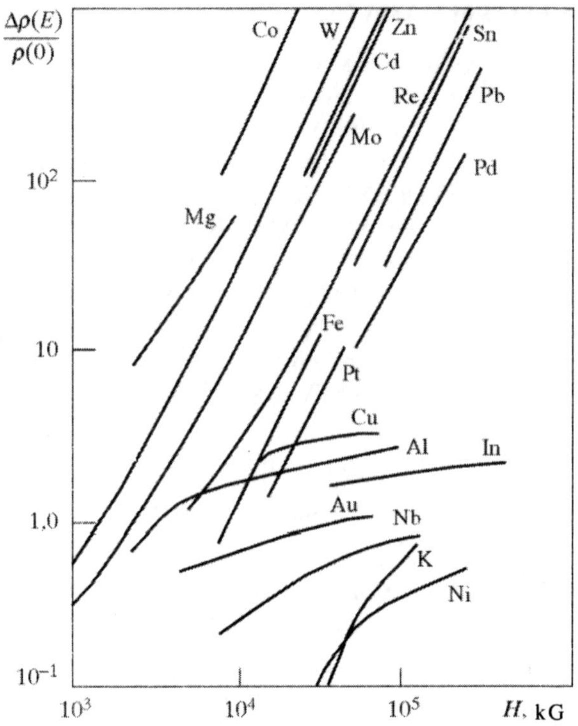

Fig. 5.31. Reduced Kohler diagram showing the transverse magnetoresistance of some metals in the high-field region [448].

over the Fermi surface. The result (5.104) may be interpreted in terms of competition between the circular motion under the influence of the Lorentz force and the straightforward motion in the electric field on the path $\bar{l}$. The former is characterized by the orbit radius

$$CH^2 = \left(\bar{l}/r_c\right)^2, \qquad BH^2 = \frac{\pi^2}{3}\left(\bar{l}/r_c\right)^2\left(\frac{k_B T}{m^*\bar{v}^2}\right)^2. \tag{5.105}$$

The occurrence of the ratio $\bar{l}/r_c \sim H/\rho$ illustrates the Kohler rule. In weak fields ($\bar{l} \ll r_c$) the increase of ρ is determined by that the spiral motion of an electron in the field H_z results in decreasing of the path in the x-direction between collisions. At large fields,

$$\Delta\rho/\rho = \frac{B}{C} = \frac{\pi^2}{3}\left(\frac{k_B T}{m^*\bar{v}^2}\right)^2. \tag{5.106}$$

Although the result (5.106) is in agreement with the law (5.101), the value of the coefficient $B \sim (T/E_F)^2$ is considerably smaller than the experimental one, and the strong temperature dependence is in fact not observed. The difficulties of the above simple theory are eliminated when one includes a dispersion of the relaxation time $\tau(\mathbf{k})$. Then the average value

$$\overline{\tau^n} = \sum_{\mathbf{k}} \frac{\partial n_{\mathbf{k}}}{\partial E_{\mathbf{k}}} v_{\mathbf{k}}^2 \tau^n(\mathbf{k}) \Big/ \sum_{\mathbf{k}} \frac{\partial n_{\mathbf{k}}}{\partial E_{\mathbf{k}}} v_{\mathbf{k}}^2 \tag{5.107}$$

differs from $\overline{\tau}^n$, and we obtain the non-zero effect at $T = 0$:

$$\Delta\rho/\rho = Q^2\left(\frac{eH}{m^* c}\right)^2, \qquad Q = \left[\overline{\tau^3}\,\overline{\tau} - \left(\overline{\tau^2}\right)^2\right] \Big/ \left(\overline{\tau^2}\right)^2. \tag{5.108}$$

Since

$$\overline{\tau^3}\,\overline{\tau} \geq \left(\overline{\tau^2}\right)^2,$$

$\Delta\rho/\rho$ is positive.

The temperature dependence of $\Delta\rho/\rho$ may be calculated provided that the factorization

$$\tau(\mathbf{k}) = \varphi(T)\chi(\mathbf{k}) \tag{5.109}$$

is possible (this may take place, e.g., at high temperatures). Then we derive

$$\Delta\rho/\rho \sim \sigma^2(T)H^2, \tag{5.110}$$

with $\sigma(T)$ being the conductivity. Thus $\Delta\rho/\rho$ is expected to increase with lowering T. The relation (5.110) is in agreement with the Kohler rule. It should be noted that the temperature dependence of (5.110) (a decrease with ρ) contradicts to expressions (5.101), (5.104) for a single group of current carriers, which yield an increase as ρ^2 with increasing T. Thus the experimental T-dependences of $\Delta\rho/\rho$-effect correspond to essentially anisotropic relaxation time or to existence of several current groups.

Consistent calculation of the parameter Q, which is defined in (5.108), is a very complicated problem. More convenient is the simple estimation from experimental data with the use of

$$Q = \frac{1}{H}(\Delta\rho/\rho)^{1/2}\frac{\rho}{R_0}, \tag{5.111}$$

where R is the Hall coefficient. For most metals, the value of Q varies from 1 to 4. An important exception are semimetals As, Sb, Bi where $Q \sim 10^2$–10^3. The giant $\Delta\rho/\rho$-effect in these substances is connected with their anomalous electronic structure [10]. Some transition metals also possess large value, e.g., for zirconium $Q = 20$, which may be connected with strong anisotropy for hcp crystals.

Higher-order approximations which take into account the **k**-dependence of electron velocity permit to obtain the longitudinal $\Delta\rho/\rho$-effect. The corresponding correction to the distribution function is given by [1]

$$f_{\mathbf{k}}^{(3)} = -\tau^3 \frac{e^3}{\hbar^2 c^2} EH^2 \frac{\partial n_{\mathbf{k}}}{\partial \varepsilon_{\mathbf{k}}} \left[v_y \left(\frac{\partial v_y}{\partial k_x} \frac{\partial v_z}{\partial k_z} - \frac{\partial v_z}{\partial k_y} \frac{\partial v_x}{\partial k_z} + \right. \right.$$
$$\left. \left. + v_y \frac{\partial^2 v_z}{\partial k_x^2} - v_x \frac{\partial^2 v_z}{\partial k_x \partial k_y} \right) + x \leftrightarrow y \right]. \qquad (5.112)$$

According to experiments by Kapitza (see [1,10]), the longitudinal effect may be comparable with the transverse one.

In 50s–60s, investigations of galvanomagnetic effects were widely applied to reconstruction of the Fermi surface shape of metals. In strong magnetic fields where

$$\omega_c \tau = eH\tau / (m^* c) \gg 1,$$

three types of the $\Delta\rho/\rho$-effect behaviour were found

1. $\Delta\rho/\rho$ is saturated at arbitrary orientation of **H** in the crystal
2. $\Delta\rho/\rho$ is not saturated at arbitrary orientation of **H**
3. $\Delta\rho/\rho$ is saturated for some orientations of **H** and continues to increase in strong fields for other orientations.

The type of behaviour depends on that whether a given metal is compensated, i.e. the number of electrons equals to that of holes, $n_e = n_h$. Evidently, the compensation situation is impossible for odd number of conduction electrons per atom. Among transition elements, the compensated metals are Ti, Cr, Mo, W, Re, Fe, Os, Ni, Pd, Pt, and Sc, V, Nb, Ta are uncompensated (for other d-metals the data in [10, 443] are absent).

As follows from (5.92), at $\omega_c^{e,h}\tau \gg 1$ the off-diagonal conductivities do not depend on τ and m^*, and the Hall fields are given by

$$E_y^{\mathrm{e,h}} = -\frac{1}{n_{\mathrm{e,h}}ec} j_x H_z, \tag{5.113}$$

so that

$$\sigma_{yx} = \sigma_{yx}^{\mathrm{e}} + \sigma_{yx}^{\mathrm{h}} = (n_{\mathrm{h}} - n_{\mathrm{e}})\frac{ec}{H}. \tag{5.114}$$

At $n_{\mathrm{e}} = n_{\mathrm{h}}$ the quantity σ_{yx} vanishes and

$$\sigma_{xx} \simeq ne^2\left(\frac{\tau_{\mathrm{e}}}{m_{\mathrm{e}}} + \frac{\tau_{\mathrm{h}}}{m_{\mathrm{h}}}\right)\frac{1}{\omega_{\mathrm{c}}^{\mathrm{e}}\omega_{\mathrm{c}}^{\mathrm{h}}\tau_{\mathrm{e}}\tau_{\mathrm{h}}}, \tag{5.115}$$

so that $\rho_{xx} \sim H^2$. Thus, in the case of a compensated metal $\Delta\rho/\rho$ does not exhibit saturation.

In a similar way one may explain the existence of peculiar field directions in a crystal where saturation is absent [10, 144]. To this end we have to consider orbits which are open in some direction (these correspond to an open orbit in the **k**-space). Then the field **H** will not influence such electron states and we obtain again $\rho_{\mathrm{xx}} \sim H^2$. The linear Kapitza law in strong fields may be connected with the anisotropy of $\Delta\rho/\rho$-effect in single crystals. Such a behaviour is obtained after averaging over orientations with account of open orbits [144].

The above classification may be changed by magnetic ordering because of lifting spin degeneracy. In particular, the compensation may become violated for partial occupation numbers with a given spin projection.

5.7 ANOMALOUS TRANSPORT EFFECTS IN FERROMAGNETIC METALS

5.7.1 The extraordinary Hall effect

According to (5.91), the Hall resistivity in ferromagnetic crystals is given by

$$\rho_H = E_y/j_x = R_0 H + R_1 M = R_0 B + 4\pi R_{\mathrm{s}} M, \tag{5.116}$$

where $B = H + 4\pi M$ is the 'true' macroscopic field in the substance,

$$R_1 = 4\pi(R_0 + R_{\mathrm{s}}), \qquad R_{\mathrm{s}} = R_1/4\pi - R_0. \tag{5.117}$$

The presence of the peculiar Hall coefficient R_{s} in ferromagnets was established in first experimental investigations of the Hall effects (see [265, 384]). Already in 1881 (the 'normal' effect was discovered in 1879) Hall found the influence of magnetization on the transverse electric field, which occurred in an external magnetic field. When

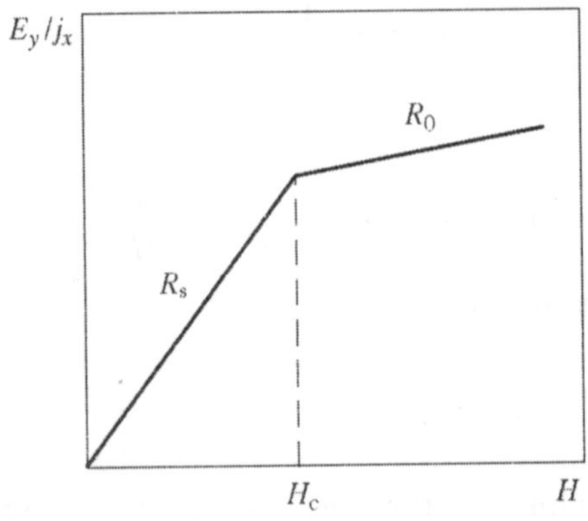

Fig. 5.32. The schematic behaviour of Hall resistivity as a function of the field H in ferromagnetic metals. The dependence $\rho_H(B)$, which demonstrates explicitly the separation of the normal and spontaneous Hall effect, has the same form.

measuring the electric field as a function of magnetic induction B in nickel, he noted that the slope of the linear increase changed after magnetic saturation.

A typical field dependence $\rho_H(H)$ in a ferromagnet is shown in Fig.5.32. At $0 < H < H_c$ (H_c is the field where magnetic domains become fully oriented) the magnetization increases rapidly from 0 to M_s, so that $\rho_H(H_c) = 4\pi (R_s + R_0) M_s$ and the slope of the curve is determined by the quantity $4\pi (R_0 + R_s)$. At $H > H_c$ the slope is determined by the usual coefficient R_s and weak high-field susceptibility ('paraprocess'). The coefficient R_s determined in such a way turns out to exceed R_0 by several orders of magnitude, which may be seen explicitly from the break of the plot $\rho_H(B)$. This demonstrates wittingly the existence of the extraordinary Hall effect.

Besides that, attempts to define the Hall coefficient in ferromagnets in the standard form $\rho_H = RB$ resulted in quite strange temperature dependences of the Hall coefficient (in particular, in a jump at the Curie point). Therefore Pugh [449] proposed to express the Hall field in terms of the magnetization. Kikoin [450] investigated the temperature dependence of the Hall field of nickel in a wide region below and above T_C. The results yielded convincing evidences for existence of the spontaneous Hall effect. Its temperature dependence was described by the expression

$$R_1(T) = a[M^2(0) - M^2(T)]. \tag{5.118}$$

It was also found that the relation with the resistivity $R(\rho) \sim \rho^n$ takes place in iron group metals for varying temperature and impurity con-

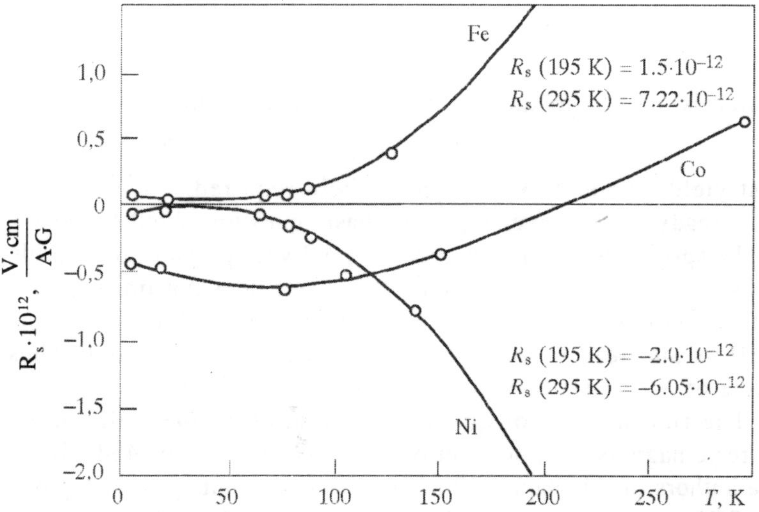

Fig. 5.33. The temperature dependences of the spontaneous Hall coefficient in iron group metals [451].

centration where n = 1.2–2 (see [394,457]). Experimental data on the temperature dependence of R_s in iron group metals are shown in Fig.5.33.

The extraordinary contribution to the Hall effect in ferromagnets occurs even in the paramagnetic phase where the magnetization is given by $M = \chi H$ and

$$R = R_0 + \chi R_1. \tag{5.119}$$

Despite the small value of $\chi \sim 10^{-3}$, the addition to R_0, may be noticeable. The extraordinary (anomalous) Hall effect is also observed in antiferromagnets and paramagnets. Its value is especially large in compounds with high values of χ, including Kondo lattices and heavy fermion systems (Chapter 6) [451, 452].

The first theoretical consideration of the anomalous Hall effect (AHE) was carried out by Rudnitskii [453]. He demonstrated that the simplest supposition about the deviation of magnetized conduction electrons in the field induced by electric current may not explain the value of the effect, leading to quantities which are smaller by three orders of magnitude than the experimental ones. Further, he put forward the idea to explain the effect by the spin-orbit interaction. The corresponding energy

$$\mathcal{E}_{so} = \langle \mathsf{H}_{so} \rangle, \qquad \mathsf{H}_{so} = \lambda \mathbf{ls} = \lambda [\mathbf{rp}] \mathbf{s} \tag{5.120}$$

is proportional to the magnetization $\langle S^z \rangle = M$ and yields the force

$$\mathbf{F}_{\mathrm{so}} = \frac{\partial}{\partial \mathbf{r}} \mathcal{E}_{\mathrm{so}} \sim [\mathbf{pM}],$$

which is similar to the Lorentz force. The estimate

$$\mathcal{E}_{\mathrm{so}} = \mu_{\mathrm{B}} H_{\mathrm{so}} \sim 10^{-13}\,\mathrm{erg}, \qquad H_{\mathrm{so}} \sim 10^{7}\ \mathrm{Oe}$$

just yields the effective magnetic field required.

Already in the paper [453] the basic question about the averaging of the spin-orbit interaction over a crystal with possible zero result was considered. Indeed, the periodic SOI itself does not provide the Hall effect, but one has to introduce inhomogeneities which scatter current carriers and lead to an asymmetry of y and $-y$ directions. Thus the value of R_{s} correlates with the electrical resistivity.

The first attempt of quantum calculation of the Hall coefficient in ferromagnets with account of SOI was carried out [454]. However, the authors of this paper did not take into account symmetry properties of SOI matrix elements. In fact, no linear in SOI corrections to the electron distribution function exist, so that the result of the calculation [454] should vanish.

The absence of the AHE within the lowest order approximation in the transport equation was demonstrated by Karplus and Luttinger [455]. To obtain a non-zero effect, they considered dynamical corrections to the electron energy in the electric field (i.e. to the field term) owing to interband matrix elements of SOI and of velocity. The corresponding off-diagonal conductivity does not depend on the scattering mechanism. Since

$$R_{\mathrm{s}} \approx \frac{1}{4\pi} R_{1} \approx -\frac{1}{4\pi M} \frac{\sigma_{yx}(M)}{\sigma_{xx}^{2}}, \tag{5.121}$$

we obtain

$$R_{\mathrm{s}} = \alpha \rho_{xx}^{2}, \tag{5.122}$$

where the constant α does not depend on temperature. The result (5.122) was widely used to fit experimental data.

Being important for the development of the theory, the calculations [455] still had incomplete and rather artificial character. Indeed, they did not take into account corrections to the collision term which have lower order in the scattering amplitude and may yield larger contributions. Thus the theory of AHE required a more consistent consideration.

A step in this direction was made in papers by Kohn and Luttinger [458] where a new method of obtaining transport equations with the use of equations of motion for the density matrix was proposed. In [459] this method was applied for calculation of AHE owing to scattering of magnetically polarized current carriers by impurity centres. Unlike [457], contributions owing to collision terms ('skew scattering') occurred in both the lowest and next orders with respect to the scattering amplitude. The final result has the form

$$R_s^i = \alpha\rho_i + \beta\rho_i^2, \tag{5.123}$$

where ρ_i is the impurity resistivity of a sample. The ratio of the first term in (5.123) to the second one equals $E_F/(3n_i\bar{\phi})$ where n_i is the concentration of impurity atoms, $\bar{\phi}$ is the impurity potential. At small n_i this ratio is large, so that almost linear dependence $R_s(\rho_i)$ should be observed.

Detailed derivation of the ρ-linear term is considered in Appendix M.1. The expression for the coefficient α in (5.123) reads

$$\alpha = -\frac{\mu_B^2 k_F^3}{18\pi\Delta^2}\frac{\rho_{\text{eff}}\bar{\varphi}}{M(0)}, \tag{5.124}$$

where the effective charge density ρ_{eff} is defined by (M.52) and includes the spin-orbital parameter, Δ is the splitting of energy levels, which is of order of bandwidth. Some peculiarities of this expression should be discussed. The sign of R_s is determined by signs of not only current carrier charge and SOI, but also the potential $\bar{\phi}$. Therefore this sign may be reversed by changing the sign of impurity charge. The influence of impurities on the Hall coefficient is stronger than that on resistivity since R_s is proportional to the third power of scattering amplitude.

The results (5.123) cannot be easily extended by replacing ρ_i to the total resistivity (such a replacement may be made only in the second term). Thus a concrete consideration of various scattering mechanisms was needed. As discussed above, the explanation of AHE is based on the spin-orbital interaction which results in occurrence of transverse contribution to current even for isotropic scattering. Since SOI is linear in magnetization, we have to calculate corrections to the distribution function which are linear in SOI. On the other hand, for non-degenerate wavefunctions (i.e. for quenched orbital momenta) the operator H_{so} has only off-diagonal matrix elements, so that linear corrections to the electron energy are absent. Therefore the transport equation in the Born approximation, which depends only on the electron energy and the squared scattering amplitude, does not yield AHE. Thus we have to consider higher-order transport equations for various scattering mechanisms (Appendix M).

The phonon mechanism was considered by Irkhin and Shavrov

[460]. The lowest-order transport equation which yields the phonon scattering contribution to AHE is the equation of the second order in the perturbation Hamiltonian H′. According to (M.72), the corresponding expression for the spontaneous Hall coefficient reads

$$R_s^{ph} = -\frac{2}{3} g \mu_B^2 \frac{e^2 n\hbar}{\Delta^2} \rho_{eff} t \left\langle \frac{1}{m^*} \right\rangle \frac{\rho^2}{M(0)}, \quad (5.125)$$

where t is the number of subbands. Introducing the effective spin orbital field

$$\mu_B H_{so} = \mathcal{E}_{so} = -\frac{\pi}{3} \mu_B^2 \rho_{eff}, \quad (5.126)$$

we derive

$$R_s^{ph} = 2t \frac{\mathcal{E}_{so}}{\Delta^2} \frac{e^2 n\hbar}{m^*} \frac{\rho_{ph}^2}{M(0)}, \quad (5.127)$$

Thus the relation $R_s^{ph} \sim \rho_{ph}^2$ takes place. The expression (5.127) may be rewritten in the form

$$R_s^{ph} = \pm 2t \frac{\hbar}{\tau_{ph}} \frac{\mathcal{E}_{so}}{\Delta^2} \frac{\rho_{ph}}{M(0)}, \quad (5.128)$$

so that the sign of R_s is determined by the sign of m^*: $R_s < 0$ for electrons and $R_s > 0$ for holes. Putting $\varepsilon_{so} \sim 10^{-14}$ erg, $\Delta \sim 10^{-1}$ erg, $\rho \sim 10^{-5}\ \Omega \cdot$ cm, $\tau \sim 10^{-13}$ s we estimate $R_s \sim 10^{-13}\ \Omega \cdot$ cm/G. Further theoretical investigations of the phonon mechanisms were performed in papers [461, 462]. The role of two-phonon scattering processes was investigated in [463].

The square dependence $R_s(\rho)$ was also found in the paper [464] for the scattering by spin inhomogeneities; the principal linear term was not found because of too simple decoupling of spin correlators. Kondo [465] considered the latter problem in the framework of the s–d exchange model with account of the proper SOI among localized electrons. He did not derive the transport equation, but used the equation by Kohn and Luttinger [458] for the impurity scattering. Thus the inelastic part of the scattering was not taken into account. The result by Kondo for d-metals reads

$$R_s^{mag} \sim \frac{\lambda}{\Delta} \frac{(m^*)^{5/2}}{E_F^{1/2} \hbar^4 e^2} G_2 \left(G_0^2 - \frac{4}{3} G_0 G_1 \right) \frac{\langle (S^z - \langle S^z \rangle)^3 \rangle}{\langle S^z \rangle}, \quad (5.129)$$

where G are the s–d exchange 'Slater' integrals of the type (K.5

Δ is the energy difference for magnetic d-electrons. Main shortcoming of the calculation [465] was using d-electron states with unquenched orbital momenta. At the same time, the unquenching of orbital momenta in d-metals is connected with the same SOI which is responsible for AHE.

This point was taken into account by Abelskii and Irkhin [466] who considered within the two-band s–d model two types of SOI: 'proper' d–d interaction λ and the interaction s-spin–d-orbit λ' (see Appendix L). The magnetized d-electrons were supposed to be itinerant and described in the tight-binding approximation with a small bandwidth, so that their orbital momenta were almost quenched. The result of this paper for high temperatures for spin S = 1/2 reads (see (M.99))

$$R_s^{\mathrm{mag}} = \frac{9\pi}{64}\left(\frac{1}{E_{\mathrm{F}}}\right)^2 \frac{\rho_\lambda}{M(0)}\left(\frac{1}{4} - \langle S^z\rangle^2\right), \tag{5.130}$$

where

$$\rho_\lambda = \lambda_{\mathrm{eff}} \frac{m^*}{e^2 n\hbar}, \qquad \lambda_{\mathrm{eff}} \approx \lambda \frac{I^{(1)}}{\Delta E}\frac{2}{3}\overline{l_z} + \lambda'.$$

Taking into account the expression for the spin-disorder resistivity (5.57) this result may be presented in the simple form

$$R_s^{\mathrm{mag}} = \pm\frac{3}{16}\frac{\lambda_{\mathrm{eff}}}{E_{\mathrm{F}}}\frac{\rho_{\mathrm{mag}}}{M(0)}, \tag{5.131}$$

where the + and − signs correspond to electron and hole conductivity. Thus we obtain a simple connection between ρ_{mag} and R_s.

The contribution of the electron-electron scattering to the extraordinary Hall effect was calculated in [467]. The result reads

$$R_s^{\mathrm{ee}} \sim \lambda\left(\frac{T}{E_{\mathrm{F}}}\right)^4. \tag{5.132}$$

Separation of various contributions to AHE may be carried out by investigating the dependence $R_s(T)$ at crossing the Curie point, since in the far paramagnetic region the magnetic scattering mechanism is saturated. Another possible way is considering the dependence of R_s on the magnetic field in the ferromagnetic phase [468]. Since the magnetic field suppresses spin disorder, $d\rho_{\mathrm{mag}}/dH < 0$ in the paraprocess region. Therefore the signs of R_s and dR_s/dH should be opposite provided that magnetic mechanism dominates. According to [468], for nickel the signs of both R_s and dR_s/dH are negative. This may be explained by the large phonon contribution. However, this does not quite agree with the dependence $R_s(T)$ above T_C (the rapid increase of R_s is not observed).

An attempt to formulate a new picture of AHE was made by Berger [469] who considered the 'side-jump' scattering of the electron wave packet under influence of various mechanisms. The universal result $R_s \sim \rho^2$ is obtained in such an approach. It was demonstrated later [470] that the side-jump scattering is just another formulation of the Luttinger's skew scattering (corrections to the field term).

On the whole, experiments at high temperatures are in agreement with the above theoretical results. However, the measurements at low T do not yield a linear dependence between ln R_s and ln ρ. Thus a special consideration of the spin-wave region, where the mean field approximation is inapplicable, is required. The calculation by Kagan and Maksimov [471] yielded the result $R_s \sim T^4$. In the paper [472] another contribution, which was due to energy dependence of the distribution function, was found. It turns out that, because of cancelation of the lowest-order terms, such contributions to AHE (unlike resistivity) are the most important ones. The final result of [472] reads (see Appendix M.3)

$$R_s = \pm\frac{3\pi}{512}\frac{\hbar I^2}{e^2 k_F M(0)}\left(k\frac{dE_{\mathbf{k}}}{dk}\right)^{-2}_{k=k_F}\left[A\left(\frac{T}{T_C}\right)^4 + B\left(\frac{T}{T_C}\right)^3\right], \quad (5.133)$$

where the constants

$$A = 1.1\lambda\bar{I}\frac{\bar{I}^{(1)}}{\Delta E}, \qquad B = 0.8\lambda' \quad (5.134)$$

are determined by 'proper' and 'improper' SOI respectively. One can see that a simple connection between the Hall coefficient and resistivity is violated at low temperatures and the dependence $R_s(T)$ may be rather complicated. This is confirmed by experimental data on Fe, Co, Ni (Fig.5.33) and diluted ferromagnetic alloys [473], which demonstrate non-monotonous temperature dependences of R_s whereas ρ behaves monotonously.

A number of calculations of the anomalous Hall effect with account of realistic band structure (in particular, the Fermi surface topology and anisotropy) were performed [474, 475]. The results show the strong influence of the Fermi surface details on the anomalous Hall effect at low temperatures.

The Hall effect in 4f-metals merits a special consideration. Due to the variety of magnetic structures and complicated phase diagrams in rare earths, the role of various factors in the Hall effect may be

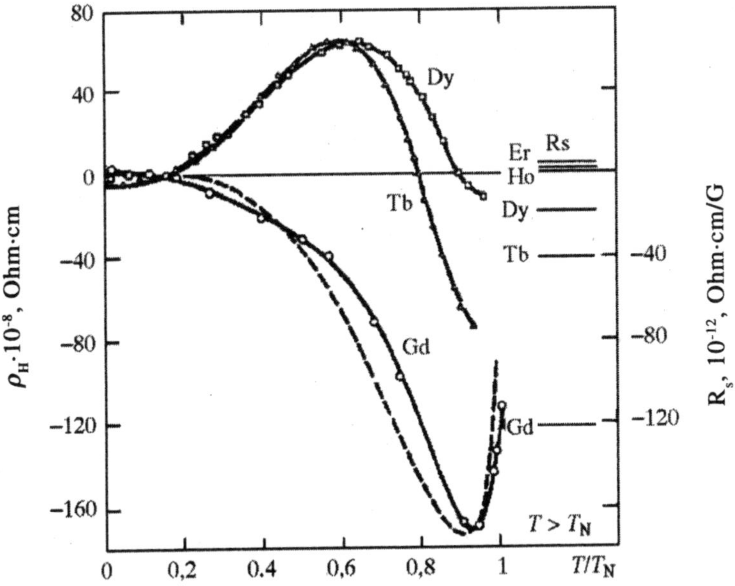

Fig. 5.34. The extraordinary Hall resistivities of gadolinium, terbium and dysprosium as a function of T/T_N. The dashed curve is the theoretical plot obtained from the results by Maranzana. Above T_N, the values of R_s for heavy rare earths are shown which are temperature independent [15].

investigated here. In particular, the influence of antiferromagnetism and strong magnetic anisotropy becomes important. On the other hand, the situation is somewhat simplified in comparison with *d* metals because of a distinct separation between current carriers (*s*, *p*, *d* electrons) and magnetic *f*-electrons, which permits to separate proper and improper spin-orbital interaction effects.

The normal and anomalous Hall effects in 4*f*-metals were extensively investigated starting from 60s. Detailed investigations were performed for heavy rare earths (a review is given the monograph [15]). The temperature dependences of the Hall coefficient for heavy rare-earth metals are shown in Fig.5.34. Main distinctive features of the anomalous Hall effect in comparison with 3*d*-metals are as follows

1. The dependence $R_s(T)$ has a more complicated non-monotonous character. However, the temperatures of a maximum (minimum) do not coincide with points of transitions between magnetic structures. A change in $R_s(T)$ sign takes place for Tb and Dy at $T \sim 0.8$–$0.9T_N$.

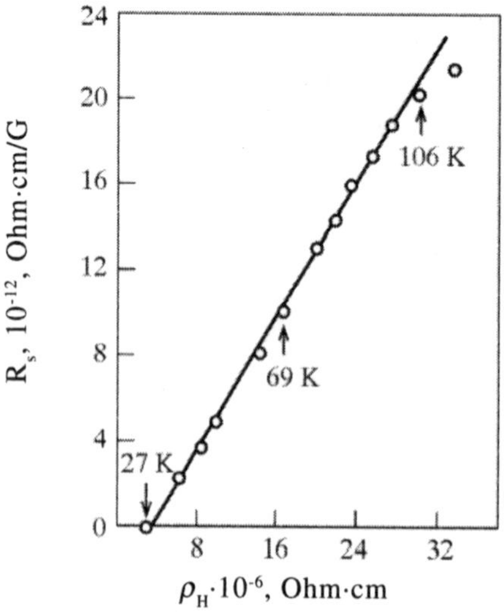

Fig. 5.35. The extraordinary Hall coefficient of dysprosium with the magnetic field along the *a*-axis as a function of spin-disorder resistivity [15].

2. A good proportionality between R_s and magnetic resistivity takes place (Fig.5.35).
3. R is highly anisotropic: the values in different crystallographic directions vary by several times.
4. The dependence $R_s(T)$ is not influenced by ferro-antiferro transitions.

It would be instructive to compare the R_s behavior in AFM region with data on other antiferromagnets, e.g., Cr and Mn [443]. However, this comparison is hampered by that the latter data are insufficient to separate the normal and anomalous contributions. Theory of AHE in rare-earths is based on the above-discussed general theory but some extra factors should be taken into account.

The terms with the vector products $[\mathbf{k}, \mathbf{k}']$, which arise from the matrix elements of the conduction electron orbital momenta $(\mathbf{l})_{\mathbf{kk}'}$, describe the anisotropic electron scattering. Such terms correspond to the coupling of conduction electron current in the external electric field to the momentum J and yield therefore the anomalous Hall effect. The Hall coefficient is proportional to $A(g-2)$, which corresponds to the interaction of electron orbital momenta with the localized orbital momenta $(2-g)\mathbf{J}$ (see (B.20)). This picture is different from that in *d*-metals where the anomalous Hall effect is due to weak spin-orbit coupling. For *f*-electrons, this coupling is strong (of order

1 eV), which enables us to consider only one J-multiplet, so that the spin-orbit coupling constant does not enter explicitly the results. It is worthwhile to mention that one of first papers, devoted to derivation of the Hamiltonian of the type (K.4), was the paper by Kondo [465] on the theory of the anomalous Hall effect (which preceeded to the famous papers on the Kondo effect).

In the antifferromagnetic region one has to determine the saturation magnetization in terms of high-field susceptibility $\chi_s = dM/dH$. Then one obtains the following equations for the Hall resistivity ρ_H

$$\rho_H(H_{dm}) = 4\pi(R_0 + R_s)M_s, \tag{5.135}$$

$$\frac{\partial \rho_H}{\partial H} = R_0 + 4\pi[R_s + R_0(1-N)]\chi_s, \tag{5.136}$$

where $H_{dm} = 4\pi NM_s$, N is the demagnetization factor. Thus one can separate the normal and anomalous Hall coefficients by measuring M_s, χ_s, N and ρ_H(H). It should be noted that in the AFM region χ is not saturated up to $H \sim 40$ kOe.

Although the spin-disorder scattering mechanism [465, 466] yields the behaviour $R_s(T) \sim \rho_{mag}(T)$ which is really observed at low temperatures, such simple theory does not explain changes in the $R_s(T)$ sign. Attempts to do this were performed by a number of authors [476–478].

The idea by Maranzana [476] was using higher-order terms in the s–f Hamiltonian derived by Kondo (see also Appendix K). Besides the main anisotropic-scattering term

$$i\lambda_1(2-g)[\mathbf{k},\mathbf{k}']J_\nu c^\dagger_{\mathbf{k}\alpha}c_{\mathbf{k}'\alpha} = i\lambda_1[\mathbf{k},\mathbf{k}']L_\nu c^\dagger_{\mathbf{k}\alpha}c_{\mathbf{k}'\alpha}, \tag{5.137}$$

the terms were considered, which had the structure

$$i\lambda_2\langle \mathbf{J}_\nu \mathbf{J}_\nu\rangle\langle[\mathbf{k},\mathbf{k}']\sigma_{\alpha\beta}\rangle c^\dagger_{\mathbf{k}\alpha}c_{\mathbf{k}'\beta}, \tag{5.138}$$

the scalar product of tensors being defined by

$$\langle AB\rangle\langle CD\rangle = \frac{1}{2}\left\{(AC)(BD) + (AD)(BC) - \frac{1}{3}(AB)(CD)\right\}. \tag{5.139}$$

As a result, higher powers of momentum operators occur, which result after averaging in new functions of magnetization. In particular, there exist the contribution to R_s which is proportional to the second derivative of the Brillouin function,

$$M_S = \langle(S^z - \langle S^z\rangle^3)\rangle = -J^3 B''_J\,(y) = -J^3 f_1\left(\frac{T}{T_N}\right), \tag{5.140}$$

where

$$y = \frac{3\langle J^z \rangle}{J+1}\frac{T_N}{T} + \frac{g_J \mu_B H}{k_B T}. \qquad (5.141)$$

Therefore this contribution is non-monotonous near T_N. Giovannini [477] introduced a still more complicated function

$$M_4 = J^4\{B_J'''(y) + 2B_J''(y) + 2[B_J'(y)]^2\} \equiv J^4 f_2(T/T_N)/B_J(y), \quad (5.142)$$

which contains the higher-order derivatives of B_J. This led to the dependence of the form

$$\rho_H = C_1 f_1(T/T_N) + C_2 f_2(T/T_N). \qquad (5.143)$$

The constants $C_1 \sim \lambda_1$ and $C_1 \sim \lambda_2$ were used as fitting parameters. For Tb and Dy $C_2/C_1 \sim 30$, and 9 respectively, whereas $C_2 = 0$ for gadolinium. The value of C_1 for Gd corresponds to the unreasonably large spin-orbital constant $\lambda \cong 0.5$ eV. The values of R_s in paramagnetic regions, which are obtained from the estimations of C_1 and C_2, turn out to be not quite satisfactory. It should be also noted that the C_2-term corresponds to a rather higher order in the s–f Hamiltonian and hardly may give a dominating contribution. Therefore, despite a qualitative explanation of the $R_s(T)$ behaviour, the mechanism discussed is debatable.

Somewhat later Fert [478] considered the influence of the side-jump scattering which corresponds to corrections to the field term in the transport equation. This mechanism yields some renormalizations of the coefficients at the Brillouin function derivatives. However, the value of AHE in gadolinium remains unexplained.

The anomalous Hall effect in heavy-fermion systems is discussed in the papers [479].

5.7.2 Magnetoresistivity in the presence of spontaneous magnetization

The $\Delta\rho/\rho$-effect in ferromagnets has important peculiarities. Its value may be of order 10^2, which is much greater than in usual metals, and have both positive and negative sign (Fig.5.36). The Kohler rule is usually not satisfied.

An important circumstance of the situation in a magnetized sample is possibility of the non-zero effect in the absence of the external magnetic field. The spontaneous effects are masked in multidomain samples where the average magnetization is zero. The single magnetic domain forms in the fields above the field of technical saturation H_c. In the low-field region (below H_c) the effect is due to a change in relative volumes of domains with $\mathbf{M} \parallel \mathbf{j}$ and $\mathbf{M} \perp \mathbf{j}$:

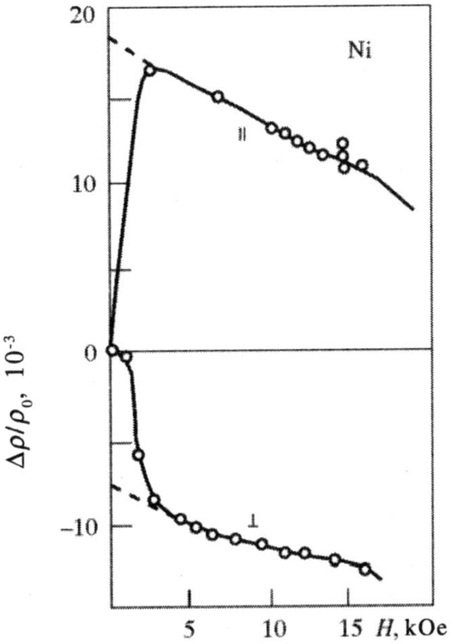

Fig. 5.36. The field dependence of $\Delta\rho/\rho$ in ferromagnetic nickel for parallel and perpendicular orientations of current and magnetic field [265].

$$\rho(H)=\frac{1}{V}(V_{\parallel}\rho_{\parallel}+V_{\perp}\rho_{\perp}). \tag{5.144}$$

For $H = 0$ we have

$$\rho=\frac{1}{3}\rho_{\parallel}+\frac{2}{3}\rho_{\perp}, \tag{5.145}$$

so that

$$\Delta\rho_{\parallel}(H_c)/\Delta\rho_{\perp}(H_c)=-2 \tag{5.146}$$

(Akulov's rule for even transport effects [265]).

Above the technical saturation $H > H_c$, the field dependence of resistivity is considerably weaker and determined by the paraprocess (the field dependence of magnetization $M = M_s + \chi H$). The dependence $\rho(H)$ is owing to suppression of spin disorder, which results in a decrease of exchange scattering of current carriers. This effect should take place also in the paramagnetic region $T > T_C$, resulting in a negative contribution to $\Delta\rho/\rho$. Experimental data are often described by the equation

$$\Delta\rho/\rho = a(M_s^2 - M^2(H)), \tag{5.147}$$

so that

$$\Delta\rho/\rho = -a_1 H - a_2 H^2. \tag{5.148}$$

Thus we obtain the linear $\Delta\rho/\rho$-effect. It should be noted that similar field dependences of $\Delta\rho/\rho$ may take place also in antiferromagnetic metals. For example, the behaviour $\Delta\rho/\rho \sim H^{3/2}$ was found in Fe_3Pt [480].

As it is clear from the above consideration, the $\Delta\rho/\rho$-effect in a single domain is determined by the difference of $\rho_\perp$ and $\rho_\parallel$, i.e. by dependence of resistivity on the angle between vectors **j** and **M**. This dependence turns out by one-two order of magnitude stronger than in the usual $\Delta\rho/\rho$-effect. Therefore the Lorentz force does not explain the effect quantitatively. The most natural relevant microscopical mechanism is, as well as for the extraordinary Hall effect, the spin- orbital interaction. The extraordinary $\Delta\rho/\rho$-effect, which is quadratic in M, occurs in the second order of perturbation theory in this interaction.

Unlike the Hall effect, the microscopic theory of the $\Delta\rho/\rho$-effect in ferromagnets is not developed in detail, and the whole physical picture is still absent. Some calculations with account of different scattering mechanisms and the spin-orbital interaction were performed starting from 50s [265]. Smit [481] and Marsocci [482] investigated the Mott s–d transition mechanism. Kondo [465] treated the scattering by magnetic inhomogeneities (as well as for the Hall effect, the picture of unquenched orbital momenta was used).

Vu Dinh Ky [483] considered the transport equation for the impurity scattering. In the second order in SOI, corrections, which are proportional to $[\mathbf{kM}]^2$, occur both in collision and scattering terms. They yield the resisitivity anisotropy required. The final result is rather cumbersome and may be represented in the form

$$\Delta\rho/\rho \sim \lambda^2 M^2, \tag{5.149}$$

so that $\Delta\rho/\rho$ decreases with temperature and vanishes above the Curie point. A numerical estimation was made in [483] by using the comparison with the anomalous Hall effect. Since the Hall resistivity $\rho_H = 4\pi R_s M$ occurs in the first order in λ, but contains an extra power of the impurity potential $\bar{\phi}$, we have

$$\Delta\rho/\rho_H \sim H_{so}/\bar{\phi}. \tag{5.150}$$

Putting for nickel $H_{so} \sim 10^{-13}$ erg and $\bar{\phi} \sim 10^{-14}$ erg one obtains $\Delta\rho/\rho \sim 0.1$ in a rough agreement with the experimental data

($\rho_H/\rho \sim 0.5\%$, $\Delta\rho/\rho \sim 3\%$).

The phenomenological consideration of magnetoresistivity in ferromagnetic metals may be performed within the two-current model with strongly different currents $j_\downarrow$ and $j_\downarrow$ [436].

5.7.3 Magnetooptical effects

Magnetooptical (MO) effects in ferromagnetic transition metals are closely related to galvanomagnetic effects. Experimentally, the Faraday and Kerr polarization plane rotation angles in ferromagnets are by several orders larger than in paramagnetic metals. They are proportional to magnetization rather than to magnetic field and strongly decrease above the Curie point.

Microscopic mechanisms of the large MO effects are connected with spin-orbital interaction. In particular, the Faraday effect is analogous to the high-frequency extraordinary Hall effect. Although the spontaneous Hall effect is determined in the static limit by the magnetization, separation of magnetic and electric characterisitics at high frequencies becomes impossible, so that the similarity with magnetooptical effects occurs. Proportionality of the Hall field and the Faraday rotation was established already in 1893 by Kundt [484].

From the phenomenological point of view, polarization plane rotation is due to gyrotropy, i.e. presence of antisymmetric contribution, in dielectric and magnetic permeability tensors $\varepsilon_{\alpha\beta}(\omega)$ and $\mu_{\alpha\beta}(\omega)$. In the optical region the magnetic gyrotropy is connected mainly with anisotropy of gyroelectric properties, the q-dependence of $\varepsilon(\mathbf{q}, \omega)$ playing an important role.

First physical explanation of MO effects in ferromagnets was given by Hulme [485] and Kittel [486], and the theory of the frequency dependence was developed by Argyres [487] and Cooper [488]. An account of electron scattering permitted to extend the theory to the low-frequency region [489,490] and the case of low temperatures [491,381].

Consider a simple theory of the MO Kerr effect. At reflection from the magnetic medium with the complex refraction factor $\tilde{n} = n + ik$ and off-diagonal conductivity σ_{xy}, the light with the frequency ω changes its polarization by the Kerr angle

$$\theta_K = \frac{4\pi}{\omega}(A\,\mathrm{Im}\,\sigma_{xy} + B\,\mathrm{Re}\,\sigma_{xy})/(A^2 + B^2), \tag{5.151}$$

where

$$A = n^3 - 3nk^2 - n, \qquad B = -k^3 + 3n^2k - k.$$

At small damping $k \ll n$ the value of θ_K is determined mainly by Im σ_{xy}. In the simplest case of a cubic structure with the magnetization vector which is parallel to the (001) plane, the Argyres formula takes the form [316]

$$\mathrm{Im}\,\sigma_{xy} = \frac{\pi}{\omega} \sum_{\mathbf{k},m\neq m'} [F^{xy}_{m'm\uparrow}(\mathbf{k}) n_{\mathbf{k}m'\uparrow}(1-n_{\mathbf{k}m\uparrow})\delta(\omega-\omega_{mm'\uparrow}(\mathbf{k})) -$$

$$- F^{xy}_{m'm\downarrow}(\mathbf{k}) n_{\mathbf{k}m'\downarrow}(1-n_{\mathbf{k}m\downarrow})\delta(\omega-\omega_{mm'\downarrow}(\mathbf{k}))], \qquad (5.152)$$

where m is the band index,

$$\omega_{mm'\sigma}(\mathbf{k}) = \varepsilon_{\mathbf{k}m\sigma} - \varepsilon_{\mathbf{k}m'\sigma}$$

is the interband transition frequency, $n_{\mathbf{k}m\sigma}$ is the Fermi distribution function,

$$F^{xy}_{m'm\sigma}(\mathbf{k}) = 2i \sum_{m''} \left[\frac{(l^z_{m''m})^*}{\omega_{m'm''\sigma}} p^x_{m''m} p^y_{mm'} + \frac{l^z_{m''m}}{\omega_{m'm''\sigma}} p^x_{m''m'} p^y_{m'm} \right], \qquad (5.153)$$

$$p^\alpha_{m'm} = \langle \mathbf{k}m'\sigma | -i\frac{\partial}{\partial x_\alpha} | \mathbf{k}m\sigma \rangle,$$

$$l^z_{m'm} = \langle \mathbf{k}m'\sigma | \xi l_z | \mathbf{k}m\sigma \rangle,$$

l_z is the orbital momentum operator z-projection,

$$\xi = \frac{2}{rc^2}\frac{\partial V_{\mathrm{eff}}}{\partial r},$$

with V_{eff} being the effective potential for conduction electrons.

The formula (5.152) demonstrates a strong dependence of the Kerr angle on electronic structure and magnetic ordering. Thus MO effects are promising from the point of view of comparison with band structure calculations. Comparison of magnetooptical properties with results of band calculations is performed, e.g., for nickel [381], Fe–Co alloys [492] and gadolinium [493].

An almost total compensation of the first and second term in the square brackets of (5.152) takes place provided that the spectrum e depends weakly on σ. At the same time, the effect is large in strongly ferromagnetic case. In particular, for half-metallic ferromagnets (Sect.4.4) at $\omega < \delta_\sigma$ (δ_σ is the gap for the spin projection σ) the corresponding term in (5.152) vanishes, so that one may expect large values of Kerr rotation. Indeed, for the system $NiMnSn_{1-x}Sb_x$ the intensity of peaks in the frequency dependence of θ_K decreases sharply with increasing x, i.e. as the Fermi level goes out of the gap [494]. According to (5.153), the angle θ_K is proportional to the spin-orbit coupling, i.e. increases for heavy elements. Therefore one may expect that the HMF which contain platinum should have larger values of θ_K. Indeed, giant values $\theta_K \cong 0.15°$ (for the red light), which exceed considerably the values for NiMnSb, were observed the compound PtMnSb [307, 308] (the results of the calculation are given in [316]). Note, however, that according to [495] main difference between electronic structures of HMF's PtMnSb and NiMnSb, which results in smaller value of θ_K in the latter compounds, is connected not so much with the spin-orbit matrix element values, as with a shift of some energy levels owing to 'scalar' relativistic effects (velocity dependence of mass and the Darvin correction). In this sense, the simplest assumption about the direct connection between the spin-orbit coupling strength and the Kerr rotation is not quite adequate.

Record values of θ_K might be observed in the ferromagnetic phase of the compound UNiSn, which should also have a half-metallic structure [315, 316]. However, experimentally this turns out to be antiferromagnetic [496, 497] (see discussion in Sect.6.6). Nevertheless, investigation of isostructural compounds containing actinides (e.g., UCoSn, UPdSn) is of interest from this point of view.

The polar, meridional and equatorial Faraday and Kerr effects, as well as the gyrotropic effect (change of reflected light intensity with magnetization) in ferromagnetic metals were studied systematically in [498–500]. In the case where the magnetization is perpendicular to the plane of the ingoing light wave it is posssible to determine simultaneously real and imaginary part of the tensors ε and μ [500]. Both intraband (indirect) and interband transitions turn out to play a role. The latter are important near resonance absorption frequencies, so that the frequency dependence of magnetooptical effects is a $1/\omega$-hyperbola with peaks owing to interband transitions. A simple expression for the intraband contribution to off-diagonal magnetic permeability may be presented in the form [499]

$$\mu_{xy}(\omega) = -\frac{\hbar}{2m^* c^2} i\bar{\kappa} \frac{\omega_p^2}{\omega}, \tag{5.154}$$

where ω_p is the plasma frequency, $\bar{\kappa}$ is an averaged (over the Fermi surface) dimensionless parameter which determines the correction to electron quasimomentum owing to spin-orbital interaction. For $\omega \sim 10^{14} s^{-1}$ one has $\mu_{xy} \sim 10^{-6}$–10^{-4} in agreement with experimental data.

It is instructive to establish a correlation between temperature dependences of magnetooptical and galvanomagnetic effects and compare quantitatively the corresponding microscopic SOI parameters. In the paper [489], the relation between the extraordinary Hall coefficient

R_s and the MO parameters defined by

$$\mathbf{j} = \sigma_1(\omega)\mathbf{E} + \sigma_2(\omega)\left[\frac{\mathbf{M}}{\tilde{M}}\mathbf{E}\right],$$

$$\sigma_2(\omega)/\sigma_1(\omega) \equiv iq, \tag{5.155}$$

was obtained in the form

$$4\pi R_s M = -\frac{\sigma_2(0)}{\sigma_1^2(\omega)}. \tag{5.156}$$

Using the expression for the one-band conductivity

$$\sigma_1(\omega) = \sigma_1(0)\frac{\gamma}{\gamma + i\omega}, \tag{5.157}$$

where $\gamma = 1/\tau$ is the relaxation rate we obtain

$$\mathrm{Re}\,q = -4\pi R_s M\sigma_1(0)\frac{\omega}{\gamma}, \qquad \mathrm{Im}\,q = 4\pi R_s M\sigma_1(0), \tag{5.158}$$

so that the signs of Re q and Im q are opposite and determined by the sign of R_s, which agrees with the data [501, 451]. At the same time, in the case of the normal MO effect the signs of Re q and Im q coincide. The formulas (5.159) provide the correct order of magnitude for Im q. Since

$$R_s \sim [\sigma_1(0)]^{-2}, \qquad \gamma \sim [\sigma_1(0)]^{-1}, \tag{5.159}$$

we obtain from (5.158) the temperature dependences

$$\mathrm{Re}\,q \sim M(T), \qquad \mathrm{Im}\,q \sim \frac{M(T)}{\sigma(T)}. \tag{5.160}$$

Their verification is hampered by that the experimental T-dependences [501] correspond to the resonance region. Therefore investigation of the long-wave region would be of interest.

MO effects in the X-ray region seem also to be promising to investigate the band structure. In particular, the magnetic X-ray dichroism (MXD) effect is discussed in Sect.2.5.

5.7.4 Thermomagnetic effects

Besides above-discussed electric, thermoelectric and galvanomagnetic effects, there exist a number of effects owing to combined action of the fields **E**, **H**, and grad T [7, 8, 265]. Although not investigated now in detail, the thermomagnetic effects (TME) may be of interest since they provide an additional information about microscopic transport mechanisms in solids.

Difficulties in studying TME increase for transition metals, especially in the case of magnetic ordering. First paper in this direction have demonstrated that TME are described by the same concepts and are determined by the same microscopic parameters as galvanomagnetic effects. In a number of cases one can perform a separation of 'normal' and 'spontaneous' TME. Therefore a correlation between TME and galvanomagnetic effects should exist which may be used to verify values of the microscopic parameters and separate different scattering mechanisms. Some attempts of this kind were made (see [475]).

At present, most investigated TME are the Nernst–Ettingshausen and Righi–Leduc effects (see also Sect.5.1). The transverse Nernst-Ettingshausen effect is the occurrence of the electric power E_y in the presence of the magnetic field H_z and temperature gradient in the x-direction. Similar to the Hall effect, we have

$$E_y^{NE} = (Q_0 B_z + 4\pi Q_s M_z)\mathrm{grad}_x T, \tag{5.161}$$

where Q_0 and Q_s are the normal and spontaneous Nernst–Ettingshausen coefficients. The general expression (5.1) under the conditions

$$\mathbf{j} = 0, \qquad \frac{\partial T}{\partial y} = \frac{\partial T}{\partial z} = 0, \tag{5.162}$$

yields

$$E_y \simeq \frac{\lambda_{xx}\sigma_{yx} - \lambda_{yx}\sigma_{xx}}{\sigma_{xx}^2}\frac{\partial T}{\partial x}. \tag{5.163}$$

The experimental dependence $Q_s(T)$ turn out to be stronger than $Q_0(T)$. This dependence is described by the empirical equation [502]

$$Q_s(T) = -T(\alpha + \beta\rho), \tag{5.164}$$

where α is determined by impurities. A derivation of the formula (5.164) with the use of the density-matrix appoach was performed by Kondorsky [503] by analogy with the Hall effect. The case of alloys was considered in papers [504]. The result (5.164) was confirmed also within the 'side-jump' mechanism approach by Berger [204].

The spontaneous Righi–Leduc effect (occurrence of grady^T in the presence of grad$_x T$ and M_z) and the Ettingshausen effect (occurence of grad$_y T$ in the presence of electric current j_x and magnetization M_z) were also investigated [475]. The normal Righi–Leduc coefficient is known to be expressed in terms of the Hall coefficient and conductivity,

$$A_0 = \sigma R_0 = \frac{e\tau}{m^* c}. \tag{5.165}$$

Search of a similar relation for the spontaneous coefficient is of interest.

Chapter 6

THE KONDO EFFECT AND PROPERTIES OF ANOMALOUS *d*- AND *f*-COMPOUNDS

In this Chapter we consider physics of some types of 4*f*- and 5*f*-compounds which exhibit anomalous electronic properties. Here belong so-called heavy-fermion, Kondo lattice and intermediate valence systems; in some aspects close is physics of some *d*-systems, in particular of copper-oxide high-T_c superconductors where strong electron correlation effects in CuO_2 planes take place. We shall demonstrate applications of many-electron models to the description of unusual physical phenomena in these substances. Of course, we do not claim to cover completely this topic which is very wide and rapidly develops, but consider some selected questions which are determined by the authors' scientific interests.

Most exotic properties are characteristic for heavy-fermion compounds. They possess giant values of effective electron mass, which are manifested most brightly in the huge linear specific heat. In a somewhat arbitrary definition of heavy-fermion systems, the boundary value of $\gamma = 400$ mJ/mol · K^2 was established. Besides that, large paramagnetic susceptibility at low temperatures and large coefficient at the T^2-term in resistivity are observed.

An especial interest in the heavy-fermion compounds was stimulated by the discovery of unconventional superconductivity in $CeCu_2Si_2$, UBe_{13}, UPt_3. The superconducting state is characterized by an isotropic (non-zero angular momentum) pairing and is possibly not caused by electron-phonon interaction [505, 506]; often superconductivity coexists with antiferromagnetic ordering.

The properties of the 'classical' heavy-fermion systems $CeAl_3$, $CeCu_6$, $CeCu_2Si_2$, UBe_{13}, UPt_3, U_2Zn_{17}, UCd_{11}, $NpBe_{13}$ are considered in detail in the review [507]. For recent years, a number of ternary Ce-based compounds with huge (of order 1 J/mol · K^2 and larger)

Table 6.1. Electronic specific heat and magnetic characteristics (paramagnetic susceptibility, Neel or Curie (in brackets) point, paramagnetic Curie temperature and the ground state moment) of some anomalous rare-earth and actinide compounds (heavy-fermion, Kondo lattice and intermediate valence systems).

	γ	$\chi(0)$	$T_N(T_C)$	M_s	$-\theta$	Ref.
	mJ/mol·K^2	10^{-3} emu/mol	K	μ_B	K	
$CeAl_3$	1620	36	1.2?	0.3	46	507–511
$CeAl_2$	135	44	3.8	0.89[a]	32	512
$CeCu_2Si_2$	1100	7	0.8		164	507, 513, 514
$CeCu_6$	1450	27			45	507, 515
$CeCu_2$	90		3.5			516
$CeCu_5$	100		3.9			517
$CeCu_2Ge_2$	100		4.1	1	18	518
$CeIn_3$	260	11	10	0.6	50	512
$CeInCu_2$	1200	20	1.6	0.1	20	519
$CePb_3$	1000[b]		1.1	0.1		520
$CeCu_4Al$	2000(<1K)				25	520
$CeCu_4Ga$	1900	20			26	520
$CeCu_3Al_2$	540(1.6K)	29			40	520
$CeCu_3Ga_2$	730(1.5K)					520
$CeAl_2Ga_2$	80		8.5	1.2	18	521
$CeInPt_4$	2500				225	520
CePtSi	800				47	520
$CePtSi_2$	1700 (1.2K)				17	520
$CePt_2Sn_2$	3500[c]		0.88		25	520
CePtIn	700				73	520
CePdIn	330		1.8			520
CePdSn			7.5		68	522
CePdSb			(17)	1.2	-10	520
$CeRu_2Si_2$	350	17			54	520
$CeZn_{11}$	2500		2			523
$CeRh_3B_2$	16		(115)	0.37	370	284
UPt_3	450	7	5.0	0.02	200	524
URu_2Si_2	180		17.5	0.03	65	525, 526
UBe_{13}	1100	15			53–70	507, 527, 528
$UZn_{8.5}$	535	12.5	9.7	0.8	105	507, 526, 529
UCd_{11}	800	45	5.0		23	507, 526
$UAgCu_4$	310		18			520, 530
UNiAl	164		19	0.8		531
UNi_2Al_3			5.2	0.24		532
UPd_2Al_3	150		14	0.85	47	533
UPdIn	280		20[d]			534, 535
U_2PtSi_3	400		(8)			536
YbCuAl	260	25.5			34	512
YbAs[f]	270		0.6	0.82		537, 538
YbP			0.4	0.79		537
YbSb			0.3	0.63		537
YbSi			1.5	0.2		539

Table 6.1. *(Cont.)*

	γ	$\chi(0)$	$T_N)$ (T_C)	M_s	$-\theta$	Ref.
$YbPdCu_4$	200		0.8			520
YbNiSn	300		(5.5)	0.4[e]	65	540
YbNiSb[f]	175		0.8		13	541
Sm_4As_3			(160)	2.5?		542
Sm_4Sb_3			(16)	1.5?		542
TmS			8.9	4.0		512
TmSe			3	1.7		512
$PrCu_2Si_2$	225		21	2.5		543
NpBe	900	56	3.4		42	507
$NPAl_2$	193		(57)			544

[a] Maximum value in modulated-moment structure.
[b] $\gamma \simeq$ 200 above the Neel temperature.
[c] Above the Neel temperature.
[d] Small canted ferromagnetic moment 0.3 μ_B below 7 K
[e] Canted ferromagnetic moment.
[f] Low carrier concentration.

values of γ, e.g., 2.5 J/mol · K^2 for $CeInPt_4$, 1.2 J/mol · K^2 for $CeInCu_2$, were synthesized. Besides that many Ce, Yb and U-based systems possess 'moderate' value of γ (of order 100 mJ/mol·K^2). Data on electronic specific heat and magnetic properties of some anomalous rare earth and actinide compounds, and corresponding bibliography are given in Table 6.1 (see also the reviews [512, 520, 545–547]).

As well for transition metals, the ratio of the coefficient at the T^2-term in resistivity to γ^2 is universal, but has a magnitude about 25 times larger: $A/\gamma^2 \sim 10^{-5} \mu\Omega \cdot$ cm (mol · K/mJ)^2. This correlation is seen in Fig.6.1 [548]. For comparison, the data on d-systems with large γ (A15 structure compounds which exhibit superconductivity with moderately high T_c) are also shown.

Modern de Haas–van Alphen investigations yielded the possibility to observe directly some bands with large effective masses [288, 549, 550]. Thus the substances under consideration provide an extremely interesting example of strong renormalization of electron characteristics owing to interelectron correlations. Standard band structure calculations of heavy-fermion systems usually greatly underestimate the values of $N(E_F)$. A satisfactory agreement may be achieved in a semi-phenomenological way by introducing large phase shifts corresponding to the strong resonance scattering of electron states at the Fermi level (see [550, 551]).

The simplest theoretical model describing the formation of heavy-fermion state is the s–f exchange model. It should be stressed that, unlike the case of systems with strong Hubbard correlations (Sect.4.6),

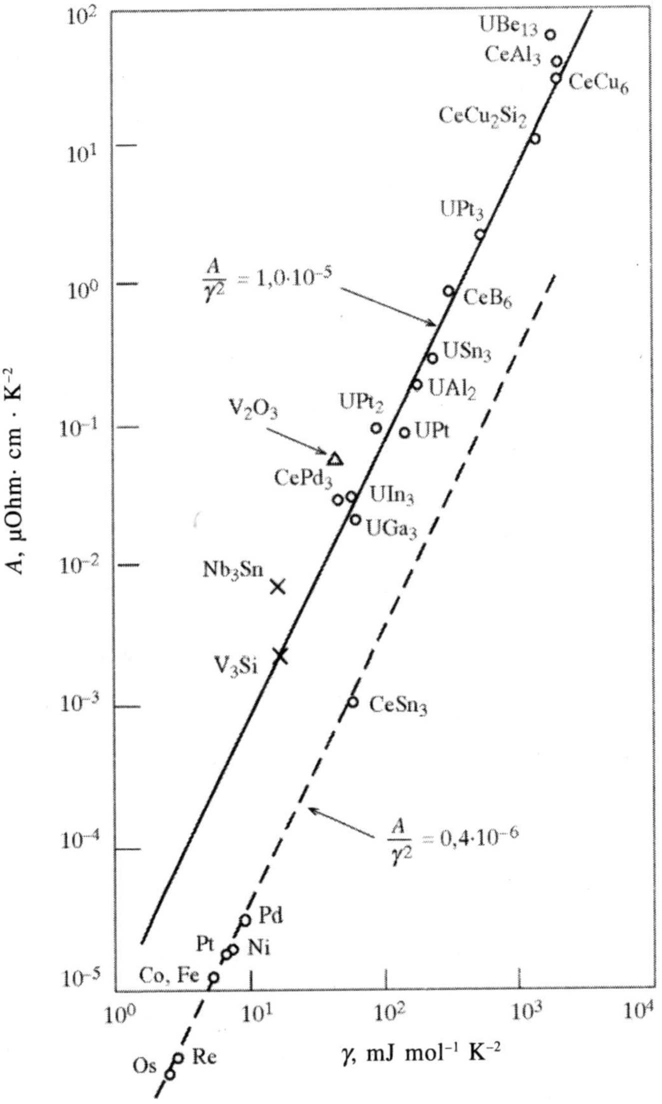

Fig. 6.1. Coefficient at the T^2-term A vs γ^2 for anomalous rare earth and actinide compounds, and some d-systems [410].

the bare interaction between current carriers and localized moments, which leads to the anomalous behaviour, is rather weak. However, owing to resonance character of s–f scattering near the Fermi level, the effective interaction in the many-electron system tends to infinity. Thus we deal with an essentially many-particle problem. In the next Section, we start the consideration of this problem from the case of one magnetic $d(f)$-impurity.

6.1 THE ONE-CENTRE KONDO EFFECT

It is believed now that the main cause for anomalous properties of heavy-fermion systems is the Kondo effect. This effect was first discussed in connection with the problem of resistivity minimum in diluted alloys of transition metals. Even in 'pure' samples of copper, gold and zinc, an increase of resistivity was observed at low temperatures below 10–20 K. It was established experimentally that this phenomenon is closely related to the presence of a small amount (10^{-2}–10^{-3}%) of impurities of transition metals (Cr, Fe, Mn), which retain magnetic moment in the host metal. Such a strong effect cannot be explained within simple one-electron approximations for impurity resistivity. Kondo [552] demonstrated that in the third order of perturbation theory the s–d exchange interaction of conduction electrons with localized moments results in a singular ln T -correction to resistivity owing to many-body effects (Fermi statistics). When combined with the usual low temperature T^5-contribution owing to electron-phonon scattering, this correction does yield the minimum of resistivity. Minimizing the expression

$$\rho = Ac\ln T + BT^5, \tag{6.1}$$

with c being the impurity concentration we obtain, $T_{\min} \sim c^{1/5}$, i.e a weak c-dependence.

Consider the occurrence of Kondo anomalies in the s–d exchange model with one impurity atom

$$\mathsf{H} = \sum_{\mathbf{k}\sigma} t_{\mathbf{k}} c^{\dagger}_{\mathbf{k}\sigma} c_{\mathbf{k}\sigma} - I \sum_{\mathbf{kk}'\sigma\sigma'} (\mathbf{S}\sigma_{\sigma\sigma'}) c^{\dagger}_{\mathbf{k}\sigma} c_{\mathbf{k}'\sigma'}. \tag{6.2}$$

The lowest-order matrix element of the elastic s–d scattering ($t_{\mathrm{k}} = t_{\mathrm{k}'} = E_{\mathrm{F}}$) do not differ from those of the usual potential impurity scattering:

$$\langle \mathbf{k}' \uparrow | T | \mathbf{k} \uparrow \rangle = -IS^z. \tag{6.3}$$

In the second order of perturbation theory, two types of scattering processes contribute to the matrix element (6.3):

1. An electron passes from the state $|\mathbf{k} \uparrow\rangle$ into the state $|\mathbf{k}' \uparrow\rangle$. The

intermediate state $|\mathbf{k}'\sigma\rangle$ should be empty.

2. An electron from the occupied state $|\mathbf{k}'\sigma\rangle$ passes into the state $|\mathbf{k}'\uparrow\rangle$, and then an electron from the state $|\mathbf{k}\uparrow\rangle$ passes into the state $|\mathbf{k}'\sigma\rangle$. The sign of this contribution is opposite to that of the first contribution because of the antisymmetry of the many-electron wave function.

The whole expression for the second order contribution reads

$$\langle \mathbf{k}'\uparrow|T|\mathbf{k}\uparrow\rangle^{(2)} = I^2(S^zS^z + S^+S^-)\sum_{\mathbf{k}'}\frac{1-n_{\mathbf{k}'}}{t_{\mathbf{k}}-t_{\mathbf{k}'}} - $$
$$+I^2(S^zS^z + S^-S^+)\sum_{\mathbf{k}'}\frac{n_{\mathbf{k}'}}{t_{\mathbf{k}}-t_{\mathbf{k}'}} =$$
$$= I^2S(S+1)\sum_{\mathbf{k}'}\frac{1}{t_{\mathbf{k}}-t_{\mathbf{k}'}} + I^2S^z\sum_{\mathbf{k}'}\frac{1-2n_{\mathbf{k}'}}{t_{\mathbf{k}}-t_{\mathbf{k}'}}, \qquad (6.4)$$

where we have used the commutation relation $[S^+, S^-] = 2S^z$; similar expressions may be obtained for other matrix elements $\langle \mathbf{k}\sigma|T|\mathbf{k}'\sigma'\rangle$. The first term in the right-hand side of (6.4) yields only a small correction to the potential scattering. At the same time, the second term, which occurs because of non-commutativity of spin operators and contains Fermi distribution functions, contains a large logarithmic factor which diverges as E approaches the Fermi energy:

$$-\sum_{\mathbf{k}'}\frac{1-2n_{\mathbf{k}'}}{t_{\mathbf{k}}-t_{\mathbf{k}'}} = \int \rho(E')\frac{1-2f(E')}{E'-E}dE' \approx 2\rho\ln\frac{W}{\max\{|E|,T\}}, \qquad (6.5)$$

where W is of the order of conduction band width, E is referred to the Fermi level, $\rho(E)$ is the bare density of states of conduction electrons with one spin projection, $\rho = \rho(E_F)$. The total contribution of magnetic scattering to resistivity is obtained after averaging of the squared matrix elements over the localized spin projections:

$$\rho_{sd} = \rho_{sd}^{(0)}\left(1 - 4I\rho\ln\frac{W}{T}\right), \qquad \rho_{sd}^{(0)} \sim I^2S(S+1) \qquad (6.6)$$

(at calculating resistivity, $|E| \sim T$). Thus, the singular Kondo contribution occurs in the third order in I.

Similar perturbation calculations may be performed for other physical properties [552]. The magnetic susceptibility is diminished by logarithmic corrections of the second order:

$$\chi = \frac{S(S+1)}{3T}\left(1 - 4I^2\rho^2 \ln\frac{W}{T}\right), \tag{6.7}$$

The s–d contribution to specific heat occurs in the fourth order [560]

$$C_{sd}(T) = 16\pi^2 S(S+1) I^4 \rho^4 \left(1 - 8I\rho \ln\frac{W}{T}\right). \tag{6.8}$$

The logarithmic term in (6.6) results in an resistivity increase with lowering T for $I<0$. This sign of the s–d exchange parameter takes place for magnetic impurities in noble metals where the effective s–d exchange is in fact due to combined action of the s–d hybridization and Coulomb interaction (see Appendix N):

$$I = V^2\left(\frac{1}{\Delta} - \frac{1}{\Delta+U}\right), \tag{6.9}$$

where V is hybridization matrix element, Δ is the position of the d-level calculated from E_F, U is the one-site Coulomb interaction.

The resistivity increase is suppressed at very low temperatures by magnetic ordering of impurities owing to long-range RKKY-interaction among them (in the ordered phase, the orientation of spins becomes fixed and the scattering becomes ineffective). For d-impurities the third-order correction (6.6) is in most cases sufficient to describe experimental data since at not too small c the higher order contributions of perturbation theory are small up to the magnetic ordering temperature. On the other hand, rare-earth impurities (e.g., Ce, Yb, Sm, Tm in Y or La-based hosts) may be considered as isolated ones up to $c \sim 1\%$; even at larger concentrations the interaction among them does not necessarily result in usual magnetic ordering, but leads to the formation of 'dense' Kondo systems [545]. Therefore the problem of accurate treatment of many-electron effects owing to s–$d(f)$ exchange interaction at low temperatures (the Kondo problem) occurs.

Summing up the leading logarithmic terms yields

$$\rho_{sd} = \rho_{sd}^{(0)}\left(1 + 2I\rho \ln\frac{W}{T}\right)^{-2}. \tag{6.10}$$

In the case of 'ferromagnetic' s–d exchange $I>0$ this 'parquet' approximation [14] solves the Kondo problem. However, in more important case $I<0$ this approximation yields a divergence of resistivity at the temperature

$$T_K = W \exp \frac{1}{2I\rho}, \tag{6.11}$$

which is called the Kondo temperature.

Unlike the critical temperature of a ferromagnet or superconductor, the Kondo temperature does not correspond to a phase transition, but is just a characteristic energy scale for a crossover between high- and low-temperature regions. The consideration of the region $T < T_K$ is a very difficult and beautiful mathematical problem. The case $T \ll T_K$ was investigated within phenomenological Fermi-liquid theory [553, 10] and analytical renormalization group methods [554, 555]. The numerical solution was obtained by Wilson with the use of renormalization group approach [556]. Finally, under some simplifying approximations (which reduce the problem to one dimension) the exact solution of the one-impurity *s–d* model was obtained by Andrei and Wiegmann with the use of the Bethe ansatz [557, 558].

It turns out that at $T \to 0$ the effective (renormalized) *s–d* interaction becomes infinitely strong, so that the impurity magnetic moment is totally compensated (screened) by conduction electrons. Strictly speaking, in the usual *s–d* model with zero orbital momentum (6.2) such a compensation occurs only for $S = 1/2$, and for a general S the Kondo effect results in the contraction of the impurity spin, $S \to S - 1/2$. However, in a realistic situation of degenerate electron bands, the number of 'scattering channels' for conduction electrons is sufficient to provide the screening.

The resistivity tends at $T \to 0$ to a finite unitary limit (which corresponds to the maximum possible phase shift of $\pi/2$), the corrections at low T being proportional to $(T/T_K)^2$ [14,552,558]:

$$\rho_{sd} = \frac{3}{\pi}(\rho v_F e)^{-2}\left(1 - \frac{\pi^2 T^2}{T_K^2} + O\left(\frac{T}{T_K}\right)^4\right). \tag{6.12}$$

The specific heat of the system has a maximum at $T \sim T_K$ and behaves linearly at $T \to 0$:

$$C_{sd}(T) = \frac{\pi}{3}\frac{T}{T_K}\left(1 + O\left(\frac{T}{T_K}\right)^2\right), \tag{6.13}$$

which is reminiscent of electronic specific heat with $E_F \to \pi T_K$.

The magnetic entropy at $T = 0$,

$$\mathcal{S}(0) = R \ln(2S+1),$$

is removed owing to the magnetic moment screening rather than to magnetic ordering. The magnetic susceptibility

$$\chi = \frac{(g\mu_B)^2}{2\pi T_K}\left(1 - O\left(\frac{T^2}{T_K^2}\right)\right) \tag{6.14}$$

demonstrates the Pauli behaviour (in contrast with the Curie law (6.7) at $T > T_K$) and is greatly enhanced, as well as specific heat. These results may be described in terms of a narrow many-particle Abrikosov–Suhl resonance at the Fermi level with a width of order T_K and a height of order $1/T_K$, so that T_K plays the role of the effective degeneracy temperature. Thus a new Fermi-liquid state is formed which is characterized by large many-electron renormalizations.

The interpolation formula for $\chi(T)$ may be presented in a form of the Curie–Weiss law with a negative paramagnetic Curie temperature, $|\theta| \sim T_K$. It should be noted in this connection that the difference between transition metal impurities, which retain magnetic moment in a given host, from 'non-magnetic' ones has quantitative rather than qualitative nature. The second case may be viewed as that of high T_K of order 10^2–10^4 K, sometimes higher than the melting point (in the case of usual Pauli susceptibility $T_K \to E_F$). Similar considerations may be applied to pure substances where local magnetic moments do not exist at low temperatures (although concrete theoretical models may be quite different). For enhanced Pauli paramagnets like Pd, Pt, UAl_2, where the Curie–Weiss law holds at high temperatures, one introduces, instead of the Kondo temperature, the so-called spin-fluctuation temperature.

6.2 THE KONDO TEMPERATURE FOR *d*-IMPURITIES

Due to the strong exponential dependence on the model parameters, the Kondo temperature varies in a large interval from 10^{-2} to 10^4 K. Values of T_K for transition metal impurities in copper and gold [559, 560], which are determined from anomalies of various physical properties (resistivity, thermopower, specific heat, magnetic susceptibility) are presented in Fig.6.2. One can see a characteristic V-shape dependence in the $3d$-series with a sharp minimum in the middle of the series ($n = 5$). In the papers [559] these data were interpreted within the Schrieffer model

$$\mathcal{H}_{sd} = -\frac{I}{n}\sum_{\mathbf{kk}'\sigma}\mathbf{S}\sigma_{\sigma\sigma'}\sum_m c^\dagger_{\mathbf{k}lm\sigma}c_{\mathbf{k}'lm\sigma'}. \tag{6.15}$$

[561] which is rather artificial since it does not take into account scattering by orbital degrees of freedom, despite they should play

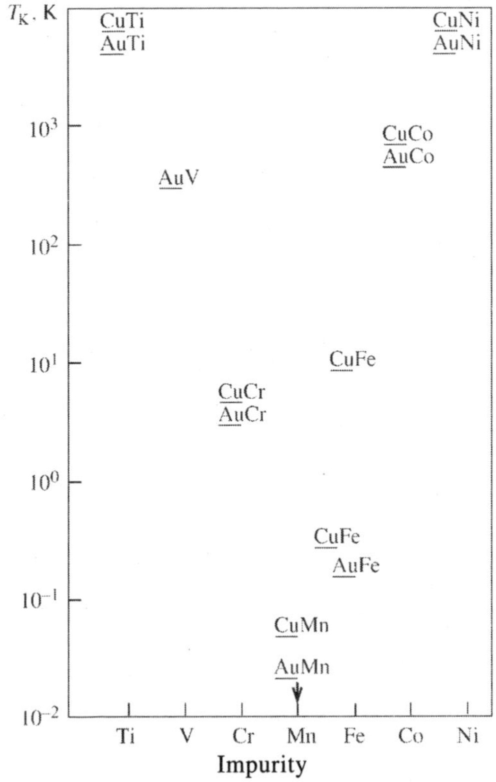

Fig. 6.2. Experimental data on the Kondo temperature for *d*-impurities in copper and gold [559, 560].

an important role [562]. The model (6.15) yields the following n-dependence of T_K

$$T_K = W \exp\left(-\frac{n}{2|I|\rho}\right). \tag{6.16}$$

Consider the calculation of the Kondo temperature with account of the orbital scattering for the localized degenerate d-level (the configuration d^n, $n<5$) at neglecting intraconfiguration splitting. Then the dependence on quantum numbers of ME terms vanishes, so that the Hamiltonian of the one-impurity model takes the form

$$\mathsf{H} = \sum_{\mathbf{k}m} t_{\mathbf{k}} c^{\dagger}_{\mathbf{k}m} c_{\mathbf{k}m} - I \sum_{\mathbf{k}\mathbf{k}'mm'm_i} \left[X(\{m_1 \dots m_{n-1}, m'\}, \{m_1 \dots m_{n-1}, m\}) - \right.$$

$$\left. - \frac{1}{2[l]} \delta_{mm'} \right] c^{\dagger}_{\mathbf{k}m} c_{\mathbf{k}'m'}, \qquad (6.17)$$

where all the indices in the sets $\{m_i\}$ ($m_i = 1, 2, \dots, 2[l]$ include both spin and orbital indices, $[l] = 2l + 1$, $l = 2$ for d-electrons) are different, X-operators mark all the possible transitions, the second term in the brackets is subtracted to exclude the potential scattering, $I < 0$. For $l = 0$ the model (6.17) is reduced to the usual s–d exchange model (6.2) with $S = 1/2$.

The Kondo temperature is determined from the pole of the T-matrix defined by

$$\langle\langle c_{\mathbf{k}'m} | c^{\dagger}_{\mathbf{k}m} \rangle\rangle_E = \frac{\delta_{\mathbf{k}\mathbf{k}'}}{E - t_{\mathbf{k}}} + \frac{T_{\mathbf{k}\mathbf{k}'}(E)}{(E - t_{\mathbf{k}})(E - t_{\mathbf{k}'})}. \qquad (6.18)$$

Write down the equation of motion

$$(E - t_{\mathbf{k}'}) \langle\langle c_{\mathbf{k}'m} | c^{\dagger}_{\mathbf{k}'m} \rangle\rangle_E = \delta_{\mathbf{k}\mathbf{k}'} - I \sum_{\mathbf{p}} \Gamma_{\mathbf{k}\mathbf{p}m}(E); \qquad (6.19)$$

$$\Gamma_{\mathbf{k}\mathbf{q}m}(E) = \sum_{m_1 \dots m_{n-1} m'} \left\langle\left\langle \left[X(\{m_1 \dots m_{n-1}, m'\}, \{m_1 \dots m_{n-1}, m\}) - \right.\right.\right.$$

$$\left.\left.\left. - \frac{1}{2[l]} \delta_{mm'} \right] c_{\mathbf{q}m} \right| c^{\dagger}_{\mathbf{k}m} \right\rangle\right\rangle_E. \qquad (6.20)$$

Performing the simplest decoupling in the equation of motion for Γ (which is an analogue of Nagaoka decoupling, see [552]) we derive

$$(E - t_{\mathbf{q}}) \Gamma_{\mathbf{k}\mathbf{q}m}(E) = -I \left\{ 1 - [l]^{-1} (2[l] - n + 1)^{-1} + \frac{1}{4} [l]^{-2} + \right.$$

$$+ (n-2) \sum_{m'm''\mathbf{p}} \left\langle c^{\dagger}_{\mathbf{p}m''} c_{\mathbf{q}m'} \left[X(\{m_1 \dots m_{n-1}, m'\}, \{m_1 \dots m_{n-1}, m\}) - \right.\right.$$

$$\left.\left.\left. - \frac{1}{2[l]} \delta_{m'm''} \right] \right\rangle \right\} \sum_{\mathbf{k}} \langle\langle c_{\mathbf{k}'m} | c^{\dagger}_{\mathbf{k}m} \rangle\rangle_E -$$

$$- I \left\{ (2[l] - 2n + 2) n_{\mathbf{q}} - n - 1 - \frac{1}{2[l]} \right\} \sum_{\mathbf{p}} \Gamma_{\mathbf{k}\mathbf{p}m}(E). \qquad (6.21)$$

Solving the equation (6.21) we obtain the estimate for T_K from the divergence in the 'parquet' approximation, which corresponds to the second Born approximation for resistivity,

$$T_{\mathrm{K}} = W \exp\left(-\frac{1}{2[l]-2n+2} \frac{1}{|I|\rho} \right). \qquad (6.22)$$

For the second half of the d-series $n > 5$ the degeneracy of the configuration d^n is smaller than that of d^{n-1}. Thus we have to consider the situation where the level d^n lies above the Fermi level, i.e. $\Delta > 0$ (otherwise, the Kondo effect is absent). Then we may pass to the hole representation and reproduce the result (6.22) with the replacement $n \to 2[l] - n$. The formula (6.22) may be fitted to experimental data of Fig.6.2 with $|I|\rho = 1/16$, whereas using (6.16) yields the unreasonably large value $|I|\rho = 1/4$.

The dependence on ME term quantum numbers, which is neglected in above consideration, seems to be important since the distance between different terms in free atoms is of order of several eV. An account of ME term splitting is performed in Appendix N within the degenerate Anderson model. The result has the form

$$T_{\mathrm{K}} = W \exp\left(-\left(\frac{[S][L]}{[S'][L']} - 1 \right)^{-1} \left(n^{1/2} G^{SL}_{S'L'} \right)^{-2} \frac{1}{|I|\rho} \right), \qquad I = \frac{v_l^2(\mathbf{k}_{\mathrm{F}})}{\Delta}. \qquad (6.23)$$

Although the general picture of ion levels (especially in a crystal field) is very complicated, the occurrence of the squared fractional parentage coefficients $(n^{1/2} G^{SL}_{S'L'})^{-2}$ in (6.23) is expected to lead to a further sharpening of the dependence $T_{\mathrm{K}}(n)$. Indeed, the n-dependence of fractional parentage coefficients on the average has a minimum in a middle of the d-series. Such a dependence is due to that the total number of ME terms is maximum near $n = 5$ by combinatorial reasons, and for a given n the values of G^2 satisfy the sum rules (A.8). However, direct use of the formula (6.23), which yields a strongly oscillating dependence $T_{\mathrm{K}}(n)$ in contradiction with experimental data, probably overestimates the role of ME term effects.

At derivation of (6.23) we have restricted ourselves to the case of LS-splitting and neglected crystal-field (CF) effects which may be very important. In particular, CF results in that $L = 0$ for some d-impurities (e.g., for V, Ni [555]). The CF splitting may be taken into consideration in a similar way, but this requires more cumbersome calculations with the use of Clebsh–Gordan and fractional parentage coefficients for a point group (these may be found, e.g., in [565]). At accurate estimations of T_{K}, the influence of several groups of degenerate levels (e.g., corresponding to different atomic or crystal-field split terms) should be also taken into account. The expression for two level groups (Δ_1 with degeneracy N_1 and Δ_2 with degeneracy N_2) has the form [565]

$$T_K = T_K^{(0)} \left(\frac{T_K^{(0)}}{\Delta_2 - \Delta_1} \right)^{N_2/N_1}, \tag{6.24}$$

where

$$T_K^{(0)} = W \exp\left(\frac{\Delta_1}{\rho V^2 (N_1 + N_2)} \right).$$

However, such calculations require a detailed information on the electronic structure of Kondo impurities.

The role of variation of the interconfiguration splitting should be also discussed. It was supposed above that the effective s–d parameters, i.e. the values of v, Δ and ρ do not depend on the configuration d^n. However, it is well known that in the many-electron picture these configurations possess different stability (e.g., the value of Δ should be related to atomic ionization potentials). In particular, the spherically symmetric configuration d^5 is rather stable which may result in a large value of $|\Delta|$ thereby lowering T_K for manganese.

6.3 SPIN DYNAMICS AND ELECTRONIC PROPERTIES OF KONDO LATTICES

The anomalous rare-earth and actinide compounds are classified as concentrated Kondo systems or Kondo lattices since the formation of low-temperature Kondo state gives most natural explanation of their unusual properties (large values of γ and $\chi(0)$). Most of these compounds exhibit ln T-contribution in resistivity at high temperatures, but have a metallic ground state with $\rho(T \to 0) \sim T^2$. However, examples of insulating Kondo lattices are known too. The system CeNiSn possesses at low temperatures extremely small energy gap of order of a few K [566, 567]. Partial replacing of Ni by Cu results in heavy-fermion metallic behaviour [568]. A similar situation takes place for the narrow-gap compound CeRhSb [569]: the system $Ce_{1-x}Th_xRhSb$ demonstrates at $x \sim 0.4$ the huge value $\gamma > 1$ J/mol$\cdot$K^2 [570]. Narrow energy gaps are found also for NdBiPt [571], $Ce_3Bi_3Pt_4$ [572].

The picture of the insulating Kondo lattice is also used sometimes for the narrow-gap intermediate valence semiconductors SmB_6, SmS (golden phase) [545], which are discussed in Sect. 6.5. The formation of the insulating Kondo state may be described in terms of the coherent

Kondo scattering, the Abrikosov–Suhl resonance being transformed into the narrow many-electron gap.

Consider the Kondo effect manifestations for the periodic lattice of localized f-moments within the framework of the s–f exchange model (G.2). This case differs from that of a single Kondo impurity by the presence of intersite exchange interactions and, consequently, of spin dynamics which tends to weaken usual Kondo divergences and also leads to some new effects.

We apply perturbation theory in the s–f exchange parameter, the Heisenberg f–f interaction being taken into account exactly. The calculation of the second-order contribution to the electron self-energy yields [367]

$$\Sigma_{\mathbf{k}}^{(2)}(E) = I^2 \sum_{\mathbf{q}} \int K_{\mathbf{q}}(\omega) \left(\frac{1 - n_{\mathbf{k}+\mathbf{q}}}{E - t_{\mathbf{k}+\mathbf{q}} + \omega} + \frac{n_{\mathbf{k}+\mathbf{q}}}{E - t_{\mathbf{k}+\mathbf{q}} - \omega} \right) d\omega, \qquad (6.25)$$

where $K_{\mathbf{q}}(\omega)$ is the spectral density of the localized spin subsystem, defined by

$$\langle \mathbf{S}_{-\mathbf{q}}(t) \mathbf{S}_{\mathbf{q}} \rangle = \int_{-\infty}^{+\infty} K_{\mathbf{q}}(\omega) e^{i\omega t} d\omega. \qquad (6.26)$$

One can see that in the presence of spin dynamics Kondo-like divergences in the self-energy arise already in the second order. They are formally due to the Fermi functions:

$$\sum_{\mathbf{q}} \frac{n_{\mathbf{k}+\mathbf{q}}}{E - t_{\mathbf{k}+\mathbf{q}} \pm \overline{\omega}} \simeq \rho \ln \frac{W}{\max\{|E|, T, \overline{\omega}\}}, \qquad (6.27)$$

where ω is the characteristic spin-fluctuation frequency. As follows from the spectral representation for the commutator Green's functions (E.18),

$$K_{\mathbf{q}}(\omega) = -\frac{1}{\pi} N_{\mathrm{B}}(\omega) \operatorname{Im} \chi_{\mathbf{q}\omega},$$

$$\chi_{\mathbf{q}\omega} = \langle\langle \mathbf{S}_{\mathbf{q}} | \mathbf{S}_{-\mathbf{q}} \rangle\rangle_{\omega}, \qquad \operatorname{Im} \chi_{\mathbf{q}\omega} = -\operatorname{Im} \chi_{\mathbf{q}-\omega}. \qquad (6.28)$$

Then in the classical limit $\omega \ll T$ one has

$$K_{\mathbf{q}}(\omega) = K_{\mathbf{q}}(-\omega) = -\frac{1}{\pi} \frac{T}{\omega} \operatorname{Im} \chi_{\mathbf{q}\omega},$$

so that the terms with the Fermi functions cancel out mutually. However, in the quantum case $\Sigma(E)$ varies sharply in an energy region of order ω near E_F. This leads to an appreciable renormalization of the Green's function residue Z (see (G.53)) and, consequently, of the electronic effective mass m^* and specific heat. These renormalizations vanish at $T \gg \omega$. In particular, the result (6.25) with $I \to U$ yields the spin-fluctuation (paramagnon) renormalization in the Hubbard model [573].

Expressions, similar to (6.25) may be also obtained in other situations. For

$$K_{\mathbf{q}}(\omega) \sim [1 - f(\Delta_{cf})]\delta(\omega + \Delta_{cf}) + f(\Delta_{cf})\delta(\omega - \Delta_{cf}) \qquad (6.29)$$

the formula (6.25) describes effects of the interaction with crystalline-electric-field excitations [263], Δ_{cf} being the CF level splitting. In the case of the electron-phonon interaction I is replaced by the corresponding matrix element, and

$$K_{\mathbf{q}}(\omega) = [1 + N_B(\omega_{\mathbf{q}})]\delta(\omega + \omega_{\mathbf{q}}) + N_B(\omega_{\mathbf{q}})\delta(\omega - \omega_{\mathbf{q}}). \qquad (6.30)$$

A spectral density of the form (6.29) with Δ, which depends weakly on $\mathbf{q}$, corresponds to localized spin fluctuations. The renormalization of m^* owing to such fluctuations is much stronger than that owing to 'soft' paramagnons because of smallness of fluctuation phase volume in the latter case.

Thus the definition of the Kondo effect in systems exhibiting dynamics is non-trivial. The condition $Z \ll 1$, characteristic for the Kondo lattices, may be satisfied not only due to the usual Kondo effect (the formation of the Abrikosov–Suhl resonance at $T < T_K$), but also due to the interaction with low-energy spin- or charge-density fluctuations.

The result $m^* \sim 1/\bar{\omega}$, which follows from (6.25), (6.27), does not change in form when account is taken of higher orders terms even for arbitrarily small $\bar{\omega}$. This problem is investigated in [574] for a simple model which describes the interaction with local excitations of a two-level system ($\bar{\omega} = \Delta$). After collecting all the singular terms of the type under consideration, we obtain in the $(n+1)$-th order the singular factors $\ln^n |(E \pm \bar{\omega})/\bar{\omega}|$, i.e. the divergences are shifted from E_F and the cutoff parameter is $\bar{\omega}$ rather than the bandwidth W. Besides that, all the singularities are cancelled at $\bar{\omega} \to 0$ due to the factors of the type $\tanh(\bar{\omega}/2T)$.

Now we consider the true Kondo divergences, corresponding to another sequence of singular terms which describes spin-flip processes and starts from the third order in the s–f parameter. These divergences do not vanish in the absence of dynamics and do yield at $E \to 0$ the factors $\ln(W/\max\{\bar{\omega}, T\})$. The corresponding contribution to the imaginary part of the self-energy with account of spin dynamics reads

$$\mathrm{Im}\,\Sigma_{\mathbf{k}}^{(3)}(E) = 2\pi I^3 \rho(E) \int \sum_{\mathbf{q}} K_{\mathbf{q}}(\omega) \frac{n_{\mathbf{k+q}}}{E - t_{\mathbf{k+q}} - \omega} d\omega \tag{6.31}$$

(the real part of the singular contribution is absent [367]). The quantity (6.31) determines the damping of one-particle state and, consequently, the relaxation rate $\tau^{-1}(E)$. Comparing (6.31) with (6.4) we see that spin dynamics results in a smearing of logarithmic term in resistivity. Using, e.g., the simplest diffusion approximation for the spectral density,

$$K_{\mathbf{q}}(\omega) = \frac{S(S+1)}{\pi} \frac{D_s \mathbf{q}^2}{\omega + (D_s \mathbf{q}^2)^2}, \tag{6.32}$$

with D_s being the spin diffusion constant, we obtain

$$\delta\tau^{-1}(E) = 4\pi I^3 \rho^2 S(S+1) \ln \frac{E^2 + \overline{\omega}^2}{W^2}, \tag{6.33}$$

where $\overline{\omega} = 4D_s k_F^2$. Thus in resistivity

$$\ln T \to \frac{1}{2} \ln(T^2 + \overline{\omega}) \approx \ln(T + a\overline{\omega}), \quad a \sim 1. \tag{6.34}$$

A similar replacement occurs in other physical properties (e.g., in paramagnetic susceptibility and specific heat). In this connection, shifts in $\rho(T)$ and $C(T)$ curves with changing content of components in heavy-fermion systems were discussed in [367].

Now we discuss the thermoelectric power $\alpha(T)$ in Kondo lattices. At moderately high (as compared to T_K) temperatures $\alpha(T)$ is usually large and has an extremum (a maximum at $\alpha > 0$, a minimum at $\alpha < 0$). Large Kondo contributions to $\alpha(T)$ correspond to the anomalous odd contribution to $\tau^{-1}(E)$ [552], which should arise, by analytical properties of $\Sigma(E)$, from the logarithmic singularity in $\mathrm{Re}\Sigma(E)$ [367]. Although such singularity is absent in the usual Kondo problem, it occurs in the presence of the potential scattering V which leads to emergence of complex factors

$$1 + V \sum_{\mathbf{k}} (E - t_{\mathbf{k}} + i0)^{-1}, \tag{6.35}$$

which 'mix' Im Σ and ReΣ in the incoherent regime. Then spin dynamics leads to the replacements

$$\ln \frac{|E|}{W} \to \frac{1}{2} \ln \frac{E^2 + \overline{\omega}^2}{W^2}, \qquad \mathrm{sign}\, E \to \frac{2}{\pi} \arctan \frac{E}{\overline{\omega}}, \tag{6.36}$$

in Im Σ and ReΣ respectively, and the anomalous contribution to $\alpha(T)$ reads

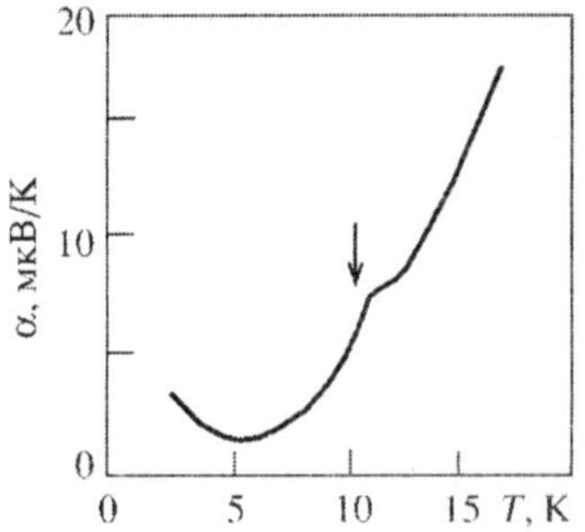

Fig. 6.3. Anomaly of thermoelectric power in CeIn at the Neel point [578].

$$\alpha(T) \sim \frac{I^3 V}{e\rho(T)} \int \frac{E}{T} \frac{\partial f(E)}{\partial E} \arctan \frac{E}{\bar{\omega}} dE \sim \frac{I^3 V}{e\rho(T)} \frac{T}{\max\{T, \bar{\omega}\}}. \tag{6.37}$$

Thus the quantity $\bar{\omega}$ plays the role of a characteristic fluctuating magnetic field which is introduced in [552] to describe thermoelectric power of diluted Kondo systems.

Near the magnetic ordering point T_M, the quantity $\bar{\omega}$ contains non-analytic contributions, proportional to $(T - T_M)^{1-\alpha}$ (α is the critical index for specific heat). Therefore, the Fisher–Langer transport prop-

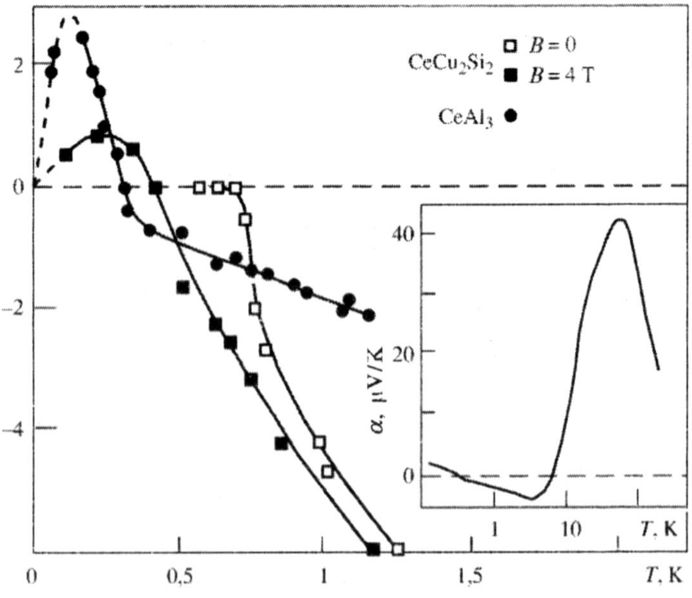

Fig. 6.4. The temperature dependences of thermoelectric power of $CeCu_2Si_2$ and $CeAl_3$; inset shows the dependence for $CeAl_3$ between 0.15 and 350 K on a logarithmic scale [579].

erties anomalies near magnetic phase transitions [575] should be enhanced in Kondo magnets. This is confirmed by experimental data on Ce_3Al [576], UCu_5 [577], $Ce_{1-x}La_xIn_3$ [578] where appreciable breaks of thermopower at the Neel point were observed (Fig.6.3).

6.4 GROUND STATE OF THE KONDO LATTICES

The calculations of the previous Section correspond to the temperature region $T \gg T_K$. As well as in the case of one Kondo impurity, below T_K the perturbation theory regime passes into the strong coupling regime where magnetic momnets are suppressed and a new Fermi-liquid state with greatly enhanced electronic effective mass occurs.

Besides the Kondo temperature, one introduces sometimes the second energy scale – the coherence temperature T_{coh}, which corresponds to onset of coherent Kondo scattering by different lattice sites. This is usually small in comparison with T_K. The picture of the coherent state formation enables one to treat experimental data on low-temperature anomalies of thermoelectric power in heavy-fermion systems [545,579,580]. With decreasing T below the above-discussed high-temperature extremum, $\alpha(T)$ often changes its sign, has an extremum again and vanishes linearly at $T \to 0$ (see Fig. 6.4). Such a behaviour may be attributed to occurrence of a pseudogap with reversing the sign of the quantity $dN(E)/dE$ at the Fermi level, which determines the $\alpha(T)$ sign (Sect.5.4). Besides that, formation of the coherent state is indicated by the positive magnetoresistivity and a sharp negative peak in the Hall coefficient.

To describe the formation of the singlet Kondo state in the strong coupling region we may use the simplest Hamiltonian of the SU (N) Anderson-lattice model

$$\mathsf{H} = \sum_{\mathbf{k}m} t_{\mathbf{k}} c^{\dagger}_{\mathbf{k}m} c_{\mathbf{k}m} + \Delta \sum_{im} X_i(mm) + V \sum_{\mathbf{k}m} [c^{\dagger}_{\mathbf{k}m} X_{\mathbf{k}}(0m) + X_{-\mathbf{k}}(m0) c_{\mathbf{k}m}] \tag{6.38}$$

($m = 1,\ldots, N$). This model is convenient at describing the inter-configuration transitions f^0–f^1 (cerium, $J = 5/2$) or f^{14}–f^{13} (ytterbium, $J = 7/2$) and is treated often within the $1/N$-expansion. A more realistic model of s–f hybridization with inclusion of two (generally speaking, magnetic) configurations is investigated in Appendix N. The model (6.38) may be mapped by a canonical transformation, which excludes the hybridization, onto the Coqblin–Schrieffer model

$$\mathsf{H}_{CS} = \sum_{\mathbf{k}m} t_{\mathbf{k}} c^{\dagger}_{\mathbf{k}m} c_{\mathbf{k}m} - I \sum_{imm'} X_i(mm') c^{\dagger}_{\mathbf{k}m'} c_{\mathbf{k}m}, \tag{6.39}$$

$$I = \frac{V^2}{\Delta}.$$

However, it is easier to treat the Kondo-lattice state in the strong-coupling regime directly in the model (6.38). To avoid difficulties owing to complicated commutation relations for the X-operators, the representation may be used [581,551]

$$X_i(m0) = f_{im}^\dagger b_i^\dagger, \qquad X_i(m'm) = f_{im'}^\dagger f_{im}, \qquad X_i(00) = b_i^\dagger b_i, \quad (6.40)$$

where $f^\dagger$ are the Fermi operators, $b^\dagger$ are the auxiliary (slave) Bose operators. This representation enables one to satisfy the needed commutation relations (A.35). At the same time, according to (A.28), one has to require fulfilment of the subsidiary condition

$$\sum_m X_i(mm) + X_i(00) = \sum_m f_{im}^\dagger f_{im} + b_i^\dagger b_i = 1. \quad (6.41)$$

Then the parameter $\langle b_i \rangle$ renormalizes the hybridization matrix elements.

A description of the crossover to the coherent regime was constructed within a modified SU (N) Anderson model [582]. The temperature dependence of effective hybridization parameter was obtained in the form

$$V_{\text{eff}}^2 \sim \langle b_i^\dagger b_i \rangle \sim \varphi(T),$$

$$\varphi(T) = \left(N + e^{-T_K/T} + 1\right)^{-1} = \begin{cases} 1, & T \ll T_{\text{coh}}, \\ O(1/N), & T_{\text{coh}} \ll T \ll T_K \end{cases} \quad (6.42)$$

with the coherence temperature $T_{\text{coh}} = T_K/\text{In } N$.

Below we describe a simple approach to the problem of the ground state of Kondo lattices, which uses directly the X-operator representation [367]. (An alternative treatment of the Kondo-lattice state, which starts directly from the s–f exchange model with $S = 1/2$ and uses the pseudofermion representation for spin operators, is performed in Appendix O and Sect.6.6.) Consider the retarded Green's function for the localized level states

$$G_{\mathbf{k}m}^f(E) = \langle\langle X_{\mathbf{k}}(0m) | X_{-\mathbf{k}}(m0) \rangle\rangle_E. \quad (6.43)$$

Write down the Dyson equation in the form

$$G_{\mathbf{k}m}^f(E) = R\left[E - \Delta - \Sigma_{\mathbf{k}}^f(E)\right]^{-1}, \quad (6.44)$$

where

$$R = N_0 + N_m = 1 - \frac{N-1}{N} n_f. \quad (6.45)$$

The equation of motion for $G^f_{\mathbf{k}m}(E)$ has the form

$$(E-\Delta)G^f_{\mathbf{k}m}(E) = R\left(1+V\langle\langle c_{\mathbf{k}m} \mid X_{-\mathbf{k}}(m0)\rangle\rangle_E + \right.$$
$$\left. +V\sum_{m'\neq m}\langle\langle X_{\mathbf{q}}(m'm)c_{\mathbf{k}-\mathbf{q}m'} \mid c^\dagger_{\mathbf{k}m}\rangle\rangle_E\right). \tag{6.46}$$

Here we have carried out the decoupling of the term which corresponds to processes without change of m and yields the formation of the hybridization gap, the corresponding spectrum being given by

$$(E-t_{\mathbf{k}})(E-\Delta) = V^2R. \tag{6.47}$$

Further we neglect for simplicity the influence of this gap, which is possible provided that the latter lies well below the Fermi level (besides that, these contributions are formally small in the inverse degeneracy of the f-level, $1/N$).

The terms with $m \neq m'$ contribute to Kondo divergences. Carrying out decouplings in the equations for the Green's function in the right- hand side of (6.46) we obtain

$$\Sigma^f_{\mathbf{k}}(E) = (N-1)V^2\sum_{\mathbf{q}}\frac{n_{\mathbf{q}}}{E-t_{\mathbf{q}}} \approx (N-1)\rho V^2 \ln\frac{W}{|E|}. \tag{6.48}$$

Here we have applied again the approximation (6.5).

Then the Green's function (6.44) has the pole, which is exponentially close to the Fermi level,

$$E = \Delta^* = T_{\mathrm{K}} \approx W\exp\left(-\frac{|\Delta|}{N\rho V^2}\right). \tag{6.49}$$

Near this pole we may expand

$$G^f_{\mathbf{k}m}(E) \approx \frac{Z}{E-\Delta^*}, \tag{6.50}$$

where the residue of the Green's function, determining the inverse effective mass, is

$$Z = R\left(1-\frac{\partial\Sigma^f_{\mathbf{k}}(E)}{\partial E}\right)^{-1}_{E=\Delta^*} \approx \frac{R}{N}\frac{T_{\mathrm{K}}}{\rho V^2} \ll 1. \tag{6.51}$$

Thus the pole (6.49) determines a characteristic low-energy scale – the Kondo temperature. Note that one has to neglect, in spirit of the $1/N$-expansion, unity in comparison with N to obtain the result which is correct in the case $N=2$, which corresponds to the s–f model with $S=1/2$. Similar calculations may be performed within the models with realistic atomic configurations (N.9), (N.5) to obtain the results (N.23), (N.24). These formulas have a simple physical meaning: we have in the exponent the ratio of degeneracies of the multiplets Γ_n and Γ_{n-1}.

The results for the Kondo temperature in the present approximation is valid also for the case of one magnetic impurity. Then the value of T_K obtained determines the position of the Abrikosov–Suhl resonance. Exponential dependence on external parameters (in particular, on the f-level position Δ) makes difficult to establish experimentally reliable correlations of T_K with ME quantum numbers. However, the expression (N.23) enables one to explain very low values of T_K, which are observed for Tm impurities (these are not obtained within the large-N approach [565]). Indeed, in this case both the configurations Γ_n and Γ_{n-1} are magnetic: $J=7/2$ for Tm^{2+} and $J=6$ for Tm^{3+}, so that the ratio $(J-J')/(2J'+1)$ in the exponent of (N.23) is small.

In a periodic lattice of f-moments (anomalous rare earth and actinide compounds), the Green's function of conduction electrons has the form

$$\langle\langle c_{\mathbf{k}m} | c^{\dagger}_{\mathbf{k}m} \rangle\rangle_E = \left[E - t_{\mathbf{k}} - V^2 G^f_{\mathbf{k}m}(E) \right]^{-1}. \tag{6.52}$$

As follows from comparison of (6.52) with (6.47), (6.50), the effective hybridization parameter near E_F is estimated as $V^* \sim VZ^{1/2} \sim (T_K/\rho)^{1/2}$. Thus, instead of the Abrikosov–Suhl resonance, a Kondo gap (or pseudogap), which has a width of order of T_K, and corresponding density-of-states peaks occur near the Fermi level, the latter being, generally speaking, shifted from the gap. Such a picture of the energy spectrum is confirmed for metalic heavy-fermion compounds, by direct far-infrared and point-contact spectroscopy data [583].

6.5 INTERMEDIATE VALENCE SYSTEMS

A number of rare-earth elements (Ce, Sm, Eu, Tm, Yb and possibly Pr) do not possess a stable valence, but vary it in different compounds. In some systems these elements may produce so-called mixed (or intermediate) valent state which is characterized by non- integer number of f-electrons per atom. Such a situation may occur provided that the configurations $4f^n(5d6s)^m$ and $4f^{n-1}(5d6s)^{m+1}$ are nearly degenerate, so that inter-configuration fluctuations are strong. In metallic systems, this corresponds to the

f-level located near the Fermi energy, f-states being hybridized with conduction band states.

The intermediate valence (IV) state is characterized by the single line in Moessbauer experiments (the time scale of these measurements is about 10^{-11}s), which has an intermediate position. On the other hand, in X-ray experiments (the time is about 10^{-11}s) two lines are seen, which correspond to configurations f^n and f^{n-1}. A peculiar feature of the transition into IV state is also the change of the lattice parameter to a value which is intermediate between those for corresponding integer-valent states. Such transformations (e.g., under pressure) are as a rule sharp (first-order) transitions. Besides that, IV compounds possess at low temperatures substantially enhanced electronic specific heat and magnetic susceptibility. At high T, $\chi(T)$ obeys the Curie–Weiss law with the effective moment which is intermediate between the the values for the corresponding atomic configurations.

Examples of IV compounds are metallic compounds $YbAl_2$, YbZn, $YbCu_2Si_2$, $YbAgCu_4$, CeN, $CeBe_{13}$, $CeSn_3$, $EuCu_2Si_2$ (see reviews [512, 520, 584]) and narrow-gap semiconductors TmSe, SmB_6, the "golden" phase of SmS (under pressure above 6 kbar), YbB_{12}.

The metallic cerium in the α-phase was earlier considered as an IV system [584] which seemed to be confirmed by the volume change at the γ–α transition under pressure. However, it is believed now that this transition is in fact the Mott–Hubbard transition in the f-band (delocalization of 4f-electrons due to overlap of f-states at different sites) without a considerable valence change [585]. This is confirmed by both spectroscopic data [586] and band calculations [587]. Besides that, the energy difference of the f^0 and f^1 configurations in metallic cerium is too high for explanation of small heats for formation of metallic Ce alloys. According to paper [588], the 'tetravalent' state of cerium (without localized f-electrons) should be attributed to f^2 rather than to f^0 configuration. As follows from atomic calculations (see [81]), a collapse of f-electrons in cerium takes place at small variations of the atomic potential. This phenomenon is accompanied by a strong decrease of f-bandwidth.

For d-metals and their compounds, one can hardly use the term 'intermediate valence' since the hybridization and, consequently, the widths of d-peaks are as a rule large, and valence fluctuations are too rapid. Thus we shall observe in any experiment the ion states with non-integer number of d-electrons.

The 5f-ions in actinide compounds demonstrate a great variety of valence states – from 1+ to 7+ [371], hybridization being also rather strong. Only the compounds with ion binding have relatively stable valence configurations. Thus the situation for actinides is more close to $3d$-elements than to rare earths. Delocalization of $5f$-electrons and analogy between intermediate valence in actinides and cerium is discussed in [81, 589].

The 'homogeneous' intermediate valence should be distinguished from the inhomogeneous case where lattice sites corresponding to different valences are inequivalent and the inter-configurational fluctuations are very slow. Such a situation takes place, e.g., in the compound Sm_3S_4 which is characterized by static charge ordering (similar to the well-known case of magnetite Fe_3O_4). The inhomogeneous mixed-valent state demonstrates two distinct lines not only in X-ray, but also in in Moessbauer spectra. This case is opposite to rapid valence fluctuations in d-metals. In a number of systems, e.g., EuM_2Si_2 (M=Fe, Co, Ni, Pd, Cu), $EuPd_2P_2$ [584], a transition from inhomogeneous mixed-valent state to the homogeneous one is observed with increasing temperature.

Consider simplest theoretical models for description of the IV state. The Hamiltonian of the spinless Falicov–Kimball model reads

$$\mathsf{H} = \sum_{\mathbf{k}} \left[t_{\mathbf{k}} c_{\mathbf{k}}^{\dagger} c_{\mathbf{k}} + \Delta f_{\mathbf{k}}^{\dagger} f_{\mathbf{k}} + V (c_{\mathbf{k}}^{\dagger} f_{\mathbf{k}} + f_{\mathbf{k}}^{\dagger} c_{\mathbf{k}}) \right] + G \sum_{i} f_i^{\dagger} f_i c_i^{\dagger} c_i, \tag{6.53}$$

where G is the parameter of on-site d–f Coulomb repulsion, we neglect for simplicity the **k**-dependence of hybridization. This Hamiltonian enables one to take simply into account strong on-site f–f repulsion (in the spinless model, doubly-occupied states are forbidden by the Pauli principle) and is convenient at description of valence phase transitions, the interaction G being important for many-electron 'exciton' effects. The Falicov–Kimball model may be generalized by inclusion of Coulomb interaction at different sites, which permits to describe charge ordering.

Unlike the model (6.53), the periodic Anderson model takes into account spin degrees of freedom and permits to describe magnetic properties, but includes explicitly the Hubbard repulsion. At neglecting orbital degeneracy the Anderson Hamiltonian has the form

$$\mathsf{H} = \sum_{\mathbf{k}\sigma} \left[t_{\mathbf{k}} c_{\mathbf{k}\sigma}^{\dagger} c_{\mathbf{k}\sigma} + \Delta f_{\mathbf{k}\sigma}^{\dagger} f_{\mathbf{k}\sigma} + V (c_{\mathbf{k}\sigma}^{\dagger} f_{\mathbf{k}\sigma} + f_{\mathbf{k}\sigma}^{\dagger} c_{\mathbf{k}\sigma}) \right] + U \sum_{i} f_{i\uparrow}^{\dagger} f_{i\uparrow} f_{i\downarrow}^{\dagger} f_{i\downarrow}. \tag{6.54}$$

The intermediate valent situation corresponds to the situation where the width of f-peak owing to hybridization, $\Gamma = \pi V^2 \rho$, is small in comparison with the distance $\Delta = |\varepsilon_F - E_F|$. On the contrary, the Kondo-lattice (heavy-fermion) state may be considered as the nearly integral limit of the IV state (the valence change does not exceed of a few percents). In a sense, IV systems may be treated as Kondo lattices with high values of T_K [545]. Unlike the Kondo lattice state, not only spin, but also charge fluctuations play an important role in the IV state. A renormalization group study of the Anderson–Falicov model is presented in [732].

The excitation spectrum in IV systems may be obtained via diagonalizing the Hamiltonian (6.53) in the simplest Hartree-Fock approximation which corresponds to the picture of 'exciton condensate' (formation of electron-hole pairs) [590]. This is achieved, similar to the superconductivity theory, by the Bogoliubov transformation

$$c_{\mathbf{k}}^{\dagger} = \cos\frac{\theta_{\mathbf{k}}}{2}\alpha_{\mathbf{k}}^{\dagger} + \sin\frac{\theta_{\mathbf{k}}}{2}\beta_{-\mathbf{k}},$$

$$f_{\mathbf{k}}^{\dagger} = \cos\frac{\theta_{\mathbf{k}}}{2}\beta_{-\mathbf{k}} - \sin\frac{\theta_{\mathbf{k}}}{2}\alpha_{\mathbf{k}}^{\dagger}. \tag{6.55}$$

Substituting (6.55) into (6.53) we derive

$$\langle \mathrm{H} \rangle = \sum_{\mathbf{k}}\left[t_{\mathbf{k}} n_{\mathbf{k}}^{\alpha} + \Delta(1 - n_{\mathbf{k}}^{\beta})\right] + \frac{1}{2}\sum_{\mathbf{k}}{}'(t_{\mathbf{k}} - \Delta)(1 - \cos\theta_{\mathbf{k}}) +$$

$$+ V\sum_{\mathbf{k}}{}'\sin\theta_{\mathbf{k}} + \frac{1}{4}G\left[1 - \left(\sum_{\mathbf{k}}{}'\cos\theta_{\mathbf{k}}\right)^2\right] - \frac{1}{4}G\left(\sum_{\mathbf{k}}{}'\sin\theta_{\mathbf{k}}\right)^2, \tag{6.56}$$

where the fourth and fifth terms correspond to the Hartree and Fock

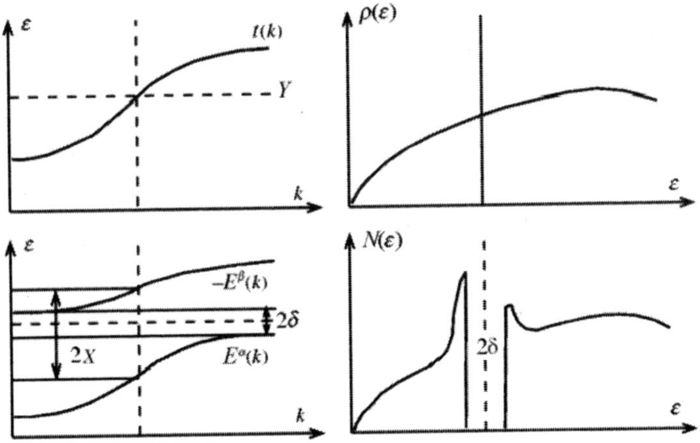

Fig. 6.5. The energy spectrum and density of states in the effective hybridization model of an intermediate valent system.

decoupling, $n_{\mathbf{k}}^{\alpha,\beta} = f(E_{\mathbf{k}}^{\alpha,\beta} \mp \zeta)$ are the Fermi distribution functions for 'electron' and 'hole' excitations,

$$\sum_{\mathbf{k}}' \ldots \equiv \sum_{\mathbf{k}} (1 - n_{\mathbf{k}}^{\alpha} - n_{\mathbf{k}}^{\beta}) \ldots$$

The one-particle energies are given by

$$E_{\mathbf{k}}^{\alpha,\beta} = \frac{\delta \langle \mathrm{H} \rangle}{\delta n_{\mathbf{k}}^{\alpha,\beta}} = \frac{1}{2}\left[E_{\mathbf{k}}^{\alpha,\beta} \pm (t_{\mathbf{k}} + \Delta) \right], \tag{6.57}$$

where

$$E_{\mathbf{k}} = \left[X^2 + (t_{\mathbf{k}} + Y)^2 \right]^{1/2},$$

$$X = 2V - G \sum_{\mathbf{k}}' \sin\theta_{\mathbf{k}}, \qquad Y = \Delta + G \sum_{\mathbf{k}}' \cos\theta_{\mathbf{k}}.$$

The hybridization picture of the electron spectrum (Fig.6.5) is confirmed by optical spectra and other data for IV semiconductors SmB_6, SmS and YbB_{12} [591] and by investigation of transport properties for TmSe [592].

In the case of the semiconductor, the chemical potential is determined from the condition

$$\sum_{\mathbf{k}} (n_{\mathbf{k}}^{\alpha} - n_{\mathbf{k}}^{\beta}) = 0. \tag{6.58}$$

The renormalized hybridization parameter X determines the 'direct' energy gap

$$|X| = \min_{\mathbf{k}} (E_{\mathbf{k}}^{\alpha} - E_{\mathbf{k}}^{\beta}), \tag{6.59}$$

which is observed in optical transitions. Unlike the case of usual semiconductors, extrema of valence and conduction band do not coincide. The width of the hybridization gap is estimated in terms of X and bandwidth W as $\delta \sim X^2/W$. The quantity Y is the position of f-level in the Hartree–Fock approximation. Varying (6.56) with respect to $\theta_{\mathbf{k}}$ we obtain

$$\sin\theta_{\mathbf{k}} = -\frac{X}{E_{\mathbf{k}}}, \qquad \cos\theta_{\mathbf{k}} = \frac{t_{\mathbf{k}} - Y}{E_{\mathbf{k}}},$$

and the equations for X and Y take the form

$$X = 2V \left(1 - G \sum_{\mathbf{k}}' \frac{1}{E_{\mathbf{k}}} \right)^{-1}; \tag{6.60}$$

$$Y = \Delta + G\left(1 - \sum_{\mathbf{k}}{}' \frac{t_{\mathbf{k}} - Y}{E_{\mathbf{k}}}\right). \tag{6.61}$$

These equations have, generally speaking, several IV solutions with various values of the energy gap. Consider the case $T = 0$. For a narrow-gap IV state with $|X| \ll W$ the quantity Y equals approximately to the Fermi energy and the equation (6.60) takes the form

$$X = 2V\left(1 - 2\lambda \ln \frac{aW}{|X|}\right)^{-1}, \quad \lambda = G\rho, \quad a \sim 1. \tag{6.62}$$

The solution to

$$X_1 = 2V\left(1 - 2\lambda \ln \frac{aW}{|V|}\right)^{-1} \tag{6.63}$$

describes the state with a hybridization gap, which is renormalized by correlation effects. It should be noted that the logarithmic divergences in the narrow-gap state are cut at the width of the energy gap, so that, unlike metallic Kondo lattices, the strong coupling regime is not achieved.

At $\lambda \geq 1$ the expression (6.63) yields the unique narrow-gap solution. In this case, there exist also two broad-band solution, which are approximately (at neglecting V) determined from the equation

$$G \sum_{\mathbf{k}}{}' \frac{1}{E_{\mathbf{k}}} = 1, \tag{6.64}$$

describing 'excitonic' gap states owing to Coulomb interaction.

In the case $\lambda \ll 1$ equation (6.62) has additional narrow-gap excitonic solutions provided that

$$|V| < V_c = aW\lambda \exp\left(-\frac{1}{2\lambda} - 1\right). \tag{6.65}$$

For $V \ll V_c$ they are given by

$$|X_{2,3}| = aW \exp\left(-\frac{1}{2\lambda}\right) \pm \frac{|V|}{\lambda}. \tag{6.66}$$

The corresponding total energy is lower than for the state with the hybridization gap (6.63). Thus, for a given valence there are metastable insulator states. One can assume that in real IV semiconductors $V \sim V_c$ so that the value of the gap is determined by a combined effect of hybridization and many-electron (exciton) effects. Phase transitions between states with different gaps were observed in the system $TmSe_{1-x}Te_x$ [593].

It should be noted that the hybridization gap can in principle be obtained in band structure calculations. Such calculations with account of relativistic effects were performed for SmS [594].

The expression for the density of states for the spectrum (6.57) in terms of the bare DOS $\rho(t)$ for the s-band has the form

$$N^{\alpha,\beta}(E)=\left(1+\frac{X^2}{4(E\mp Y)^2}\right)\rho\left(\frac{X^2}{4(Y\mp E)}\pm E\right). \tag{6.67}$$

Near the tops of the bare band and of the lower 'hole' hybridization band we have respectively

$$\rho(T)=A(t_{\max}-t)^{1/2}; \tag{6.68}$$

$$N^{\beta}=\tilde{A}(E_{\max}-E)^{1/2}, \qquad \tilde{A}=A\left(\frac{2t_{\max}}{X}\right)^2\approx A\frac{W}{\delta}. \tag{6.69}$$

Taking into account the renormalization of the chemical potential (calculated from the band top) we obtain for the electronic specific heat enhancement in a highly doped semiconductor

$$\frac{\tilde{\gamma}}{\gamma}=\frac{N(\tilde{\xi})}{\rho(\xi)}=\frac{\tilde{A}}{A}\left(\frac{\tilde{\xi}}{\xi}\right)^{1/2}=\left(\frac{\tilde{A}}{A}\right)^{2/3}=\left(\frac{W}{\delta}\right)^{2/3}. \tag{6.70}$$

Thus the states near the gap possess large effective mass. This explains the large value of the linear specific heat ($\gamma=145$ mJ/mol $\cdot$ K^2) [595] in the 'golden phase' of SmS where the current carriers (heavy holes) lie near the lower edge of the gap.

To consider finite temperature behaviour we treat the simplest case where the conduction band is symmetric and f-level lies exactly at the conduction band centre, so that the mean number of f-electrons equals to 1/2 and the chemical potential does not depend on temperature. Then the equation (6.61) takes the form

$$X=\frac{2V}{1-GL}, \qquad L=\sum_{\mathbf{k}}\frac{1-2n_{\mathbf{k}}}{E_{\mathbf{k}}}. \tag{6.71}$$

In the narrow-gap state we have

$$L=\int_{-W}^{W}\frac{\rho(E)}{(E^2+X^2)^{1/2}}\tanh\frac{(E^2+X^2)^{1/2}+E}{4T}dE\approx\rho\int_{-W/2T}^{W/2T}\frac{\tanh x}{x}dx. \tag{6.72}$$

At high temperatures, the energy gap logarithmically decreases with T:

$$X(T)\approx\frac{2V}{\lambda\ln(T/T^*)}, \quad T\gg\delta(T=0),T^*, \tag{6.73}$$

with

$$T^{*}=\frac{\pi}{4}\frac{W}{1.14}\exp\left(-\frac{1}{\lambda}\right), \qquad (6.74)$$

being of order of the direct excitonic gap. Such a behaviour explains qualitatively data on the IV semiconductor SmB_6 where an appreciable temperature dependence of the gap is observed in ESR experiments [596]. At low temperatures, the corrections are exponentially small, as well as in the case of the superconducting gap.

To calculate the electronic specific heat in the narrow-gap insulating state we write down the expression for the entropy of quasiparticles α and β

$$\mathcal{S}=-\sum_{\mathbf{k},j=\alpha,\beta}\left[n_{\mathbf{k}}^{j}\ln n_{\mathbf{k}}^{j}+(1-n_{\mathbf{k}}^{j})\ln(1-n_{\mathbf{k}}^{j})\right].$$

Then we get

$$C(T)=T\frac{\partial S}{\partial T}=-\frac{1}{T}\sum_{\mathbf{k}j}\frac{\partial n_{\mathbf{k}}^{j}}{\partial E_{\mathbf{k}}^{j}}\left(1-\frac{T}{2}\frac{\partial}{\partial T}\right)(E_{\mathbf{k}}^{j})^{2}. \qquad (6.75)$$

At low temperatures $T\ll\delta(0)$ specific heat is exponentially small,

$$C(T)\approx\frac{2\rho}{T}X^{2}(0)\exp\left(\frac{\delta}{T}\right). \qquad (6.76)$$

At high temperatures $\delta\ll T<|X|$ we have

$$C(T)=\frac{\rho}{T}X^{2}(T)\left[1-\frac{2}{\ln(T^{*}/T)}\right]+\frac{\pi^{2}}{3}\rho T. \qquad (6.77)$$

The first term in (6.77) dominates over the usual linear term. Thus $C(T)$ should have a maximum at $T\sim\delta$. Such a behaviour of $C(T)$ is reminiscent of the Schottky anomaly for localized states, but because of hybridization the dependence $1/T^2$ transforms into $1/T$.

The magnetic properties of the IV state may be investigated after including spin variables within the model (6.54) [597]. The corresponding energy spectrum in the magnetic field is given by

$$E_{\mathbf{k}\sigma}^{\alpha,\beta}(H)=E_{\mathbf{k}\sigma}^{\alpha,\beta}(0)-\frac{\sigma}{2}H\left[\mu_{s}+\mu_{f}\pm(\mu_{s}-\mu_{f})\frac{t_{\mathbf{k}}-E_{\mathrm{F}}}{E_{\mathbf{k}}}\right], \qquad (6.78)$$

where μ_s and μ_f are magnetic moments of s- and f-electrons. The field dependence of the energy gap turns out to be simple

$$\delta(H)=\frac{1}{2}\left[\min_{\mathbf{k}}E_{\mathbf{k}\uparrow}^{\alpha}(H)+\min_{\mathbf{k}}E_{\mathbf{k}\uparrow}^{\beta}(H)\right]\approx\delta(0)-\mu_{f}H. \qquad (6.79)$$

One can see that at $\delta=\mu_f H$ the gap vanishes and a transition into metallic state takes place. Such a transition was observed in YbB_{12} at $H \sim 200$ kOe [598].

Magnetic susceptibility is given by

$$\chi(T) = \lim_{H \to 0} \sum_{\sigma} \frac{\sigma(\mu_f \langle f_\sigma^\dagger f_\sigma \rangle + \mu_s \langle c_\sigma^\dagger c_\sigma \rangle)}{H}. \qquad (6.80)$$

At high temperatures $T \gg \delta$, χ is determined by the contribution of transitions between hybridization subbands

$$\Delta\chi(T) = \frac{1}{2} \sum_{\mathbf{k}} \left\{ \left(-\frac{\partial n_{\mathbf{k}}^{\alpha}}{\partial E_{\mathbf{k}}} \right) \left[\mu_f \left(1 - \frac{t_{\mathbf{k}} - E_{\mathrm{F}}}{E_{\mathbf{k}}} \right) + \mu_s \left(1 + \frac{t_{\mathbf{k}} - E_{\mathrm{F}}}{E_{\mathbf{k}}} \right) \right]^2 + \right.$$

$$\left. + \left(-\frac{\partial n_{\mathbf{k}}^{\alpha}}{\partial E_{\mathbf{k}}} \right) \left[\mu_f \left(1 + \frac{t_{\mathbf{k}} - E_{\mathrm{F}}}{E_{\mathbf{k}}} \right) + \mu_s \left(1 - \frac{t_{\mathbf{k}} - E_{\mathrm{F}}}{E_{\mathbf{k}}} \right) \right]^2 \right\} \approx \frac{\mu_f^2}{T} \rho W. \qquad (6.81)$$

Thus we obtain, owing to strong energy dependence of density of states (narrow peaks), a Curie-like behaviour with a non-integer magnetic moment. This mechanism of temperature dependence was also discussed for transition d-metals [599]. At $T \ll \delta$ the contribution (6.81) is exponentially small, and $\chi(T)$ is determined by intrasubband transition contribution which is given by

$$\chi(0) = 2(\mu_s - \mu_f)^2 \rho. \qquad (6.82)$$

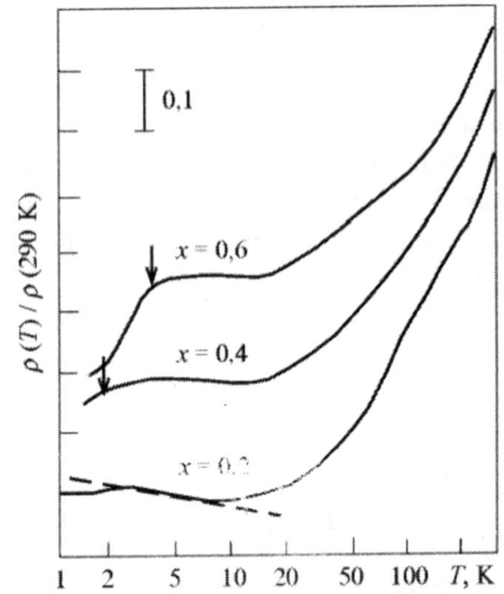

Fig. 6.6. Temperature dependences of resistivity for the system $Ce_x La_{1-x} Ge_2$ (which is ferromagnetic at $0.4 < x < 1$), the logarithmic scale being used. Arrows show the value of T_C [601].

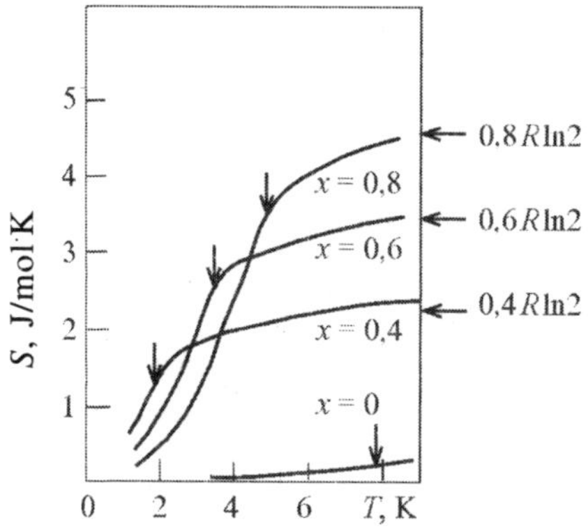

Fig. 6.7. Temperature dependences of magnetic entropy for the system $Ce_xLa_{1-x}Ge_2$. Arrows show the value of T_C [601].

Note that at $\mu_s=\mu_f$ expression (6.82) vanishes since the ground state is singlet.

We see that $\chi(T)$ should also have a maximum at $T\sim\delta$, characteristic energy scale δ playing thereby the role of the 'Kondo temperature'. Such a maximum is weakly pronounced in SmB_6 and SmS [512] and clearly observed in YbB_{12} [600], the susceptibility at low T being masked by Van Vleck contribution of the Sm^{2+} ion or by paramagnetic impurities.

6.6 MAGNETIC ORDERING IN KONDO LATTICES AND HEAVY-FERMION COMPOUNDS

It was traditionally believed for many years that the competition of the intersite RKKY exchange interaction and the Kondo effect should result in the formation of either the usual magnetic ordering with large atomic magnetic moments (as in elemental rare-earth metals) or the non-magnetic Kondo state with suppressed magnetic moments.

However, experimental investigations of last years have convincingly demonstrated that magnetic ordering and pronounced spin fluctuations are widely spread among heavy-fermion systems and other anomalous 4*f*- and 5*f*-compounds, which are treated usually as concentrated Kondo systems. Data on magnetic properties of such systems are presented in the Table 6.1.

The class of 'Kondo' magnets is characterized by the following features [601]

1. Logarithmic temperature dependence of resistivity at $T>T_K$

characteristic of Kondo systems (Fig.6.6).

2. Small value of the magnetic entropy at the ordering point, in comparison with the value $R\ln(2S+1)$, which corresponds to the usual magnets with localized moments (Fig.6.7). This phenomenon is connected with the suppression of magnetic specific heat owing to the Kondo effect (see Sect.6.1).
3. The ordered magnetic moment M_s is small in comparison with the 'high-temperature' moment μ_{eff} determined from the Curie constant. The latter has as a rule 'normal' value, which is close to that for the corresponding rare-earth ion (e.g., $\mu_{eff} \simeq 2.5\mu_B$ for Ce^{3+} ion). Such a behaviour is reminiscent of weak itinerant magnets (see Sect.4.4).
4. Paramagnetic Curie temperature θ is as rule negative (even for ferromagnets) and exceeds appreciably in absolute value the magnetic ordering temperature. This behaviour is due to the large single-site Kondo contribution to the paramagnetic susceptibility ($\chi(T=0) \sim 1/T_K$). The most bright example is the Kondo ferromagnet $CeRh_3B_2$ with $T_C = 115$ K, $\theta = -370$ K [284] with moderate $\gamma = 16$ mJ/(mol · K^2) (Large value of T_C in this compound, which exceeds even that for $GdRh_3B_2$, $T_C = 105$K, is not typical and probably connected with strong d–f hybridization.)

There exist numerous examples of systems (ferromagnets CePdSb, $CeSi_x$, Sm_3Sb_4, Ce_4Bi_3, $NpAl_2$, antiferromagnets $CeAl_2$, TmS, CeB_6, $UAgCu_4$; some experimental data and bibliography are presented in Table 6.1) where 'Kondo' anomalies in thermodynamic and transport properties coexist with magnetic ordering, the saturation moment M_s being of order of $1\mu_B$.

As for heavy-fermion systems themselves, the situation is more complicated. There exist unambiguous evidences for the antiferromagnetism in UCd_{11} and U_2Zn_{17} with the same order of magnitude of M_s [507]. For UPt_3 and URu_2Si_2, $M_s \simeq 2-3\cdot10^{-2}\mu_B$ [524–526]. Antiferromagnetic ordering with very small M_s was also reported for $CeAl_3$ [509, 510], UBe_{13} [527], $CeCu_2Si_2$ [513, 514]. Indications of possible magnetic transition at 2mK were obtained for $CeCu_6$ [515]. However, the data for $CeAl_3$ and UBe_{13} were not confirmed in papers [511] and [528] respectively. Generally, a characteristic feature of heavy fermion magnets is high sensitivity of M_s to external factors such as pressure and doping by a small amount of impurities. For example, UBe_{13} becomes an antiferromagnet with an appreciable M_s under the pressure $P > 23$ kBar; on the contrary, $CeAl_3$ becomes paramagnetic above $P = 3$ kBar [510]. The moment in UPt_3 increases

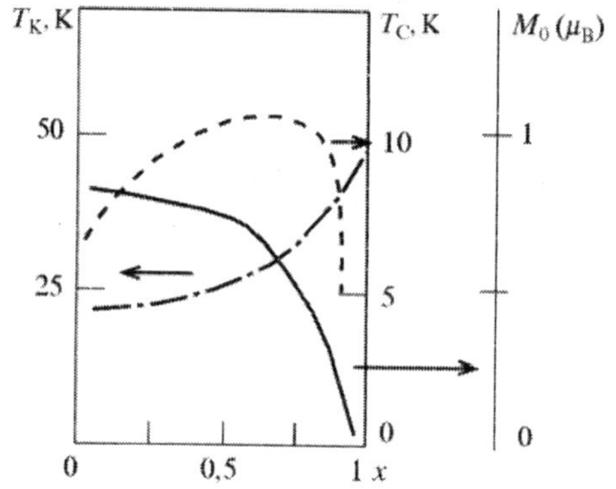

Fig. 6.8. Concentration dependences of saturation moment M_0 (solid line), T_C (broken line) and T_K (schematically, dash-dotted line) in the system $CeNi_xPd_{1-x}$ [601].

up to values of order $1\mu_B$ upon adding 5% of Pd instead of Pt or 5%0 of Th instead of U [602]. A number of heavy-fermion systems demonstrates metamagnetic transitions in weak magnetic fields with a sharp increase of magnetic moment [603]. The 'marginal' situation in the magnetic state of heavy-fermion systems is discussed in [604] by using experimental data on their critical behaviour.

The problem of magnetic ordering in the Kondo lattices was investigated in a number of theoretical works [605–612]. The roles of the Kondo effect and and the intersite RKKY interaction are determined by the relation of the two energy scales: the Kondo temperature $T_K = W \exp(1/2I\rho)$ which determines the crossover from the free-moment regime to the strong coupling region, and $T_{RKKY} \sim I^2\rho$. The latter quantity is of the order of magnetic ordering temperature T_M in the absence of the Kondo effect. The ratio T_K/T_M may vary depending on external parameters and alloy composition. As an example, Fig.6.8 shows concentration dependences of the saturation magnetization M_0, T_K and T_C in the alloy $CeNi_{1-x}Pd_x$.

In the non-magnetic case, $T_{RKKY} \sim \bar{\omega}$ with $\bar{\omega}$ being a characteristic spin-fluctuation frequency. For most compounds under consideration, $T_K > T_{RKKY}$. However, there exist also anomalous cerium and uranium-based magnets with $T_K \ll T_N$, e.g., $CeAl_2Ga_2$ [521], $UAgCu_4$ (T_N= 18 K, T_K=3 K [530]). This case is close to that of usual rare-earth magnets with the Kondo effect almost suppressed by magnetic ordering.

To describe the formation of magnetic Kondo state, we consider Kondo perturbation corrections to magnetic characteristics with account of spin dynamics. Calculation of magnetic susceptibility [367] yields (cf. (6.7))

$$\chi = \frac{S(S+1)}{3T}(1-4I^2L), \tag{6.83}$$

where

$$L = \frac{1}{S(S+1)}\sum_{\mathbf{pq}}\int K_{\mathbf{p-q}}(\omega)\frac{n_{\mathbf{p}}(1-n_{\mathbf{q}})}{(t_{\mathbf{q}}-t_{\mathbf{p}}-\omega)^2}d\omega \tag{6.84}$$

and the spin spectral density is defined by (6.26). A simple estimation of the integral in (6.84) yields

$$\chi = \frac{S(S+1)}{3T}\left(1-2I^2\rho^2\ln\frac{W^2}{T^2+\bar{\omega}^2}\right), \tag{6.85}$$

where the quantity in square brackets describes the suppression of the effective moment. The Kondo corrections to the magnetic moment in the ferro- and antiferromagnetic state are obtained by using the standard spin-wave result

$$\delta\bar{S} = -\sum_{\mathbf{q}}\langle b_{\mathbf{q}}^{\dagger}b_{\mathbf{q}}\rangle \tag{6.86}$$

and substituting zero-point corrections to magnon occupation numbers

$$\delta\langle b_{\mathbf{q}}^{\dagger}b_{\mathbf{q}}\rangle = 2I^2S\sum_{\mathbf{k}}\frac{n_{\mathbf{k}\downarrow}(1-n_{\mathbf{k-q}\uparrow})}{(t_{\mathbf{k}\downarrow}-t_{\mathbf{k-q}\uparrow}-\omega_{\mathbf{q}})^2}, \tag{6.87}$$

$$\delta\left\{\begin{matrix}\langle b_{\mathbf{q}}^{\dagger}b_{\mathbf{q}}\rangle\\ \langle b_{\mathbf{q}}^{\dagger}b_{-\mathbf{q}}^{\dagger}\rangle\end{matrix}\right\} = 2I^2S\sum_{\mathbf{k}}\left(\frac{n_{\mathbf{k}}(1-n_{\mathbf{k-q}})}{(t_{\mathbf{k}}-t_{\mathbf{k-q}}-\omega_{\mathbf{q}}^2}\pm\frac{n_{\mathbf{k}}(1-n_{\mathbf{k+Q-q}})}{(t_{\mathbf{k}}-t_{\mathbf{k+Q-q}})^2-\omega_{\mathbf{q}}^2}\right) \tag{6.88}$$

respectively (see Appendix G). The integration in both the cases gives

$$\delta\bar{S}/S = -2I^2\rho^2\ln\frac{W}{\bar{\omega}}. \tag{6.89}$$

The obtained corrections to the ground state moment occur in any conducting magnets including pure 4f-metals. However, in the latter case they should be small (of order of 10^{-2}). On the other hand, it would be interesting to search them in rare-earth compounds with high values of ρ.

To obtain a self-consistent picture for a magnet with appreciable Kondo renormalizations, we have to calculate the corrections to characteristic spin-fluctuation frequencies $\bar{\omega}$. In the paramagnetic phase, we can use the estimation from the second-order correction to the dynamic susceptibility

$$\omega_{\mathbf{q}}^2 = (\dot{S}_{-\mathbf{q}}^z, \dot{S}_{\mathbf{q}}^z)/(S_{-\mathbf{q}}^z, S_{\mathbf{q}}^z), \tag{6.90}$$

with

$$(A,B) \equiv \int_0^{1/T} \langle \exp(\lambda H) B \exp(-\lambda H) B \rangle d\lambda. \tag{6.91}$$

The calculation yields [608]

$$\omega_{\mathbf{q}}^2 = \frac{4}{3} S(S+1) \sum_{\mathbf{p}} (J_{\mathbf{q-p}} - J_{\mathbf{p}})^2 [1 - 4I^2 L(1-\alpha_{\mathbf{q}})]. \tag{6.92}$$

Here L is defined by (6.85),

$$\alpha_{\mathbf{q}} = \sum_{\mathbf{R}} J_{\mathbf{R}}^2 \left(\frac{\sin k_{\mathrm{F}} R}{k_{\mathrm{F}} R} \right)^2 [1 - \cos \mathbf{qR}] \Big/ \sum_{\mathbf{R}} J_{\mathbf{R}}^2 [1 - \cos \mathbf{qR}]. \tag{6.93}$$

Since $0 < \alpha_{\mathbf{q}} < 1$, the Kondo effect results in decreasing $\bar{\omega}(T)$ as T is lowered. In the approximation of nearest neighbours (with the distance d) for $J(\mathbf{R})$, the value of α does not depend on $\mathbf{q}$:

$$\alpha_{\mathbf{q}} = \alpha = \left(\frac{\sin k_{\mathrm{F}} d}{k_{\mathrm{F}} d} \right)^2. \tag{6.94}$$

The corrections to the spin-wave frequency in ferromagnetic and antiferromagnetic phase owing to magnon-magnon interactions are obtained by using the results of Appendix G, (E.5), (E.13), (6.87), (6.88). Then we obtain

$$\delta\omega_{\mathbf{q}} / \omega_{\mathbf{q}} = -4I^2 \rho^2 a \ln \frac{W}{\bar{\omega}}, \tag{6.95}$$

where the factor a depends on the type of magnetic ordering.

The above perturbation theory results permit a qualitative description of the magnetic Kondo-lattice state with a small magnetic moment. Suppose we lower the temperature starting from the paramagnetic state. As we do it, the magnetic moment is 'compensated', but, in contrast with the one-impurity situation, the degree of compensation is determined by $(T^2 + \bar{\omega}^2)^{1/2}$ rather than T. At the same time, $\bar{\omega}$ itself decreases according to (6.92). This process cannot be described analytically in terms of perturbation theory. However, one needs to take $\bar{\omega} \sim T_{\mathrm{K}}$ at $T < T_{\mathrm{K}}$ if he has in view the establishment of an universal energy scale of the order of T_{K}. The latter fact is confirmed by a large body of experimental data on quasi-elastic neutron scattering in Kondo systems, which demonstrate that at low T the typical "central-peak" width $\Gamma \sim \bar{\omega}$ is of the same order of magnitude as the Fermi degeneracy temperature determined from thermodynamic and transport properties, i.e. T_{K}. Thus the process of the magnetic moment compensation terminates somewhere at the boundary of the strong-coupling region and results in the state with a finite (although possibly small) saturation moment M_s.

A quantitative consideration of the Kondo lattice magnetism problem may be performed within renormalization group approach in the simplest form of Anderson's "poor man scaling" [554]. The above expressions, obtained from perturbation theory, enable one to write down the renormalization group equations for the effective s–f parameter and $\bar{\omega}$ [612]. This is achieved by considering the integrals over **k** with the Fermi functions in the Kondo corrections to electron self-energies (G.33), (G.73) and spin-fluctuation frequencies. To construct a scaling procedure, one picks out the contributions from the energy layer $C < E < C + \delta C$ near the Fermi level $E_F = 0$. For example, in the case of ferromagnet we have for the effective splitting in the electron spectrum

$$2I_{\text{eff}} S = 2IS - \left[\Sigma_{\mathbf{k}\uparrow}^{\text{FM}}(E_F) - \Sigma_{\mathbf{k}\downarrow}^{\text{FM}}(E_F)\right]_{k=k_F} \tag{6.96}$$

Using (G.34) we derive

$$\delta I_{\text{eff}} = I^2 \sum_{C<t_{\mathbf{k+q}}<C+\delta C} \left(\frac{1}{t_{\mathbf{k+q}} + \omega_{\mathbf{q}}} + \frac{1}{t_{\mathbf{k+q}} - \omega_{\mathbf{q}}} \right) = \frac{\rho I^2}{\bar{\omega}} \delta C \ln \left| \frac{C - \bar{\omega}}{C + \bar{\omega}} \right|, \tag{6.97}$$

where

$$\bar{\omega} = 4Dk_F^2,$$

D is the spin-wave stiffness. Introducing the dimensionless coupling parameters

$$g = -2I\rho, \qquad g_{\text{eff}}(C) = -2I_{\text{eff}}(C)\rho, \tag{6.98}$$

we obtain the system of renormalization group equations. In the nearest-neighbour approximation for the intersite Heisenberg exchange interaction it has a simple form

$$\begin{aligned} \partial g_{\text{eff}}(C)/\partial C &= -\Phi, \\ \partial \ln \bar{\omega}_{\text{eff}}(C)/\partial C &= \Phi/2, \\ \partial \ln \bar{S}_{\text{eff}}(C)/\partial C &= \Phi/2, \end{aligned} \tag{6.99}$$

with

$$\Phi = \Phi(C, \bar{\omega}_{\text{eff}}(C)) = [g_{\text{eff}}^2(C)/C]\phi(\bar{\omega}_{\text{eff}}(C)/C). \tag{6.100}$$

The scaling function for para-, ferro- and antiferromagnetic phases has the form

$$\phi(x)=\begin{cases} x^{-1}\arctan x, & \text{PM}, \\ \frac{1}{2x}\ln\left|\frac{1+x}{1-x}\right|, & \text{FM}, \\ -x^{-2}\ln|1-x^2|, & \text{FM}. \end{cases} \tag{6.101}$$

In all the cases $\phi(0)=1$, which guarantees the correct limit of one Kondo impurity (Sect.6.1).

The results of investigation of the equations (6.99)–(6.101) [612] are as follows. Depending on the relation between the one-impurity Kondo temperature and the bare spin-fluctuation frequency, three regimes are possible at $I<0$:

1. the strong coupling regime where g_{eff} diverges at [illegible] ne C is roughly determined by the condition
$$\bar{\omega}<T_{\text{K}}=W\exp(-1/g). \tag{6.102}$$
Then $I_{\text{eff}}(C\to 0)=\infty$, so that all the conduction ele[illegible]s are bound into singlet states and spin dynamics is suppressed.
2. the regime of a 'Kondo' magnet with an appreciable, but not total compensation of magnetic moments. This is realized in the interval $T_{\text{K}}<\bar{\omega}<AT_{\text{K}}$ (A is a numerical factor of order unity), which corresponds to a small interval $\delta g\sim g^2$. In this interval, the renormalized values of magnetic moment and spin-fluctuation frequency, $S_{\text{eff}}(0)$ and $\bar{\omega}_{\text{eff}}(0)$, increase from zero to approximately the bare values.
3. the regime of 'usual' magnets with small logarithmic corrections to the ground state moment (see (6.89)), which occurs at $\bar{\omega}>AT_{\text{K}}$. Note that the same situation takes place in the case of 'ferromagnetic' s–f exchange interaction $I>0$.

High sensitivity of the magnetic state to external factors, which was discussed above, is explained by that in the case 2 the magnetic moment changes strongly at small variations of the bare coupling constant. Of course, the quantitative description should be different for the realistic long-range behaviour of the RKKY-interaction. The renormalization of the latter may be not described by the single constant α.

Outwardly, the described mechanism of the formation of magnetic state with small M_s differs radically from the ordinary mechanism for weak itinerant ferromagnets which are assumed to be located in the immediate vicinity of Stoner instability. Recall, however, that both the energy spectrum of new Fermi quasiparticles and the ef-

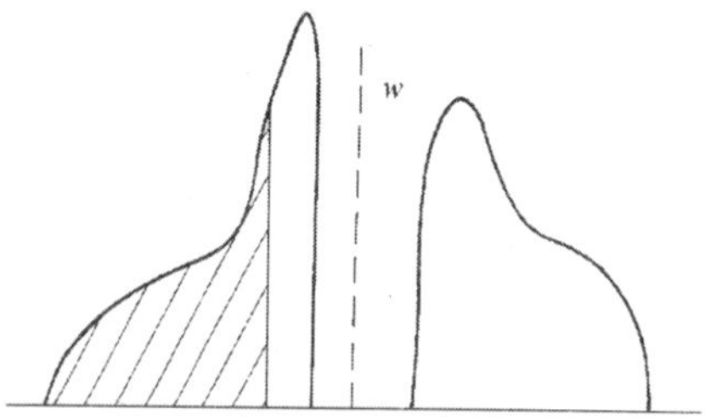

Fig. 6.9. The density of states picture in the mean-field approximation for the s-f exchange model (the non-magnetic phase) [608].

fective interaction among them suffer strong renormalizations. Therefore the inapplicability of the Stoner criterion with bare parameters for Kondo magnets is practically evident.

Since a continuous transition exists between the highly-correlated Kondo lattices and the 'usual' itinerant-electron systems (in particular, Pauli paramagnets may be viewed as systems with high T_K of order of the Fermi energy), the question arises about the role which many-electron effects play in the 'classical' weak itinerant magnets like $ZrZn_2$. It may turn out that the closeness of the ground state to the Stoner instability point, i.e. the smallness of M_s, in the latter systems is not due to accidental bare values of $N(E_F)$ and Stoner parameter, but determined by their renormalization. In this context, it would be of interest to describe weak itinerant magnets not from the 'band' side, but from the side of local magnetic moments which are nearly compensated. As it is customary now to treat UPt_3, $CeSi_x$ and $CeRh_3B_2$ as weak itinerant magnets (see, e.g., [613]), the second approach appears already by far less natural than the first. From the formal point of view, the calculations in the Hubbard model (Appendix G), which describes itinerant electron systems, are similar to those in the *s–f* exchange model if one postulates the existence of local moments.

The simple analytical description of crossover between the high-temperature region $T > T_K$, where perturbation theory holds, and the strong-coupling regime is hardly possible. Therefore it is important to discuss approximations which work at $T < T_K$. A special mean-field approximation for the ground state of magnetic Kondo lattices is considered in Appendix O. This approach exploits the pseudofermion representation for spin operators and reduces the *s–f* exchange model to an effective hybridization model.

The corresponding energy spectrum contains narrow DOS peaks owing to the pseudofermion contribution (Fig.6.9). It should be noted that *f*-pseudofermions become itinerant in the situation under consideration. Delocalization of *f*-electrons in heavy-fermion systems, which is confirmed by observation of large electron mass in de Haas–van Alphen experiments, is not simply understandable in the *s–f* exchange model (unlike the situation in the Anderson model with *f*-states near the Fermi level with bare *s–f* hybridization included). This delocalization is analogous to the occurrence of the Fermi excitation branch in the resonating valence bond (RVB) theory of high-T_C superconductors (see Sect.6.8).

As discussed above in Sect.6.4, the hybridization form of electron spectrum with the presence of DOS peaks is confirmed by numerous experimental investigations of Kondo lattices, including heavy-fermion systems. As for ferromagnetic Kondo systems, of interest are the results on the temperature dependences of magnetization in Sm_4Sb_3 and Sm_4As_3 [542], which turn out to be non-monotonous. Such a behaviour (temperature-induced ferromagnetism) may be explained by the sharp energy dependence of DOS in the hybridization model [614].

Because of the dependence of the effective hybridization on spin projection, there exist, generally speaking, several ferromagnetic solutions. For the constant bare DOS, only the saturated ferromagnetic state turns out to be stable (remember that the same situation takes place in the Wohlfarth model, i.e. the Stoner model with the rectangle DOS). One may assume that in more general models (e.g., for large degeneracy of electron bands) the role of this dependence is not so important, so that it may be neglected. Then the criterion of ferromagnetism $J \sim T_K$ reduces, roughly speaking, to the Stoner criterion with the replacement of the intrasite interaction parameter by the intersite exchange parameter J, the effective density of states at the Fermi level $N(E_F) \sim 1/T_K$ being large because of hybridization DOS peaks. Thus magnetism of Kondo lattices has features of both localized spin magnets (essential role of intersite exchange interaction) and of itinerant ones (non-integer value of the magnetic moment and its connection with the DOS structure).

The influence of spin-fluctuation corrections to the mean-field approximation was investigated in [608]. Calculating the contribution to the saturation magnetization from the fluctuations of the Heisenberg interaction by analogy with (6.88) we obtain

$$\delta\overline{S} \sim -\frac{J}{T_K}\ln\frac{T_K}{J}. \qquad (6.103)$$

(T_K plays the role of a characteristic energy scale in the electron spectrum). Thus at $J \sim T_K$ we have $\delta\overline{S} \sim S$ and formation of a state with a small moment is possible.

To consider antiferromagnetic ordering of the Kondo lattices one has to pass to the local coordinate system (see (E.8)). Then the mean-field Hamiltonian of the f-subsystem in the pseudofermion representation takes the form

$$\mathsf{H}_f = -J_{\mathbf{Q}}\bar{S}\sum_{\mathbf{k}}(f^{\dagger}_{\mathbf{k}+\mathbf{Q}\uparrow}f_{\mathbf{k}\downarrow} + f^{\dagger}_{\mathbf{k}\downarrow}f_{\mathbf{k}+\mathbf{Q}\uparrow}). \tag{6.104}$$

Unlike the case of a ferromagnet, one may neglect the σ-dependence of the hybridization since corrections owing to spin polarization have the structure $(J_{\mathbf{Q}}\bar{S})^2/(t_{\mathbf{k}+\mathbf{Q}}-t_{\mathbf{k}})$ and are proportional to $(J_{\mathbf{Q}}\bar{S})^2/W$. Thus the criterion of antiferromagnetism has the usual form

$$J_{\mathbf{Q}}\chi_{\mathbf{Q}} > 1, \tag{6.105}$$

with $\chi_{\mathbf{Q}}$ being the non-enhanced ($\mathsf{H}f \to 0$) staggered susceptibility of the f-subsystem in the effective hybridization model (O.2). Using the Bogoliubov transformation (O.8) yields

$$\chi_{\mathbf{q}} = \sum_{\mathbf{k}}\left[v^2_{\mathbf{k}}v^2_{\mathbf{k}+\mathbf{q}}\frac{n^{\alpha}_{\mathbf{k}} - n^{\alpha}_{\mathbf{k}+\mathbf{q}}}{E^{\alpha}_{\mathbf{k}} - E^{\alpha}_{\mathbf{k}+\mathbf{q}}} + u^2_{\mathbf{k}}u^2_{\mathbf{k}+\mathbf{q}}\frac{n^{\beta}_{\mathbf{k}} - n^{\beta}_{\mathbf{k}+\mathbf{q}}}{E^{\beta}_{\mathbf{k}} - E^{\beta}_{\mathbf{k}+\mathbf{q}}} - 2u^2_{\mathbf{k}}v^2_{\mathbf{k}+\mathbf{q}}\frac{n^{\beta}_{\mathbf{k}} - n^{\alpha}_{\mathbf{k}+\mathbf{q}}}{E^{\beta}_{\mathbf{k}} - E^{\alpha}_{\mathbf{k}+\mathbf{q}}}\right]. \tag{6.106}$$

Owing to the hybridization peaks, the contribution from the intersubband transitions (the last term in brackets of (6.106)) turns out to be large: $\chi_{\mathbf{Q}} \sim 1/T_{\mathrm{K}}$. Thus antiferromagnetism occurs at

$$J_{\mathbf{Q}} > \nu T_{\mathrm{K}}, \tag{6.107}$$

where the constant ν of the order of unity is determined by the band structure.

The electron spectrum of antiferromagnetic Kondo lattices is disturbed by both hybridization (Kondo) gap and an antiferromagnetic gap. Consider some experimental examples. For the intermediate-valent semiconductor TmSe the gap at E_{F} seems to be of a hybridization nature since there exist indication of its retaining in the paramagnetic state [592]. As discussed in Sect.6.4, the insulator Kondo state with very small (of order of 3 K) gap occurs in the compound CeNiSn which is non-magnetic (however, recently static spin correlations with extremely small local moment of order $10^{-3}\mu_{\mathrm{B}}$ were found in this substance at $T<0.13$ K [615]). The system YbNiSn has a small canted ferromagnetic moment (see Table 6.1), the absence of gap being probably connected with weakness of d–f hybridization in ytterbium compounds.

The compound UNiSn turns out to be an antiferromagnet [496, 497], in contradiction witn the band structure calculations [315] which yield a half-metallic ferromagnetic structure. A transition from metallic antiferromagnetic state to semiconductor paramagnetic one takes place at 47 K with increasing T (in contrast with the usual picture of the temperature-induced metal-insulator transition [25]), the gap in the semiconducting state being rather small (of order of 10 K). Most simple explanation of this phenomenon is that the emergence of the sublattice magnetization results in a shift of the Fermi level outside the energy gap. The gap may have either Kondo nature (as in CeNiSn) or usual band origin. The latter situation takes place in the systems TiNiSn, ZrNiSn where the gap in the C_{1b} lattice has the 'vacansion' nature [566]. (This situation is reminiscent of the Heusler aloys RMnSb, see Sect.4.5; recently, a hypothesis about the formation of the half-metallic ferromagnetic state in TiCoSn with $T_C = 143$ K was put forward [616].) However, the gap in these compounds is large (of order of 10^3 K). Therefore the Kondo origin of the gap in UNiSn seems to be more probable. Then the possible explanation of the metal-insulator transition observed is that the antiferromagnetic exchange interaction suppresses the Kondo order parameter $V \sim \langle c^\dagger f \rangle$ (see Appendix O) and, consequently, the gap [508].

6.7 CURRENT CARRIERS IN A TWO-DIMENSIONAL ANTIFERROMAGNET

The interest in many-electron models with strong correlations has been recently greatly revived in connection with the discovery of high-T_c superconductivity in copper-oxide ceramics $La_{2-x}Sr_xCuO_4$ [617] and $YBa_2Cu_3O_{7-y}$ [618]. The current carriers in these systems move in weakly coupled CuO_2-layers and form rather narrow energy bands. A characteristic feature of these systems are presence of pronounced spin fluctuations and, for some compositions, the antiferromagnetic ordering. The important role of correlations is confirmed by that La_2CuO_4 is a typical Mott–Hubbard insulator. This compound yields also the best known example of a quasi-two-dimensional Heisenberg antiferromagnet with small magnetic anisotropy.

In the present section we do not discuss the problem of high-temperature superconductivity itself, but demonstrate the application of simple ME models to the description of current carrier states in highly correlated two-dimensional systems.

The electron states in CuO_2-planes of copper-oxide perovskites may be described by the so-called Emery model

$$\mathsf{H} = \sum_{\mathbf{k}\sigma}\left[\varepsilon p^{\dagger}_{\mathbf{k}\sigma} p_{\mathbf{k}\sigma} + \Delta d^{\dagger}_{\mathbf{k}\sigma} d_{\mathbf{k}\sigma} + V_{\mathbf{k}}(p^{\dagger}_{\mathbf{k}\sigma} d_{\mathbf{k}\sigma} + d^{\dagger}_{\mathbf{k}\sigma} p_{\mathbf{k}\sigma})\right] + \\ + U\sum_{i} d^{\dagger}_{i\uparrow} d_{i\uparrow} d^{\dagger}_{i\downarrow} d_{i\downarrow}, \quad (6.108)$$

where ε and Δ are positions of p- and d-levels for Cu and O ions respectively. The **k**-dependence of matrix elements of p–d hybridization for the square lattice is given by

$$V_{\mathbf{k}} = 2V_{pd}\left(\sin^2 k_x + \sin^2 k_y\right)^{1/2}. \quad (6.109)$$

At $|V_{pd}| \ll \varepsilon - \Delta$ the Hamiltonian (6.108) is reduced by a canonical transformation [619] to the Hubbard model with strong Coulomb repulsion and the effective Cu–Cu transfer integrals

$$t_{\mathrm{eff}} = \frac{V_{pd}^2}{\varepsilon - \Delta}. \quad (6.110)$$

At present, a large number of models for high-T_{C} superconductors are developed which take into account formation of several hybridized narrow and wide bands with orbital degeneracy. Here we confine ourselves to a simple consideration of the current carriers within the s–d exchange model [620].

An important property of two-dimensional (2D) Heisenberg magnets is the absence of long-range order at finite temperatures since it is destroyed by long-wavelength fluctuations. At the same time, the strong short-range order with large correlation length ξ persists up to the temperatures of order of the intersite exchange parameter J. Unlike purely 2D Heisenberg magnets, quasi-2D compounds possess finite values of magnetic ordering temperature because of a weak interlayer coupling J' and/or easy-axis magnetic anisotropy which are estimated as

$$T_{\mathrm{M}} = 4\pi |J| S^2 \Big/ \ln \frac{|J|}{\max\{|J'|, |J^z - J^{\perp}|\}}. \quad (6.111)$$

Thus the ordering temperature is small, which is reminiscent to weak itinerant magnets. In such a situation, a consistent perturbation theory can be developed. Formulas, which are more exact than (6.111) and yield a quantitative agreement with experimental data, were obtained in [728].

Experimental data on layered perovskites (including La–Cu–O systems) demonstrate a pronounced short-range order above the Neel point. A similar situation takes place in frustrated three-dimensional (3D) antiferromagnets where long-range magnetic ordering is also partially suppressed (see Sect.6.8).

Recent developments in the theory of two-dimensional Heisenberg antiferromagnets have provided a simple and successful description of their thermodynamic properties. Unlike the usual mean-field approximation, the self-consistent spin-wave theories (SSWT), based on the non-linear representations of Schwinger bosons [622, 623] or the Dyson–Maleev ideal magnons [624], yield a smooth transition from the ordered state at $T=0$ to the finite-temperature state with a strong short-range order, the correlation length ξ being exponentially large at low T. The short-range order parameter is described as an anomalous average of Bose operators, and the long-range ordering as Bose condensation.

It is clear from the physical point of view that the electron spectrum in 2D systems at low T does not change its form in comparison with the ordered state and should be determined by short-range order (the situation is reminiscent of spontaneous spin splitting above the Curie point in strong itinerant ferromagnets). To obtain a quantitative description we calculate the electron Green's function. First we treat the broad band case with the use of perturbation expansion in the s–d exchange parameter I. Substituting the result for the spin spectral density $K_{\mathbf{q}}(\omega)$ in SSWT (see (P.18), (P.22)) in the expression for electron self-energy (6.25) we obtain

$$\Sigma_{\mathbf{k}}(E)=\frac{I^2\overline{S}_{\mathrm{eff}}^2}{E-t_{\mathbf{k+Q}}}+I^2\overline{S}_{\mathrm{eff}}\sum_{q,|\mathbf{q}-\mathbf{Q}|>\xi^{-1}}\left(\frac{1-\phi_{\mathbf{q}}}{1+\phi_{\mathbf{q}}}\right)^{1/2}\times$$
$$\times\left(\frac{1-n_{\mathbf{k+q}}+N_{\mathrm{B}}(\omega_{\mathbf{q}})}{E-t_{\mathbf{k+q}}-\omega_{\mathbf{q}}}+\frac{n_{\mathbf{k+q}}+N_{\mathrm{B}}(\omega_{\mathbf{q}})}{E-t_{\mathbf{k+q}}+\omega_{\mathbf{q}}}\right),\tag{6.112}$$

with

$$\phi_{\mathbf{q}}=\frac{1}{2}(\cos q_x+\cos q_y),$$

which has the same form as the self-energy of the usual 3D antiferromagnet at $T\ll T_{\mathrm{N}}$ (G.69). The first term in (6.112) describes the formation of antiferromagnetic gap (the effective "sublattice magnetization" $\overline{S}_{\mathrm{eff}}(T)$ is determined by the singular contribution to the spin correlation function and has the linear T-dependence (P.21)), and the second terms corresponds to the interaction with magnons.

Consider the peculiarities of the electron spectrum near the band bottom in the case of a single current carrier. Summing up higher orders corrections of perturbation series (i.e. replacing energy denominators by the exact Green's functions) [520] we derive the self-consistent equation

$$\Sigma_{\mathbf{k}}(E) = \Phi_{\mathbf{k}}(E) - I^2\overline{S}_{\mathrm{eff}}^2/(E - t_{\mathbf{k+Q}} - \Phi_{\mathbf{k+Q}}(E)), \tag{6.113}$$

where

$$\Phi_{\mathbf{k}}(E) = I^2 \sum_{|\mathbf{q}-\mathbf{Q}|>\xi^{-1}} \int K_{\mathbf{q}}(\omega) G_{\mathbf{k+q}}(E+\omega) d\omega. \tag{6.114}$$

For $T = 0$ we obtain

$$\Phi_{\mathbf{k}}(E) = I^2\overline{S}_{\mathrm{eff}} \sum_{\mathbf{q}} \left(\frac{1-\phi_{\mathbf{q}}}{1+\phi_{\mathbf{q}}}\right)^{1/2} G_{\mathbf{k+q}}(E-\omega_{\mathbf{q}}). \tag{6.115}$$

To solve the equation (6.113) one can use the 'dominant pole' approximation [625]

$$G_{\mathbf{k}}(E) = \frac{Z_{\mathbf{k}}}{E - \tilde{E}_{\mathbf{k}}} + G_{\mathrm{inc}}(\mathbf{k}, E), \tag{6.116}$$

where G_{inc} is the incoherent contribution to the Green's function,

$$Z_{\mathbf{k}} = \left(1 - \frac{\partial}{\partial E}\mathrm{Re}\,\Sigma_{\mathbf{k}}(E)\right)^{-1}_{E=\tilde{E}_{\mathbf{k}}} \tag{6.117}$$

is the residue at the pole near the band bottom, corresponding to the spectrum of new quasiparticles,

$$\tilde{E}_{\mathbf{k}} \simeq t_{\min} + Z_{\mathbf{k}}(t_{\mathbf{k}} - t_{\min}) \simeq t_{\min} + Z|t|k^2. \tag{6.118}$$

Substituting (6.116) into (6.115) and performing integration over q we obtain the estimation in the 2D case

$$Z^{-1} - 1 \sim I^2/|Jt|. \tag{6.119}$$

Thus, as $|I|$ increases, the spectral weight passes into the incoherent contribution and undamped quasiparticles become heavy, so that at $I^2 \gg J|t|$ we have the 'heavy-fermion' situation with $m^*/m = Z^{-1} \gg 1$. In the 3D case the divergence is weaker, and corrections to effective mass contain a logarithmic factor:

$$Z^{-1} - 1 \sim \frac{I^2 S}{t^2} \ln\left|\frac{t}{JS}\right|. \tag{6.120}$$

The terms with the Bose functions in (6.112) yield corrections to (6.119) which are proportional to $T/\xi J$, i.e. exponentially small. Therefore the picture of the electron spectrum holds at finite $T \ll J$.

It should be noted that similar results are obtained in the case of the interaction with acoustic phonons if we replace

$$I^2\overline{S}_{\text{eff}}\left(\frac{1-\phi_{\mathbf{q}}}{1+\phi_{\mathbf{q}}}\right)^{1/2}\to\Lambda^2 q,$$

with Λ being the electron–phonon interaction constant. Then the estimation for the residue renormalization reads

$$Z^{-1}-1\sim\Lambda^2/\overline{\omega}|t|\sim 1. \qquad (6.121)$$

It is instructive to perform a comparison with the case of an usual paramagnet without strong antiferromagnetic correlations, so that the electron-spin interaction matrix element is constant at $q\to 0$ rather than proportional to $q^{1/2}$. In such a situation we may use the spin diffusion approximation (6.32). Then quasiparticles turn out to possess a strong damping

$$\Gamma_{\mathbf{k}}=-I^2\sum_{\mathbf{q}}\operatorname{Im}\int K_{\mathbf{q}}(\omega)\frac{Z_{\mathbf{k+q}}}{\tilde{E}_{\mathbf{k}}-\tilde{E}_{\mathbf{k+q}}+\omega}d\omega, \qquad (6.122)$$

At small k we have

$$\Gamma_{\mathbf{k}}\sim\frac{I^2JS^3}{Zt^2}\begin{cases}|\ln k|, & D=2,\\ 1, & D=3.\end{cases} \qquad (6.123)$$

Using the dominant pole approximation (6.116) with $\tilde{E}_{\mathbf{k}}\to\tilde{E}_{\mathbf{k}}-i\Gamma_{\mathbf{k}}$ we obtain

$$Z^{-1}-1\sim\begin{cases}|t/JS|^{1/2}, & D=2,\\ |I^2S/tJ|^{1/2}, & D=3.\end{cases} \qquad (6.124)$$

Although the residue of the damped quasiparticles may still be small, it is difficult to separate them from the background of the incoherent contribution.

Now we treat the s–d model with strong correlations $|I|\to\infty$, which includes as a particular case the Hubbard model with $U\to\infty$ (Appendix I). Consider the Green's function

$$G_{\mathbf{k}\alpha\sigma}(E)=\langle\langle g_{\mathbf{k}\alpha\sigma}|g^{\dagger}_{\mathbf{k}\alpha\sigma}\rangle\rangle_E,\quad \alpha=\operatorname{sign}I,$$

where the operators g are defined in (I.4). The result of calculation with account of spin fluctuations has the form [520]

$$G_{\mathbf{k}\alpha}(E)=\left[E\left(\Psi_{\mathbf{k}\alpha}(E)-\frac{\overline{S}^2_{\text{eff}}t_{\mathbf{k+Q}}/(2S+1)^2}{E-\Psi_{\mathbf{k+Q}}(E)t_{\mathbf{k+Q}}}\right)^{-1}-t_{\mathbf{k}}\right]^{-1}, \qquad (6.125)$$

with

$$\Psi_{\mathbf{k}\alpha}(E) = P_\alpha + \sum_{|\mathbf{q}-\mathbf{Q}|>\xi^{-1}} \frac{t_{\mathbf{k}+\mathbf{q}}}{(2S+1)^2} \int K_{\mathbf{q}}(\omega)\Psi^{-1}_{\mathbf{k}+\mathbf{q},\alpha}(E) G_{\mathbf{k}+\mathbf{q},\alpha}(E+\omega)d\omega,$$

$$P_+ = \frac{S+1}{2S+1}, \qquad P_- = \frac{S}{2S+1}. \tag{6.126}$$

When neglecting spin fluctuations, $\Psi_\alpha = P_\alpha$ and the electron spectrum contains two quasiparticle subbands, as well as in the three-dimensional antiferromagnet (cf. (I.15)):

$$E^{1,2}_{\mathbf{k}\alpha} = \frac{P_\alpha}{2}(t_{\mathbf{k}} + t_{\mathbf{k}+\mathbf{Q}}) \pm \left(\frac{P^2_\alpha}{4}(t_{\mathbf{k}} - t_{\mathbf{k}+\mathbf{Q}})^2 + \frac{\overline{S}^2_{\text{eff}}}{(2S+1)^2} t_{\mathbf{k}} t_{\mathbf{k}+\mathbf{Q}} \right)^{1/2}. \tag{6.127}$$

In the nearest-neighbour approximation ($t_{\mathbf{k}+\mathbf{Q}} = -t_{\mathbf{k}}$) we have

$$E^{1,2}_{\mathbf{k}} = \pm \frac{t_{\mathbf{k}}}{2S+1} \begin{cases} \left[S^2 - \overline{S}^2_{\text{eff}}(T)\right]^{1/2}, & \alpha = -, \\ \left[(S+1)^2 - \overline{S}^2_{\text{eff}}(T)\right]^{1/2}, & \alpha = +. \end{cases} \tag{6.128}$$

The second term in (6.126) (spin-fluctuation corrections) leads to qualitative changes in the spectrum near the band bottom. To solve the system (6.125), (6.126) at $T = 0$ we employ again the dominant pole approximation

$$G_{\mathbf{k}\alpha}(E) = \Psi_{\mathbf{k}\alpha}\left[\frac{Z_{\mathbf{k}}}{E - \tilde{E}_{\mathbf{k}}} + G_{\text{inc}}(\mathbf{k}, E) \right]. \tag{6.129}$$

The estimate for the residue is analogous to (6.119), I^2 being replaced by $(t/2S)^2$ which is typical for the strong-coupling limit. Then we obtain

$$Z^{-1} - 1 \sim \begin{cases} |t/JS|^{1/2}, & D = 2, \\ S^{-1} \ln |t/JS|^{1/2}, & D = 3. \end{cases} \tag{6.130}$$

Thus a narrow quasiparticle band with the width of order of $|J|$ is formed near the bare band bottom in the 2D case. This result was firstly obtained analytically by Kane et al [625] and confirmed by numerical calculations in the t–J model [626].

We see that strong electron-spin interaction in two-dimensional systems may result in heavy electron mass formation near the band bottom even in the case of one current carrier. Simultaneous consideration of this effect and many-electron Kondo divergences is an interesting physical problem. In this connection we mention the system $Y_{1-x}Pr_xBa_2Cu_3O_7$ where increasing x results in suppression of superconductivity and heavy-fermion behavior [627] and the high-T_c superconductor $Nd_{2-x}Ce_xCuO_4$ which possesses a logarithmic term in the temperature dependence of resistivity [628]. Recently [629], heavy-fermion behaviour with very large γ was found in the latter system at $T < 0.3$ K.

Investigation of electron spectrum in a 2D antiferromagnetic metal was performed in Ref.[433]. It was demonstrated that a behavior, which is similar to the marginal Fermi liquid [645] (the linear in $|E - E_F|$ dependence of the electron damping) takes place in some interval near the Fermi energy due to intersubband transitions. This results in anomalous temperature dependences of thermodynamic and transport properties at not too low temperatures. A similar picture takes place in the 3D case for the 'nesting" situation where $t_{\mathbf{k+Q}} - E_F = E_F - t_{\mathbf{k}}$ Interaction with lattice degrees of freedom (strong polaron effects) should also play an important role in the formation of electron spectrum of high-T_c superconductors. The investigation of the interaction among electrons and ions in a double-well potential was performed by Yu and Anderson [530] with application to A15-superconductors with moderate T_c. A pseudo-Kondo lattice model, which treats the interaction of current carriers with strongly anharmonic displacements of oxygen atoms (which may be described as two-level systems), is developed in papers [405, 442].

6.8 SPIN-LIQUID STATE IN SYSTEMS WITH SPIN AND CHARGE DEGREES OF FREEDOM

The consideration of the ground state of Kondo lattices (Sect.6.6, Appendix O) uses essentially the idea of physical reality of auxiliary pseudofermions that arise when localized-spin operators are 'dismantled'. The pseudofermion representation was applied by Coleman and Andrei [711] to describe the spin-liquid state in the two-dimensional periodic s–f model. Similar concepts were extensively used in Anderson's theory of resonating valence bonds (RVB) for copper-oxide high-T_c superconductors [631]. This state is characterized by absence of long-range magnetic ordering and formation of singlet bonds between neighbour spins on the square lattice. As well as in quantum chemistry (e.g., for the benzol molecule), the bonds can move (resonate) in the crystal, so that the ground state is a superposition of the corresponding wavefunctions.

Anderson put forward the idea of separating spin and and charge degrees of freedom by using the representation of slave Bose and

Fermi operators

$$c_{i\sigma}^{\dagger} = X_i(\sigma,0) + \sigma X_i(2,-\sigma) = s_{i\sigma}^{\dagger}e_i + \sigma d_i^{\dagger}s_{i-\sigma}, \qquad (6.131)$$

where $s_{i\sigma}^{\dagger}$ are creation operators for neutral fermions (spinons) and $e_i^{\dagger}$, $d_i^{\dagger}$ for charged spinless bosons (holons). The physical sense of such excitations may be explained as follows [632]. Consider the lattice with one electron per site with strong Hubbard repulsion, so that each site is neutral. In the ground RVB state each site takes part in one bond. When a bond becomes broken, two uncoupled sites occur which possess spins of 1/2. The corresponding excitations (spinons) are uncharged. On the other hand, the empty site (hole) in the system carries the charge, but not spin.

The requirement of the Fermi commutation relations for electron operators yields

$$e_i^{\dagger}e_i + d_i^{\dagger}d_i + \sum_{\sigma} s_{i\sigma}^{\dagger}s_{i\sigma} = 1. \qquad (6.132)$$

In the half-filled case only spinon excitations with the kinetic energy of order of $|J|$ are present. At doping the system by holes, there occur the current carriers which are described by holon operators $e_i^{\dagger}$. In the simplest gapless version, the Hamiltonian of the system for a square lattice may be presented as

$$\mathcal{H} = \sum_{\mathbf{k}}(4t\phi_{\mathbf{k}} - \zeta)e_{\mathbf{k}}^{\dagger}e_{\mathbf{k}} + 4\sum_{\mathbf{k}}(\Delta + t\delta)\phi_{\mathbf{k}}(s_{\mathbf{k}\sigma}^{\dagger}s_{-\mathbf{k}-\sigma}^{\dagger} + s_{\mathbf{k}\sigma}s_{-\mathbf{k}-\sigma}) + \ldots, \qquad (6.133)$$

with Δ being the RVB order parameter, which is determined by anomalous averages of the spinon operators, $\delta = \langle e^{\dagger}e\rangle$ the hole concentration. Thus a spin-liquid state with long-range magnetic order suppressed, a small energy scale J, and a large linear term in specific heat ($\gamma \sim 1/|J|$), which is owing to existence of the spinon Fermi surface, can arise in a purely spin systems without conduction electrons. Experimentally, some data indicate presence of a T-linear term in the insulating phase of copper-oxide systems.

Later, more complicated versions of the RVB theory, which use topological consideration and analogies with the fractional quantum Hall effect, were developed (see, e.g., [633]). These ideas led to rather unusual and beautiful results. For example, it was shown that spinons may obey fractional statistics, i.e. the wavefunction of the system acquires a complex factor at permutation of two quasiparticles.

Here we demonstrate the suppression of long-range magnetic order at $T = 0$ and occurrence of the spin-liquid state within a simple spin-wave treatment of a two-dimensional Heisenberg antiferromagnet [601, 609]. To this end we write down the correction to sublattice magnetization due to zero-point vibrations (see (E.14))

$$\delta\overline{S} = -\frac{1}{2}\sum_{\mathbf{q}}[S(4J_{\mathbf{Q}} - J_{\mathbf{Q+q}} - J_{\mathbf{Q-q}} - 2J_{\mathbf{q}})/\omega_{\mathbf{q}} - 1]. \tag{6.134}$$

At $q \to 0$ we have

$$\omega_{\mathbf{q}}^2 \simeq S^2(J_{\mathbf{Q}} - J_0)\left[\alpha q^2 + \beta f(\phi) q^4\right], \tag{6.135}$$

where $\beta > 0$ and $f(\phi)$ is a positive polar-angle function. For $\alpha \to 0$ (frustration situation) we find

$$\overline{S} = S - a\ln\frac{\beta}{\alpha}, \tag{6.136}$$

$$a = \frac{1}{16\pi^2}(J_{\mathbf{Q}} - J_0)^{1/2}\int_0^{2\pi}\frac{1}{[\beta f(\phi)]^{1/2}}d\phi,$$

so that $\overline{S} = 0$ in some regions of parameters

$$\alpha < \beta\exp(-S/a). \tag{6.137}$$

Thus, unlike the Kondo lattices (cf. (6.89), (6.103)), destruction of magnetic ordering is due to frustrations of the f–f exchange interaction itself rather than to Kondo screening of magnetic moments. One may assume that similar frustration effects (e.g., large next-neighbour exchange interactions or presence equilateral nearest-neighbour triangles) may result in total or partial destruction of magnetic moment in some three-dimensional systems. Such Heisenberg systems, which possess developed spin fluctuations, reduced magnetic moments and strong short-range order above the Neel point [634] are reminiscent in a sense of itinerant magnets.

There exist a number of experimental data which indicate realization of a spin-liquid-type state in three-dimensional d- and f-systems [635, 609]. Consider some examples.

The compound YMn_2 has a frustrated AFM structure and pronounced short-range order above T_N [636]. In the system $Y_{1-x}Sc_xMn_2$ with $x = 0.03$ the long-range order is destroyed, while linear term in specific heat reaches very large value, $\gamma = 140$ mJ/mol · K^2 (a similar situation occurs under pressure) [637]. This value exceeds by ten times the result of band calculations and is a record for d-systems.

The Mott insulator NiS_{2-y} [638] exhibits magnetic ordering, but one may talk of a suppression of the magnetism in the sense that, similar to magnetic Kondo lattices, $T_N = 45$ K is small in comparison with $|\theta| \simeq 1500$ K. The slope of the phase diagram line indicates a high entropy value for the insulator phase [44], which is characteristic of a spin-liquid-type state. In metallic $NiS_{2-x}Se_x$ near the transition

($x \simeq 0.5$), the value of $\gamma = 30$ mJ/mol · K^2) is rather large.

The most striking of the spin-liquid state realization is probably the intermediate valence semiconductor Sm_3Se_4 [639], a system with not only spin, but also charge degrees of freedom, where the ions Sm^{2+} and Sm^{3+} are distributed over crystallographically equivalent sites in the Th_3P_4-lattice. This situation may be described by an effective (generally speaking, anisotropic) pseudospin Hamiltonian for the charge degrees of freedom [635]. By contrast with isostructural compounds like Eu_3S_4, there is no indication of charge ordering up to $T = 0$ in Sm_3Se_4, and γ has at $T < 1$K the giant value 4.5 J/(mol · K^2).

In the system $Yb_4As_{3-x}P_4$ with the anti-Th_3P_4 structure, where charge ordering occurs near 300K, γ increases from 200 to 400 mJ/(mol K^2) as x is varied from 0 to 0.3, the current carrier concentration n (of order of 10^{-3} per atom) remaining practically unchanged [640]. Thus we deal with a new class of heavy-fermion systems with extremely small concentration of current carriers. Here belong also the compound YbAs with $n \sim 10^{-2}$, $\gamma = 270$ mJ/(mol · K^2) [538], and possibly YbNiSb [541], YbBiPt [571, 641]. In the latter system, γ reaches the value of 8 J/(mol · K^2).

Although in most cases there are no serious grounds in doubting that the Kondo effect is the major cause of anomalous behaviour of f-systems, which were discussed above, the role of f–f interaction in the formation of low-energy spectrum should be also investigated. Apparently, an intermediate state between the usual magnetic state and the spin liquid arises in some compounds. In Kondo lattices with a small number of carriers, the exchange interaction frustrations can be even more important than the Kondo effect. Thus, the semimetal CeSb (a classical example of a system with competing exchange interactions) displays a complicated magnetic phase diagram–the "devil's staircase" [642]. According to [643], the system $Ce_{0.8}La_{0.2}Sb$ possesses large electronic specific heat. The 'frustrated' picture of magnetism, that is reminiscent of spin-glass state (absence of a marked phase transition) is discussed for $CeAl_3$ [508]. The problem of the role of frustrations and of the Kondo effect in the formation of complicated magnetic structures and reduced magnetic moments is also of interest for rare-earth metals (Sect.4.7).

To conclude this Chapter, we discuss briefly some modern concepts in the theory of systems with strong correlations. The idea of unusual excitation spectrum in correlated metallic systems was extensively developed by Anderson [644]. He put forward the concept about the inadequacy of the Fermi-liquid picture and formation of the Luttinger liquid state. The latter is characterized by the absence of simple poles for the one-electron Green's function, i.e. of usual single-particle Fermi excitations). The transition into the Luttinger liquid is connected with the phase shift at the Fermi level owing to strong electron scattering.

A phenomenological 'marginal Fermi-liquid' theory of high-T_c superconductors, which yields close results, was proposed by Varma et al [645]. The non-Fermi liquid behaviour (unusual power-law temperature dependence of electronic specific heat, resistivity etc.) is found now for a number of f-electron systems [646]. To explain this, a number of mechanisms were proposed. In particular, a competition of the Kondo effect and intersite exchange interactions with a peculiar spin dynamics was considered in [733]. The power-law divergence of magnetic susceptibility, $\chi(T) \sim T^{-\alpha}$, is observed in low-dimensional organic conductors; the analogy with the behaviour of yttrium-based ceramics was discussed in [647].

In the Luttinger model [648], which was developed for the one-dimensional electron systems, the bare spectrum contains two linear branches

$$E_k^{1,2} = \pm v_F k \quad (k > 0), \tag{6.138}$$

which is reminiscent of relativistic models in quantum electrodynamics (the vacuum of the states with $k < 0$ is excluded). The Luttinger Hamiltonian may be 'bosonized' by introducing the collective excitation operators ($n = 1, 2$; $q > 0$)

$$\rho_n(q) = \sum_k c^\dagger_{k+q,n} c_{kn}, \qquad \rho_n(-q) = \rho_n^\dagger(q), \tag{6.139}$$

so that

$$b_q^\dagger = \left(\frac{2\pi}{N|q|}\right)^{1/2} \begin{cases} \rho_1(q), & q > 0, \\ \rho_2(q), & q < 0, \end{cases} \tag{6.140}$$

(the summation in (6.139) goes over both positive and negative k, N is the number of lattice sites). The operators (6.140) turn out to satisfy the Bose commutation relations, which is connected with account of the 'vacuum' states in (6.139) [649]. Charge and spin operators may be introduced by

$$a_k^\dagger = \frac{1}{\sqrt{2}}(b_{k\uparrow}^\dagger + b_{k\downarrow}^\dagger), \qquad s_k^\dagger = \frac{1}{\sqrt{2}}(b_{k\uparrow}^\dagger - b_{k\downarrow}^\dagger). \tag{6.141}$$

In the representation (6.140), both kinetic energy (6.138) and interaction Hamiltonians contain quadratic terms only. Then the one-electron Green's function is calculated exactly and turns out to have the singularities of the form

$$G_k^{1,2}(E) \sim (E \mp v_F k)^{\alpha-1}, \tag{6.142}$$

where the parameter $\alpha > 0$ is determined by the electron interaction. Then the electron distribution function at $T=0$ has, instead of a jump, a power-law behaviour at the 'Fermi surface',

$$\langle c_{kn}^\dagger c_{kn} \rangle - \frac{1}{2} \sim | k \mp k_F |^\alpha \operatorname{sign}(k \mp k_F). \tag{6.143}$$

Such a behaviour may be derived rigorously for the one-dimensional Hubbard model which permits rigorous consideration (in this case, in contradiction with the Landau theory, an arbitrarily small interaction leads to a reconstruction of the ground state, e.g. to the metal-insulator transition for a half-filled band). However, generalization of these results to two-dimensional systems of interacting electrons is a very difficult problem. As discussed by Anderson [644], the violation of the Fermi-liquid picture may be described in terms of the Hubbard splitting: the states in the upper Hubbard subband correspond to the anomalous forward scattering. Modern treatment of the Fermi-liquid state and of its instabilities with the use of the 'Bose' representation is given, e.g., in [650].

Appendix A

MANY-ELECTRON CREATION OPERATORS FOR ATOMIC CONFIGURATIONS AND HUBBARD'S OPERATORS

In this Appendix we consider an operator description of many-electron systems with large intrasite Coulomb correlations.

At passing to the standard second quantization representation, the many-electron (ME) wave function of a crystal $\Psi(x_1 \cdots x_N)$ ($x=\mathbf{r}_i s_i$), s_i is the spin coordinate) is chosen in the form of a linear combination of the Slater determinants. These are constructed from the one-electron wave functions $\psi_\lambda(x)$ ($\lambda=\{\nu\gamma\}$, ν are the indices of lattice sites and γ are the one-electron quantum number sets):

$$\Psi(x_1 \ldots x_N) = \sum_{\lambda_1 \ldots \lambda_N} c(\lambda_1 \ldots \lambda_N) \Psi_{\lambda_1 \ldots \lambda_N}(x_1 \ldots x_N), \tag{A.1}$$

where

$$\Psi_{\lambda_1 \ldots \lambda_N}(x_1 \ldots x_N) = (N!)^{-1/2} \sum_P (-1)^P P \prod_i \psi_{\lambda_i}(x_i), \tag{A.2}$$

with P being all the possible permutations of x_i. The expansion (A.1) is exact provided that the system of functions ψ_λ is complete [651]. The second quantization representation is introduced by using the one-electron occupation numbers n_λ as new variables:

$$\Psi(x_1 \ldots x_N) = \sum_{\{n_\lambda\}} c(\ldots n_\lambda \ldots) \Psi_{\{n_\lambda\}}(x_1 \ldots x_N). \tag{A.3}$$

Then $c(\ldots n_\lambda \ldots)$ plays the role of a new wave function. The Fermi one-electron creation and annihilation operators are defined by

$$a_\lambda c(\ldots n_\lambda \ldots) = (-1)^{\eta_\lambda} n_\lambda c(\ldots n_\lambda - 1 \ldots),$$

$$a_\lambda^\dagger c(\ldots n_\lambda \ldots) = (-1)^{\eta_\lambda} (1 - n_\lambda) c(\ldots n_\lambda + 1 \ldots), \tag{A.4}$$

with

$$\eta_\lambda = \sum_{\lambda'>\lambda} n_{\lambda'}, \qquad a_\lambda^\dagger a_\lambda = \hat{n}_\lambda .$$

Now we try to generalize this approach by introducing quantum numbers of some electron groups. In particular, we may combine electrons at a given lattice site ($\Lambda = \{\nu\Gamma\}$, Γ_i are ME terms) to obtain

$$\Psi(x_1 \ldots x_N) = \sum_{\{N_\lambda\}} c(\ldots N_\lambda \ldots)\Psi_{\{N_\lambda\}}(x_1 \ldots x_N). \tag{A.5}$$

In the case of equivalent electron configuration l^n, the ME wave function of the electron group is constructed as (see [20])

$$\Psi_{\Gamma_n}(x_1 \ldots x_N) = \sum_{\Gamma_{n-1},\gamma} G^{\Gamma_n}_{\Gamma_{n-1}} C^{\Gamma_n}_{\Gamma_{n-1},\gamma} \Psi_{\Gamma_{n-1}}(x_1 \ldots x_{n-1})\psi_\gamma(x_n), \tag{A.6}$$

where C are the Clebsh–Gordan coefficients. In the case of LS-coupling we use the notation

$$C^{\Gamma_n}_{\Gamma_{n-1},\gamma} \equiv C^{L_n M_n}_{L_{n-1}M_{n-1},\, lm} C^{S_n \mu_n}_{S_{n-1}\mu_{n-1},\frac{1}{2}\sigma}, \tag{A.7}$$

and the summation over $\gamma=\{lm\sigma\}$ (we omit for brevity the principal quantum number) stands for the summation over one-electron orbital projection m and spin projection σ, but not over l. The case of jj-coupling, where strong spin-orbit coupling should be taken into account in the first place, is applicable for $5f$ actinide compounds. Here we have $\gamma=j\mu$, $\Gamma=JM$ with $j=l+1/2$, so that the consideration is formally more simple. The quantities

$$G^{\Gamma_n}_{\Gamma_{n-1}} \equiv G^{S_n L_n \alpha_n}_{S_{n-1}L_{n-1}\alpha_{n-1}}$$

are the fractional parentage coefficients (α are additional quantum numbers which distinguish different terms with the same S, L, e.g., the Racah's seniority; we will omit α for brevity where it is possible). They do not depend on momentum projections, and the quantities $(G^{\Gamma_n}_{\Gamma_{n-1}})^2$ have the meaning of fracture of the term Γ_{n-1} at formation of the term Γ_n (at $n<2$, $G\equiv 1$). The fractional parentage coefficients satisfy the orthogonality relations [20,32]

$$\sum_{\{S'L'\alpha'\}} G^{SL\alpha}_{S'L'\alpha'} G^{SL\alpha''}_{S'L'\alpha'} = \delta_{\alpha\alpha''}; \tag{A.8a}$$

$$n\sum_{\{SL\alpha\}} [S][L] G^{SL\alpha}_{S'L'\alpha'} G^{SL\alpha}_{S'L'\alpha''} = (2[l]+1-n)[S'][L']\delta_{\alpha'\alpha''}, \tag{A.8b}$$

where

$$[A] \equiv 2A+1.$$

The relation (A.8b) is obtained from (A.8a) after passing to the hole representation in the atomic shell. In the case of *jj*-coupling

$$[S][L] \to [J], \qquad 2[l] \to [j]. \tag{A.9}$$

The recurrence relation (A.6) enables one to obtain the ME functions with an arbitrary electron number *n*. The fractional parentage and Clebsh–Gordan coefficients ensure the antisymmetry of the function (A.6) with respect to the permutation of electron coordinates.

If the added electron belongs to another shell, we can write down

$$\Psi_{\Gamma_n}(x_1 \ldots x_n) = n^{-1/2} \sum_{i,\Gamma_{n-1},\gamma} (-1)^{n-i} C^{\Gamma_n}_{\Gamma_{n-1},\gamma} \times \\ \times \Psi_{\Gamma_{n-1}}(x_1 \ldots x_{i-1}, x_{i-1} \ldots x_{n-1}) \psi_\gamma(x_i) \tag{A.10}$$

(unlike the case of equivalent electrons, antisymmetrization is required). Note, however, that introducing ME functions and operators, which describe several electron shells, is advisable in the solid state theory only provided that the interaction between the shells is large in comparison with the band energies (e.g., in the narrow-band *s–d* exchange model, see Appendix I).

The wave function of the the whole crystal (A.5) may be now obtained as an antisymmetrized product of ME functions for electron groups.

By analogy with (A.6), (A.10) we can introduce ME creation operators for electron groups [652]. For equivalent electrons

$$A^\dagger_{\Gamma_n} = n^{-1/2} \sum_{\Gamma_{n-1},\gamma} G^{\Gamma_n}_{\Gamma_{n-1}} C^{\Gamma_n}_{\Gamma_{n-1},\gamma} a^\dagger_\gamma A^\dagger_{\Gamma_{n-1}}. \tag{A.11}$$

At adding an electron from another shell

$$A^\dagger_{\Gamma_n} = \sum_{\Gamma_{n-1},\gamma} C^{\Gamma_n}_{\Gamma_{n-1},\gamma} a^\dagger_\gamma A^\dagger_{\Gamma_{n-1}}. \tag{A.12}$$

Antisymmetry of the functions

$$|\Gamma_n\rangle = A^\dagger_{\Gamma_n} |0\rangle \tag{A.13}$$

is provided by anticommutation of the Fermi operators. Extra factors of $(1/n)^{1/2}$ in (A.11), (A.12) in comparison with (A.6), (A.10) respectively arise at passing from the *x*-representation to second quantization one.

In the particular case of two equivalent electrons

$$A_\Gamma^\dagger = \frac{1}{\sqrt{2}} \sum_{\substack{m_1 m_2 \\ \sigma_1 \sigma_2}} C^{LM}_{lm_1, lm_2} C^{S\mu}_{\frac{1}{2}\sigma_1, \frac{1}{2}\sigma_2} a^\dagger_{lm_2\sigma_2} a^\dagger_{lm_1\sigma_1}. \tag{A.14}$$

One can see from (A.14) that only the terms with even $S+L$ are possible [20]: as follows from the properties of the Clebsh–Gordan coefficients, A_Γ is identically zero if $S+L$ is odd.

Unlike the Fermi operators, the commutation relation for the operators A are complicated. For example, we obtain from (A.14)

$$[A_{\Gamma'}, A_\Gamma^\dagger] = \delta_{\Gamma\Gamma'} + 2 \sum_{\gamma_1\gamma_2\gamma_3} C^{\Gamma}_{\gamma_1\gamma_3} C^{\Gamma'}_{\gamma_2\gamma_3} a^\dagger_{\gamma_2} a_{\gamma_1}.$$

Representing the operator products in terms of one-electron Fermi operators, performing the pairings of the latter and using the orthogonality relations for the Clebsh–Gordan and fractional parentage coefficients, we obtain

$$\langle 0 | A_{\Gamma'} A_\Gamma^\dagger | 0 \rangle = \delta_{\Gamma\Gamma'}, \qquad A_{\Gamma'_n} A^\dagger_{\Gamma_n} | 0 \rangle = \delta_{\Gamma_n \Gamma'_n} | 0 \rangle. \tag{A.15}$$

However, for $m < n$

$$A_{\Gamma'_m} A^\dagger_{\Gamma_n} | 0 \rangle \neq 0.$$

Therefore the operators (A.11), (A.12) are convenient only for the treatment of configurations with a fixed number of electrons (e.g., in the Heitler–London model). At the same time, they are insufficient at considering problems with electron transfer between different shells or sites because of the 'non-orthogonality' for different n. In such situations, it is suitable to define new ME creation operators, which contain projection factors [653, 654]

$$\tilde{A}_\Gamma^\dagger = A_\Gamma^\dagger \prod_\gamma (1 - \hat{n}_\gamma). \tag{A.16}$$

Formally, the product in (A.16) goes over all the relevant one-electron states γ. Howewer, because of the identity $a^\dagger_\gamma \hat{n}_\gamma = 0$, it is sufficient to retain only those γ which do not enter the corresponding operator products in A_Γ. It should be noted that introducing the ME operators, which depend on all the one-electron quantum numbers (both for occupied and empty states), is in a sense the next step in the quantum-field description after the usual second quantization.

We obtain, instead of (A.15), the operator identities

$$\tilde{A}_\Gamma \tilde{A}_\Gamma^\dagger = \delta_{\Gamma\Gamma'} \prod_\gamma (1 - \hat{n}_\gamma); \tag{A.17}$$

$$\tilde{A}_\Gamma \tilde{A}_{\Gamma'} = \tilde{A}^\dagger_{\Gamma'} \tilde{A}^\dagger_\Gamma = 0 \quad (|\Gamma\rangle \neq 0) \tag{A.18}$$

(after reducing to the normal form, the terms with Fermi operators in the left-hand side of (A.17) are cancelled by the factors $(1-\hat{n}_\gamma)$). Thus we may pass to the representation of ME occupation numbers N_Γ at a given site:

$$\tilde{A}_\Gamma |\Gamma'\rangle = \delta_{\Gamma\Gamma'} |0\rangle, \qquad \tilde{A}^\dagger_\Gamma |\Gamma'\rangle = \delta_{\Gamma' 0} |\Gamma\rangle,$$

$$\tilde{A}^\dagger_\Gamma \tilde{A}_\Gamma = \hat{N}_\Gamma, \qquad \hat{N}_\Gamma |\Gamma'\rangle = \delta_{\Gamma\Gamma'} |\Gamma\rangle, \qquad \sum_\Gamma \hat{N}_\Gamma = 1. \tag{A.19}$$

Note that, unlike (A.4), only one of the numbers N_Γ is non-zero, and the commutation relation for the operators (A.16) differ considerably from those for the Fermi operators:

$$[\tilde{A}_\Gamma, \tilde{A}^\dagger_{\Gamma'}]_\pm = \hat{N}_0 \delta_{\Gamma\Gamma'} \pm \tilde{A}^\dagger_{\Gamma'} \tilde{A}_\Gamma, \tag{A.20}$$

with

$$\hat{N}_0 = \prod_\gamma (1 - \hat{n}_\gamma) = \prod_\Gamma (1 - \hat{N}_\Gamma).$$

Therefore, at practical calculations, it is convenient to pass from the ME creation and annihilation operators to X-operators, which possess more simple properties.

For different Γ and Γ', the product

$$X(\Gamma, \Gamma') = \tilde{A}^\dagger_\Gamma \tilde{A}_{\Gamma'} \tag{A.21}$$

transforms the state Γ' into the state Γ. Such operators were firstly introduced by Hubbard [31] in the axiomatic way as generalized projection operators:

$$X(\Gamma, \Gamma') = |\Gamma\rangle\langle\Gamma'|, \tag{A.22}$$

where $|\Gamma\rangle$ are the exact eigenstates of the intraatomic Hamiltonian.

Using the introduced operators of electron configurations enables one to obtain explicit expressions for X-operators in terms of one-electron operators. In particular,

$$X(\Gamma, 0) = \tilde{A}^\dagger_\Gamma, \qquad X(\Gamma, \Gamma) = \hat{N}_\Gamma. \tag{A.23}$$

For example, we consider the simplest case of s-electrons where $\gamma = \sigma = \pm(\uparrow, \downarrow)$, $\Gamma = 0, \sigma, 2$ and $|0\rangle$ with $|0\rangle$ being the empty state (hole) and $|2\rangle$ the doubly-occupied singlet state (double) on a site. Then we have

$$X(0,0) = (1 - \hat{n}_\uparrow)(1 - \hat{n}_\downarrow), \qquad X(2,2) = \hat{n}_\uparrow \hat{n}_\downarrow,$$

$$X(\sigma, \sigma) = \hat{n}_\sigma (1 - \hat{n}_{-\sigma}), \qquad X(\sigma, -\sigma) = a^\dagger_\sigma a_{-\sigma},$$

$$X(\sigma,0) = a_\sigma^\dagger(1-\hat{n}_{-\sigma}), \qquad X(2,\sigma) = -\sigma a_{-\sigma}^\dagger \hat{n}_\sigma. \tag{A.24}$$

As follows from (A.17), the simple multiplication rules, postulated by Hubbard [31], hold

$$X(\Gamma,\Gamma')X(\Gamma'',\Gamma''') = \delta_{\Gamma'\Gamma''}X(\Gamma,\Gamma'''). \tag{A.25}$$

For each number of electrons n, the sum rule is satisfied

$$\sum_{\Gamma_n} X(\Gamma,\Gamma) = \frac{1}{n!}\sum_{\gamma_i \neq \gamma_j} \hat{n}_{\gamma_n}\hat{n}_{\gamma_{n-1}}\ldots\hat{n}_{\gamma_1}\prod_{\gamma\neq\gamma_i}(1-\hat{n}_\gamma) \equiv X_n, \tag{A.26}$$

which may be verified by direct multiplication

$$X(\Gamma',\Gamma_m)X_n = \delta_{nm}X(\Gamma',\Gamma_m). \tag{A.27}$$

Finally, one can prove the completeness relation

$$\sum_\Gamma X(\Gamma,\Gamma) = \sum_n X_n = 1. \tag{A.28}$$

An arbitrary operator $\hat{O}$ acting on the electrons at a given site i, is expressed in terms of X-operators as

$$\hat{O} = \left(\sum_{\Gamma\Gamma'}\int \Psi_\Gamma^*(x_1\ldots x_n)\hat{O}\Psi_{\Gamma'}(x_1\ldots x_n)dx_1\ldots dx_n\right)\tilde{A}_\Gamma^\dagger\tilde{A}_{\Gamma'} =$$
$$= \sum_{\Gamma\Gamma'}\langle\Gamma|\hat{O}|\Gamma'\rangle X(\Gamma,\Gamma'). \tag{A.29}$$

It follows from (A.11) that the matrix elements of the one-electron Fermi operators read

$$\langle\Gamma_n|a_\gamma^\dagger|\Gamma_{n-1}\rangle = n^{1/2}G_{\Gamma_{n-1}}^{\Gamma_n}C_{\Gamma_{n-1},\gamma}^{\Gamma_n}. \tag{A.30}$$

Then we obtain the representation [32]

$$a_\gamma^\dagger = \sum_n n^{1/2}\sum_{\Gamma_n\Gamma_{n-1}} G_{\Gamma_{n-1}}^{\Gamma_n}C_{\Gamma_{n-1},\gamma}^{\Gamma_n}X(\Gamma_n,\Gamma_{n-1}). \tag{A.31}$$

In particular, for s-electrons one gets

$$a_\sigma^\dagger = X(\sigma,0) + \sigma X(2,-\sigma). \tag{A.32}$$

Using (A.31) and the orthogonality relations for Clebsh–Gordan and fractional parentage coefficients (A.8) we obtain

$$\sum_\gamma a_\gamma^\dagger a_\gamma = \sum_{n\Gamma_n} nX(\Gamma_n,\Gamma_n),$$

$$\sum_\gamma a_\gamma a_\gamma^\dagger = \sum_{n\Gamma_n}(2[l]-n)X(\Gamma_n,\Gamma_n). \tag{A.33}$$

The fulfilment of Fermi commutation relations is ensured by the identity

$$n\sum_{\Gamma_{n-1}} G^{\Gamma_n}_{\Gamma_{n-1}} G^{\Gamma'_n}_{\Gamma_{n-1}} C^{\Gamma_n}_{\Gamma_{n-1},\gamma'} C^{\Gamma_n}_{\Gamma_{n-1},\gamma} +$$

$$+(n+1)\sum_{\Gamma_{n-1}} G^{\Gamma_{n+1}}_{\Gamma_n} G^{\Gamma_{n+1}}_{\Gamma'_n} C^{\Gamma_{n+1}}_{\Gamma_n,\gamma} C^{\Gamma_{n+1}}_{\Gamma'_n,\gamma'} = \delta_{\gamma\gamma'}\delta_{\Gamma_n\Gamma'_n}. \quad (A.34)$$

After multiplying by $C^{\gamma''}_{\gamma\gamma'}$, and summing over momentum projections, this identity may be expressed in terms of $6j$-symbols and used to obtain recurrence relations for fractional parentage coefficients [32].

Up to now we have considered the algebra of ME operators at one lattice site. General commutation relations for X-operators at sites ν and ν' read

$$[X_\nu(\Gamma,\Gamma'), X_\nu(\Gamma'',\Gamma''')]_\pm = \delta_{\nu\nu'}\{X(\Gamma,\Gamma''')\delta_{\Gamma'\Gamma''} \pm X(\Gamma'',\Gamma')\delta_{\Gamma\Gamma'''}\}, \quad (A.35)$$

where the plus sign corresponds to the case where both X-operators have the Fermi type, i.e. change the number of electrons by an odd number, and minus sign to all the other cases. For the Fourier-transforms of X-operators, equation (A.35) takes the form

$$[X_{\mathbf{k}}(\Gamma,\Gamma'), X_{-\mathbf{k}'}(\Gamma'',\Gamma''')]_\pm = X_{\mathbf{k}-\mathbf{k}'}(\Gamma,\Gamma''')\delta_{\Gamma'\Gamma''} \pm X_{\mathbf{k}-\mathbf{k}'}(\Gamma'',\Gamma')\delta_{\Gamma\Gamma'''}. \quad (A.36)$$

The Hubbard operators may be introduced not only for the eigenstates of the Hamiltonian with spherical symmetry, but also in the presence of a strong crystal field. In such a situation, we have to use the irreducible representations of the point group and corresponding Clebsh–Gordan coefficients [564].

Appendix B

ANGULAR MOMENTUM OPERATORS AND DOUBLE IRREDUCIBLE TENSOR OPERATORS

In a number of problems of solid state physics, it is convenient to pass into the angular-momentum operator representation. The first example of such approach was introducing spin operators for $S=1/2$ by Dirac

$$a_{\uparrow}^{\dagger}a_{\downarrow}=S^{+},\qquad a_{\downarrow}^{\dagger}a_{\uparrow}=S^{-},\qquad \frac{1}{2}(a_{\uparrow}^{\dagger}a_{\uparrow}-a_{\downarrow}^{\dagger}a_{\downarrow})=S^{z},\tag{B.1}$$

to represent the electron exchange Hamiltonian for singly-occupied s-states in the form of a scalar product:

$$-\sum_{\sigma\sigma'}a_{1\sigma}^{\dagger}a_{1\sigma'}a_{2\sigma}^{\dagger}a_{2\sigma'}=-\left(\frac{1}{2}+2\mathbf{S}_1\mathbf{S}_2\right).\tag{B.2}$$

This yielded a basis for the Heisenberg model of magnetism (Sect.4.1), which was also applied for arbitrary S. However, for $S>\frac{1}{2}$ one has to consider electrons in orbit-degenerate states and, consequently, to treat orbital momenta.

Consider the connection of spin and orbital momentum operators for an atomic shell (1.7) with Hubbard's operators. The momentum operators are diagonal in LS-terms:

$$\langle SL\mu M\alpha|L_q|S'L'\mu'M'\alpha'\rangle=\delta_{SL\alpha,S'L'\alpha'}\delta_{\mu\mu'}\sqrt{L(L+1)}C^{LM}_{LM',1q};\tag{B.3}$$

$$\langle SL\mu M\alpha|S_q|S'L'\mu'M'\alpha'\rangle=\delta_{SL\alpha,S'L'\alpha'}\delta_{MM'}\sqrt{S(S+1)}C^{S\mu}_{S\mu',1q},\tag{B.4}$$

where we have introduced the cyclic components of a vector

$$A_{\pm1}=\mp\frac{1}{\sqrt{2}}A^{\pm}=\mp\frac{1}{\sqrt{2}}(A^{x}\pm iA^{y}),\qquad A_0=A^{z}.$$

In the presence of spin-orbit coupling, we have to construct the functions with definite total angular momentum **J**=**L**+**S**. In the Russel–Saunders approximation we have

$$|SLJM_J\rangle = \Psi_{SLJM_J} = \sum_{\mu M} C^{JM_J}_{S\mu,LM} \, |SL\mu M\rangle, \tag{B.5}$$

and the corresponding ME creation operator reads

$$A^{\dagger}_{SLJM_J} = \sum_{\mu M} C^{JM_J}_{S\mu,LM} A^{\dagger}_{SL\mu M}. \tag{B.6}$$

The matrix elements of the total momentum operator within the multiplet with a given J are

$$\langle JM \,|\, J_q \,|\, JM' \rangle = \sqrt{J(J+1)} C^{JM}_{JM',1q}. \tag{B.7}$$

Standard components of the vector $\mathbf{I} = \mathbf{S}, \mathbf{L}, \mathbf{J}$ are expressed as

$$\begin{aligned} I^{+} &= \sum_{M} [(I-M)(I+M+1)]^{1/2} X(M+1,M), \\ I^{-} &= \sum_{M} [(I-M+1)(I+M)]^{1/2} X(M-1,M), \\ I^{z} &= \sum_{M} MX(M,M). \end{aligned} \tag{B.8}$$

At passing from exchange Hamiltonians, which are expressed in terms of ME operators, into momentum representation, we need the inverse transformations from X-operators to momentum operators. To obtain them, we can write down expressions for all the powers $(I^{\alpha})^k$ with k=0...2I (higher powers are not linearly independent) and solve this system of $(2I+1)^2$ equations, which is very cumbersome. It is more convenient to use the irreducible tensor operators

$$I^{(k)}_q = \sum_{MM'} C^{IM}_{IM',kq} X(IM, IM') \quad (q = -k \ldots k), \tag{B.9}$$

so that

$$\begin{aligned} S_q &= \sqrt{S(S+1)} S^{(1)}_q, \\ L_q &= \sqrt{L(L+1)} L^{(1)}_q. \end{aligned} \tag{B.10}$$

The operators (B.9) are connected with the Stevens equivalent operators in the crystal field theory (see Sect.1.3). Using the orthogonality relations for the Clebsh–Gordan coefficients, it is easy to obtain the inverse relations required

$$X(IM, IM') = \sum_{kq} C^{IM}_{IM',kq} I^{(k)}_q. \tag{B.11}$$

The commutation rules for the operators (B.9) have the form

$$[I_q^{(k)}, I_{q'}^{(k)}]_- =$$
$$= \sum_{k''q''} ((-1)^{k''} - (-1)^{k+k'}) \begin{Bmatrix} k & k' & k'' \\ I & I & I \end{Bmatrix} ([K''][I])^{1/2} C_{kq,k'q'}^{k''q''} I_{q''}^{(k'')},$$

where $\{\dots\}$ being the $6j$-symbols. Explicit expressions for the irreducible tensor operators in terms of usual momentum operators are tabulated [41, 43]. Besides that, one can apply at analytical calculations the recurrence relation

$$(-1)^k ([k-p][p][I])^{1/2} \begin{Bmatrix} k-p & p & k \\ I & I & I \end{Bmatrix} I^{(k)} = [I^{(k-p)} \times I^{(p)}]^{(k)}, \tag{B.12}$$

where the tensor product of rank c is defined by

$$[A^{(a)} \times B^{(b)}]_\gamma^{(c)} = \sum_{\alpha\beta} C_{a\alpha,b\beta}^{c\gamma} A_\alpha^{(a)} B_\beta^{(b)}. \tag{B.13}$$

To express X-operators, connecting terms with different L, S, J, one can use 'hyperbolic' operators which change the value of the momentum itself [656]. They satisfy the commutation relations

$$[K^\pm, K^z] = \mp K^\pm, \qquad [K^+, K^-] = -2K^z \tag{B.14}$$

and have the following non-zero matrix elements

$$\langle JM \mid K^+ \mid J-1M \rangle = [(J+M)(J-M)]^{1/2},$$

$$\langle JM \mid K^z \mid JM \rangle = J + \frac{1}{2}. \tag{B.15}$$

This problem is discussed also by Popov and Loginov [657].

At passing from the LS-representation to the J-representation for the multiplet with a given J, we have to express the products $L^{(k)}S^{(\kappa)}$ in terms of operators $J^{(p)}$. Using (A.11), (B.6) and summing the product of four Clebsh–Gordan coefficients with the use of a $9j$-symbol we get

$$S_\xi^{(\kappa)} L_q^{(k)} = \sum_{p\rho} \begin{Bmatrix} J & S & L \\ J & S & L \\ p & \kappa & k \end{Bmatrix} ([S][L][J][p])^{1/2} C_{\kappa\xi,kq}^{p\rho} J_\rho^{(p)}. \tag{B.16}$$

Note that $(p+\kappa+k)$ is even (in the opposite case, the $9j$-symbol in (B.15) is zero due to its symmetry properties). For $k=0$ or $\kappa=0$ the $9j$-symbol is simplified, so that we obtain

$$S^{(\kappa)}=(-1)^{J+S+L+\kappa}([S][J])^{1/2}\begin{Bmatrix} J & S & L \\ S & J & \kappa \end{Bmatrix} J^{(\kappa)}; \tag{B.17}$$

$$L^{(k)}=(-1)^{J+S+L+k}([L][J])^{1/2}\begin{Bmatrix} J & L & S \\ L & J & k \end{Bmatrix} J^{(k)}. \tag{B.18}$$

In particular, at $\kappa=1$ or $k=1$, substituting explicit values of 6j-coefficients yields

$$\mathbf{S}=(g-1)\mathbf{J}; \tag{B.19}$$

$$\mathbf{L}=(2-g)\mathbf{J}, \tag{B.20}$$

where

$$g=1+\frac{(\mathbf{LS})}{J^2}=1+\frac{J(J+1)-S(S+1)-L(L+1)}{2J(J+1)}.$$

is the Lande factor. Thus the well-known de Gennes formula (B.19) (see Sect.4.7) is obtained from the general relation (B.16).

Now we discuss another approach to description of the Fermi operator products. Since matrix elements of the Fermi creation operators $a^\dagger_{lm\sigma}$ (A.10) are proportional to the Clebsh–Gordan coefficients, these operators may be considered as $2(2l+1)$ components of a double tensor operator acting in the spin and orbital spaces. At calculating matrix elements of products of Fermi operators at one site, it is convenient to introduce for a given shell the double irreducible tensor operators with the components

$$W^{(\kappa k)}_{\xi q}=\sum_{mm'\sigma\sigma'} C^{lm}_{lm',kq}C^{1/2\sigma}_{1/2\sigma',\kappa\xi}a^\dagger_{lm\sigma}a_{lm'\sigma'} \tag{B.21}$$

(we use the definition, slightly different from that in [32]). Any operator of the form

$$\hat{F}=\sum_i \hat{f}_i,$$

where $\hat{f}$ is an arbitrary one-electron operator, may be represented in terms of the operators (B.21). In particular,

$$W^{(00)}=\sum_{m\sigma}a^\dagger_{lm\sigma}a_{lm\sigma}=\hat{n}. \tag{B.22}$$

is the number-of-particle operator for the shell. The cyclic components of the total spin and orbital momentum operators (1.7) are given by (cf.(B.10))

$$S_\xi=\frac{\sqrt{3}}{2}W^{(10)}_\xi, \tag{B.23}$$

$$L_q = \sqrt{l(l+1)}W_q^{(01)}, \tag{B.24}$$

and the inner scalar product

$$\frac{1}{2}\sqrt{3l(l+1)}\sum_q (-1)^q W_{q,-q}^{(11)} = \sum_i \mathbf{s}_i \mathbf{l}_i \tag{B.25}$$

is proportional to the operator of spin-orbit coupling.

The operators (B.21) for $k+\kappa>1$, unlike the operators (B.23), (B.24), have non-zero matrix elements between terms with different S and L and are, generally speaking, not reduced to products of the momentum operators (B.9). However, the connection may be established within a concrete term with given L, S (or J). Calculating the matrix element of the product of Fermi operators with account of (3.10) we obtain in agreement with the Wigner–Eckart theorem

$$\begin{aligned} &\langle SL\mu M\alpha | W_{\xi q}^{(\kappa k)} | S'L'\mu'M'\alpha'\rangle = \\ &\qquad = ([S][L])^{-1/2} C_{L'M',kq}^{LM} \langle SL\alpha | W^{(\kappa k)} | S'L'\alpha'\rangle, \end{aligned} \tag{B.26}$$

where the reduced matrix elements have the form

$$\begin{aligned} &\langle SL\alpha \| W^{(\kappa k)} \| S'L'\alpha'\rangle = n(2[l][S][S'][L][L'])^{1/2} \times \\ &\times \sum_{\{\bar{S}\bar{L}\bar{\alpha}\}} G_{\bar{S}\bar{L}\bar{\alpha}}^{SL\alpha} G_{\bar{S}\bar{L}\bar{\alpha}}^{S'L'\alpha'} \begin{Bmatrix} \frac{1}{2} & \kappa & \frac{1}{2} \\ S & \bar{S} & S' \end{Bmatrix} \begin{Bmatrix} l & k & l \\ L & \bar{L} & L' \end{Bmatrix} (-1)^{1/2+l+S+L+\bar{S}+\bar{L}+\kappa+k}; \end{aligned} \tag{B.27}$$

$$\begin{aligned} &\langle SL\alpha \| W^{(0k)} \| S'L'\alpha'\rangle = n\delta_{SS'}([l][S][L][L'])^{1/2} \times \\ &\qquad \times \sum_{\{\bar{S}\bar{L}\bar{\alpha}\}} G_{\bar{S}\bar{L}\bar{\alpha}}^{SL\alpha} G_{\bar{S}\bar{L}\bar{\alpha}}^{S'L'\alpha'} \begin{Bmatrix} l & k & l \\ L & \bar{L} & L' \end{Bmatrix} (-1)^{l+L+\bar{L}+k+1}. \end{aligned} \tag{B.28}$$

They can be also taken from the tables available (see, e.g., [10, 659]). Thus we have for a given term

$$W^{(\kappa k)} = ([S][L])^{1/2} \langle SL\alpha \| W^{(\kappa k)} \| SL\alpha\rangle S_\xi^{(\kappa)} L_q^{(k)}. \tag{B.29}$$

Comparing (B.27), (B.29) with (B.3), (B.4), (B.23), (B.24) one can see that the double irreducible tensor operator approach yields a summation of fractional parentage coefficients at $k=0$, $\kappa=1$ or $k=1$, $\kappa=0$ [660]. We shall see in Appendix D that this leads to a simplification of some terms in the exchange Hamiltonian. In particular, substituting the values of $6j$-symbols at $SL\alpha = S'L'\alpha'$, we derive

$$\sum_{\{\overline{S}\overline{L}\overline{\alpha}\}} \left(G^{SL\alpha}_{\overline{S}\overline{L}\overline{\alpha}}\right)^2 (-1)^{S-\overline{S}+1/2}[S]^{-1} = \frac{1}{n} \quad (S \neq 0); \tag{B.30}$$

$$\sum_{\{\overline{S}\overline{L}\overline{\alpha}\}} \left(G^{SL\alpha}_{\overline{S}\overline{L}\overline{\alpha}}\right)^2 \overline{L}(\overline{L}+1) = l(l+1) + \left(1 - \frac{2}{n}\right) L(L+1) \quad (L \neq 0). \tag{B.31}$$

The double irreducible tensor operator formalism may be generalized by considering triple tensor operators $a^{(qls)}$ $(q = s = 1/2)$ with the components

$$a^{(qls)}_{1/2m\sigma} = a^{\dagger}_{lm\sigma}, \qquad a^{(qls)}_{-1/2m\sigma} = (-1)^{l+s-m-\sigma} a_{lm\sigma}, \tag{B.32}$$

and constructing from them the operators

$$T^{(\Lambda\kappa k)}_{\lambda\xi r} = \sum_{\substack{\varphi\varphi'\sigma\sigma' \\ mm'}} C^{q\varphi}_{q\varphi',\Lambda\lambda} C^{S\sigma}_{S\sigma',\kappa\xi} C^{lm}_{lm',kr} a^{(qls)}_{\varphi m\sigma} a^{(qls)}_{\varphi' m'\sigma'}. \tag{B.33}$$

In particular, for κ=0, k=0, Λ=1 we obtain the quasispin operator with the components

$$Q^{+} = \frac{1}{2} \sum_{m\sigma} (-1)^{l+1/2-m-\sigma} a^{\dagger}_{lm\sigma} a^{\dagger}_{l-m-\sigma},$$

$$Q^{z} = -\frac{1}{2}\left(2l+1 - \sum_{m\sigma} a^{\dagger}_{lm\sigma} a_{lm\sigma}\right), \tag{B.34}$$

which satisfy the usual commutation relations for spin operators. Note that the operator Q^{+} coincides to a numerical factor with the creation operator (2.14) for the two-electron singlet zero-L state ^1S and adds such a pair to configuration l^{n-2} . The value of quasispin for a given term sequence equals to maximum value of Q^{z}:

$$Q = \frac{1}{2}(2l+1-v), \tag{B.35}$$

where v is the Racah's seniority quantum number, $(n-v)/2$ being the number of closed electron pairs with zero spin and orbital momenta in the given ME term. Using the Wigner–Eckart theorem in the quasispin space enables one to establish additional symmetry of the ME problem (in particular, to present the dependences of reduced matrix elements and fractional parentage coefficients on the electron number in terms of v) [32]. This formalism may be useful in the solid state theory at considering charge fluctuations. Further applications of the group theory to the classification of many-electron states are discussed in the book [65].

Appendix C

HAMILTONIAN OF A CRYSTAL WITH MANY-ELECTRON ATOMS

To present derivation of some many-electron (ME) models we treat the general Hamiltonian of the ME system in a crystal

$$\mathsf{H} = \sum_i \left(-\frac{\hbar^2}{2m}\Delta_{\mathbf{r}_i} + V(\mathbf{r}_i) \right) + \frac{1}{2}\sum_{i\neq j} \frac{e^2}{|\mathbf{r}_i - \mathbf{r}_j|}, \tag{C.1}$$

where $V(\mathbf{r})$ is the periodic crystal potential. Further we pass to the second quantization representation. To this end one has to use orthogonal wave functions. However, the atomic wave functions

$$\varphi_{lm\sigma}(x) = \varphi_{lm}(\mathbf{r})\chi_\sigma(s) = R_l(r)Y_{lm}(\hat{\mathbf{r}})\chi_\sigma(s) \tag{C.2}$$

(s is the spin coordinate, R_l is the radial wavefunction, Y is the spherical function, $\hat{\mathbf{r}}=(\theta, \phi)$) do not satisfy this condition at different sites ν. The non-orthogonality problem is one of most difficult in the magnetism theory [656, 661]. Here we use the orthogonalization procedure developed by Bogolyubov [651]. To lowest order in the overlap of atomic wavefunctions the orthogonalized functions read

$$\psi_{\nu lm}(\mathbf{r}) = \varphi_{\nu lm}(\mathbf{r}) - \frac{1}{2}\sum_{\nu'\neq\nu}\sum_{l'm'} \varphi_{\nu' l' m'}(\mathbf{r}) \int \varphi^*_{\nu' l' m'}(\mathbf{r}')\varphi_{\nu lm}(\mathbf{r}) d\mathbf{r}'. \tag{C.3}$$

The non-orthogonality corrections may be neglected at considering intrasite interactions and two-site Coulomb (but not exchange) matrix elements.

Then we obtain the many-electron Hamiltonian of the polar model [651, 662, 663] in the general case of degenerate bands

$$\begin{aligned} \mathsf{H} = &\sum_{\nu lm\sigma} \varepsilon_l a^\dagger_{\nu lm\sigma} a_{\nu lm\sigma} + \sum_{\nu_1\neq\nu_2}\sum_{l_i m_i \sigma} \beta_{\nu_1\nu_2}(l_1m_1, l_2m_2) a^\dagger_{\nu_1 l_1 m_1\sigma} a_{\nu_2 l_2 m_2 \sigma} + \\ &+ \frac{1}{2}\sum_{\nu_i l_i m_i \sigma_1\sigma_2} I_{\nu_1\nu_2\nu_3\nu_4}(l_1m_1, l_2m_2, l_3m_3, l_4m_4) \times \\ &\times a^\dagger_{\nu_1 l_1 m_1 \sigma_1} a^\dagger_{\nu_2 l_2 m_2 \sigma_2} a_{\nu_4 l_4 m_4 \sigma_2} a_{\nu_3 l_3 m_3 \sigma_1}. \end{aligned} \tag{C.4}$$

Here we have used the orthogonality of spin wave functions $\chi_\sigma(s)$,

$$\varepsilon_l = \int_0^\infty r^2 R_l^2(r)\left[-\frac{1}{2mr^2}\frac{d}{dr}\left(r^2\frac{d}{dr}\right)+\frac{l(l+1)}{2mr^2}+v(r)\right]dr \tag{C.5}$$

are the one-electron levels in the central potential of a given site $v(\mathbf{r})$ (we neglect the influence of potentials of other atoms, i.e. crystal-field effects),

$$\beta_{\nu_1\nu_2}(l_1m_1,l_2m_2) = \int \psi^*_{\nu_1l_1m_1}(\mathbf{r})\left(-\frac{\hbar^2}{2m}\Delta+V(\mathbf{r})\right)\psi_{\nu_2l_2m_2}(\mathbf{r})d\mathbf{r} \tag{C.6}$$

are the transfer matrix elements between the sites ν_1 and ν_2,

$$I_{\nu_1\nu_2\nu_3\nu_4}(l_1m_1,l_2m_2,l_3m_3,l_4m_4) =$$

$$= \int \psi^*_{\nu_1l_1m_1}(\mathbf{r})\psi^*_{\nu_2l_2m_2}(\mathbf{r}')\frac{e^2}{|\mathbf{r}-\mathbf{r}'|}\psi_{\nu_3l_3m_3}(\mathbf{r})\psi_{\nu_4l_4m_4}(\mathbf{r}')d\mathbf{r}d\mathbf{r}' \tag{C.7}$$

are the matrix elements of electrostatic interaction. Consider the electrostatic interactions between two atomic shells on the same lattice site $\nu_1=\nu_2=\nu_3=\nu_4$. We use the standard expansion

$$\frac{1}{|\mathbf{r}-\mathbf{r}'|} = \sum_{p=0}^{\infty}\frac{4\pi}{[p]}\frac{r_<^p}{r_>^{p+1}}\sum_{q=-p}^{p}Y_{pq}(\hat{\mathbf{r}})Y_{pq}(\hat{\mathbf{r}}') \tag{C.8}$$

and the expression for the matrix element of a spherical function

$$\int_0^\pi \sin\theta d\theta \int_0^{2\pi} d\varphi Y^*_{l_1m_1}(\hat{\mathbf{r}})Y_{l_2m_2}(\hat{\mathbf{r}})Y_{l_3m_3}(\hat{\mathbf{r}}) =$$

$$= \left(\frac{[l_2][l_3]}{4\pi[l_1]}\right)^{1/2} C^{l_10}_{l_20,l_30}C^{l_1m_1}_{l_2m_2,l_3m_3} \equiv \tilde{C}^{l_1m_1}_{l_2m_2,l_3m_3}. \tag{C.9}$$

The quantity $\tilde{C}$ vanishes if $l_1+l_2+l_3$ is odd, and for $l_1+l_2+l_3=2g$ one has

$$C^{l_10}_{l_20,l_30} = (-1)^{l_1+g}[l_1]^{1/2}g!\{(2g+1)\}^{-1/2}\prod_{i=1}^{3}\frac{\{(2g-2l_i)!\}^{1/2}}{(g-l_i)!}.$$

Carrying out the integration, we obtain for the Coulomb term ($l_1=l_3$, $l_2=l_4$)

$$\mathsf{H}_{\text{coul}}(l_1l_2) = \sum_p C^{l_10}_{l_10,p0}C^{l_20}_{l_20,p0}F^{(p)}(l_1l_2)(W_1^{(0p)}W_2^{(0p)}), \tag{C.10}$$

where

$$F^{(p)}(l_1l_2) = e^2\int r_1^2r_2^2\frac{r_<^p}{r_>^{p+1}}R^2_{l_1}(r_1)R^2_{l_2}(r_2)dr_1dr_2, \tag{C.11}$$

are the Slater parameters, the irreducible tensor operators W are given by (B.21) and the scalar product is defined by

$$(A^{(a)}B^{(a)}) = \sum_{\alpha}(-1)^{\alpha} A^{(a)}_{\alpha} B^{(a)}_{-\alpha} \equiv (-1)^{a}[a]^{1/2}[A^{(a)} \times B^{(a)}]^{(0)}. \qquad \text{(C.12)}$$

The Hamiltonian (C.10) may be expressed in terms of many-electron X-operators with the use of (B.29) (see Appendix D).

For the exchange integral we obtain

$$I_{\nu}(l_1m_1, l_2m_2, l_3m_3, l_4m_4) = \sum_{pq} \frac{4\pi}{[p]} \tilde{C}^{l_1m_1}_{l_2m_3,pq} \tilde{C}^{l_2m_2}_{l_1m_4,pq} G^{(p)}(l_1l_2),$$

where

$$G^{(p)}(l_1l_2) = e^2 \int r_1^2 r_2^2 R_{l_1}(r_1) R_{l_2}(r_2) \frac{r_<^p}{r_>^{p+1}} R_{l_2}(r_1) R_{l_1}(r_2) dr_1 dr_2. \qquad \text{(C.13)}$$

Transforming the product of Clebsh–Gordan coefficients

$$\sum_q C^{l_1m_1}_{l_2m_3,pq} C^{l_2m_2}_{l_1m_4,pq} = \sum_{k\mu} \begin{Bmatrix} l_1 & l_2 & p \\ l_2 & l_1 & k \end{Bmatrix} C^{l_1m_1}_{l_1m_4,pq} C^{l_2m_2}_{l_2m_3,pq} (-1)^{k-\mu}, \qquad \text{(C.14)}$$

we derive

$$\mathsf{H}_{\text{exch}}(l_1l_2) = -\frac{1}{2} \sum_{kp\kappa} \begin{Bmatrix} l_1 & l_2 & p \\ l_2 & l_1 & k \end{Bmatrix} (-1)^k [k][\kappa] \left(\frac{[l_2]}{[l_1]} \right)^{1/2} \times$$

$$\times \left(C^{l_1 0}_{l_2 0, p0} \right)^2 G^{(p)}(l_1l_2) \left(W_1^{(\kappa k)} W_2^{(\kappa k)} \right). \qquad \text{(C.15)}$$

For k=0 we obtain from (B.22), (B.23) the intra-atomic exchange Hamiltonian

$$\sum_{\kappa} \mathsf{H}^{(\kappa 0)}_{\text{exch}}(l_1l_2) = -\frac{1}{2} \sum_p \left(\frac{[l_2]}{[l_1]} \right)^{1/2} \left(C^{l_1 0}_{l_2 0, p0} \right)^2 G^{(p)}(l_1l_2)[n_1n_2 + 4(\mathbf{S}_1\mathbf{S}_2)], \qquad \text{(C.16)}$$

and for κ=0, k=1 from (B.24) the orbital exchange Hamiltonian

$$\mathsf{H}^{(01)}_{\text{exch}}(l_1l_2) = -\frac{3}{4} \sum_p \left(C^{l_1 0}_{l_2 0, p0} \right)^2 \frac{l_1(l_1+1) + l_2(l_2+1) - p(p+1)}{l_1(l_1+1) l_2(l_2+1)[l_1]} \times$$

$$\times G^{(p)}(l_1l_2)(\mathbf{L}_1\mathbf{L}_2). \qquad \text{(C.17)}$$

In the case of electrostatic interaction between electrons of the same shell, F=G, so that we may represent the Hamiltonian in both forms (C.10) and (C.15). Thus

$$\mathsf{H}(l) = \frac{1}{2} \sum_p \left(C^{l_1 0}_{l_2 0, p0} \right)^2 F^{(p)}(ll)[(W^{(0p)}W^{(0p)}) - n], \qquad \text{(C.18)}$$

where the second term in square brackets arises because of the condition $i \neq j$ in (C.1). In the ME representation, the Hamiltonian (C.18) takes the quasidiagonal form. Calculating the matrix elements of the scalar product we obtain

$$\mathsf{H}(l) = \sum_{SL\mu M} \sum_{\alpha\alpha'} E_{SL}^{\alpha\alpha'} X(SL\mu M\alpha, SL\mu M\alpha'),$$

with

$$E_{SL}^{\alpha\alpha'} = \frac{1}{2} \sum_{SLM\mu} \sum_{p} \left(C_{l_2 0, p0}^{l_1 0} \right)^2 F^{(p)}(ll) \{ [S]^{-1} [L]^{-1} \times$$

$$\times \sum_{\bar{L}\bar{\alpha}} \langle SL\alpha \| W^{(0p)} \| S\bar{L}\bar{\alpha} \rangle \langle SL\alpha' \| W^{(0p)} \| S\bar{L}\bar{\alpha} \rangle, \quad \text{(C.19)}$$

and the reduced matrix elements are given by (B.27). If several ME terms with the same values of L, S are present (which is typical for d- and f-electrons), additional diagonalization is required. Retaining in (C.19) the contribution with $p=0$ only we obtain

$$E_{\Gamma} = \frac{1}{2} n(n-1) F^{(0)}(ll). \quad \text{(C.20)}$$

The contributions with $p=2, 4, \ldots$ yield the dependence of term energy on the ME quantum numbers S, L according to the Hund rules.

Expressions (C.10), (C.16), (C.18) reduce the problem of electrostatic interaction between electrons to calculation of the Slater integrals. These may be considered as parameters, which should be taken from experimental data (such a procedure is used often in the atomic spectroscopy). In this case, the wave functions, which enter the Slater integrals, are to be calculated in a self-consistent way from the corresponding integro-differential equations [20]. Consider the one-band Hamiltonian of a crystal in the many-electron representation. Besides the on-site Coulomb repulsion and the intersite electron transfer (which are taken into account in the Hubbard model) we take into account the electron transfer owing to matrix elements of electrostatic interaction (C.7) with $\nu_1 \neq \nu_3$, $\nu_2 = \nu_4$. We confine ourselves to the terms with $\nu_1 = \nu_2$ or $\nu_2 = \nu_3$ (three-site terms are smaller due to decrease of the Coulomb interaction with distance). Using (B.21), (C.19) and taking into account corrections owing to the non-orthogonality we obtain

$$\mathsf{H} = \sum_{\nu\Gamma} E_{\Gamma} X_{\nu}(\Gamma, \Gamma) +$$

$$+ \sum_{\nu_1 \neq \nu_2} \sum_{\Gamma_n \Gamma_{n-1} \Gamma_{n'} \Gamma_{n'-1}} B_{\nu_1\nu_2}(\Gamma_n \Gamma_{n-1}, \Gamma_{n'} \Gamma_{n'-1}) X_{\nu_1}(\Gamma_n \Gamma_{n-1}) X_{\nu_2}(\Gamma_{n'} \Gamma_{n'-1}), \quad \text{(C.21)}$$

where

$$B_{\nu_1\nu_2}(\Gamma_n\Gamma_{n-1},\Gamma_{n'}\Gamma_{n'-1}) = (nn')^{1/2} G^{\Gamma_n}_{\Gamma_{n-1}} G^{\Gamma_{n'}}_{\Gamma_{n'-1}} \times$$
$$\times \sum_{\gamma_1\gamma_2} C^{\Gamma_n}_{\Gamma_{n-1},\gamma_1} C^{\Gamma_{n'}}_{\Gamma_{n'-1},\gamma_2} \delta_{\sigma_1\sigma_2} \beta_{\nu_1\nu_2}(lm_1, lm_2) +$$
$$+[n(n-1)n'(n'-1)]^{1/2} \left[\sum_{\overline{\Gamma}_{n-1}\overline{\Gamma}_{n-2}} G^{\Gamma_n}_{\overline{\Gamma}_{n-1}} G^{\overline{\Gamma}_{n-1}}_{\overline{\Gamma}_{n-2}} G^{\Gamma_{n-1}}_{\overline{\Gamma}_{n-2}} G^{\Gamma_{n'}}_{\Gamma_{n'-1}} \times \right.$$
$$\times \sum_{\gamma_1\cdots\gamma_4} G^{\Gamma_n}_{\overline{\Gamma}_{n-1},\gamma_1} G^{\overline{\Gamma}_{n-1}}_{\overline{\Gamma}_{n-2},\gamma_2} G^{\Gamma_{n-1}}_{\overline{\Gamma}_{n-2},\gamma_3} G^{\Gamma_{n'}}_{\overline{\Gamma}_{n'-1},\gamma_4} \delta_{\sigma_1\sigma_3} \delta_{\sigma_2\sigma_4} \times$$
$$\times I_{\nu(1)\nu_2\nu_3\nu_4}(lm_1, lm_2, lm_3, lm_4) +$$
$$\left. + \{\Gamma_n, \Gamma_{n-1}\} \leftrightarrow \{\Gamma_{n'}, \Gamma_{n'-1}\} \right]. \tag{C.22}$$

are the many-electron transfer integrals, e.g., for the s-band we have

$$\mathsf{H} = U \sum_{\nu} X_{\nu}(2,2) +$$
$$+ \sum_{\nu_1\nu_2\sigma} \{\beta^{(00)}_{\nu_1\nu_2} X_{\nu_1}(\sigma,0) X_{\nu_2}(0,\sigma) + \beta^{(22)}_{\nu_1\nu_2} X_{\nu_1}(2,\sigma) X_{\nu_2}(\sigma,2) +$$
$$+ \sigma\beta^{(02)}_{\nu_1\nu_2} X_{\nu_1}(\sigma,0) X_{\nu_2}(-\sigma,2) + X_{\nu_1}(2,-\sigma) X_{\nu_2}(0,\sigma)\}, \tag{C.23}$$

where $U=I_{\nu\nu\nu\nu}=F^{(0)}(00)$ is the Hubbard parameter,

$$\beta^{(00)}_{\nu_1\nu_2} = \beta_{\nu_1\nu_2},$$
$$\beta^{(22)}_{\nu_1\nu_2} = \beta_{\nu_1\nu_2} + 2I_{\nu_1\nu_1\nu_2\nu_1},$$
$$\beta^{(02)}_{\nu_1\nu_2} = \beta^{(20)}_{\nu_1\nu_2} = \beta_{\nu_1\nu_2} + I_{\nu_1\nu_1\nu_2\nu_1} \tag{C.24}$$

are the transfer integral for holes and doubles, and the integral of the double-hole pair creation; according to (C.3),

$$I_{\nu_1\nu_1\nu_2\nu_1} = I^{(0)}_{\nu_1\nu_1\nu_2\nu_1} - \frac{U}{2} \int \varphi_{\nu_2}(\mathbf{r}) \varphi_{\nu_1}(\mathbf{r}) d\mathbf{r}, \tag{C.25}$$

with $I^{(0)}$ being calculated for the atomic functions φ. It should be noted that the dependence of the transfer integrals on the atomic ME terms may be more complicated if we use at solving the atomic problem the approaches which are more complicated than in Appendix A. For example, the general Hartree–Fock approximation (see [20]) yields the radial one-electron wave functions which depend explicitly on atomic term. In some variational approaches of the many-electron atom theory (see [664] the ME wavefunction is not many-electron atom theory (see [664] the ME wavefunction is not factorized into one-electron ones. Then the transfer integrals are to be calculated with the use of ME wave functions:

$$B_{\nu_1\nu_2}(\Gamma_n\Gamma_{n-1},\Gamma_{n'}\Gamma_{n'-1}) = \int\prod \Psi^*_{\nu_1\Gamma_n}\Psi^*_{\nu_2\Gamma_{n'}} \times$$
$$\times\sum_i \left(-\frac{\hbar^2}{2m}\Delta_{\mathbf{r}_i} + V(\mathbf{r}_i)\right)\Psi_{\nu_1\Gamma_{n-1}}\Psi_{\nu_2\Gamma_{n'-1}}dx_i. \quad \text{(C.26)}$$

In particular, for the s-band the integrals (C.24) may be different even at neglecting interatomic Coulomb interactions and non-orthogonality. Besides that, the many-configuration approach, which takes into account the interaction of different electron shells, is sometimes required.

To analyze the m-dependence of two-site matrix elements, it is convenient to transform them into one-site ones by expanding the wave functions $\varphi_{\nu_2}(\mathbf{r})$ about the first atom [665]. Passing to the Fourier transforms

$$\varphi_{lm}(\mathbf{r}) = \int \tilde{R}_l(k)Y_{lm}(\hat{\mathbf{k}})e^{i\mathbf{kr}}d\mathbf{k}, \quad \text{(C.27)}$$

(with $\tilde{R}_l(\mathrm{k})$ being the Fourier transforms of radial functions) and expanding plane waves in spherical harmonics

$$e^{i\mathbf{kr}} = 4\pi\sum_{\lambda\mu} i^\lambda j_\lambda(\mathbf{kr})Y^*_{\lambda\mu}(\hat{\mathbf{k}})Y_{lm}(\hat{\mathbf{r}}), \quad \text{(C.28)}$$

where $j_\lambda(x)$ are the spherical Bessel function, we obtain

$$\varphi_{\nu_2 lm}(\mathbf{r}) \equiv \varphi_{\nu_1 lm}(\mathbf{r}+\boldsymbol{\rho}) = 4\pi\sum_{\lambda\mu\eta\xi}\tilde{C}^{\eta\xi}_{lm,\lambda\mu}R_{l\eta\lambda}(\mathbf{r},\boldsymbol{\rho})Y^*_{\lambda\mu}(\hat{\boldsymbol{\rho}})Y_{\eta\xi}(\hat{\mathbf{r}}). \quad \text{(C.29)}$$

with

$$R_{l\eta\lambda}(\mathbf{r},\boldsymbol{\rho}) = \int_0^\infty k^2 R_l(k)i^{\lambda+\eta}j_\eta(\mathbf{kr})j_\lambda(\mathbf{k}\boldsymbol{\rho})dk. \quad \text{(C.30)}$$

As well as in (C.3), we retain only the crystal potential at the site ν_1 is retained; in this approximation, non-orthogonality corrections to one-electron transfer integrals are absent [651]. Then, substituting (C.29) into (C.6) we pick out the dependence of transfer matrix elements on magnetic quantum numbers [660]

$$\beta_{\nu_1\nu_2}(l_1m_1,l_2m_2) = 4\pi\sum_{\lambda\mu}\tilde{C}^{l_1m_1}_{l_2m_2,\lambda\mu}\bar{\beta}_{\nu_1\nu_2}(l_1l_2\lambda)Y_{\lambda\mu}(\boldsymbol{\rho}_{\nu_1\nu_2}), \quad \text{(C.31)}$$

where

$$\bar{\beta}_{\nu_1\nu_2}(l_1l_2\lambda) = \int_0^\infty r^2 R_{l_1}(r)v(r)R_{l_2l_1\lambda}(r,\boldsymbol{\rho}_{\nu_1\nu_2})dr. \quad \text{(C.32)}$$

integrals with $l_1=l_2$) λ is even, and for odd l_1-l_2 (in particular, for the matrix elements of s–p, p–d and d–f hybridization) the angle dependence with $\lambda=1$ emerges. The Coulomb contributions to (C.22) turn out to have qualitatively the same anisotropy as (C.31) yields.

In the case of strong intraatomic Coulomb interaction we may retain in the Hamiltonian (C.21) only two lower ME terms, $\Gamma=\{SL\}$ and $\Gamma=\{S'L'\}$. Then the intraatomic Hamiltonian yields a constant energy shift only and may be omitted. The transfer Hamiltonian may be represented through the spin and orbital momentum operators corresponding to the term Γ_n. To this end, we pick out from X-operators one Fermi operator with the use of (A.21), (A.17), (A.11). Substituting (C.31) and transforming products of Clebsh–Gordan coefficients we get

$$\mathsf{H} = (4\pi)^{1/2}\left(G^{\Gamma_{n+1}}_{\bar{\Gamma}_n}\right)^4 \sum_{\nu_i m_i \sigma_i \sigma} \sum_{k_1 k_2 \lambda p q k} \bar{\beta}_{\nu_1\nu_2}(ll\lambda)[k_1][k_2]([k][p])^{1/2} \times$$

$$\times [L']^2[L]^{-1} \begin{Bmatrix} L & L & k_1 \\ l & l & k_2 \end{Bmatrix} \begin{Bmatrix} k_1 & l & l \\ k_2 & l & l \\ k & p & \lambda \end{Bmatrix} (-1)^q C^{l0}_{l0,\lambda 0} C^{lm_1}_{lm_2,pq} \times$$

$$\times a^\dagger_{\nu_1 l m_1 \sigma_1} P_{\nu_1\sigma_1\sigma} \left[Y^{(\lambda)}(\hat{\boldsymbol{\rho}}) \times \left[L^{(k_1)}_{\lambda_1} \times L^{(k_2)}_{\lambda_2} \right]^{(\lambda)} \right]^{(p)}_{-q} P_{\nu_2\sigma\sigma_2} a_{\nu_2 l m_2 \sigma_2}, \qquad \text{(C.33)}$$

where

$$\hat{P}_\nu = \frac{1}{2[S]}([S'] + (-1)^{S-S'+1/2} 2(\mathbf{S}_\nu \boldsymbol{\sigma})).$$

Here σ are the Pauli matrices, $Y^{(\lambda)}$ is the irreducible tensor with the components $Y_{\lambda\mu}$, the vector product of irreducible tensor operators is defined by (B.14).

Appendix D

INTERATOMIC ELECTROSTATIC INTERACTION AND DERIVATION OF THE HEISENBERG HAMILTONIAN

In this Appendix we treat, starting from the general many-electron Hamiltonian of a crystal (C.1), the 'direct' Coulomb and exchange interaction between two degenerate atomic shells at different sites. The consideration of this mechanism enables one to establish general features of Hamiltonian of magnetic ion systems with unquenched orbital momenta.

Putting in (C.7) $v_1l_1=v_3l_3$, $v_2l_2=v_4l_4$, making the expansion (C.29), transforming the product of Clebsh–Gordan coefficients and using the multiplication formula for spherical harmonics

$$Y_{\lambda_1\mu_1}(\hat{\rho})Y_{\lambda_2\mu_2}(\hat{\rho}) = \sum_{\lambda\mu}\tilde{C}^{\lambda\mu}_{\lambda_1\mu_1,\lambda_2\mu_2}Y_{\lambda\mu}(\hat{\rho}), \qquad \text{(D.1)}$$

we obtain, similar to (C.10), the Hamiltonian of the intersite Coulomb interaction [660]

$$\begin{aligned} \mathrm{H}_{\text{coul}}(v_1v_2) = (4\pi)^{1/2}\sum_{pb\lambda}\sum_{\lambda_1\lambda_2\eta_1\eta_2} F^{(p)}(l_1l_2\eta_1\eta_2\lambda_1\lambda_2)[b][\lambda_1][\lambda_2][l_2]^{1/2}\times \\ \times(-1)^{\lambda_1}C(l_1pl_1,l_2\lambda_1\eta_1,\eta_2p\eta_1,l_2\lambda_2\eta_2,\lambda_1\lambda_2\lambda)\times \\ \times\begin{Bmatrix} l_1 & \lambda_1 & \eta_1 \\ l_2 & \lambda_2 & \eta_2 \\ l & \lambda & p \end{Bmatrix}\left([W_1^{(0p)}\times W_2^{(ab)}]^{(\lambda)}Y^{(\lambda)}(\hat{\rho})\right), \end{aligned} \qquad \text{(D.2)}$$

where

$$\begin{aligned} F^{(p)}(l_1l_2\eta_1\eta_2\lambda_1\lambda_2) = e^2\int_0^\infty r_1^2r_2^2R^*_{l_1\eta_1\lambda_1}(r_2,\rho)\frac{r_<^p}{r_>^{p+1}}\times \\ \times R_{l_1\eta_1\lambda_1}(r_2,\rho)R^2_{l_1}(r_1)dr_1dr_2. \end{aligned} \qquad \text{(D.3)}$$

is a generalized Slater integral for two-site interaction, and the notation is used for brevity

$$C(a_1,b_1,c_1,a_2,b_2,c_2,\ldots)=\prod_i C^{c_i 0}_{a_i 0,b_i 0}. \qquad \text{(D.4)}$$

One can see from (C.9) that p, $\lambda_1+\lambda_2$ and λ are even (b is also even because of time-reversal invariance), $p\le 2l_1$, $b\le 2l_2$, $\lambda\le 2(l_1+l_2)$. Further we pass to the many-electron representation. According to (B.29), for a given LS-term we have to substitute

$$W_i^{(0k)}=([S_i][L_i])^{-1/2}\langle S_iL_i \,\|\, W_i^{(0k)} \,\|\, S_iL_i\rangle L_i^{(k)}, \qquad \text{(D.5)}$$

with the reduced matrix elements being given by (B.28). The transition to usual vectors is performed with the use of equation (B.13) and factorization of spherical harmonics according to (D.1). For example,

$$(L_1^{(2)}L_2^{(2)})=20\sqrt{\frac{5}{3}}\prod_{i=1,2}[L_i]^{1/2}(2L_i-1)^{-1}(2L_i+3)^{-1}\times$$

$$\times\left\{\frac{1}{3}L_1(L_1+1)L_2(L_2+1)-(\mathbf{L}_1\mathbf{L}_2)^2-\frac{1}{2}(\mathbf{L}_1\mathbf{L}_2)\right\}, \qquad \text{(D.6)}$$

$$(L_1^{(2)}Y^{(2)}(\hat{\rho}))=\frac{5}{\sqrt{2\pi}}\frac{[L_1]^{1/2}}{(2L_1-1)(2L_1+3)}\left\{\frac{1}{3}L_1(L_1+1)-\frac{(\mathbf{L}_1\boldsymbol{\rho})^2}{\rho^2}\right\}. \qquad \text{(D.7)}$$

In the classical limit (L_1, $L_2 \gg 1$) we obtain the expansion

$$\mathsf{H}_{\text{coul}}(v_1v_2)=\sum_{\alpha\beta\gamma}Q_{\alpha\beta\gamma}\frac{(\mathbf{L}_1\boldsymbol{\rho})^{\alpha}(\mathbf{L}_2\boldsymbol{\rho})^{\beta}(\mathbf{L}_1\mathbf{L}_2)^{\gamma}}{\rho^{\alpha+\beta}}, \qquad \text{(D.8)}$$

where the coefficients Q are linear combinations of the Slater integrals (D.3), the powers of L_1, L_2 and $\hat{\rho}$ in each summand are even and $\alpha+\gamma\le 2l_1$, $\beta+\gamma\le 2l_2$. Anisotropic terms, which depend on orientation of only one orbital momentum in the lattice, are present. Vector products do not appear since they are transformed into scalar ones:

$$([\mathbf{L}_1\times\mathbf{L}_2]\boldsymbol{\rho})^2=\det\begin{pmatrix} L_1(L_1+1) & (\mathbf{L}_1\mathbf{L}_2) & (\mathbf{L}_1\boldsymbol{\rho}) \\ (\mathbf{L}_1\mathbf{L}_2) & L_2(L_2+1) & (\mathbf{L}_2\boldsymbol{\rho}) \\ (\mathbf{L}_1\boldsymbol{\rho}) & (\mathbf{L}_2\boldsymbol{\rho}) & \rho^2 \end{pmatrix}. \qquad \text{(D.9)}$$

For concrete electron configurations, the total number of terms in the series (D.7) is small. So, maximum powers of L_1 and L_2 do not exceed 2 for p-electrons and 4 for d-electrons.

At considering the intersite direct exchange interaction [660, 666, 667] ($\nu_1 l_1 = \nu_4 l_4$, $\nu_2 l_2 = \nu_3 l_3$), which is small in the overlap of atomic wave functions at different sites, one has to take into account corrections owing to non-orthogonality since they have the same (second) order of smallness. Calculating the integral (C.6) for the functions (C.3), and expanding the wavefunctions, that arise, about the site ν with the use of (C.29), we obtain the intersite exchange Hamiltonian [660]

$$\mathsf{H}_{\text{exch}}(\nu_1\nu_2) = -\frac{1}{2}\sum_{k_1k_2\lambda\kappa}(4\pi)^{1/2}[\kappa]^{3/2}(-1)^{\kappa} I(l_1l_2k_1k_2\lambda)\times \\ \times\Big([W_1^{(\kappa k_1)}\times W_2^{(\kappa k_2)}]^{(0\lambda)}Y^{(\lambda)}(\rho_{\nu_1\nu_2})\Big), \qquad \text{(D.10)}$$

where effective exchange parameters are given by

$$I(l_1l_2k_1k_2\lambda) = \frac{1}{2}\Bigg[\sum_{p\eta_1\eta_2\lambda_1\lambda_2}\Big\{G^{(p)}(l_1l_2\eta_1\eta_2\lambda_1\lambda_2) - \\ -\mathrm{Re}[G^{(p)}(l_1l_2\eta_1\eta_2\lambda_2)Z(l_2\eta_2\lambda_2)] + \frac{1}{2}G^{(p)}(l_1\eta_1\eta_2)Z(l_1\eta_1\lambda_1)Z^*(l_2\eta_2\lambda_2)\Big\}\times \\ \times(-1)^{l_1-l_2}[k_1][k_2][\lambda_1][\lambda_2]\left(\frac{[l_2][\eta_1]}{[\lambda]}\right)^{1/2}\times \\ \times C(\eta_1pl_1,l_1p\eta_2,l_2\lambda_2\eta_2,l_2\lambda_1\eta_1,\lambda_1\lambda_2\lambda)\begin{Bmatrix}\eta_1 & \eta_2 & k_1\\ l_1 & l_1 & p\end{Bmatrix}\begin{Bmatrix}l_2 & \lambda_1 & \eta_1\\ l_2 & \lambda_2 & \eta_2\\ k_2 & \lambda & k_1\end{Bmatrix} + \\ +\frac{1}{4}\sum_{p\lambda_i\Lambda_i\eta_i a}F^{(p)}(l_1l_2\eta_1\eta_2\lambda_2\lambda_3\lambda_4)Z(l_1\eta_1\lambda_1)Z(l_2\eta_3\lambda_3)\times \\ \times(-1)^{\eta_3}\prod_{i=1}^{3}[\lambda_i][k_1][k_2][a]\left(\frac{[\Lambda_1][\Lambda_2][\eta_1][\eta_3]}{[\lambda]}\right)^{1/2}\times \\ \times C(l_1\lambda_1\eta_1,\eta_3pl_1,\eta_1\lambda_4\eta_4,l_2\lambda_3\eta_3,\eta_2p\eta_4,l_2\lambda_2\eta_2,\lambda_1\lambda_4\Lambda_1,\Lambda_1\lambda_3\Lambda_2,\Lambda_2\lambda_2\lambda)\times \\ \times\begin{Bmatrix}l_1 & \eta_4 & \Lambda_1\\ \lambda_4 & \lambda_1 & \eta_1\end{Bmatrix}\begin{Bmatrix}l_2 & a & \Lambda_2\\ \Lambda_1 & \gamma_3 & \eta_3\end{Bmatrix}\begin{Bmatrix}l_1 & \eta_3 & \Lambda_1\\ l_1 & \Lambda_1 & \eta_4\\ k & a & \eta_2\end{Bmatrix}\begin{Bmatrix}l_2 & \lambda_2 & \eta_2\\ l_2 & \Lambda_2 & a\\ k & \lambda & k_1\end{Bmatrix}\Bigg] + \\ +\frac{1}{2}[\{\nu_1l_1k_1\}\leftrightarrow\{\nu_2l_2k_2\}]^*, \qquad \text{(D.11)}$$

with k_1+k_2 and λ being even, $k_1 \leq 2l_1$, $k_2 \leq 2l_2$, $\lambda \leq 2(l_1+l_2)$, $\kappa \leq 1$. Here we have introduced the two-site Slater integrals

$$G^{(p)}(l_1 l_2 \eta_1 \eta_2 \lambda_1 \lambda_2) = e^2 \int r_1^2 r_2^2 R_{l_1}(r_1) R^*_{l_2 \eta_2 \lambda_2}(r_2, \rho) \frac{r_<^p}{r_>^{p+1}} \times$$
$$\times R_{l_2 \eta_1 \lambda_2}(r_1, \rho) R_{l_1}(r_2) dr_1 dr_2; \qquad \text{(D.12)}$$

$$G^{(p)}(l_1 l_2 \eta_1 \eta_2 \lambda_1) = e^2 \int r_1^2 r_2^2 R_{l_1}(r_1) R_{\eta_2}(r_2) \frac{r_<^p}{r_>^{p+1}} \times$$
$$\times R_{l_2 \eta_1 \lambda_1}(r_1, \rho) R_{l_1}(r_2) dr_1 dr_2; \qquad \text{(D.13)}$$

$$G^{(p)}(l \eta_1 \eta_2) = e^2 \int r_1^2 r_2^2 R_l(r_1) R_{\eta_2}(r_2) \frac{r_<^p}{r_>^{p+1}} R_{\eta_1}(r_1) R_l(r_2) dr_1 dr_2; \qquad \text{(D.14)}$$

$$F^{(p)}(l_1 l_2 \eta_1 \eta_2 \eta_3 \eta_4 \lambda_2 \lambda_3 \lambda_4) = e^2 \int r_1^2 r_2^2 R_{l_1}(r_1) R^*_{l_2 \eta_2 \lambda_2}(r_2, \rho) \frac{r_<^p}{r_>^{p+1}} \times$$
$$\times R_{l_3 \eta_3 \lambda_3}(r_1, \rho) R_{\eta_1 \eta_4 \lambda_4}(r_2, \rho) dr_1 dr_2, \qquad \text{(D.15)}$$

and the non-orthogonality integrals

$$Z(l \eta \lambda) = \int_0^\infty r^2 R_l(r) R_{l \eta \lambda}(r, \rho) dr. \qquad \text{(D.16)}$$

Putting $k_1 = k_2 = \lambda = 0$ we obtain, similar to (C.16), the spin Hamiltonian

$$\sum_\kappa \mathsf{H}^{(\kappa 0)}_{\text{exch}}(\nu_1 \nu_2) = -\frac{1}{2} I(l_1 l_2 000)[n_1 n_2 + 4(\mathbf{S}_1 \mathbf{S}_2)]. \qquad \text{(D.17)}$$

Putting $k_1 = k_2 = 1$, $\kappa = 0$ we obtain the orbital exchange Hamiltonian

$$\mathsf{H}^{(01)}_{\text{exch}}(\nu_1 \nu_2) = -\frac{1}{2}(l_1(l_1+1) l_2(l_2+1))^{-1/2} \times$$
$$\times \left\{ I(l_1 l_2 110)(\mathbf{L}_1 \mathbf{L}_2) + (8\pi)^{-1/2} I(l_1 l_2 112) \left[\frac{1}{3}(\mathbf{L}_1 \mathbf{L}_2) - \frac{(\mathbf{L}_1 \rho)(\mathbf{L}_2 \rho)}{\rho^2} \right] \right\}. \qquad \text{(D.18)}$$

For all the other contributions, we have to pass to the operators $S^{(\kappa)}$ and $L^{(k)}$ for a concrete term with the use of (B.29). Then we obtain the expansion

$$\mathsf{H}_{\text{exch}}(\nu_1 \nu_2) = -\sum_{\alpha\beta\gamma\kappa} I_{\alpha\beta\gamma\kappa} \frac{(\mathbf{L}_1 \rho)^\alpha (\mathbf{L}_2 \rho)^\beta (\mathbf{L}_1 \mathbf{L}_2)^\gamma (\mathbf{S}_1 \mathbf{S}_2)^\kappa}{\rho^{\alpha+\beta}} \qquad \text{(D.19)}$$

with even $\alpha+\beta$, $\alpha+\gamma \leq 2l_1$, $\beta+\gamma \leq 2l_2$, $\kappa = 0,1$. It should be stressed that the powers of orbital momentum operators in (D.7), (D.19) are restricted in the microscopic treatment not only by $2L$ (as follows from kinematic relations), but also by $2l$. Higher-order terms in spin operators, e.g., biquadratic exchange ones, do not occur in (D.19) since the electron spin s equals to 1/2. Such terms may be obtained from the higher-order corrections owing to non-orthogonality [656, 668]. However, they are considerably smaller in the overlap of wave functions at different sites.

The above treatment of the two-site problem is, generally speaking, insufficient for obtaining the exchange Hamiltonian of a crystal. However, neglect of multi-site terms is justified in the nearest-neighbor approximation. Corrections to the two-centre approximation are especially important if some nearest neighbour form equilateral triangles (e.g., for the fcc and hcp lattices).

It should be noted that the electrostatic interaction may be investigated by direct using the representation of ME wave functions [655]. We obtain

$$\mathsf{H}(\nu_1\nu_2) = \int \prod_{ij} \Psi^*_{\nu_1\Gamma_1}(\{x_i\})\Psi^*_{\nu_2\Gamma_2}(\{x'_j\}) \sum_{ij} \frac{e^2}{|\mathbf{r}_i - \mathbf{r}_j|} \times$$
$$\times (1 - P_{ij})\Psi_{\nu_1\Gamma_3}(\{x_i\})\Psi_{\nu_2\Gamma_4}(\{x'_j\}) X_{\nu_1}(\Gamma_1,\Gamma_3) X_{\nu_2}(\Gamma_2,\Gamma_4) dx_i dx'_j, \tag{D.20}$$

where P_{ij} is the permutation operator. Such an approach does not yield immediately the results (D.17), (D.18) and requires a summation of fractional parentage coefficients. However, this approach permits to take into account the dependence of radial wavefunctions on ME quantum numbers which occurs, e.g., in the Hartree–Fock approximation and is therefore more general.

In the case of strong (in comparison with crystal field) spin-orbital interaction the orbital and spin momenta are coupled into the total momentum J (the Russel–Saunders scheme, appropriate for rare-earth ions). In the Coulomb term, the spin-orbital coupling results in the replacement $L_i \to J_i$ according to (B.18). For the exchange term, one has to make some additional manipulations. Using (B.16) and performing summation of Clebsh–Gordan coefficients, we derive the result (D.10) with the replacement

$$\left[W_1^{(\kappa k_1)} W_2^{(\kappa k_2)}\right]^{(0\lambda)} \to [\kappa]^{-1/2} \sum_{p_1 p_2} \left[J_1^{(p_1)} \times J_2^{(p_2)}\right]^{(\lambda)} \begin{Bmatrix} p_1 & p_2 & \lambda \\ k_1 & k_2 & \kappa \end{Bmatrix} \times$$
$$\times \prod_{i=1,2} [p_i]([S_i][L_i][J_i])^{1/2} \begin{Bmatrix} J_i & S_i & L_i \\ J_i & S_i & L_i \\ p_i & \kappa & k_i \end{Bmatrix} \langle S_i L_i \| W_i^{(\kappa k_i)} \| S_i L_i \rangle. \tag{D.21}$$

Then the expansion of the exchange Hamiltonian reads

$$\mathsf{H}_{\text{exch}}(\nu_1\nu_2) = -\sum_{\alpha\beta\gamma} I_{\alpha\beta\gamma} \frac{(\mathbf{J}_1\rho)^\alpha (\mathbf{J}_2\rho)^\beta (\mathbf{J}_1\mathbf{J}_2)^\gamma}{\rho^{\alpha+\beta}}, \tag{D.22}$$

with $\alpha+\gamma \leq 2l_1+1$, $\beta+\gamma \leq 2l_2+1$, $\alpha+\beta$ being odd.

In the limit of strong spin-orbital coupling under consideration, antisymmetric Dzyaloshinsky-Moriya-type exchange interaction of the form $K_{12}[\mathbf{J}_1 \times \mathbf{J}_2]$ may be obtained with account of crystal-field effects only. Unlike the case of weak spin-orbital coupling considered by Moriya [669], where the antisymmetric exchange is determined by matrix elements of orbital momenta, the components of the pseudovector K_{ij} are given by matrix elements of electrostatic interaction in the local coordinate system [666].

Besides the 'potential' exchange (D.10), we have to consider the 'kinetic' exchange interaction. We consider the degenerate Hubbard model (C.21) with large Coulomb interaction. Then in the ground state the electron states at all sites correspond to the same SL-term, so that perturbation theory is applicable. The kinetic exchange occurs in the second order in the electron transfer

$$\begin{aligned}\tilde{\mathsf{H}} = 2\sum_{\nu_i m_i \sigma_i} \beta_{\nu_1\nu_2}(lm_1, lm_2)\beta_{\nu_2\nu_1}(lm_3, lm_4) \times \\ \times \sum_{\Gamma_n^{(i)}\Gamma_{n-1}\Gamma_{n+1}} (2E_{\Gamma_n} - E_{\Gamma_{n+1}} - E_{\Gamma_{n-1}})^{-1} \times \\ \times \langle \Gamma_{n+1} | a^\dagger_{\nu_1 lm_1\sigma_1} | \Gamma_{n-1} \rangle \langle \Gamma_{n-1} | a_{\nu_1 lm_4\sigma_2} | \Gamma'_n \rangle \langle \Gamma''_n | a_{\nu_2 lm_2\sigma_1} | \Gamma_{n+1} \rangle \times \\ \times \langle \Gamma_{n+1} | a^\dagger_{\nu_2 lm_3\sigma_2} | \Gamma'''_n \rangle X_{\nu_1}(\Gamma_n, \Gamma'_n) X_{\nu_2}(\Gamma''_n, \Gamma'''_n),\end{aligned} \tag{D.26}$$

where sum over $\Gamma^{(i)}$ stands for the sum over momentum projections of the term $\Gamma_n = \{SL\mu M\}$. In particular, for the s-band we obtain the standard result for the Anderson's kinetic exchange (4.8).

Generally speaking, in the problem of kinetic exchange for degenerate bands we cannot use the double irreducible tensor operators (B.21) since the denominators in (D.26) depend on ME quantum numbers S, L of virtual states. This may be, however, made if we take into account the dependence of E_Γ on the number of electrons only (C.20) to obtain

$$\tilde{\mathsf{H}} = \sum_{\nu_1\nu_2} \left\{ \frac{\overline{\beta}^2_{\nu_1\nu_2}(ll0)}{F^{(0)}(ll)} \right\} \{ n_1 n_2 + 4(\mathbf{S}_1\mathbf{S}_2) - [l](n_1+n_2) \}. \tag{D.27}$$

The interaction (D.27) is antiferromagnetic and dominates as a rule over the potential exchange interaction (D.10). It should be noted that the kinetic exchange interaction survives even in the limit $F^{(2)} = U \to \infty$ owing to non-orthogonality corrections (the second term in (C.25)).

In the general case we have to apply directly the ME operator approach. Substituting the expressions for matrix elements of Fermi operators (A.30), taking into account the m dependence of transfer integrals (C.31) and performing summation of the Clebsh–Gordan coefficients we obtain [660]

$$\tilde{\mathsf{H}} = (4\pi)^{1/2} n(n+1) \sum_{\substack{\nu_i k_i \lambda_i \lambda \\ \{S_{n\pm1}, L_{n\pm1}\}}} \left(G_{\Gamma_{n-1}}^{\Gamma_n} G_{\Gamma_n}^{\Gamma_{n+1}} \right)^2 \frac{\bar{\beta}_{\nu_1\nu_2}(ll\lambda_1)\bar{\beta}_{\nu_1\nu_2}(ll\lambda_2)}{E_{\Gamma_{n+1}} + E_{\Gamma_{n-1}} - 2E_{\Gamma_n}} \times$$

$$\times C(l\lambda_1 l, l\lambda_2 l, \lambda_1\lambda_2\lambda)[l][\lambda_1][\lambda_2][k_1][k_2][L_{n+1}][\lambda]^{-1/2}(-1)^{L_{n+1}-L_{n-1}} \times$$

$$\times \begin{Bmatrix} L & L & k_1 \\ l & l & L_{n-1} \end{Bmatrix} \begin{Bmatrix} L & L & k_2 \\ l & l & L_{n+1} \end{Bmatrix} \begin{Bmatrix} k_1 & k_2 & \lambda \\ l & l & \lambda_1 \\ l & l & l_2 \end{Bmatrix} \times$$

$$\times ([L_1^{(k_1)} \times L_2^{(k_2)}]^{(\lambda)} Y^{(\lambda)}(\hat{\boldsymbol{\rho}}_{1,2})) \times$$

$$\times (-1 + 4(-1)^{S_{n+1}-S_{n-1}}([S_{n+1}][S_{n-1}])^{-1}(\mathbf{S}_1\mathbf{S}_2)), \qquad \text{(D.28)}$$

with λ_1, λ_2, λ and k_1+k_2 being even, λ_1, $\lambda_2 \leq 2l$, k_1, k_1, $k \leq 2l$, $2L$. Passing to the usual vectors, we obtain the multipole expansion in the same form as (D.19). The Hamiltonian (D.28) contains only bilinear terms in spin operators. The biquadratic exchange may be obtained in the fourth order of perturbation theory [661, 668–670] which also corresponds to higher order corrections in the overlap parameter.

The sign of contributions of virtual configurations Γ_{n-1} and Γ_{n+1} to the effective exchange parameter ($k_1=k_2=0$) is determined by their spins. The coupling is antiferromagnetic if $S_{n+1}=S_{n-1}=S_n\pm\frac{1}{2}$ and ferromagnetic if $S_{n+1}-S_{n-1}=\pm1$.

Similar rules for the coupling between orbital momenta ($k_1=k_2=1$, $\lambda_1=\lambda_2=0$) are obtained after substituting explicit values of $6j$-coefficients in (D.28). The orbital exchange interaction is 'antiferromagnetic' if both the differences

$$\Delta_\pm = L(L+1) + l(l+1) - L_{n\pm1}(L_{n\pm1}+1) \qquad \text{(D.29)}$$

have the same sign and 'ferromagnetic' in the opposite case.

In real situations, the form of exchange Hamiltonians is strongly modified by crystal field (CF) which quenches, at leat partially, orbital momenta. Even in the case of intermediate CF, one has to consider, instead of the many-electron *SL*-terms, the corresponding irreducible representations of point groups. Besides that, the overlap between partially occupied $d(f)$-shells and, consequently, the direct exchange is as a rule small, so one has to take into account more complicated 'superexchange' mechanisms via non-magnetic atoms [661]. The kinetic exchange may be treated as a particular case of the superexchange interaction (the indirect interaction via valence band). The case of a narrow-band metal or semiconductor with non-integer band filling where exchange interaction is mediated by current carriers (the 'double exchange' situation) [668] may be described within the Hubbard and s–d exchange models with strong correlations. The corresponding Hamiltonians (D.28), (I.10) describe the interaction of electrons with spin and orbital degrees of freedom. In this case the exchange interaction is not reduced to a Heisenberg form.

Appendix E

SPIN WAVES IN THE HEISENBERG MAGNETS AND THE GREEN'S FUNCTION METHOD

The exponential behaviour of magnetization at low temperatures in the mean-field approximation (4.20) contradicts experimental data on both insulator and metallic ferromagnets. At $T \ll T_C$ the excitation spectrum and thermodynamics of the Heisenberg model may be investigated in detail. This is achieved by passing from spin operators S^z, $S^{\pm} = S^x \pm iS^y$ to Bose operators which correspond to weakly interacting excitations in the ferromagnet – spin waves (magnons). The simplest way to do this is using the Holstein–Primakoff representation

$$S^z = S - b^\dagger b,$$

$$S^+ = (2S)^{1/2}\left(1 - \frac{1}{2S}b^\dagger b\right)^{1/2} b, \qquad S^- = (2S)^{1/2} b^\dagger \left(1 - \frac{1}{2S}b^\dagger b\right)^{1/2}. \tag{E.1}$$

In fact, this representation is well defined only on the basis of physical states where spin deviations (magnon occupation numbers) on a site do not exceed $2S$. Unlike the representation (E.1), the Dyson–Maleev representation of ideal bosons

$$S^z = S - b^\dagger b, \qquad S^+ = (2S)^{1/2}\left(1 - \frac{1}{2S}b^\dagger b\right) b, \qquad S^- = (2S)^{1/2} b^\dagger \tag{E.2}$$

does not contain irrational functions of Bose operators, but is not-Hermitian. However, this lead to errors which are exponentially small at low temperatures.

Performing expansion of (E.1) in the quasiclassical parameter $1/2S$ we reduce the Heisenberg Hamiltonian (4.9) to the form

$$\mathsf{H} = \sum_{\mathbf{q}} \omega_{\mathbf{q}} b^\dagger_{\mathbf{q}} b_{\mathbf{q}} + \frac{1}{2}\sum_{\mathbf{pqr}} (J_{\mathbf{q}} + J_{\mathbf{p}} - 2J_{\mathbf{q-p}}) b^\dagger_{\mathbf{q}} b^\dagger_{\mathbf{r}} b_{\mathbf{p}} b_{\mathbf{q+r-p}} + \cdots,$$

$$\omega_{\mathbf{q}} = 2S(J_0 - J_{\mathbf{q}}). \tag{E.3}$$

The first term of the Hamiltonian (E.3) describes the system of magnons with the frequencies $\omega_{\mathbf{q}}$, and the second term the dynamical interaction (magnon–magnon scattering). At small q for a cubic lattice we have

$$\omega_{\mathbf{q}} = Dq^2, \qquad D = \frac{1}{3}J_0 S \approx T_C/(S+1). \tag{E.4}$$

The quantity D is called the spin-wave stiffness constant. The temperature dependent correction to the spin-wave spectrum owing to magnon–magnon interaction is obtained by decoupling the second term of (E.3) with the use of the Wick theorem

$$\delta\omega_{\mathbf{q}} = 2\sum_{\mathbf{p}}(J_{\mathbf{q}} + J_{\mathbf{p}} - J_{\mathbf{p-q}} - J_0)N_{\mathbf{p}}, \qquad N_{\mathbf{q}} = \langle b_{\mathbf{q}}^{\dagger} b_{\mathbf{q}} \rangle \tag{E.5}$$

and is proportional to $T^{5/2}$. The temperature dependence of magnetization is given by the Bloch law

$$\delta\langle S^z \rangle = -\sum_{\mathbf{p}} N_{\mathbf{p}} = -\frac{v_0}{2\pi^2}\int_0^{\infty}\frac{q^2}{\exp(Dq^2/T)-1}dq =$$

$$= -\zeta\left(\frac{3}{2}\right)\frac{v_0}{8\pi^{3/2}}\left(\frac{T}{D}\right)^{3/2}, \tag{E.6}$$

where v_0 is the lattice cell volume, $\zeta(x)$ is the Riemann zeta-function.

Now we consider the case of an antiferromagnet. We treat a general case of the spiral structure, characterized by the wavevector $\mathbf{Q}$. The corresponding classical spin configuration is written as

$$\langle S_i^x \rangle = S\cos\mathbf{Q}R_i, \qquad \langle S_i^y \rangle = S\sin\mathbf{Q}R_i, \qquad \langle S_i^z \rangle = 0. \tag{E.7}$$

In particular, for the usual 'chess' ordering in the simple cubic lattice one has $\mathbf{Q}=(\pi,\pi,\pi)$, and in the case of ferromagnetic planes with alternating magnetization direction $\mathbf{Q}=(0,0,\pi)$. It is convenient to introduce the local coordinate system where spins at each site are directed along the new local z-axis:

$$S_i^x \to S_i^z\cos\mathbf{QR}_i - S_i^y\sin\mathbf{QR}_i,$$

$$S_i^y \to S_i^z\sin\mathbf{QR}_i + S_i^x\cos\mathbf{QR}_i, \qquad S_i^z \to -S_i^x. \tag{E.8}$$

Passing from the spin operators in the local coordinate system to the operators of spin deviations with the use of (E.1) we represent the Heisenberg Hamiltonian as

$$\mathsf{H} = \sum_{\mathbf{q}}\left[C_{\mathbf{q}} b_{\mathbf{q}}^{\dagger} b_{\mathbf{q}} + \frac{1}{2}D_{\mathbf{q}}(b_{\mathbf{q}} b_{-\mathbf{q}} + b_{\mathbf{q}}^{\dagger} b_{-\mathbf{q}}^{\dagger})\right] + \ldots,$$

$$C_{\mathbf{q}} = S\left[2J_{\mathbf{Q}} - \frac{1}{2}(J_{\mathbf{Q+q}} + J_{\mathbf{Q-q}}) - J_{\mathbf{q}}\right],$$

$$D_{\mathbf{q}} = S\left[\frac{1}{2}(J_{\mathbf{Q+q}} + J_{\mathbf{Q-q}}) - J_{\mathbf{q}}\right]. \tag{E.9}$$

The Hamiltonian (E.9) can be diagonalized by the Bogoliubov–Tyablikov transformation

$$b_{\mathbf{q}} = u_{\mathbf{q}}\beta_{\mathbf{q}} - v_{\mathbf{q}}\beta^{\dagger}_{-\mathbf{q}}, \tag{E.10}$$

$$u_{\mathbf{q}}^2 = \frac{1}{2}\left(\frac{C_{\mathbf{q}}}{\omega_{\mathbf{q}}} + 1\right), \qquad v_{\mathbf{q}}^2 = \frac{1}{2}\left(\frac{C_{\mathbf{q}}}{\omega_{\mathbf{q}}} - 1\right),$$

to obtain

$$\mathsf{H} = \text{const} + \sum_{\mathbf{q}} \omega_{\mathbf{q}} \beta^{\dagger}_{\mathbf{q}} \beta_{\mathbf{q}} + \cdots, \tag{E.11}$$

where the spin-wave spectrum reads

$$\omega_{\mathbf{q}} = (C_{\mathbf{q}}^2 - D_{\mathbf{q}}^2)^{1/2} = 2S\left\{\left[J_{\mathbf{Q}} - \frac{1}{2}(J_{\mathbf{Q+q}} + J_{\mathbf{Q-q}})\right][J_{\mathbf{Q}} - J_{\mathbf{q}}]\right\}^{1/2}. \tag{E.12}$$

The spin-wave frequency tends to zero at both $q \to 0$ and $\mathbf{q} \to \mathbf{Q}$, the q-dependence being linear, unlike the case of a ferromagnet (we do not discuss here magnetic anisotropy which results in a disturbance of spin-wave spectrum at small q, $|\mathbf{q} - \mathbf{Q}|$ [15, 16]). Using the Wick theorem for the quartic terms, which are neglected in the Hamiltonian (E.9), we may obtain corrections to $C_{\mathbf{q}}$ and $D_{\mathbf{q}}$ owing to magnon–magnon interaction

$$\delta C_{\mathbf{q}} = -\frac{1}{2}\sum_{\mathbf{p}}(4J_{\mathbf{Q}} + 2J_{\mathbf{Q+q-p}} + 2J_{\mathbf{Q-q-p}} - 2J_{\mathbf{p}} - 2J_{\mathbf{Q}} - $$
$$- J_{\mathbf{Q+q}} - J_{\mathbf{Q-q}} - 2J_{\mathbf{Q+p}})\langle b^{\dagger}_{\mathbf{p}} b_{\mathbf{p}}\rangle - \sum_{\mathbf{p}}(J_{\mathbf{Q+p}} - J_{\mathbf{p}})\langle b^{\dagger}_{\mathbf{p}} b_{-\mathbf{p}}\rangle,$$

$$\delta D_{\mathbf{q}} = -\frac{1}{2}\sum_{\mathbf{p}}(J_{\mathbf{Q+q}} + 2J_{\mathbf{Q+p}} + J_{\mathbf{Q-q}} - 2J_{\mathbf{q}} - 2J_{\mathbf{p}})\langle b^{\dagger}_{\mathbf{p}} b_{\mathbf{p}}\rangle +$$
$$+ \sum_{\mathbf{p}}(J_{\mathbf{Q+p}} + J_{\mathbf{p}})\langle b^{\dagger}_{\mathbf{p}} b^{\dagger}_{-\mathbf{p}}\rangle. \tag{E.13}$$

The low-temperature behaviour of the sublattice magnetization is given by

$$\overline{S} = \langle S^z \rangle = S - \langle b^{\dagger} b\rangle = S - \sum_{\mathbf{q}}[u_{\mathbf{q}}^2 N_{\mathbf{q}} + v_{\mathbf{q}}^2(1 + N_{\mathbf{q}})], \tag{E.14}$$

so that

$$\delta\bar{S}(T) \sim -(T/T_{\mathrm{N}})^2.$$

The spin-wave approach enables one to obtain an interpolation description of the Heisenberg ferromagnet in the whole temperature region. This was made by Tyablikov [357] within the Green's function method. Define the anticommutator (commutator) double-time retarded Green's function for operators A and B by [671]

$$\langle\langle A | B \rangle\rangle_E^{\pm} = \int_{-\infty}^{0} \langle [e^{i\mathrm{H}t} A e^{-i\mathrm{H}t}, B]_{\pm} \rangle e^{iEt}\, dt, \quad \mathrm{Im}\, E > 0. \tag{E.15}$$

The Green's function (E.15) satisfies the equation of motion

$$E\langle\langle A | B \rangle\rangle_E^{\pm} = \langle [A,B]_{\pm} \rangle + \langle\langle [A,\mathrm{H}\,] | B \rangle\rangle_E^{\pm} \tag{E.16a}$$

or

$$E\langle\langle A | B \rangle\rangle_E^{\pm} = \langle [A,B]_{\pm} \rangle + \langle\langle A | [\mathrm{H}\,, B \rangle\rangle_E^{\pm} \tag{E.16b}$$

and is thereby expressed in terms of more complicated Green's functions. As one can see from (E.16), it is convenient to use the commutator Green's function in the case of Bose-type operators A, B (where the commutator $[A, B]$, which stands in the average, is a 'simpler' operator), and the anticommutator functions in the case of Fermi-type operators. In the cases of free bosons and fermions, where the Hamiltonian has a diagonal form, the equations (E.16) are closed and we find

$$\langle\langle b_{\mathbf{q}} | b_{\mathbf{q}}^{\dagger} \rangle\rangle_{\omega}^{-} = \frac{1}{\omega - \omega_{\mathbf{q}}}, \qquad \langle\langle c_{\mathbf{k}} | c_{\mathbf{k}}^{\dagger} \rangle\rangle_{E}^{+} = \frac{1}{E - E_{\mathbf{k}}}. \tag{E.17}$$

Calculation of the Green's function permits to restore the corresponding thermodynamical averages by using the spectral representation

$$\langle BA \rangle = -\frac{1}{\pi} \int_{\infty}^{\infty} \mathrm{Im}\, \langle\langle A | B \rangle\rangle_{E+i\delta}^{\pm} \frac{1}{e^{E/T} \pm 1} dE. \tag{E.18}$$

In particular, we obtain from (E.17)

$$\langle b_{\mathbf{q}}^{\dagger} b_{\mathbf{q}} \rangle = N_{\mathrm{B}}(\omega_{\mathbf{q}}), \qquad \langle c_{\mathbf{k}}^{\dagger} c_{\mathbf{k}} \rangle = f(E_{\mathbf{k}}). \tag{E.19}$$

In the general case, the infinite sequence of equations (E.16) may be 'decoupled' by reducing higher-order Green's functions to simpler ones. In the case of interacting quasiparticles the Green's function are expressed by the Dyson equations

$$\begin{aligned} \langle\langle b_{\mathbf{q}} | b_{\mathbf{q}}^{\dagger} \rangle\rangle_{\omega} &= [\omega - \omega_{\mathbf{q}} - \Pi_{\mathbf{q}}(\omega)]^{-1}, \\ \langle\langle c_{\mathbf{k}} | c_{\mathbf{k}}^{\dagger} \rangle\rangle_{E} &= [E - E_{\mathbf{k}} - \Sigma_{\mathbf{k}}(E)]^{-1}. \end{aligned} \tag{E.20}$$

The real part of the self-energy Σ (or of the polarization operator Π) yields the energy shift, and the imaginary part determines the quasiparticle damping. Provided that the interaction Hamiltonian contains a small parameter, the method of equations of motion permits to construct the perturbation expansion in a convenient form. In particular, applying both equations (E.16) we obtain the expression for the self-energy

$$\Sigma_{\mathbf{k}}(E) = \Lambda_{\mathbf{k}} + \langle\langle [c_{\mathbf{k}}, \mathsf{H}_{\mathrm{int}}] - \Lambda_{\mathbf{k}} c_{\mathbf{k}} \,|\, [\mathsf{H}_{\mathrm{int}}, c^{\dagger}_{\mathbf{k}}] - \Lambda_{\mathbf{k}} c^{\dagger}_{\mathbf{k}} \rangle\rangle^{\mathrm{irr}}_{E},$$

$$\Lambda_{\mathbf{k}} = \langle \{ [c_{\mathbf{k}}, \mathsf{H}_{\mathrm{int}}], c^{\dagger}_{\mathbf{k}} \} \rangle, \tag{E.21}$$

where the symbol 'irr' means that the divergent contributions, containing the denominators $(E-\varepsilon_{\mathbf{k}})^n$, should be omitted at further calculations of the irreducible Green's function (E.21). In this book we apply the method of double-time retarded Green's functions to various many-electron models describing highly-correlated *d*- and *f*-systems. In transition metal theory, this technique turns out to be very useful since we meet with operators which do not possess simple commutation relations, so that standard diagram expansions [27] are inapplicable. Here belong many-electron *X*-operators of the Fermi and Bose type (Appendix A) and spin operators (Appendix B). Below we describe briefly the derivation of the Tyablikov equation for magnetization in the Heisenberg model for spin $S = 1/2$ (the general case is discussed in Appendix F within framework of the Hubbard *X*-operator approach). Write down the equation of motion for the commutator transverse spin Green's function

$$(\omega - H)\langle\langle S^+_{\mathbf{q}} \,|\, S^-_{-\mathbf{q}} \rangle\rangle_{\omega} = 2\langle S^z \rangle + 2\sum_{\mathbf{p}} (J_{\mathbf{p}-\mathbf{q}} - J_{\mathbf{p}}) \langle\langle S^z_{\mathbf{p}} S^+_{\mathbf{q}-\mathbf{p}} \,|\, S^-_{-\mathbf{q}} \rangle\rangle_{\omega}.$$

Performing the simplest decoupling at different lattice sites we derive

$$\langle\langle S^+_{\mathbf{q}} \,|\, S^-_{-\mathbf{q}} \rangle\rangle_{\omega} = \frac{2\langle S^z \rangle}{\omega - \omega_{\mathbf{q}}}, \qquad \omega_{\mathbf{q}} = 2\langle S^z \rangle (J_0 - J_{\mathbf{q}}) + H. \tag{E.22}$$

Presence of the factor $2\langle S^z \rangle$ in comparison with (E.17) is connected with non-Bose commutation relation for spin operators. The result (E.22) is valid for arbitrary S. For $S = 1/2$, using (E.18) and the identity

$$S^z_i = \frac{1}{2} - S^-_i S^+_i, \tag{E.23}$$

we obtain the self-consistent equation for $\langle S^z \rangle$

$$\langle S^z \rangle = \frac{1}{2}\left[1 + 2\sum_{\mathbf{q}} N_{\mathrm{B}}(\omega_{\mathbf{q}})\right]^{-1},$$

which may be transformed to the form

$$\langle S^z \rangle = \frac{1}{2}\left[\sum_{\mathbf{q}} \coth \frac{2\langle S^z \rangle (J_0 - J_{\mathbf{q}}) + H}{T}\right]^{-1}. \tag{E.24}$$

The equation (E.24) differs from the mean-field equation (4.14) by the presence of dispersion (q-dependence) of excitation spectrum. It describes satisfactorily thermodynamics of the Heisenberg model at both high and low temperatures, although higher-order terms in the low-T expansion do not quite agree with the rigorous Dyson's results (in particular, the spurious T^3-terms in magnetization occurs). Numerous attempts to improve the spin-wave region description by using more complicated decoupling procedures (see [357]) resulted in fact in deterioration of the interpolation. The values of ferromagnetic and paramagnetic Curie temperature in the Tyablikov approximation for arbitrary S value are given by

$$T_C = \frac{S(S+1)}{3}\left(\sum_{\mathbf{q}} \frac{1}{J_0 - J_{\mathbf{q}}}\right)^{-1}; \tag{E.25}$$

$$\theta = \frac{S(S+1)}{3} J_0 \tag{E.26}$$

and are somewhat different (the mean-field result (4.18) is obtained from (E.25) by averaging the denominator over $\mathbf{q}$, i.e. in the lowest order in $1/z$). The corresponding expressions for an antiferromagnet are obtained from (E.25), (E.26) by the replacement $J_0 \to J_{\mathbf{Q}}$, $\theta \to -\theta$. To end this Appendix, we discuss one more method in the theory of Heisenberg magnets [672]. To take into account kinematical interactions, this approach introduces auxiliary pseudofermion operators c into the Dyson–Maleev representation (E.2):

$$S^z = S - b^\dagger b - (2S+1)c^\dagger c, \qquad S^- = (2S)^{1/2} b^\dagger,$$

$$S^+ = (2S)^{1/2}\left(1 - \frac{1}{2S} b^\dagger b\right) b - 2(2S+1)(2S)^{1/2} b c^\dagger c. \tag{E.27}$$

The distribution function of the dispersionless 'fermions' turns out to be

$$\langle c_i^\dagger c_i \rangle = -N_B((2S+1)\tilde{H}), \qquad \tilde{H} = H + 2J_0 \langle S^z \rangle. \tag{E.28}$$

We see that the pseudofermion operators exclude non-physical states, so that an extrapolation to high temperatures becomes possible. In particular, neglecting dispersion of spin-wave excitations we obtain from (E.27), (E.28)

$$\langle S^z \rangle = S - N_B(\tilde{H}) + (2S+1)N_B((2S+1)\tilde{H}) = S - B_S(S\tilde{H}/T). \qquad (E.29)$$

Thus we have rederived the mean-field equation for the magnetization (4.15).

Appendix F

HUBBARD OPERATOR APPROACH IN THE HEISENBERG MODEL

Using the Hubbard operators yields a possibility to obtain in a simple way main results of the theory of Heisenberg magnets. In particular, this formalism enables one to take into account the strong single-ion magnetic anisotropy in the zeroth order approximation [673–677]. The Hamiltonian of the Heisenberg model with an arbitrary single-site anisotropy has the form

$$\mathcal{H} = -\sum_{ij} J_{ij}\mathbf{S}_i\mathbf{S}_j + \mathcal{H}_a = -\sum_{\mathbf{q}} J_{\mathbf{q}}\mathbf{S}_{\mathbf{q}}\mathbf{S}_{-\mathbf{q}} + \mathcal{H}_a; \tag{F.1}$$

$$\mathcal{H}_a = \sum_i \varphi(S_i^x, S_i^y, S_i^z, H), \tag{F.2}$$

where $J_{\mathbf{q}}$ are the Fourier transforms of exchange parameters, H is the external magnetic field, φ is an arbitrary function. Note that, as follows from the results of Appendix C, the Hamiltonan (F.1) with $\mathbf{S} \to \mathbf{L}$ describes the system of interacting orbital momenta in the intermediate crystal field, $\mathcal{H}_a$ having the sense of the crystalline electric field Hamiltonian. Thus the anisotropic Heisenberg model may be applied to the problem of quenching of orbital momenta. First, we consider the easy-axis ferromagnet where

$$\mathcal{H}_a = -\sum_i \varphi(S_i^z) - H\sum_i S_i^z. \tag{F.3}$$

We use the representation of momentum operators in terms of Hubbard operators (Appendix B)

$$S_i^{\pm} = \sum_{M=-S}^{S} \gamma_S(\pm M) X_i(M \pm 1, M), \qquad S_i^z = \sum_{M=-S}^{S} M X(M,M),$$

$$\gamma_S(M) \equiv [(S-M)(S+M+1)]^{1/2}. \tag{F.4}$$

Then the anisotropy Hamiltonian takes the diagonal form

$$\mathrm{H}_a = -\sum_{iM}[\varphi(M) + HM]X_i(M,M). \tag{F.5}$$

It is convenient to introduce the commutator Green's functions

$$G_{\mathbf{q}}(\omega) = \langle\langle S_{\mathbf{q}}^+ | S_{-\mathbf{q}}^- \rangle\rangle_\omega, \qquad G_{\mathbf{q}M}(\omega) = \langle\langle X_{\mathbf{q}}(M+1,M) | S_{-\mathbf{q}}^- \rangle\rangle_\omega. \tag{F.6}$$

Write down the equation of motion

$$\begin{aligned}(\omega - H - \varphi(M+1) + \varphi(M) - 2J_0\langle S^z\rangle)G_{\mathbf{q}M}(\omega) = \\ = \gamma_S(M)(N_{M+1} - N_M)[1 - J_{\mathbf{q}}G_{\mathbf{q}}(\omega)],\end{aligned} \tag{F.7}$$

where we have carried out a simplest decoupling, which corresponds to Tyablikov's decoupling at different sites (E.22),

$$N_M = \langle X(M,M)\rangle.$$

After summing over M one obtains

$$G_{\mathbf{q}}(\omega) = \frac{\Phi_S(\omega)}{1 + J_{\mathbf{q}}\Phi_S(\omega)}, \tag{F.8}$$

where

$$\Phi_S(\omega) = \sum_M \frac{\gamma_S^2(M)(N_{M+1} - N_M)}{\omega - H - \varphi(M+1) + \varphi(M) - 2J_0\langle S^z\rangle}. \tag{F.9}$$

The structure of the expression (F.8) is reminiscent of the Green's functions in the itinerant electron magnetism theory (see Appendices G,H). The excitation spectrum is determined by the equation

$$1 + J_{\mathbf{q}}\Phi_S(\omega) = 0 \tag{F.10}$$

and contains $2S$ branches. The expressions (F.7), (F.8) enable one to calculate the occupation numbers N_M and obtain a self-consistent equation for the magnetization

$$\langle S^z\rangle = \sum_M MN_M. \tag{F.11}$$

For simplicity we restrict ourselves to analyzing this equation in the isotropic case where reduces to (E.22). Using the multiplication rule for the Hubbard operators and the spectral representation (E.18) we obtain

$$\langle S_i^- X_i(M+1,M)\rangle = \gamma_S(M)N_M = \gamma_S(M)(N_{M+1} - N_M)\sum_{\mathbf{q}} N_{\mathrm{B}}(\omega_{\mathbf{q}}). \tag{F.12}$$

Then the system for N_M reads

$$N_M = (N_{M+1} - N_M)P_S, \qquad \sum_M N_M = 1, \qquad P_S \equiv \sum_{\mathbf{q}} N_{\mathrm{B}}(\omega_{\mathbf{q}}). \tag{F.13}$$

Solving this system we derive the equation for $\langle S^z\rangle$ [675]

$$\langle S^z \rangle = SB_S(-S\ln(P_S/[1+P_S])). \tag{F.14}$$

The equation (F.14) has the form, which is somewhat different from the standard one [357], and is more convenient. Now we discuss the general case [675–677]. The Hamiltonian (F.2) may be diagonalized to obtain

$$\mathcal{H}_a | \psi_\mu \rangle = E_\mu | \psi_\mu \rangle. \tag{F.15}$$

The eigenfunctions ψ_μ are be expanded in the eigenfunctions $|M\rangle$ of the operator S^z:

$$| \psi_\mu \rangle = \sum_M c_{\mu M} | M \rangle. \tag{F.16}$$

Then the spin operators are presented in terms of the Hubbard operators as

$$S_i^+ = \sum_{\mu\mu' M} c_{\mu',M} c^*_{\mu,M+1} \gamma_S(M) X_i(\mu,\mu'); \tag{F.17}$$

$$S_i^z = \sum_{\mu\mu' M} c_{\mu',M} c^*_{\mu,M} M X_i(\mu,\mu'). \tag{F.18}$$

For $H \parallel \langle S^z \rangle$, in the simplest mean-field approximation we may take into account the intersite exchange interaction by replacing H to the effective field

$$\tilde{H} = H + 2J_0 \langle S^z \rangle. \tag{F.19}$$

Then the occupation numbers N_M are given by

$$N_M = \sum_\mu | c_{\mu M} |^2 N_\mu = \frac{\sum_\mu | c_{\mu M} |^2 \exp(-E_\mu/T)}{\sum_\mu \exp(-E_\mu/T)}. \tag{F.20}$$

Consider an easy-plane ferromagnet where

$$\mathrm{H}_a = 2D \sum_i (S_i^x)^2 - H \sum_i S_i^z, \quad D > 0. \tag{F.21}$$

The results turn out to be essentially different for integer and half-integer spins. (Such a difference is typical for quantum spin systems, see, e.g., [678].) This is due to the fact that the ground state of an ion with an integer S is singlet, and excited magnetic states are separated by an energy gap. In particular, for $S = 1$ we have [673]

$$| \psi_{1,3} \rangle = \cos\theta | \pm 1 \rangle \pm \sin\theta | \mp 1 \rangle, \tag{F.22}$$

$$E_{1,3} = D \mp (H^2 + D^2)^{1/2}, \qquad E_2 = 2D,$$

where

$$\cos 2\theta = \frac{H}{(H^2 + D^2)^{1/2}}, \qquad \sin 2\theta = -\frac{D}{(H^2 + D^2)^{1/2}}. \tag{F.23}$$

According to (F.10), (F.20) the equation for the magnetization has the form

$$\langle S^z \rangle = \frac{\cos 2\tilde{\theta} \sinh[(\tilde{H}^2 + D^2)^{1/2}/T]}{\cosh[(\tilde{H}^2 + D^2)^{1/2}/T] + \frac{1}{2}\exp(-D/T)}. \tag{F.24}$$

At zero temperature and magnetic field we obtain

$$\langle S^z \rangle = \begin{cases} [1-(D/2J_0)^2]^{1/2}, & D < 2J_0, \\ 0, & D > 2J_0. \end{cases} \tag{F.25}$$

Thus at $D > 2J_0$ the ground state $|\psi_1\rangle$ is a singlet superposition of the states $|-1\rangle$ ($\sin\theta = \cos\theta$), and ferromagnetic ordering is suppressed. Easy-plane highly-anisotropic systems exhibit also a non-trivial behaviour of magnetization at changing magnetic field [679, 680], which is reminiscent of the Hall conductivity behaviour in the situation of the quantum Hall effect [681]. For half-integer S, the ground state is ordered at arbitrary D. However, the value of magnetization is diminished due to mixing of states with different M by the anisotropy field. So, for $S = 3/2$ we have

$$\begin{aligned} |\psi_{1,4}\rangle &= \cos\theta_\pm |\pm 3/2\rangle \mp \sin\theta_\pm |\mp 1/2\rangle, \\ |\psi_{2,3}\rangle &= \cos\theta_\mp |\pm 1/2\rangle \mp \sin\theta_\mp |\mp 3/2\rangle, \end{aligned} \tag{F.26}$$

$$E_{1,4} = \frac{1}{2}(5D \mp H) \mp E_\pm, \qquad E_{2,3} = \frac{1}{2}(5D \pm H) \mp E_\mp,$$

where

$$\cos\theta_\pm = (H \pm D + E_\pm)/\left[3D^2 + (H \pm D + E_\pm)^2\right]^{1/2},$$

$$E_\pm = \left[(H \pm D)^2 + 3D^2\right]^{1/2}. \tag{F.27}$$

One can see that the lowest level $|\psi_1\rangle$ is a mixture of the states with $M = 3/2$ and $M = -1/2$. The magnetization at $T = 0$ is given by

$$\langle S^z \rangle = \frac{3}{2} - \frac{6D^2}{3D^2 + (\tilde{H} + D + \tilde{E}_+)^2}. \tag{F.28}$$

At $D \gg J$ one obtains from (F.28) $\langle S^z \rangle = 1$, and the ground state is a doublet.

For cubic anisotropic ferromagnets we have

$$\mathsf{H}_a = -\frac{1}{2}K\sum_i [(S_i^x)^4 + (S_i^y)^4 + (S_i^z)^4] - H\sum_i S_i^z. \tag{F.29}$$

At $S \le 3/2$ the single-ion cubic anisotropy does not occur by kinematical reasons. For $S = 2$ one obtains

$$|\psi_{1,2}\rangle = \cos\theta|\pm 2\rangle \pm \sin\theta|\mp 2\rangle, \qquad |\psi_{3,4}\rangle = |\pm 1\rangle, \qquad |\psi_5\rangle = |0\rangle,$$

$$E_{1,2} = C \mp E, \qquad E_{3,4} = C + K \mp H, \qquad E_5 = C - K, \tag{F.30}$$

$$C = -\frac{21}{2}K, \qquad E = \left(H^2 + K^2\right)^{1/2},$$

where

$$\cos\theta = \frac{2H + E}{\left[K^2 + (2H+E)^2\right]^{1/2}}, \qquad \sin\theta = \frac{D}{\left[K^2 + (2H+E)^2\right]^{1/2}}. \tag{F.31}$$

We treat the case where $K > 0$, so that the easy magnetization direction corresponds to the z axis. Then the ground state is $|\psi_1\rangle$. The equation for the magnetization has the form

$$\langle S^z \rangle = \frac{2\cos 2\tilde{\theta}\sinh(\tilde{\varepsilon}/T) + \exp(-K/T)\sin(\tilde{H}/T)}{\cosh(\tilde{\varepsilon}/T) + \exp(K/T)/2 + \exp(-K/T)\cos(\tilde{H}/T)}. \tag{F.32}$$

The ground state magnetization reads

$$\langle S^z \rangle = 2\left[1 - (K/8J_0)^2\right]^{1/2} \tag{F.33}$$

and vanishes at $K > 8J_0$. The ferromagnetic ordering at large K does not also occur for $S = 4$. Thus the orbital momenta $L = 2$ and $L = 4$ are quenched by the cubic crystal field. The intra-atomic orbital exchange interaction, which is required for unquenching, is determined by (F.19), (F.32). For $S = 5/2$, 3 and 7/2 the ground state turns out to be always ordered, the values of $\langle S^z \rangle$ at $T = 0$ in the strong anisotropy limit being 11/6, 3/2 and 7/6 respectively [676].

Appendix G

ELECTRON–MAGNON INTERACTION IN MAGNETIC METALS

In this Appendix we calculate the spectrum of single-particle and spin-wave excitations in metallic magnets. To this end we use many-electron models which permit to describe effects of interelectron correlations. The simplest model of such a type is the Hubbard model. In the case of a non-degenerate band its Hamiltonian reads

$$\mathsf{H} = \sum_{\mathbf{k}\sigma} t_{\mathbf{k}} c_{\mathbf{k}\sigma}^{\dagger} c_{\mathbf{k}\sigma} + \mathsf{H}_{\mathrm{int}}, \tag{G.1}$$

$$\mathsf{H}_{\mathrm{int}} = U \sum_{i} c_{i\uparrow}^{\dagger} c_{i\uparrow} c_{i\downarrow}^{\dagger} c_{i\downarrow},$$

with U being the on-site repulsion parameter. The Hubbard model was widely used to consider itinerant electron ferromagnetism since this takes into account the largest term of the Coulomb interaction – the intra-atomic one. Despite apparent simplicity, this model contains a very reach physics, and its rigorous investigation is a very difficult problem.

Besides the Hubbard model, it is sometimes convenient to use for theoretical description of magnetic metals the s–$d(f)$ exchange model. The s–d exchange model was first proposed for transition d-metals to consider peculiarities of their electrical resistivity [265]. This model postulates existence of two electron subsystems: itinerant 's-electrons' which play the role of current carriers, and localized 'd-electrons' which give the main contribution to magnetic moment. Such an assumption may be hardly justified quantitatively for d-metals, but it may be useful at qualitative consideration of some physical properties (especially of transport phenomena). At the same time, the s–f model provides a good description of magnetism in rare-earth metals and their compounds with well-localized $4f$-states.

The Hamiltonian of the s–$d(f)$ model in the simplest version has the form

$$\mathsf{H} = \sum_{\mathbf{k}\sigma} t_{\mathbf{k}} c^{\dagger}_{\mathbf{k}\sigma} c_{\mathbf{k}\sigma} - \sum_{\mathbf{q}} J_{\mathbf{q}} \mathbf{S}_{-\mathbf{q}} \mathbf{S}_{\mathbf{q}} + \mathsf{H}_{\text{int}} \equiv \mathsf{H}_{s} + \mathsf{H}_{d(f)} + \mathsf{H}_{\text{int}}, \tag{G.2}$$

$$\mathsf{H}_{\text{int}} = -I \sum_{i\sigma\sigma'} (\mathbf{S}_i \sigma_{\sigma\sigma'}) c^{\dagger}_{i\sigma} c_{i\sigma'},$$

where $\mathbf{S}_{\mathbf{q}}$ are operators for localized spins, σ are the Pauli matrices, I is the parameter of the s–$d(f)$ exchange interaction which is assumed to be contact (derivation of the s–$d(f)$ model in a more general situation is considered in Appendix K), $J_{\mathbf{q}}$ are the Fourier transforms of the exchange parameters between localized spins. In rare earth metals the latter interaction is usually the indirect RKKY exchange via conduction electrons which is due to the same s–f interaction. However, at constructing perturbation theory, it is convenient to include this interaction in the zero-order Hamiltonian.

Altough more complicated in its form, the s–d model turns out to be in some respect simpler than the Hubbard model since it permits to construct the quasiclassical expansion in the small parameter $1/2S$. Within simple approximations, the results in the s–$d(f)$ and Hubbard models differ as a rule by the replacement $I \to U$ only.

Below we perform a systematic investigation of spin-wave and electron spectra of conducting ferro- and antiferromagnets within the above models. We demonstrate similarities and differences in comparison with localized-moment insulator magnets which are described by the Heisenberg model.

G.1 FERROMAGNETS

Here we consider spin-wave theory of the Hubbard ferromagnet using the Stoner spin-split state as the zero-order approximation following mainly to paper [338]. (The limit of strong correlations is discussed in Sect.4.6 and Appendices H,J.) The simplest Hartree-Fock (Stoner) approximation in the Hubbard model, which corresponds formally to first-order perturbation theory in U, yields the electron spectrum of the form

$$E_{\mathbf{k}\sigma} = t_{\mathbf{k}} + Un_{-\sigma} = t_{\mathbf{k}} + U\left(\frac{n}{2} - \sigma\langle S^z\rangle\right) \equiv t_{\mathbf{k}\sigma}, \tag{G.3}$$

so that we have for the spin splitting

$$\Delta = U(n_{\uparrow} - n_{\downarrow}) = 2U\langle S^z\rangle, \tag{G.4}$$

and U plays the role of the Stoner parameter. The Hartree-Fock decoupling does not take into account correctly the formation of doubles, i.e. doubly occupied states at a site. (This can be made within the

Hubbard many-electron approach [28–31], see Appendix H). Note that this error does not play a role for the saturated ferromagnetic state.

Unlike the Stoner theory, the Hubbard model enables one to describe spin-wave excitations in an itinerant ferromagnet. To this end we present the interaction Hamiltonian in the form

$$\mathsf{H}_{\text{int}} = \frac{U}{2}\sum_{\mathbf{k}\sigma} c^{\dagger}_{\mathbf{k}\sigma} c_{\mathbf{k}\sigma} - \frac{U}{2}\sum_{\mathbf{q}} (S^{-}_{-\mathbf{q}} S^{+}_{\mathbf{q}} + S^{+}_{\mathbf{q}} S^{-}_{-\mathbf{q}}), \tag{G.5}$$

where we have introduced the Fourier components of spin density operators

$$S^{+}_{\mathbf{q}} = \sum_{\mathbf{k}} c^{\dagger}_{\mathbf{k}\uparrow} c_{\mathbf{k}+\mathbf{q}\downarrow}, \qquad S^{-}_{\mathbf{q}} = \sum_{\mathbf{k}} c^{\dagger}_{\mathbf{k}\downarrow} c_{\mathbf{k}+\mathbf{q}\uparrow},$$

$$S^{z}_{\mathbf{q}} = \frac{1}{2}\sum_{\mathbf{k}} (c^{\dagger}_{\mathbf{k}\uparrow} c_{\mathbf{k}+\mathbf{q}\uparrow} - c^{\dagger}_{\mathbf{k}\downarrow} c_{\mathbf{k}+\mathbf{q}\downarrow}). \tag{G.6}$$

The first term in (G.5) yields a renormalization of the chemical potential and may be omitted. Consider the spin Green's function

$$G_{\mathbf{q}}(\omega) = \langle\langle S^{+}_{\mathbf{q}} | S^{-}_{-\mathbf{q}} \rangle\rangle_{\omega}$$

Writing down the sequence of equations of motion for this we derive

$$G_{\mathbf{q}}(\omega) = 2\langle S^{z}\rangle + \sum_{\mathbf{k}} (t_{\mathbf{k}+\mathbf{q}} - t_{\mathbf{k}}) \langle\langle c^{\dagger}_{\mathbf{k}\uparrow} c_{\mathbf{k}+\mathbf{q}\downarrow} | S^{-}_{-\mathbf{q}} \rangle\rangle_{\omega}; \tag{G.7}$$

$$(\omega - t_{\mathbf{k}+\mathbf{q}} + t_{\mathbf{k}} - \Delta) \langle\langle c^{\dagger}_{\mathbf{k}\uparrow} c_{\mathbf{k}+\mathbf{q}\downarrow} | S^{-}_{-\mathbf{q}} \rangle\rangle_{\omega} = \\ = (n_{\mathbf{k}\uparrow} - n_{\mathbf{k}+\mathbf{q}\downarrow})[1 - UG_{\mathbf{q}}(\omega)] - U\sum_{\mathbf{p}} L_{\mathbf{kqp}}(\omega), \tag{G.8}$$

where we have introduced the irreducible Green's function

$$L_{\mathbf{kqp}}(\omega) = \delta\langle\langle c^{\dagger}_{\mathbf{k}\uparrow} S^{+}_{\mathbf{p}} c_{\mathbf{k}+\mathbf{q}-\mathbf{p}\uparrow} - c^{\dagger}_{\mathbf{k}+\mathbf{p}\downarrow} S^{+}_{\mathbf{p}} c_{\mathbf{k}+\mathbf{q}\downarrow} - \delta_{\mathbf{pq}} (n_{\mathbf{k}\uparrow} - n_{\mathbf{k}+\mathbf{q}\downarrow}) S^{+}_{\mathbf{q}} | S^{-}_{-\mathbf{q}} \rangle\rangle_{\omega} \tag{G.9}$$

(the symbol δ means that the Hartree–Fock decouplings have to be excluded). Substituting (G.8) into (G.7) we get

$$G_{\mathbf{q}}(\omega) = \frac{\langle S^{z}\rangle - \Omega_{\mathbf{q}}(\omega)/U}{\omega - \Omega_{\mathbf{q}}(\omega) - \pi_{\mathbf{q}}(\omega)}; \tag{G.10}$$

$$\Omega_{\mathbf{q}}(\omega) = U\sum_{\mathbf{k}} \frac{t_{\mathbf{k}+\mathbf{q}} - t_{\mathbf{k}}}{t_{\mathbf{k}+\mathbf{q}} - t_{\mathbf{k}} + \Delta - \omega} (n_{\mathbf{k}\uparrow} - n_{\mathbf{k}+\mathbf{q}\downarrow}), \tag{G.11}$$

where the self-energy π is determined by the Green's function (G.9). When neglecting π we come to the random phase approximation (RPA). Unlike the standard form

$$G_{\mathbf{q}}(\omega) = \frac{\Pi_{\mathbf{q}}(\omega)}{1 - U\Pi_{\mathbf{q}}(\omega)}, \tag{G.12}$$

with

$$\Pi_{\mathbf{q}}(\omega) = \sum_{\mathbf{k}} \frac{n_{\mathbf{k}\uparrow} - n_{\mathbf{k+q}\downarrow}}{\omega + t_{\mathbf{k}\uparrow} - t_{\mathbf{k+q}\downarrow}},$$

the representation (G.10) yields explicitly the magnon (spin-wave) pole

$$\omega_{\mathbf{q}} \simeq \Omega_{\mathbf{q}}(0) = \sum_{\mathbf{k}\sigma} A^{\sigma}_{\mathbf{kq}} n_{\mathbf{k}\sigma}, \tag{G.13}$$

where

$$A^{\sigma}_{\mathbf{kq}} = \sigma U \frac{t_{\mathbf{k+q}} - t_{\mathbf{k}}}{t_{\mathbf{k+q}} - t_{\mathbf{k}} + \sigma\Delta} \tag{G.14}$$

has the meaning of the electron–magnon interaction amplitude. Expanding in q we get

$$\omega_{\mathbf{q}} = D_{\alpha\beta} q_\alpha q_\beta,$$

where

$$D_{\alpha\beta} = \frac{U}{\Delta} \sum_{\mathbf{k}} \left[\frac{\partial^2 t_{\mathbf{k}}}{\partial k_\alpha \partial k_\beta} (n_{\mathbf{k}\uparrow} + n_{\mathbf{k}\downarrow}) - \frac{1}{\Delta} \frac{\partial t_{\mathbf{k}}}{\partial k_\alpha} \frac{\partial t_{\mathbf{k}}}{\partial k_\beta} (n_{\mathbf{k}\uparrow} - n_{\mathbf{k}\downarrow}) \right] \tag{G.15}$$

are spin-wave stiffness tensor components. For a weak ferromagnet ($\Delta \ll E_{\mathrm{F}}$, U) we derive

$$D_{\alpha\beta} = \frac{U\Delta}{4} \sum_{\mathbf{k}} \left[\frac{\partial^2 t_{\mathbf{k}}}{\partial k_\alpha \partial k_\beta} \frac{\partial^2 n_{\mathbf{k}}}{\partial t_{\mathbf{k}}^2} + \frac{1}{6} \frac{\partial t_{\mathbf{k}}}{\partial k_\alpha} \frac{\partial t_{\mathbf{k}}}{\partial k_\beta} \frac{\partial^3 n_{\mathbf{k}}}{\partial t_{\mathbf{k}}^3} \right], \tag{G.16}$$

so that $D \sim \Delta$. The magnon damping in the RPA is given by

$$\gamma^{(1)}_{\mathbf{q}}(\omega) = -\mathrm{Im}\, \Omega_{\mathbf{q}}(\omega) = \pi U \Delta \omega \sum_{\mathbf{k}} \left(-\frac{\partial n_{\mathbf{k}\uparrow}}{\partial t_{\mathbf{k}\uparrow}} \right) \delta(\omega - t_{\mathbf{k+q}\downarrow} + t_{\mathbf{k}\uparrow}); \tag{G.17}$$

$$\gamma^{(1)}_{\mathbf{q}} \equiv \gamma^{(1)}_{\mathbf{q}}(\omega_{\mathbf{q}}) \simeq \pi U \Delta \omega_{\mathbf{q}} N_\uparrow(E_{\mathrm{F}}) N_\downarrow(E_{\mathrm{F}}) \theta(\omega_{\mathbf{q}} - \omega_-), \tag{G.18}$$

with $\theta(x)$ being the step function. Here $\omega_- = \omega(q_0)$ is the threshold energy which is determined by the condition of entering into the Stoner continuum (decay into the Stoner excitations, i.e. electron–hole pairs), q_0 being the minimal (in **k**) solution to the equation

$$t_{\mathbf{k+q}_0\downarrow} = t_{\mathbf{k}\uparrow} = E_{\mathrm{F}}. \tag{G.19}$$

The quantity ω_- determines a characteristic energy scale separating two temperature regions: the contributions of spin waves (poles of the Green's function (G.10)) dominate at $T<\omega_-$, and those of Stoner excitations (its branch cut) at $T>\omega_-$.

In the case of weak ferromagnets, the contribution of the branch cut of the spin Green's function may be approximately treated as that of a paramagnon pole at imaginary ω, and we obtain

$$q_0 = k_{F\uparrow} - k_{F\downarrow}, \qquad \omega_- = D(k_{F\uparrow} - k_{F\downarrow})^2 \sim \Delta^3 \sim T_C^2/E_F. \tag{G.20}$$

Since q_0 is small, we have at small $q>q_0$, instead of (G.18),

$$\gamma_{\mathbf{q}}^{(1)}(\omega_{\mathbf{q}}) \simeq \frac{U\Delta\omega}{q}\frac{v_0}{4\pi}(m^*)^2 \equiv \frac{A}{q}. \tag{G.21}$$

The estimation (G.20) holds also for the s—$d(f)$ exchange model with the indirect RKKY-interaction where

$$D \sim T_C/S \sim I^2S/E_F. \tag{G.22}$$

The damping at very small $q<q_0$ (where (G.17) vanishes) is due to the two-magnon scattering processes. To consider these we have to calculate the function π to leading order in the fluctuating part of the Coulomb interaction. Writing down the equation of motion for the Green's function (G.9) we obtain

$$\begin{aligned}\Pi_{\mathbf{q}}(\omega) = \sum_{\mathbf{pk}}(A_{\mathbf{kq}}^{\uparrow})^2[&B(\mathbf{k}\uparrow,\mathbf{k}+\mathbf{q}-\mathbf{p}\uparrow,\omega_{\mathbf{p}}-\omega)+\\ &+B(\mathbf{k}+\mathbf{p}\downarrow,\mathbf{k}+\mathbf{q}\downarrow,\omega_{\mathbf{p}}-\omega)-\\ &-B(\mathbf{k}+\mathbf{p}\downarrow,\mathbf{k}\uparrow,\omega_{\mathbf{p}})-B(\mathbf{k}+\mathbf{q}\downarrow,\mathbf{k}+\mathbf{q}-\mathbf{p}\downarrow,\omega_{\mathbf{p}})],\end{aligned} \tag{G.23}$$

where

$$B(\mathbf{k}'\sigma',\mathbf{k}\sigma,\omega) = \frac{N_{\mathbf{p}}(n_{\mathbf{k}'\sigma'} - n_{\mathbf{k}\sigma}) + n_{\mathbf{k}'\sigma'}(1-n_{\mathbf{k}\sigma})}{\omega - t_{\mathbf{k}'\sigma'} + t_{\mathbf{k}\sigma}}.$$

The magnon damping needed is given by the imaginary part of (G.23). After some transformations we derive

$$\begin{aligned}\gamma_{\mathbf{q}}^{(2)}(\omega) = \pi\sum_{\mathbf{kp}\sigma}(A_{\mathbf{kq}}^{\uparrow})^2(n_{\mathbf{k}\sigma} - n_{\mathbf{k}+\mathbf{q}-\mathbf{p}\sigma})[N_{\mathbf{p}} - N_B(\omega_{\mathbf{p}}-\omega)]\times\\ \times\delta(\omega + t_{\mathbf{k}} - t_{\mathbf{k}+\mathbf{q}-\mathbf{p}} - \omega_{\mathbf{p}}).\end{aligned} \tag{G.24}$$

Integration for the isotropic electron spectrum gives [682, 683]

$$\gamma_{\mathbf{q}}^{(2)}(\omega) = \frac{v_0^2}{12\pi^3}\frac{q^4}{4\langle S^z\rangle^2}\sum_\sigma k_{F\sigma}^2 \begin{cases}\omega_{\mathbf{q}}/35, & T \ll \omega_{\mathbf{q}},\\ (T/4)\left(\ln(T/\omega_{\mathbf{q}}) + 5/3\right), & T \gg \omega_{\mathbf{q}}.\end{cases} \tag{G.25}$$

The real part of (G.23) describes the temperature dependence of the spin stiffness owing to two-magnon processes (besides the simplest T^2-contribution which occurs from the temperature dependence of the Fermi distribution functions in (G.10)). The spin-wave contribution connected with the magnon distribution functions is proportional to T. More interesting is the non-analytical many-electron contribution owing to the Fermi functions:

$$\delta D_{\alpha\beta} = \frac{1}{4\langle S^z\rangle^2}\sum_{\mathbf{pk}}\frac{\partial t_\mathbf{k}}{\partial k_\alpha}\frac{\partial t_\mathbf{k}}{\partial k_\beta}\left[\frac{n_{\mathbf{k}\uparrow}(1-n_{\mathbf{k-p}\uparrow})}{t_\mathbf{k}-t_\mathbf{k-p}-\omega_\mathbf{p}}+\frac{n_{\mathbf{k+p}\downarrow}(1-n_{\mathbf{k}\downarrow})}{t_\mathbf{k+p}-t_\mathbf{k}-\omega_\mathbf{p}}-\right.$$
$$\left.-\frac{n_{\mathbf{k+p}\downarrow}(1-n_{\mathbf{k}\uparrow})}{t_{\mathbf{k+p}\downarrow}-t_{\mathbf{k}\uparrow}-\omega_\mathbf{p}}-\frac{n_{\mathbf{k}\downarrow}(1-n_{\mathbf{k-p}\uparrow})}{t_{\mathbf{k}\downarrow}-t_{\mathbf{k-p}\uparrow}-\omega_\mathbf{p}}\right]. \tag{G.26}$$

Performing integration for parabolic spectra of electrons and magnons ($t_\mathbf{k}=k^2/2\mathrm{m}^*$, $\omega_\mathbf{q}=Dq^2$) yields

$$\delta D(T)=\left(\frac{\pi v_0 T}{12\langle S^z\rangle m^*}\right)^2\frac{1}{D}\left[\sum_\sigma N_\sigma^2(E_\mathrm{F})\ln\frac{T}{\omega_+}-\right.$$
$$\left.-2N_\uparrow(E_\mathrm{F})N_\downarrow(E_\mathrm{F})\ln\frac{\max(\omega_-,T)}{\omega_+}\right], \tag{G.27}$$

with

$$\omega_\pm = D(k_{\mathrm{F}\uparrow}\pm k_{\mathrm{F}\downarrow})^2, \qquad N_\sigma(E_\mathrm{F})=m^* v_0 k_\mathrm{F}/2\pi^2. \tag{G.28}$$

At $\omega_- \ll T \ll \omega_+$ we have

$$\delta D(T)=\left(\frac{\pi v_0 T}{12\langle S^z\rangle m^*}\right)^2\frac{1}{D}[N_\uparrow(E_\mathrm{F})-N_\downarrow(E_\mathrm{F})]^2\ln\frac{T}{\omega_+}. \tag{G.29}$$

It should be noted that the correction (G.27) dominates at low temperatures over the above-mentioned T^2-correction, which demonstrates an important role of corrections to the RPA approximation. Unfortunately, the $T^2\ln T$-term has not yet to be considered at analyzing magnon spectra of ferromagnetic metals. We see that temperature dependences of spin-wave characteristics in conducting magnets differ considerably from those in the Heisenberg model.

To obtain corrections to the Stoner approximation for the electron spectrum (G.3) we have to calculate the one-electron Green's function

$$G_{\mathbf{k}\sigma}(E)=\langle\langle c_{\mathbf{k}\sigma}|c^\dagger_{\mathbf{k}\sigma}\rangle\rangle_E=\left[E-t_{\mathbf{k}\sigma}-\Sigma_{\mathbf{k}\sigma}(E)\right]^{-1}. \tag{G.30}$$

One obtains for the self-energy [573]

$$\Sigma_{\mathbf{k}\sigma}(E) = -U^2 \sum_{\mathbf{q}} \int_{-\infty}^{\infty} \frac{1}{\pi} \mathrm{Im}\, \langle\langle S_{\mathbf{q}}^{\sigma} | S_{-\mathbf{q}}^{-\sigma} \rangle\rangle_{\omega} \frac{N_{\mathrm{B}}(\omega) + n_{\mathbf{k}+\mathbf{q},-\sigma}}{E - t_{\mathbf{k}+\mathbf{q},-\sigma} + \omega} d\omega. \tag{G.31}$$

Retaining only the magnon pole contribution to the spectral density (i.e. neglecting the spin-wave damping) we may put

$$-\frac{1}{\pi} \mathrm{Im}\, \langle\langle S_{\mathbf{q}}^{\sigma} | S_{-\mathbf{q}}^{-\sigma} \rangle\rangle_{\omega} = 2\sigma \langle S^z \rangle \delta(\omega - \sigma\omega_{\mathbf{q}}), \tag{G.32}$$

so that

$$\langle S_{-\mathbf{q}}^{-} S_{\mathbf{q}}^{+} \rangle = 2\langle S^z \rangle N_{\mathbf{q}}. \tag{G.33}$$

Then we get

$$\begin{aligned} \Sigma_{\mathbf{k}\uparrow}(E) &= U\Delta \sum_{\mathbf{q}} \frac{N_{\mathbf{q}} + n_{\mathbf{k}+\mathbf{q}\downarrow}}{E - t_{\mathbf{k}+\mathbf{q}\downarrow} + \omega_{\mathbf{q}}}, \\ \Sigma_{\mathbf{k}\downarrow}(E) &= U\Delta \sum_{\mathbf{q}} \frac{1 + N_{\mathbf{q}} - n_{\mathbf{k}-\mathbf{q}\uparrow}}{E - t_{\mathbf{k}-\mathbf{q}\uparrow} - \omega_{\mathbf{q}}}. \end{aligned} \tag{G.34}$$

The results (G.34) are valid in the *s–d* model ($U \to I$) to the first order in the small parameter $1/2S$ [323].Taking into account the relation

$$\langle S^z \rangle = S_0 - \sum_{\mathbf{p}} N_{\mathbf{p}}, \tag{G.35}$$

where S_0 is the saturation magnetization, one obtains for the spin-wave correction to the electron energy

$$\delta E_{\mathbf{k}\sigma}(T) = \sum_{\mathbf{q}} A_{\mathbf{kq}}^{\sigma} N_{\mathbf{q}} = \frac{v_0}{2\langle S^z \rangle} \frac{\zeta(5/2)}{32\pi^{3/2}} \left(\frac{T}{D}\right)^{5/2} \left[\frac{\partial^2 t_{\mathbf{k}}}{\partial k_x^2} - \frac{\sigma}{U\langle S^z \rangle}\left(\frac{\partial t_{\mathbf{k}}}{\partial k}\right)^2\right]. \tag{G.36}$$

The $T^{5/2}$-dependence of the electron spectrum owing to magnons is weaker than the $T^{3/2}$-dependence of the magnetization. This fact is due to vanishing of electron–magnon interaction amplitude A at zero magnon wavevector, which is connected with the symmetry of exchange interaction. Such a weakening of temperature dependence of the spin splitting was observed in iron [145]. It should be noted that the same $T^{5/2}$-dependence takes place also in a ferromagnet with the Hubbard subbands [337, 338].

The one-electron occupation numbers are obtained via the spectral representation for the anticommutator Green's function (G.30):

$$\langle c^\dagger_{\mathbf{k}\uparrow} c_{\mathbf{k}\uparrow}\rangle = f(t_\mathbf{k} + \mathrm{Re}\,\Sigma_{\mathbf{k}\uparrow}(t_{\mathbf{k}\uparrow})) + \\ + U\Delta \sum_\mathbf{p} \frac{N_\mathbf{p}(n_{\mathbf{k}+\mathbf{p}\downarrow} - n_{\mathbf{k}\uparrow}) + n_{\mathbf{k}+\mathbf{p}\downarrow}(1 - n_{\mathbf{k}\uparrow})}{(t_{\mathbf{k}+\mathbf{p}\downarrow} - t_{\mathbf{k}\uparrow} - \omega_\mathbf{p})^2}, \tag{G.37}$$

$$\langle c^\dagger_{\mathbf{k}\downarrow} c_{\mathbf{k}\downarrow}\rangle = f(t_\mathbf{k} + \mathrm{Re}\,\Sigma_{\mathbf{k}\downarrow}(t_{\mathbf{k}\downarrow})) + \\ + U\Delta \sum_\mathbf{p} \frac{N_\mathbf{p}(n_{\mathbf{k}\downarrow} - n_{\mathbf{k}-\mathbf{p}\uparrow}) + n_{\mathbf{k}\downarrow}(1 - n_{\mathbf{k}-\mathbf{p}\uparrow})}{(t_{\mathbf{k}-\mathbf{p}\uparrow} - t_{\mathbf{k}\downarrow} + \omega_\mathbf{p})^2}, \tag{G.38}$$

where the second term comes from the imaginary part of the self-energy. Retaining only the magnon contributions up to $T^{3/2}$ we get

$$\langle c^\dagger_{\mathbf{k}\sigma} c_{\mathbf{k}\sigma}\rangle \simeq n_{\mathbf{k}\sigma} \frac{S_0 + \langle S^z\rangle}{2S_0} + n_{\mathbf{k}-\sigma} \frac{S_0 - \langle S^z\rangle}{2S_0}, \tag{G.39}$$

with $\langle S^z\rangle$ defined by (G.35). Thus, despite the presence of the spin splitting, electron occupation numbers have a strong $T^{3/2}$-dependence rather than an exponential one (as in the Stoner theory). This dependence arises because of thermal magnon emission and absorption processes. Thus the conduction electron spin polarization

$$P = \frac{n_\uparrow - n_\downarrow}{n_\uparrow + n_\downarrow} \tag{G.40}$$

is equal to the relative magnetization, which is obvious for an itinerant ferromagnet. However, such a behaviour takes place also in arbitrary conducting ferromagnets, e.g., for ferromagnetic semiconductors which are describe by the *s–d(f)* exchange model [329]. Formally, the $T^{3/2}$-dependence of $P(T)$ is due to the strong temerature dependence of the residues of the electron Green's functions and to the occurence of the non-quasiparticle states in an 'alien' spin subband owing to electron-magnon scattering. The picture of the density of states is discussed in more details in Sect.4.5.

Consider corrections to the magnetization $\langle S^z\rangle$. We have

$$\langle S^z\rangle = \frac{n}{2} - \sum_\mathbf{q} \langle S^-_\mathbf{q} S^+_\mathbf{q}\rangle - \langle \hat{n}_{i\uparrow}\hat{n}_{i\downarrow}\rangle. \tag{G.41}$$

The first average involved in (G.41) is calculated from the spectral representation of the RPA Green's function (G.10):

$$\langle S^-_{-\mathbf{q}} S^+_\mathbf{q}\rangle = 2S_0 N_\mathbf{q} \quad (q < q_0), \tag{G.42}$$

$$\langle S^-_{-\mathbf{q}} S^+_{\mathbf{q}} \rangle = \frac{1}{\pi} \int_{-\infty}^{\infty} \frac{N_B(\omega)\gamma^{(1)}_{\mathbf{q}}(\omega)(\Delta - \omega)/U}{[\omega - \operatorname{Re}\Omega_{\mathbf{q}}(\omega)]^2 + [\gamma^{(1)}_{\mathbf{q}}(\omega)]^2} d\omega \quad (q > q_0). \tag{G.43}$$

Using the identity

$$N_B(t_{\mathbf{k}+\mathbf{q}\downarrow} - t_{\mathbf{k}\uparrow})(n_{\mathbf{k}\uparrow} - n_{\mathbf{k}+\mathbf{q}\downarrow}) = n_{\mathbf{k}+\mathbf{q}\downarrow}(1 - n_{\mathbf{k}\uparrow}), \tag{G.44}$$

we derive from (G.43)

$$\langle S^-_{-\mathbf{q}} S^+_{\mathbf{q}} \rangle = \sum_{\mathbf{k}} \frac{(t_{\mathbf{k}+\mathbf{q}\downarrow} - t_{\mathbf{k}\uparrow})^2 n_{\mathbf{k}+\mathbf{q}\downarrow}(1 - n_{\mathbf{k}\uparrow})}{(t_{\mathbf{k}+\mathbf{q}\downarrow} - t_{\mathbf{k}\uparrow} - \omega_{\mathbf{q}})^2 + [\gamma^{(1)}_{\mathbf{q}}(t_{\mathbf{k}+\mathbf{q}\downarrow} - t_{\mathbf{k}\uparrow})]^2}. \tag{G.45}$$

In contradiction with (G.41), (G.42), the true Bloch spin-wave contribution to magnetization should be given by (G.35) since every magnon decreases $\langle S^z \rangle$ by unity. The agreement may be restored by allowing not only the magnon pole, but also branch cut contributions. Replacing in (G.45) $n_{\mathbf{k}\sigma} \to \langle c^\dagger_{\mathbf{k}\sigma} c_{\mathbf{k}\sigma} \rangle$ and using (G.39) we obtain

$$\delta\langle S^z \rangle_{\text{sw}} = -\sum_{\mathbf{q}} N_{\mathbf{q}} \Bigg[2S_0 + \frac{1}{2S_0} \sum_{\mathbf{k}\mathbf{k}'} \frac{(t_{\mathbf{k}} - t_{\mathbf{k}'})^2 (n_{\mathbf{k}\uparrow} - n_{\mathbf{k}\downarrow})}{(t_{\mathbf{k}'\downarrow} - t_{\mathbf{k}\uparrow} - \omega_{\mathbf{k}-\mathbf{k}'})^2} (1 - n_{\mathbf{k}'\uparrow} + n_{\mathbf{k}'\downarrow}) \Bigg] \simeq -\sum_{\mathbf{q}} N_{\mathbf{q}}, \tag{G.46}$$

where we have neglected the spin splitting in the denominator.

In the semi-phenomenological manner, it is convenient to introduce 'magnon' operators which satisfy on the average the Bose commutation relations:

$$b_{\mathbf{q}} = (2S_0)^{-1/2} S^+_{\mathbf{q}}, \qquad b^\dagger_{\mathbf{q}} = (2S_0)^{-1/2} S^-_{-\mathbf{q}}. \tag{G.47}$$

Then we have

$$\delta\langle S^z \rangle = -\sum_{\mathbf{q}} \langle b^\dagger_{\mathbf{q}} b_{\mathbf{q}} \rangle = -\frac{1}{2S_0} \sum_{\mathbf{q}} \langle S^-_{-\mathbf{q}} S^+_{\mathbf{q}} \rangle. \tag{G.48}$$

Performing integration over ω in (G.43) at $T = 0$ we obtain

$$\delta\langle S^z \rangle = -\frac{1}{\pi} \sum_{\mathbf{q}} \frac{\gamma^{(1)}_{\mathbf{q}}}{\omega_{\mathbf{q}}} \ln \frac{W}{\omega_{\mathbf{q}}}, \tag{G.49}$$

with W being the bandwidth. This contribution describes zero-point decrease of the magnetization due to the ground-state magnon damping which is owing to the Stoner excitations. For parabolic electron and magnon spectra, neglecting the damping in the denominator of (G.43) we obtain at low temperatures $T < \omega_-$

$$\delta\langle S^z\rangle_e \simeq -U\Delta\sum_{\mathbf{kk}'}\frac{n_{\mathbf{k}'\downarrow}(1-n_{\mathbf{k}\uparrow})}{(t_{\mathbf{k}'\downarrow}-t_{\mathbf{k}\uparrow}-\omega_{\mathbf{k}-\mathbf{k}'})^2} =$$

$$= -\left(\frac{m^* v_0}{2\pi^2}\right)^2 \frac{U\Delta}{4D}\left[\omega_+ \ln\frac{W}{\omega_+} - \omega_- \ln\frac{W}{\omega_-} + \frac{2\pi^2}{3}T^2\left(\frac{1}{\omega_-}-\frac{1}{\omega_+}\right)\right], \qquad \text{(G.50)}$$

(G.50) with $\omega_\pm$ defined in (G.28). For a weak ferromagnet, the temperature correction to $\langle S^z\rangle$ in (G.50) is proportional to $(T/T_C)^2$, in agreement with the self-consistent renormalization theory [296, 26]. It should be stressed that the T^2-correction obtained is much larger than the Stoner contribution of the order of $(T/E_F)^2$.

An account of the damping at low T influences numerical factors in (G.50) only. At the same time, at high $T > \omega_-$ the damping in the denominator dominates at small q in the case of a weak ferromagnet. Taking into account (G.21) we obtain from (G.43)

$$\delta\langle S^-_{-\mathbf{q}} S^+_{\mathbf{q}}\rangle = \frac{\Delta}{\pi U}\int_{-\infty}^{\infty} d\omega N_B(\omega)\int_0^{\infty} dq \frac{\omega A q}{(Dq^2)^2 + A^2\omega^2/q^2} \sim \left(\frac{T}{E_F}\right)^{4/3}. \qquad \text{(G.51)}$$

Thus we get from (G.41) the $T^{4/3}$-contribution to the magnetization, which agrees with the result of the phase transition scaling theory near $T = T_C$. For a ferromagnet with well-localized magnetic moments the damping may be neglected and we derive [716]

$$\delta\langle S^z\rangle_e \sim -I^2\omega_- \ln(T/\omega_-). \qquad \text{(G.52)}$$

Consider the renormalization of electronic specific heat in an itinerant ferromagnet due to interaction with spin fluctuations. Integration in (G.34) at $T = 0$ gives

$$\operatorname{Re}\Sigma_\sigma(k_{F\sigma}, E) = -\frac{U\Delta}{\omega_+ - \omega_-} N_{-\sigma}(E_F)\sum_{\alpha=\pm}\alpha(E-\omega_\alpha)\ln\frac{|E-\omega_\alpha|}{W}. \qquad \text{(G.53)}$$

Then the inverse residue of the electron Green's function

$$Z^{-1}_{\mathbf{k}\sigma}(E) = 1 - \frac{\partial}{\partial E}\operatorname{Re}\Sigma_{\mathbf{k}\sigma}(E)$$

takes the form

$$Z^{-1}_\sigma(k_{F\sigma}, E) = 1 + \frac{U\Delta}{\omega_+ - \omega_-} N_{-\sigma}(E_F)\ln\left|\frac{E-\omega_+}{E-\omega_-}\right|. \qquad \text{(G.54)}$$

The quantity (G.54) determines the renormalization of the electron effective mass owing to the electron–magnon interaction. Thus we obtain for the coefficient at the linear term in the electronic specific heat at $T \ll \omega_-$

$$\gamma_\sigma = \gamma_\sigma^{(0)} / Z_\sigma(k_{F\sigma}, E_F) = \frac{\pi^2}{3} N_\sigma(E_F) \left[1 + \frac{U\Delta}{\omega_+ - \omega_-} N_{-\sigma}(E_F) \ln \frac{\omega_+}{\omega_-} \right]. \quad (G.55)$$

For weak itinerant ferromagnets we have

$$\ln \frac{\omega_+}{\omega_-} \simeq -2\ln(UN(E_F) - 1), \quad (G.56)$$

so that the expression (G.55) describes the paramagnon enhancement of the specific heat [297, 573] discussed in Sect.4.4. The numerical factor in (G.55) is inexact in this limit because of neglecting longitudinal spin fluctuations (see [26]). On the other hand, our consideration is not restricted to the case of weak ferromagnets. This is important since a considerable enhancement of specific heat owing to spin fluctuations is observed in a number of strong ferromagnets. For example, the experimental value of γ in the system $CeFe_{1-x}Co_xS_2$ (where a ferro–antiferro transition takes place at increasing x) in the ferromagnetic phase, $\gamma = 48$ mJ/mol·K^2 ($x = 0$) is twice the theoretical value obtained from the calculated density of states, and is considerably larger than the values in para- and antiferromagnetic phases [684].

Other thermodynamic properties may be investigated by calculating the free energy of the system. At low $T < \omega_-$ the many-electron (branch cut) contribution reads

$$\mathcal{F}_e = \frac{1}{2S_0} \sum_{\mathbf{q} > \mathbf{q}_0} \omega_\mathbf{q} \langle S_{-\mathbf{q}}^- S_\mathbf{q}^+ \rangle \simeq U\Delta \sum_{\mathbf{k}\mathbf{k}'} \frac{n_{\mathbf{k}'\downarrow}(1 - n_{\mathbf{k}\uparrow})}{t_{\mathbf{k}\uparrow} - t_{\mathbf{k}'\downarrow} + \omega_{\mathbf{k}-\mathbf{k}'}} \equiv \mathcal{F}_e(0) + \delta\mathcal{F}_e(T), \quad (G.57)$$

where

$$\mathcal{F}_e(0) = \frac{1}{\pi} \sum_\mathbf{q} \gamma_\mathbf{q}^{(1)} \ln \frac{W}{\omega_\mathbf{q}} \approx \frac{1}{8D} \left(\frac{m^* v_0}{2\pi^2} \right)^2 \left[\omega_+^2 \ln \frac{W}{\omega_+} - \omega_-^2 \ln \frac{W}{\omega -} \right], \quad (G.58)$$

$$\delta\mathcal{F}_e(T) = -\frac{U\Delta}{\omega_+ - \omega_-} N_\uparrow(E_F) \frac{\pi^2}{3} T^2 \ln \frac{\omega_+}{\max(\omega_-, T)}. \quad (G.59)$$

The spin-wave contribution to the free energy has the form, usual for the Bose excitations with the square dispersion law

$$\delta\mathcal{F}_{sw} = -\frac{2}{3} \delta \langle \mathsf{H} \rangle_{sw}, \quad (G.60)$$

$$\delta \langle \mathsf{H} \rangle_{sw} = \sum_{q < q_0} \omega_\mathbf{q} N_\mathbf{q} = \frac{3v_0}{16\pi^{3/2}} \zeta\left(\frac{5}{2}\right) \frac{T^{5/2}}{D^{3/2}}.$$

Temperature-dependent corrections to physical properties are obtained from (G.59), (G.60). Differentiating (G.59) over T we obtain

$$\delta C_{\mathrm{e}} = -\frac{\partial}{\partial T}\delta \mathcal{F}_{\mathrm{e}}(T) = U^2 \frac{2\langle S^z \rangle}{\omega_+ - \omega_-} N_\uparrow (E_{\mathrm{F}}) N_\downarrow (E_{\mathrm{F}}) \frac{2\pi^2}{3} T \ln \frac{\omega_+}{\max(\omega_-, T)}. \tag{G.61}$$

Thus at $T \gg \omega_-$ we obtain instead of (G.55) the $T \ln T$-dependence of specific heat.

Consider the local magnetic moment at a site

$$\langle S^z \rangle = \frac{3}{4}(n - 2N_2), \qquad N_2 = \langle n_{i\uparrow} n_{i\downarrow} \rangle. \tag{G.62}$$

The number of doubles N_2 may be determined by using the Hellman-Feynman theorem $N_2 = \partial \mathcal{F}/\partial U$. We obtain at $T < \omega_-$

$$\delta \langle S^z \rangle_{\mathrm{e}}(T) \sim -(T/\Delta)^2 \delta \langle S^z \rangle_{\mathrm{e}}(T) \sim -(T/T_{\mathrm{C}})^{5/2}. \tag{G.63}$$

Thus the temperature dependence of the spin-wave contribution to $\langle \mathbf{S}^2 \rangle$ is weaker than that to $\langle S^z \rangle$, which justifies neglecting the former in the above discussion of the magnetization (G.41). At high T, the local moment has the T-dependence:

$$\delta \langle S_i^2 \rangle = \delta \sum_{\mathbf{q}} \langle S_{-\mathbf{q}}^- S_{\mathbf{q}}^+ \rangle \sim (T/E_{\mathrm{F}})^{4/3}. \tag{G.64}$$

As one can see from (G.55), the enhancement of effective mass and electronic specific heat owing to spin fluctuations is absent in the half-metallic state (Sect.4.5). We shall demonstrate that the specific heat of a conducting ferromagnet may contain spin-fluctuation contributions of another nature. Write down a general expression for the specific heat in the s–d exchange model in terms of the total energy

$$C(T) = \frac{\partial \langle \mathrm{H} \rangle}{\partial T} = \frac{\partial}{\partial T} \int E f(E) N_{\mathrm{t}}(E) dE =$$

$$= \frac{\pi^2}{3} N_{\mathrm{t}}(E) T + \int E f(E) \frac{\partial N_{\mathrm{t}}(E,T)}{\partial T} dE, \tag{G.65}$$

where

$$N_{\mathrm{t}}(E) = -\frac{1}{\pi} \sum_{\mathbf{k}\sigma} \mathrm{Im}\, G_{\mathbf{k}\sigma}(E)$$

is the total density of states. The first term in the right-hand side of (G.65) yields the standard result of the Fermi-liquid theory. The second term is due to the energy dependence of the density of states. Such a dependence occurs in the conducting ferromagnet owing to non-quasiparticle (incoherent) states (Sect.4.5). Substituting (4.87) into this term we derive [338]

$$\delta C_\sigma(T) = 2\sigma I^2 \langle S^z \rangle \sum_{\mathbf{kq}} \frac{f(t_{\mathbf{k+q},-\sigma} - \sigma\omega_{\mathbf{q}})}{(t_{\mathbf{k+q},-\sigma} - t_{\mathbf{k},\sigma})^2} \frac{\partial n_{\mathbf{k+q},-\sigma}}{\partial T}. \qquad \text{(G.66)}$$

At low temperatures

$$f(t_{\mathbf{k+q},\downarrow} - \omega_{\mathbf{q}}) = 1, \qquad f(t_{\mathbf{k+q},\uparrow} + \omega_{\mathbf{q}}) = 0.$$

Thus the non-quasiparticle states with $\sigma = \downarrow$ do not contribute to linear specific heat since they are empty at T=0. In the half-metallic state the non-quasiparticle contributions (G.66) with $\sigma = \uparrow$ are present for $I < 0$ only, and we obtain

$$\delta C_\uparrow(T) = \frac{2\pi^2}{3} I^2 \langle S^z \rangle N_\downarrow(E_\mathrm{F}) T \sum_{\mathbf{k}} \frac{1}{(t_{\mathbf{k}\uparrow} - E_\mathrm{F})^2}. \qquad \text{(G.67)}$$

To avoid misunderstanding, it should be stressed that presence of such contributions to specific heat means inapplicability of the Fermi-liquid description in terms of dynamical quasiparticles only, which are determined by poles of Green functions. It may be shown rigorously that the entropy of interacting Fermi systems at low T is expressed in terms of Landau quasiparticles with the energies, determined as variational derivatives of the total energy with respect to occupation numbers [685]. Thus, even in the presence of non-pole contributions to the Green functions, the description of thermodynamics in terms of statistical quasiparticles [685] holds. (However, the quasiparticle description is insufficient for spectral characteristics, e.g., optical and emission data.) The anomalous γT-term is determined by the difference of the spectra of statistical and dynamical quasiparticles.

Similar contributions to specific heat in the Hubbard model with strong correlations are discussed in the paper [338] too. They dominate in the enhancement of specific heat for half-metallic ferromagnets and may be important, besides the effective mass enhancement (G.55), for "usual" magnets with well-defined local moments.

G.2 ANTIFERROMAGNETS

To consider the electron and magnon spectrum of a metallic antiferromagnet in the s–$d(f)$ exchange model, we have to pass to the local coordinate system according to (E.8). Then the

Hamiltonian of the s–$d(f)$ exchange interaction takes the form

$$\begin{aligned}\mathrm{H}_{sd} = -I\sum_{\mathbf{kq}}[S^x_{\mathbf{q}}(c^\dagger_{\mathbf{k+q}\downarrow}c_{\mathbf{k}\downarrow} - c^\dagger_{\mathbf{k+q}\uparrow}c_{\mathbf{k}\uparrow}) + \\ + iS^y_{\mathbf{q}}(c^\dagger_{\mathbf{k-Q}\downarrow}c_{\mathbf{k-Q}\uparrow} - c^\dagger_{\mathbf{k+q}\uparrow}c_{\mathbf{k-Q}\downarrow}) + \\ + S^z_{\mathbf{q}}(c^\dagger_{\mathbf{k+q}\uparrow}c_{\mathbf{k-Q}\downarrow} + c^\dagger_{\mathbf{k-Q}\downarrow}c_{\mathbf{k-q}\uparrow})].\end{aligned} \tag{G.68}$$

Passing to the magnon representation with the use of (E.1), (E.10) and calculating the electron self-energy to second order in I we obtain

$$\begin{aligned}\Sigma_{\mathbf{k}}(E) = \frac{I^2\overline{S}^2}{E - t_{\mathbf{k-Q}}} + \\ + \frac{1}{2}I^2 S\sum_{\mathbf{q}}\left\{(u_{\mathbf{q}} - v_{\mathbf{q}})^2\left[\frac{1 - n_{\mathbf{k-q}} + N_{\mathbf{q}}}{E - t_{\mathbf{k-q}} - \omega_{\mathbf{q}}} + \frac{n_{\mathbf{k-q}} + N_{\mathbf{q}}}{E - t_{\mathbf{k-q}} + \omega_{\mathbf{q}}}\right] + \right. \\ \left. +(u_{\mathbf{q}} + v_{\mathbf{q}})^2\left[\frac{1 - n_{\mathbf{k+q-Q}} + N_{\mathbf{q}}}{E - t_{\mathbf{k+q-Q}} - \omega_{\mathbf{q}}} + \frac{n_{\mathbf{k+q-Q}} + N_{\mathbf{q}}}{E - t_{\mathbf{k+q-Q}} + \omega_{\mathbf{q}}}\right]\right\},\end{aligned} \tag{G.69}$$

with S being the sublattice magnetization. In the mean-field approximation the electron spectrum contains two split antiferromagnetic subbands:

$$E^{1,2}_{\mathbf{k}} = \frac{1}{2}(t_{\mathbf{k}} + t_{\mathbf{k-Q}}) \pm \frac{1}{2}\left[(t_{\mathbf{k}} - t_{\mathbf{k-Q}})^2 + 4I^2\overline{S}^2\right]^{1/2}. \tag{G.70}$$

The fluctuation correction (second term in (G.68)) are important, besides the dependence $S(T)$, at calculating the temperature dependence of electron spectrum. Consider the magnon corrections owing to the Bose functions

$$\delta E_{\mathbf{k}}(T) = I^2 S\sum_{\mathbf{q}}\left[-2\frac{u^2_{\mathbf{q}} + v^2_{\mathbf{q}}}{t_{\mathbf{k}} - t_{\mathbf{k-Q}}} + \frac{(u_{\mathbf{q}} - v_{\mathbf{q}})^2}{t_{\mathbf{k}} - t_{\mathbf{k-q}}} + \frac{(u_{\mathbf{q}} + v_{\mathbf{q}})^2}{t_{\mathbf{k}} - t_{\mathbf{k+q-Q}}}\right]N_{\mathbf{q}}. \tag{G.71}$$

The corrections to the band bottom ($t_{\mathbf{k}} = t_{\min}$) owing to sublattice magnetization (the first term in square brackets) and transverse fluctuations have opposite signs. The contribution of fluctuations prevails which results in a 'blue' shift of the conduction band bottom with decreasing temperature, as observed in antiferromagnetic semiconductors [352, 686], in contrast to the 'red' shift in ferromagnetic semiconductors (cf.(G.36)). (The same situation takes place at high temperatures [686,687].) In particular, in the nearest-neighbour approximation for simple lattices where $t_{\mathbf{k+Q}} = -t_{\mathbf{k}}$ the fluctuation contribution is larger in absolute value by two times. The integration yields (z is the nearest-neighbour number)

$$\delta E_{\min}(T) = -\frac{3}{16}\frac{I^2 S}{W}\left(\frac{z}{2}\right)^{3/2}(S+1)^{3/2}\left(\frac{T}{T_N}\right)^2. \qquad (G.72)$$

The T^2-dependence of electron spectrum (the same as of the sublattice magnetization) is due to linear dispersion of spin-wave spectrum and q^{-1}-dependence of electron–magnon interaction amplitude, which are specific for antiferromagnets. Therefore, the behavior of electron spectrum in a ferrimagnet is qualitatively similar to that in ferromagnets (the $T^{5/2}$-dependence) [687].

We write down also the third-order many-electron contribution to the self-energy, which describes renormalization of the antiferromagnetic gap due to Kondo-like divergences [367] (see Chapter 6)

$$\delta\Sigma_{\mathbf{k}}^{(3)}(E) = 2I^3 S^2 \sum_{\mathbf{q}} n_{\mathbf{k}+\mathbf{q}}(E - t_{\mathbf{k}+\mathbf{q}})\frac{1}{(E - t_{\mathbf{k}+\mathbf{q}})^2 - \omega_{\mathbf{q}}^2} \times$$
$$\times\left(\frac{1}{t_{\mathbf{k}+\mathbf{q}} - t_{\mathbf{k}-\mathbf{Q}+\mathbf{q}}} - \frac{1}{E - t_{\mathbf{k}+\mathbf{Q}}}\right). \qquad (G.73)$$

To investigate the magnon spectrum, we calculate the retarded commutator Green's functions

$$\Gamma_{\mathbf{q}}(\omega) = \langle\langle b_{\mathbf{q}} | b_{\mathbf{q}}^\dagger \rangle\rangle_\omega, \qquad \overline{\Gamma}_{\mathbf{q}}(\omega) = \langle\langle b_{-\mathbf{q}}^\dagger | b_{\mathbf{q}}^\dagger \rangle\rangle_\omega.$$

Writing down the sequence of equations of motion to second order in I and performing the simplest possible decouplings we derive [716] (cf.(E.9))

$$\Gamma_{\mathbf{q}}(\omega) = \frac{\omega + C_{\mathbf{q}-\omega}}{(\omega - C_{\mathbf{q}\omega})(\omega + C_{\mathbf{q}-\omega}) + D_{\mathbf{q}\omega}^2}; \qquad (G.74)$$

$$\overline{\Gamma}_{\mathbf{q}}(\omega) = \frac{D_{\mathbf{q}\omega}}{(\omega - C_{\mathbf{q}\omega})(\omega + C_{\mathbf{q}-\omega}) + D_{\mathbf{q}\omega}^2}; \qquad (G.75)$$

$$C_{\mathbf{q}\omega} = S(J_{\mathbf{Q}+\mathbf{q},\omega}^{\mathrm{tot}} + J_{\mathbf{q}\omega}^{\mathrm{tot}} - 2J_{\mathbf{Q}0}^{\mathrm{tot}}) +$$
$$+\sum_{\mathbf{p}}[C_{\mathbf{p}}\Phi_{\mathbf{pq}\omega} - (C_{\mathbf{p}} - D_{\mathbf{p}})\Phi_{\mathbf{p}00} + \phi_{\mathbf{pq}\omega}^{+} + \phi_{\mathbf{pq}\omega}^{-}] + g_{\mathbf{q}}, \qquad (G.76)$$

$$D_{\mathbf{q}\omega} = D_{\mathbf{q}-\omega} = S(J_{\mathbf{q}\omega}^{\mathrm{tot}} - J_{\mathbf{Q}+\mathbf{q},\omega}^{\mathrm{tot}}) + \sum_{\mathbf{p}} D_{\mathbf{p}}\Phi_{\mathbf{pq}\omega} + h_{\mathbf{q}},$$

where the s–d exchange contributions of the first order in $1/2S$ correspond to the RKKY approximation

$$J_{\mathbf{q}\omega}^{\mathrm{tot}} = J_{\mathbf{q}} + J_{\mathbf{q}}^{\mathrm{RKKY}}(\omega) = J_{\mathbf{q}} + I^2 \sum_{\mathbf{k}} \frac{n_{\mathbf{k}} - n_{\mathbf{k-q}}}{\omega + t_{\mathbf{k}} - t_{\mathbf{k-q}}}, \tag{G.77}$$

$J_{\mathbf{q}}^{\mathrm{RKKY}}(\omega)$ being the ω-dependent Fourier transform of the integral of indirect exchange interaction via conduction electrons (cf.(K.1)). The function Φ, which determines the second-order corrections, is given by

$$\Phi_{\mathbf{pq}\omega} = (\phi_{\mathbf{pq}\omega}^{+} - \phi_{\mathbf{pq}\omega}^{-})/\omega_{\mathbf{p}}, \tag{G.78}$$

$$\phi_{\mathbf{pq}\omega}^{\pm} = I^2 \sum_{\mathbf{k}} \frac{n_{\mathbf{k}}(1 - n_{\mathbf{k+p-q}}) + N_{\mathrm{B}}(\pm\omega_{\mathbf{p}})(n_{\mathbf{k}} - n_{\mathbf{k+p-q}})}{\omega + t_{\mathbf{k}} - t_{\mathbf{k+p-q}} \mp \omega_{\mathbf{p}}},$$

(note that $\phi^{+}_{\mathbf{pq}\omega} = -\phi^{-}_{\mathbf{pq}-\omega}$, so that the corrections of order of $(1/2S)^2$ violate the equality $C_{\mathbf{q}\omega} = C_{\mathbf{q}-\omega}$), $\omega_{\mathbf{p}}$ is the magnon frequency to zeroth order in I and $1/2S$. We have taken into account in (G.76) the equations of motion of the type

$$(\omega - t_{\mathbf{k}} + t_{\mathbf{k+q}})\langle\langle c_{\mathbf{k+q}\uparrow}^{\dagger} c_{\mathbf{k}\uparrow} | b_{\mathbf{q}}^{\dagger}\rangle\rangle_{\omega} = I\left(\frac{S}{2}\right)^{1/2} (n_{\mathbf{k+q}} - n_{\mathbf{k}})\langle\langle b_{\mathbf{q}} | b_{\mathbf{q}}^{\dagger}\rangle\rangle_{\omega} \tag{G.79}$$

and the expressions for the static correlation functions that occurred in the equations of motion

$$I\sum_{\mathbf{k}} \langle c_{\mathbf{k-Q}\downarrow}^{\dagger} c_{\mathbf{k}\uparrow}\rangle = -S(J_{\mathbf{Q}0}^{\mathrm{tot}} - J_{\mathbf{Q}}); \tag{G.80}$$

$$I\sum_{\mathbf{k}} \langle b_{\mathbf{p}}^{\dagger}(c_{\mathbf{k-p}\uparrow}^{\dagger} c_{\mathbf{k}\uparrow} - c_{\mathbf{k-p}\downarrow}^{\dagger} c_{\mathbf{k}\downarrow})\rangle = -(2S)^{1/2}(C_{\mathbf{p}} - D_{\mathbf{p}})\Phi_{\mathbf{p}00}. \tag{G.81}$$

These are obtained by calculating the corresponding retarded Green's functions and using the spectral representation (E.8). The functions

$$g_{\mathbf{q}} = \sum_{\mathbf{p}} [(2J_{\mathbf{Q}} + 2J_{\mathbf{q-p}} - 2J_{\mathbf{p}} - J_{\mathbf{Q+q}} - J_{\mathbf{q}})\langle b_{\mathbf{p}}^{\dagger} b_{\mathbf{p}}\rangle - 2J_{\mathbf{p}}\langle b_{-\mathbf{p}} b_{\mathbf{p}}\rangle],$$

$$h_{\mathbf{q}} = \sum_{\mathbf{p}} [(J_{\mathbf{Q+q}} - J_{\mathbf{q}})\langle b_{\mathbf{p}}^{\dagger} b_{\mathbf{p}}\rangle - 2J_{\mathbf{q-p}}\langle b_{-\mathbf{p}} b_{\mathbf{p}}\rangle] \tag{G.82}$$

describe the 'direct' magnon–magnon interaction, The s–d exchange contributions to the averages in (G.82) are obtained by using (G.74), (G.75) to the first order in $1/2S$ and the spectral representation

$$\left\{\begin{matrix} \langle b_{\mathbf{q}}^{\dagger} b_{\mathbf{q}}\rangle \\ \langle b_{\mathbf{q}}^{\dagger} b_{-\mathbf{q}}^{\dagger}\rangle \end{matrix}\right\} = -\frac{1}{\pi}\int_{-\infty}^{\infty} \mathrm{Im}\left\{\begin{matrix} \Gamma_{\mathbf{q}}(\omega) \\ \overline{\Gamma}_{\mathbf{q}}(\omega) \end{matrix}\right\} N_{\mathrm{B}}(\omega)d\omega. \tag{G.83}$$

The energy denominators in (G.77), (G.78) do not take into account the band splitting which comes from AFM ordering. At the same time, it is important to separate the contributions of the transitions within and between AFM subbands. Such a separation may be performed by taking into account the AFM splitting $\Delta = 2|I|S$ (S is the sublattice magnetization) in the zero-order approximation, e.g. within perturbation theory in $1/2S$. However, the corresponding expressions are very cumbersome (see the end of this Appendix). Therefore we use the simple perturbation theory in I bearing in mind that the transitions between AFM subbands correspond to the electron quasimomentum transfer $\mathbf{q} \sim \mathbf{Q}$. Generally speaking, the intersubband contributions to the spectrum characteristics and thermodynamic properties are more singular, but in fact they should be cut off because of the AFM splitting. The corresponding threshold value of the magnon quasimomentum transfer is estimated as

$$\min|\mathbf{q} - \mathbf{Q}| = q_0 = \Delta/v_F$$

(v_F is the electron velocity at the Fermi level). This quantity determines a characteristic temperature and energy scale

$$T^* = \omega(q_0) = cq_0 \sim (\Delta/v_F)T_N, \tag{G.84}$$

with c being the magnon velocity and the magnon spectrum is given by the pole of (G.74),

$$\omega_{\mathbf{q}}^2 = \Omega_{\mathbf{q}}^2(\omega_{\mathbf{q}}) = C_{\mathbf{q}}^2(\omega_{\mathbf{q}}) - D_{\mathbf{q}}^2(\omega_{\mathbf{q}}).$$

The dependence $J_{\mathbf{q}}^{\mathrm{RKKY}}(\omega)$, which is lost in the standard method of canonical transformation [265], is important at calculating the magnon damping. The spin-wave damping owing to one-magnon decay processes, which is determined by imaginary part of (G.77), reads at small q

$$\gamma_{\mathbf{q}}^{(1)} = \pi S\left[\frac{A}{L}\omega_{\mathbf{q}} + B\,\psi(q)\right], \tag{G.85}$$

where $L = 2S(J_0 - J_{\mathbf{Q}})$ the function ψ describes entering the 'Stoner continuum', $\psi(q < q_0) = 0$, $\psi(q \gg q_0) = 1$,

$$A = cI^2 \lim_{\mathbf{q}\to 0} q \sum_{\mathbf{k}} \delta(t_{\mathbf{k}})\delta(t_{\mathbf{k}-\mathbf{q}}), \tag{G.86}$$

$$B = LI^2 \sum_{\mathbf{k}} \delta(t_{\mathbf{k}})\delta(t_{\mathbf{k}-\mathbf{Q}}), \tag{G.87}$$

$t_{\mathbf{k}}$ being referred to the Fermi level. Generally speaking, A depends on the direction of the vector $\mathbf{q}$. For an isotropic electron spectrum one has

$$A = cI^2 v_0 \left\{ 4\pi^2 \, | k^{-1} \partial t_{\mathbf{k}} / \partial k |^2_{k=k_F} \right\}^{-1}, \tag{G.88}$$

where v_0 is the lattice cell volume.

One can see that the one-magnon damping (G.85) is finite at arbitrarily small q (in contrast with the FM case), but becomes considerably larger when intersubband transitions begin to work ($q > q_0$). Thus the ratio $\gamma/\omega|_{\mathbf{q}\to 0}$ makes up about I^2S/W^2 and does not depend on wavevector and electron concentration [689]. The same situation, which is similar to the case of electron-phonon interaction, takes place for an itinerant antiferromagnet [690]. The linear dependence of damping on wavevector was observed, e.g., in the antiferromagnet $Mn_{0,9}Cu_{0,1}$ [691]. An account of relaxation of conduction electrons, which occurs at disordering, results in a change of the q-dependence. The calculation in such a situation [689] yields $\gamma \sim \omega^2$ at small q, which is in agreement with hydrodynamics.

Similar to (G.20), we obtain from (G.74), (G.77) and the spectral representation the logarithmic correction to magnon occupation numbers and sublattice magnetization (6.88), (6.89).

The damping owing to two-magnon scattering processes is determined by the imaginary part of the function (G.78). The intersubband transitions turn out to contain smaller powers of ω and T, but contribute at $\max(T, \omega_{\mathbf{q}}) > T^*$. only. Using the identity

$$n(\varepsilon)[1 - n(\varepsilon')] = N(\varepsilon - \varepsilon')[n(\varepsilon') - n(\varepsilon)]$$

and expanding in $\omega = \omega_{\mathbf{q}}$ and $\omega_{\mathbf{p}}$ we obtain from the pole of (G.77)

$$\gamma_{\mathbf{q}}^{(2)} = \frac{\pi}{2} I^2 \sum_{\mathbf{k},\mathbf{p}} \sum_{\alpha,\beta=\pm} \left(\frac{C_{\mathbf{q}} - \alpha D_{\mathbf{q}}}{\omega} \frac{C_{\mathbf{p}} + \alpha D_{\mathbf{p}}}{\omega_{\mathbf{p}}} - \beta \right) \times$$

$$\times (\omega - \beta\omega_{\mathbf{p}})[N_B(\omega_{\mathbf{p}}) - N_B(\omega_{\mathbf{p}} - \beta\omega)]\delta(t_{\mathbf{k}})\delta(t_{\mathbf{k}+\mathbf{p}-\mathbf{q}}). \tag{G.89}$$

Integration at $T \ll \omega$ with account of leading temperature corrections gives

$$\gamma_{\mathbf{q}}^{(2)} = \frac{v_0}{24\pi c^3} \left[\left(\frac{9}{5} A\omega + \tilde{B} \right) \omega^2 + 4\pi^2 (2A\omega + \tilde{B}) T^2 \right], \tag{G.90}$$

with $\tilde{B}(\omega \gg T^*) = B$, $\tilde{B}(\omega \ll T^*) = 0$. At $\omega \ll T$ we find

$$\gamma_{\mathbf{q}}^{(2)} = \frac{v_0}{2\pi c^3} \left[6\zeta(3) A T + \frac{\pi^2}{3} \tilde{B} \right] T^2, \tag{G.91}$$

with $\tilde{B}(T \gg T^*) = B$, $\tilde{B}(T \ll T^*) = 0$, $\zeta(3) \simeq 1.2$.

The non-analytic corrections to the sublattice magnetization and magnon frequency come from both inter- and intrasubband transitions. The intrasubband contribution to the magnon velocity reads [716]

$$(\delta c/c)_1 = -\frac{v_0}{3\pi c^3} A T^2 \ln\frac{\overline{\omega}}{T}. \tag{G.92}$$

For $T > T^*$ the intersubband contribution to the sublattice magnetization has the form

$$(\delta \overline{S}_e)_2 = -\frac{v_0}{\pi^2 c^3} SLBT^* \ln\frac{T}{T^*}. \tag{G.93}$$

The above results are valid also in the Hubbard model. In the case of small S the magnon damping plays an important role at calculating temperature dependences of magnetic and thermodynamic properties at not too low temperatures are determined by contribution of spin fluctuations with small $|\mathbf{q}-\mathbf{Q}|$ and, as well as for a ferromagnet, the magnon damping plays an important role. However, unlike (G.21), the damping at $\mathbf{q} \to \mathbf{Q}$ does not contain the factor of $|\mathbf{q}-\mathbf{Q}|^{-1}$ (However, such a dependence occurs in some q-region provided that the electron spectrum approximately satisfies the 'nesting' condition $t_{\mathbf{k}+\mathbf{Q}} = -t_{\mathbf{k}}$ in a large part of the Fermi surface.) Then we have to replace in the denominator of the Green's functions (G.74)

$$\frac{1}{\omega^2 - \Omega_{\mathbf{q}}^2(\omega)} \to \frac{A\omega}{\left[\omega^2 - \Omega_{\mathbf{q}}^2(\omega)\right]^2 + A^2\omega^2}, \tag{G.94}$$

which yields for a weak itinerant antiferromagnet

$$\delta \overline{S} \sim -\int \sum_{\mathbf{q}} \frac{A\omega}{\omega_{\mathbf{q}}^4 + A^2\omega^2} N_B(\omega) d\omega \sim -\left(\frac{T}{E_F}\right)^{3/2}. \tag{G.95}$$

Now we consider the electron and magnon spectrum of the Hubbard antiferromagnet with strong correlations. Occurrence of AFM ordering results in splitting the bare electron band into two Slater subbands [44] which are described by new electron operators [692, 693]

$$\begin{aligned} \alpha_{\mathbf{k}}^\dagger &= A_{\mathbf{k}} c_{\mathbf{k}+\mathbf{Q}/2\uparrow}^\dagger + B_{\mathbf{k}} c_{\mathbf{k}-\mathbf{Q}/2\downarrow}^\dagger, \\ \beta_{\mathbf{k}}^\dagger &= A_{\mathbf{k}} c_{\mathbf{k}-\mathbf{Q}/2\downarrow}^\dagger - B_{\mathbf{k}} c_{\mathbf{k}+\mathbf{Q}/2\uparrow}^\dagger, \end{aligned} \tag{G.96}$$

$$A_{\mathbf{k}}, B_{\mathbf{k}} = \frac{1}{2}\left(1 \mp \frac{\tau_{\mathbf{k}}}{E_{\mathbf{k}}}\right), \qquad E_{\mathbf{k}} = \left(\tau_{\mathbf{k}}^2 + U^2 \overline{S}^2\right)^{1/2},$$

$$\theta_{\mathbf{k}}, \tau_{\mathbf{k}} = \frac{1}{2}(t_{\mathbf{k}+\mathbf{Q}/2} \pm t_{\mathbf{k}-\mathbf{Q}/2}).$$

In the Hartree–Fock approximation, the transformation (G.96) reduces the Hubbard Hamiltonian to the diagonal form

$$\mathsf{H} = \sum_{\mathbf{k}} (E_{\mathbf{k}}^{\alpha} \alpha_{\mathbf{k}}^{\dagger} \alpha_{\mathbf{k}} + E_{\mathbf{k}}^{\beta} \beta_{\mathbf{k}}^{\dagger} \beta_{\mathbf{k}}), \tag{G.97}$$

where the one-particle energies are given by

$$E_{\mathbf{k}}^{\alpha,\beta} = \theta_{\mathbf{k}} \mp E_{\mathbf{k}}. \tag{G.98}$$

The quantity

$$\overline{S} = \sum_{\mathbf{k}} \langle c_{\mathbf{k}\uparrow}^{\dagger} c_{\mathbf{k}+\mathbf{Q}\downarrow} \rangle,$$

which determines the AFM splitting, satisfies the self-consistency equation

$$1 = \frac{U}{2} \sum_{\mathbf{k}} \frac{n_{\mathbf{k}\alpha} - n_{\mathbf{k}\beta}}{(\tau_{\mathbf{k}}^2 + U^2 \overline{S}^2)^{1/2}}. \tag{G.99}$$

If the Coulomb interaction is strong enough, the whole energy band is split, so that the gap occurs in the all directions. In particular, for one electron per atom, a metal-insulator transition takes place. Provided that the 'nesting' condition $t_{\mathbf{k}} - E_{\mathrm{F}} = E_{\mathrm{F}} - t_{\mathbf{k}+\mathbf{Q}}$ holds for a given vector $\mathbf{Q}$, the insulator state is favourable for arbitrarily small U, the gap being exponentially small [692].

To obtain the spin-wave corrections to the Hartree–Fock approximation we pass to local coordinate system for electron operators

$$d_{\mathbf{k}\sigma}^{\dagger} = \frac{1}{\sqrt{2}}(c_{\mathbf{k}+\mathbf{Q}/2\uparrow}^{\dagger} + \sigma c_{\mathbf{k}-\mathbf{Q}/2\downarrow}^{\dagger}) \tag{G.100}$$

and represent the Hubbard Hamiltonian in the form

$$\mathsf{H} = \sum_{\mathbf{k}\sigma} \left[\left(\theta_{\mathbf{k}} + \frac{U}{2} \right) d_{\mathbf{k}\sigma}^{\dagger} d_{\mathbf{k}\sigma} + \tau_{\mathbf{k}} d_{\mathbf{k}\sigma}^{\dagger} d_{\mathbf{k}-\sigma} \right] - \frac{U}{2} \sum_{\mathbf{q}} (S_{-\mathbf{q}}^{-} S_{\mathbf{q}}^{+} + S_{\mathbf{q}}^{+} S_{-\mathbf{q}}^{-}), \tag{G.101}$$

where

$$S_{\mathbf{q}}^{\sigma} = \sum_{\mathbf{k}} d_{\mathbf{k}\sigma}^{\dagger} d_{\mathbf{k}+\mathbf{q},-\sigma}, \qquad S_{\mathbf{q}}^{z} = \frac{1}{2} \sum_{\mathbf{k}\sigma} \sigma d_{\mathbf{k}\sigma}^{\dagger} d_{\mathbf{k}+\mathbf{q},\sigma}. \tag{G.102}$$

Calculating the transverse spin Green's function with antiferromagnetic gap being taken into account in the zero-order order approximation we obtain for the magnon spectrum [693]

$$\omega_{\mathbf{q}}^2 = \frac{U^4 \overline{S}^2}{4}\left[\sum_{\mathbf{k}} \frac{2}{E_{\mathbf{k}}}(n_{\mathbf{k}\alpha} - n_{\mathbf{k}\beta}) + \right.$$

$$+\left(1 - \frac{U^2 \overline{S}^2}{E_{\mathbf{k}} E_{\mathbf{k+q}}}\right)\left(\frac{n_{\mathbf{k}\alpha} - n_{\mathbf{k+q}\alpha}}{E_{\mathbf{k}}^{\alpha} - E_{\mathbf{k+q}}^{\alpha}} + \frac{n_{\mathbf{k}\beta} - n_{\mathbf{k+q}\beta}}{E_{\mathbf{k}}^{\beta} - E_{\mathbf{k+q}}^{\beta}}\right) +$$

$$\left. + 2\left(1 + \frac{U^2 \overline{S}^2}{E_{\mathbf{k}} E_{\mathbf{k+q}}}\right)\left(\frac{n_{\mathbf{k}\alpha} - n_{\mathbf{k+q}\beta}}{E_{\mathbf{k}}^{\alpha} - E_{\mathbf{k+q}}^{\beta}}\right)\right]^2 -$$

$$-\frac{U^4 \overline{S}^2}{4}\left[\sum_{\mathbf{k}} \frac{\tau_{\mathbf{k}} \tau_{\mathbf{k+q}}}{E_{\mathbf{k}} E_{\mathbf{k+q}}}\left(\frac{n_{\mathbf{k}\alpha} - n_{\mathbf{k+q}\alpha}}{E_{\mathbf{k}}^{\alpha} - E_{\mathbf{k+q}}^{\alpha}} + \frac{n_{\mathbf{k}\beta} - n_{\mathbf{k+q}\beta}}{E_{\mathbf{k}}^{\beta} - E_{\mathbf{k+q}}^{\beta}} - 2\frac{n_{\mathbf{k}\alpha} - n_{\mathbf{k+q}\beta}}{E_{\mathbf{k}}^{\alpha} - E_{\mathbf{k+q}}^{\beta}}\right)\right]^2. \quad \text{(G.103)}$$

For small U we may expand (G.103) to obtain the result (G.74)–(G.77) with $I \to U$. In the case of large U and half-filled conduction band we obtain the result (E.12) with

$$J_{\mathbf{q}} = -\frac{2}{U}\sum_{\mathbf{k}} t_{\mathbf{k}} t_{\mathbf{k+q}} \quad \text{(G.104)}$$

being the kinetic exchange integral (see Sect.5.1). Finally, in the case of $U \to \infty$ and non-integer band filling, where the current carriers result in the non-Heisenberg double exchange interaction, we obtain

$$\omega_{\mathbf{q}}^2 = \left\{J_{\mathbf{Q}}^{\mathrm{H}} - J_{\mathbf{q}}^{\mathrm{H}} + \sum_{\mathbf{k}}\left[(\Theta_{\mathbf{k+q}} - \Theta_{\mathbf{k}})n_{\mathbf{k}} + \tau_{\mathbf{k}}(\tau_{\mathbf{k+q}} + \tau_{\mathbf{k}})\frac{n_{\mathbf{k+q}} - n_{\mathbf{k}}}{\Theta_{\mathbf{k+q}} - \Theta_{\mathbf{k}}}\right]\right\} \times$$

$$\times\left\{\frac{1}{2}(J_{\mathbf{Q}}^{\mathrm{H}} - J_{\mathbf{Q+q}}^{\mathrm{H}} - J_{\mathbf{Q-q}}^{\mathrm{H}}) + \right.$$

$$\left. + \sum_{\mathbf{k}}\left[(\Theta_{\mathbf{k+q}} - \Theta_{\mathbf{k}})n_{\mathbf{k}} - \tau_{\mathbf{k}}(\tau_{\mathbf{k+q}} - \tau_{\mathbf{k}})\frac{n_{\mathbf{k+q}} - n_{\mathbf{k}}}{\Theta_{\mathbf{k+q}} - \Theta_{\mathbf{k}}}\right]\right\}, \quad \text{(G.105)}$$

where

$$n_{\mathbf{k}} = n_{\mathbf{k}\beta} = f(\Theta_{\mathbf{k}}), \qquad n_{\mathbf{k}\alpha} \equiv 1,$$

and the Heisenberg interaction J^{H} is introduced to stabilize the antiferromagnetic state.

The spin-wave corrections to electron spectrum are determined, as well as for the ferromagnet (cf.(G.13), (G.36), (G.60)) by the electron-magnon interaction amplitude

$$\frac{\delta E^i_{\mathbf{k}}}{\delta N_{\mathbf{q}}} = \frac{\delta \omega_{\mathbf{k}}}{\delta n_{\mathbf{k}i}} = \frac{\delta^2 \langle \mathrm{H}_{\mathrm{sw}} \rangle}{\delta N_{\mathbf{q}} \delta n_{\mathbf{k}i}}. \tag{G.106}$$

The corresponding correction to the free energy reads

$$\delta \mathcal{F}_{\mathrm{sw}} = -\frac{1}{3}\delta \langle \mathrm{H}_{\mathrm{sw}} \rangle = -\frac{1}{3}\sum_{\mathbf{q}} \omega_{\mathbf{q}} N_{\mathbf{q}} \sim -\left(\frac{T}{T_{\mathrm{N}}}\right)^4 \tag{G.107}$$

(the difference with (G.60) is due to the linear dispersion law of spin waves). The temperature dependence of the local moment is given by the Hellman–Feynman theorem:

$$\delta \langle \mathbf{S}_i^2 \rangle = -\frac{3}{2}\delta N_2 = -\frac{3}{2}\frac{\partial}{\partial U}\delta \mathcal{F}_{\mathrm{sw}} \sim T^4. \tag{G.108}$$

The considered Hartree–Fock-type approximation with fluctuation corrections yields the same results for electron and magnon spectra in the case of the s–d model (with the replacement $U \to I$). In particular, for $|I| \to \infty$ the electron spectrum with account of spin-wave and many-electron corrections reads [693]

$$E^{\alpha}_{\mathbf{k}} = -I(S + n_{\beta}), \qquad E^{\beta}_{\mathbf{k}} = -I(S + 1 - n_{\alpha}).$$

Thus the generalized Hatree–Fock approximation yields correct 'atomic' values $E = \pm IS$, $\pm I(S+1)$ (Appendix I) for integer values n_{α}, n_{β}.

Appendix H

THE HUBBARD MODEL WITH STRONG CORRELATIONS

The Hubbard model with a degenerate band is applicable in some cases for d-electrons in transition metals and their compounds. The corresponding Hamiltonian has the form

$$\mathsf{H} = \sum_{\mathbf{k}m\sigma} t_{\mathbf{k}} a^{\dagger}_{\mathbf{k}lm\sigma} a_{\mathbf{k}lm\sigma} + \mathsf{H}_{\text{int}}, \tag{H.1}$$

where $t_{\mathbf{k}}$ is the band energy. For simplicity, we do not take into account m-dependence of transfer integrals, i.e. we neglect crystal-field effects and retain in (C.31) the contribution with $\lambda = 0$ only. Such an approximation enables one to treat in the simplest situation effects of many-electron term structure in the electron spectrum. In the many-electron representation of Hubbard's operators (A.22) the interaction Hamiltonian takes the diagonal form

$$\mathsf{H}_{\text{int}} = \sum_{i\Gamma} E_{\Gamma} X_i(\Gamma, \Gamma), \tag{H.2}$$

where the energies E_{Γ} are given by (C.19) and do not depend on momentum projections. However, the treatment of the kinetic energy term becomes more difficult because of complicated commutation relations for X-operators (A.36).

Consider the one-electron Green's function. According to (A.31),

$$G_{\mathbf{k}\gamma}(E) = \langle\langle a_{\mathbf{k}\gamma} | a^{\dagger}_{\mathbf{k}\gamma} \rangle\rangle_E = \sum_{n\Gamma_n\Gamma_{n-1}} n^{1/2} G^{\Gamma_n}_{\Gamma_{n-1}} C^{\Gamma_n}_{\Gamma_{n-1},\gamma} \langle\langle X_{\mathbf{k}}(\Gamma_{n-1}, \Gamma_n) | a^{\dagger}_{\mathbf{k}\gamma} \rangle\rangle_E . \tag{H.3}$$

In the equation of motion for the Green's function in the right-hand side of (H.3) we perform the simplest decoupling which corresponds to the 'Hubbard-I' decoupling [28, 29, 31] at different lattice sites:

$$\begin{aligned}(E - E_{\Gamma_n} + E_{\Gamma_{n-1}}) \langle\langle X_{\mathbf{k}}(\Gamma_{n-1}, \Gamma_n) | a^{\dagger}_{\mathbf{k}\gamma} \rangle\rangle_E = \\ = n^{1/2} G^{\Gamma_n}_{\Gamma_{n-1}} C^{\Gamma_n}_{\Gamma_{n-1},\gamma} (N_{\Gamma_n} + N_{\Gamma_{n-1}}) [1 + t_{\mathbf{k}} G_{\mathbf{k}\gamma}(E)].\end{aligned}$$

Then we obtain

$$G_{\mathbf{k}\gamma}(E) = \frac{\Phi_\gamma(E)}{1 - t_\mathbf{k}\Phi_\gamma(E)}, \qquad \text{(H.4)}$$

where

$$\Phi_\gamma(E) = \sum_{n\Gamma_n\Gamma_{n-1}} n\left(G^{\Gamma_n}_{\Gamma_{n-1}} C^{\Gamma_n}_{\Gamma_{n-1},\gamma}\right)^2 \frac{N_{\Gamma_n} + N_{\Gamma_{n-1}}}{E - E_{\Gamma_n} + E_{\Gamma_{n-1}}}. \qquad \text{(H.5)}$$

Note that the expressions (H.4), (H.5) have the structure which is reminiscent of (F.8),(F.9). As well as in the Appendix F, our consideration may be easily generalized to include effects of single-site crystal field (see also [29]). If we use for E_Γ the approximation (C.20), the fractional parentage coefficients in (H.5) may be summed up and the dependence on ME quantum numbers L, S vanishes, which corresponds to the approximation [29].

In the absence of magnetic and orbital ordering the occupation numbers N_Γ in (H.5) do not depend on spin projections and we have

$$\Phi_\Gamma(E) = \sum_{n\Gamma_n\Gamma_{n-1}} \frac{n}{2[l]}([S_{n-1}][L_{n-1}])^{-1}\left(G^{\Gamma_n}_{\Gamma_{n-1}}\right)^2 \frac{N_{\Gamma_n} + N_{\Gamma_{n-1}}}{E - E_{\Gamma_n} + E_{\Gamma_{n-1}}}. \qquad \text{(H.6)}$$

The excitation spectrum is given by

$$1 - t_\mathbf{k}\Phi_\gamma(E) = 0. \qquad \text{(H.7)}$$

Thus the intersite electron transfer results in a smearing of each transition between atomic levels into an energy Hubbard subband. These subbands are separated by correlation gaps. In particular, for s-band we obtain the spectrum which contains in ferromagnetic region four subbands

$$E^{1,2}_{\mathbf{k}\sigma} = \frac{1}{2}\left\{t_\mathbf{k} + U \mp \left[(t_\mathbf{k} - U)^2 + 4t_\mathbf{k}U(N_{-\sigma} + N_2)\right]^{1/2}\right\}. \qquad \text{(H.8)}$$

For a more general model (C.23),

$$E^{1,2}_{\mathbf{k}\sigma} = \frac{1}{2}\left[\beta^{(00)}_\mathbf{k}(N_0 + N_\sigma) + \beta^{(22)}_\mathbf{k}(N_2 + N_{-\sigma}) + U\right] \mp$$
$$\mp \frac{1}{2}\left\{\left[\beta^{(00)}_\mathbf{k}(N_0 + N_\sigma) - \beta^{(22)}_\mathbf{k}(N_2 + N_{-\sigma}) - U\right]^2 + \right.$$
$$\left. + 4|\beta^{(02)}_\mathbf{k}|^2 (N_0 + N_\sigma)(N_2 + N_{-\sigma})\right\}^{1/2}. \qquad \text{(H.9)}$$

The expression (H.8), (H.9) may be also rewritten in terms of one-electron occupation numbers since

$$N_\sigma + N_2 = n_\sigma, \qquad N_\sigma + N_0 = 1 - n_{-\sigma}.$$

One can see that the dependence of the spectrum on the occupation numbers does not reduce to the constant shift of subbands, as it takes place in the Stoner approximation (G.3). The Hubbard-I spectrum has the most simple form in the case of large U where

$$E^1_{\mathbf{k}\sigma} = (1 - n_{-\sigma})t_{\mathbf{k}}, \qquad E^2_{\mathbf{k}\sigma} = t_{\mathbf{k}} n_{-\sigma} + U. \tag{H.10}$$

One can assume that some subbands are in fact ill-defined because of large damping. Indeed, it is demonstrated in Appendix J for the saturated ferromagnetic state that in higher-order approximations some energy denominators are replaced by resolvents, and the corresponding states have a non-quasiparticle nature. A similar situation takes place in the Hubbard-III approximation [30, 694, 695] where damping is finite and large at Fermi level (see (H.17), (J.24)).

Consider the Hubbard model with $U \to \infty$, $n = 1 - c < 1$ ($c = N_0 = n_0$ is the hole concentration, $N_2 = 0$) with inclusion of the external magnetic field (4.96). In the 'Hubbard-I approximation' we obtain the Green's function

$$\langle\langle X_{\mathbf{k}}(\sigma 0) | X_{-\mathbf{k}}(0\sigma)\rangle\rangle_E = \frac{c + n_\sigma}{E - \tau_{\mathbf{k}\sigma} - \sigma H/2}, \tag{H.11}$$

$$\tau_{\mathbf{k}\sigma} = (c + n_\sigma)\varepsilon_{\mathbf{k}} = [(1+c)/2 + \sigma\langle S^z\rangle]\varepsilon_{\mathbf{k}},$$

and, with the use of the spectral representation (E.18), the corresponding expressions for the occupation numbers

$$n_{\mathbf{k}\sigma} \equiv \langle X_{-\mathbf{k}}(0\sigma) X_{\mathbf{k}}(\sigma 0)\rangle = \left(\frac{1+c}{2} + \sigma\langle S^z\rangle\right) f\left(\tau_{\mathbf{k}\sigma} + \frac{1}{2}h\sigma\right). \tag{H.12}$$

In principle, the equations (H.4), (H.5) may be used to investigate the magnetic ordering in the Hubbard model. However, the Hubbard-I approximation is hardly satisfactory in this problem since it is difficult to formulate a reasonable criterion of ferromagnetism by direct using the expressions for one-electron Green's functions like (H.4), (H.11). The first attempt of this kind was made by Hubbard [28], who found no magnetic solutions for simple bare densities of states (however, the situation may change considerably in the case of degenerate d-bands [696]). The X-operator approach clarifies the causes of this failure. In particular, one can see that the approximation (H.11) violates the kinematical requirements (A.25) since it is impossible to satisfy at $\langle S^z\rangle \neq 0$ the relation

$$\sum_{\mathbf{k}} n_{\mathbf{k}\sigma} = \langle X(00)\rangle = c \tag{H.13}$$

for both spin projections σ. Besides that, as discussed above, such Stoner-like approaches do not describe the formation of LMM and are physically unsatisfactory. Therefore, we use in Sect 4.5 the approach which is based on the spin Green's function

$$G_{\mathbf{q}}(\omega) = \langle\langle X_{\mathbf{q}}(+-) | X_{-\mathbf{q}}(-+)\rangle\rangle_\omega = \langle\langle S^+_{\mathbf{q}} | S^-_{-\mathbf{q}}\rangle\rangle_\omega. \tag{H.14}$$

To calculate this we write down the sequence of equations of motion

$$(\omega - H)G_{\mathbf{q}}(\omega) = n_\uparrow - n_\downarrow + \sum_{\mathbf{k}}(\varepsilon_{\mathbf{k}-\mathbf{q}} - \varepsilon_{\mathbf{k}})\langle\langle X_{\mathbf{q}-\mathbf{k}}(0-)X_{\mathbf{k}}(+0) | X_{-\mathbf{q}}(-+)\rangle\rangle_\omega, \tag{H.15}$$

$$(\omega - \tau_{\mathbf{k}\uparrow} + \tau_{\mathbf{k}-\mathbf{q}\uparrow} - H)\langle\langle X_{\mathbf{q}-\mathbf{k}}(0-)X_{\mathbf{k}}(+0) | X_{-\mathbf{q}}(-+)\rangle\rangle_\omega = \\ = n_{\mathbf{k}\uparrow} - n_{\mathbf{k}-\mathbf{q}\downarrow} + (\varepsilon_{\mathbf{k}-\mathbf{q}} n_{\mathbf{k}-\mathbf{q}\downarrow} - \varepsilon_{\mathbf{k}} n_{\mathbf{k}\uparrow})G_{\mathbf{q}}(\omega),$$

where we have performed the simplest decoupling, which corresponds to neglecting fluctuations of hole occupation numbers. Substituting (H.15) into (H.14) we obtain the expression (4.97).

Fluctuation correction to the electron spectrum (H.4) for a ferromagnet with s-band were investigated in [337, 338] within the $1/z$-expansion, z being the nearest-neighbour number. They are expressed in terms of one-particle occupation numbers and spin and charge correlation functions. Unlike the Hubbard-I approximation, such expression enable one to obtain correctly the magnetic phase diagram and describe the saturated and non-saturated ferromagnetic state [729]. The first critical concentration of current carriers, which corresponds to the instability of the saturated ferromagnetism, makes up for various lattices about 30%.

For paramagnetic phase, the gap in the spectrum (H.8) persists for arbitrarily small U. To describe the metal–insulator transition which takes place at $U \sim W$ (with W being the bandwidth) more complicated self-consistent approximations for the one-electron Green's functions were used. First description of this type was proposed by Hubbard [30], and a more simple approximation was used by Zaitsev [697].

The Hubbard-III expression for the one-electron Green's function in the case of half-filled band may be represented in the form [695]

$$G_{\mathbf{k}}(E) = [E - t_{\mathbf{k}} - \Sigma(E)]^{-1}, \tag{H.16}$$

the electron self-energy being determined in a self-consistent way in terms of the exact resolvent

$$\Sigma(E) = \frac{U^2}{16\Psi}R(E) \Bigg/ \left[1 + \Sigma(E)R(E) + ER(E)\left(\frac{1}{4\Psi} - 1\right)\right], \tag{H.17}$$

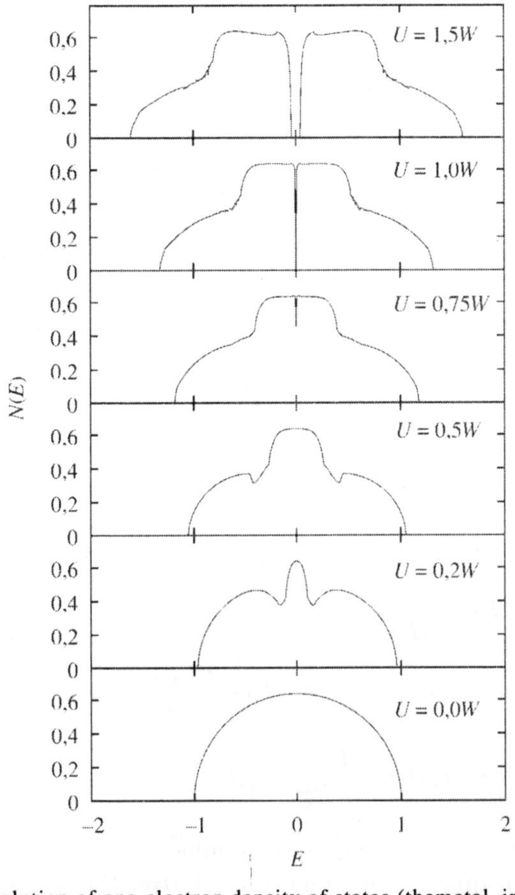

Fig. H.1. Evolution of one-electron density of states (themetal–insulator transition) with increasing interaction parameter in the Hubbard model for the semielliptic bare DOS [730].

$$R(E) = \sum_{\mathbf{k}} G_{\mathbf{k}}(E),$$

where $\Psi = 3/4$. The expression (H.17) holds also for the classical ($S \to \infty$) s–d exchange model (see Appendix I) if we put $\Psi = 1/4$, $U \to |IS|$. Then (H.17) is simplified and coincides with the coherent potential approximation (CPA) result in the disordered alloy theory [435]. Evolution of electron spectrum vs. interaction parameter is shown in Fig.H1.

Some shortcomings of the approximations [30, 697] (violence of analytical properties of the Green's functions and inconsistent description of thermodynamic quantities) are discussed in [694, 695] from the point of view of the $1/z$-expansion. Construction of a correct physical picture of the Mott–Hubbard transition has been up to now a serious physical problem. Last time, the large-d (d is the space dimensionality) approximation is widely used in this problem [705]. This approximation yields a three-peak structure of the density of states, including a Kondo peak at the Fermi level. Such approaches yield sometimes two phase transitions: at $U > U_{c1}$ the Fermi-liquid picture breaks, and at $U > U_{c2} > U_{c1}$ the system passes into the insulator state. An account of the Fermi excitations within the $1/z$-expansion was performed in Ref.[730]. Using the locator representation for the Green's functions

$$G_{\mathbf{k}}(E) = \left[F(E) - t_{\mathbf{k}}\right]^{-1} \tag{H.18}$$

enabled one to obtain correct analytical properties and reproduce the three-peak structure. The self-consistent result for the locator $F(E)$ reads

$$F(E) = \frac{b(E)}{a(E)},$$

$$a(E) = 1 + \frac{3}{4}\frac{U^2}{E^2}\sum_{\mathbf{q}} t_{\mathbf{q}} G_{\mathbf{q}}(E) + \frac{2U}{E}\sum_{\mathbf{q}} t_{\mathbf{q}} G_{\mathbf{q}}(E) n_{\mathbf{q}},$$

$$b(E) = F_0(E) + \frac{2U}{E}\sum_{\mathbf{q}} t_{\mathbf{q}}^2 G_{\mathbf{q}}(E) n_{\mathbf{q}},$$

where $n_{\mathbf{q}}$ are the exact occupation numbers.

We have seen that the one-electron (Hartree–Fock) and Hubbard approaches yield essentially different results for electron spectrum. The failure of the one-electron approach in the case of large U may be demonstrated by treatment of the case of small electron concentrations n [353]. Consider the expansion of the one-electron Green's function in the electron occupation numbers. In the one-electron representation ($f_{\mathbf{k}} = \langle a^\dagger_{\mathbf{k}\sigma} a_{\mathbf{k}\sigma}\rangle$) we obtain the equation of motion

$$(E - t_{\mathbf{k}})\langle\langle a_{\mathbf{k}\uparrow} | a^\dagger_{\mathbf{k}\uparrow}\rangle\rangle_E = 1 + U\sum_{\mathbf{k}_1\mathbf{k}_2} L_{\mathbf{k}}(\mathbf{k}_1\mathbf{k}_2 E),$$

$$L_{\mathbf{k}}(\mathbf{k}_1\mathbf{k}_2 E) = \langle\langle a^\dagger_{\mathbf{k}_1\downarrow} a_{\mathbf{k}_2\downarrow} a_{\mathbf{k}+\mathbf{k}_1-\mathbf{k}_2\uparrow} | a^\dagger_{\mathbf{k}\uparrow}\rangle\rangle_E. \tag{H.19}$$

To lowest-order in $f_{\mathbf{k}}$ we obtain the closed integral equation

$$(E - t_{\mathbf{k}+\mathbf{k}_1-\mathbf{k}_2} + t_{\mathbf{k}_1} - t_{\mathbf{k}_2})L_{\mathbf{k}}(\mathbf{k}_1,\mathbf{k}_2,E) = \delta_{\mathbf{k}_1\mathbf{k}_2} f_{\mathbf{k}} + U(1-f_{\mathbf{k}_2})\sum_{\mathbf{p}} L_{\mathbf{k}}(\mathbf{k},\mathbf{p},E).$$

Solving this we derive the expression for the electron self-energy

$$\Sigma_{\mathbf{k}}(E) = U\sum_{\mathbf{k}_1} f_{\mathbf{k}_1}\left[1 - U\sum_{\mathbf{k}_2}\frac{(1-f_{\mathbf{k}_2})(1-f_{\mathbf{k}+\mathbf{k}_1-\mathbf{k}_2})}{E - t_{\mathbf{k}+\mathbf{k}_1-\mathbf{k}_2} + t_{\mathbf{k}_1} - t_{\mathbf{k}_2}}\right]^{-1}, \tag{H.20}$$

which is finite at $U \to \infty$ [348]. At the same time, according to (E.18), (H.19) the number of doubles

$$N_2 = \langle a_{i\uparrow}^\dagger a_{i\uparrow} a_{i\downarrow}^\dagger a_{i\downarrow}\rangle = -\frac{1}{\pi}\int \mathrm{Im}\sum_{\mathbf{k}\mathbf{k}_1\mathbf{k}_2} L_{\mathbf{k}}(\mathbf{k}_1\mathbf{k}_2 E) f(E) dE =$$

$$= -\frac{1}{\pi U}\int \mathrm{Im}\sum_{\mathbf{k}}\frac{\Sigma_{\mathbf{k}}(E)}{E - t_{\mathbf{k}}} f(E) dE \tag{H.21}$$

behaves as $1/U$ in this limit. Then the Hellman-Feynman theorem $N_2 = \partial\mathcal{E}/\partial U$ yields a divergence of the ground-state energy $\mathcal{E}$

$$\mathcal{E}(U) - \mathcal{E}(0) = \int_0^\infty N_2(U) dU \sim \ln U. \tag{H.22}$$

This divergence indicates formation of the Hubbard subbands and inadequacy of the one-electron picture at large U. On the other hand, calculation in the ME representation yields [353]

$$\langle\langle X_{\mathbf{k}}(\sigma 2) | X_{-\mathbf{k}}(2\sigma)\rangle\rangle_E \simeq \sum_{\mathbf{k}_1} f_{\mathbf{k}_1}\left[E + t_{\mathbf{k}_1} - U - \sum_{\mathbf{k}_2}\frac{(t_{\mathbf{k}+\mathbf{k}_1-\mathbf{k}_2} + t_{\mathbf{k}_2})^2}{E - t_{\mathbf{k}+\mathbf{k}_1-\mathbf{k}_2} + t_{\mathbf{k}_1} - t_{\mathbf{k}_2}}\right]^{-1}, \tag{H.23}$$

so that

$$N_2 = -\frac{1}{\pi}\mathrm{Im}\sum_{\mathbf{k}}\int f(E)\,\mathrm{Im}\,\langle\langle X_{\mathbf{k}}(\sigma 2) | X_{-\mathbf{k}}(2\sigma)\rangle\rangle_E dE \simeq$$

$$\simeq \frac{1}{U^2}\sum_{\mathbf{k}\mathbf{k}_1\mathbf{k}_2}(t_{\mathbf{k}+\mathbf{k}_1-\mathbf{k}_2} + t_{\mathbf{k}_2})^2 f_{\mathbf{k}_1} f(t_{\mathbf{k}+\mathbf{k}_1-\mathbf{k}_2} - t_{\mathbf{k}_1} + t_{\mathbf{k}_2}) \simeq \frac{n^2}{U^2} t_{\min}. \tag{H.24}$$

Thus we obtain the correct asymptotic $N_2 \sim 1/U^2$.

Appendix I

NARROW-BAND *s–d* EXCHANGE MODEL AND *t–J* MODEL

At considering the electron transfer in narrow degenerate bands, one can use, besides the Hubbard model, the *s–d* exchange model with strong correlations. This model corresponds to the case where the current carriers do not belong to the same energy band where magnetic moments are formed. Such a situation takes place in some magnetic semiconductors and insulators [668].

Unlike the degenerate Hubbard model, the standard Hamiltonian of the *s–d* exchange model (G.2) does not include orbital degrees of freedom. In the case of large *s–d* exchange parameter $|I|$ it is convenient to pass to the atomic representation [698–700]. Substituting the values of the Clebsh–Gordan coefficients, corresponding to the coupling of momenta S and 1/2, we obtain the eigenfunctions of H_{sd}:

$$|M\rangle \equiv |SM\rangle|0\rangle, \qquad |M2\rangle \equiv |SM\rangle|2\rangle, \tag{I.1}$$

$$|\mu\pm\rangle = \left(\frac{S\pm\mu+1/2}{2S+1}\right)^{1/2} |S,\mu-\frac{1}{2}\rangle|\uparrow\rangle \pm \left(\frac{S\mp\mu+1/2}{2S+1}\right)^{1/2} |S,\mu+\frac{1}{2}\rangle|\downarrow\rangle, \tag{I.2}$$

where $|\mu\alpha\rangle$ are the singly-occupied states with the total on-site spin $S+\alpha/2$ and its projection μ. Then H_{sd} takes the diagonal form

$$\mathsf{H}_{sd} = -IS\sum_{\mu=-S-1/2}^{S+1/2}\sum_i X_i(\mu+,\mu+) + I(S+1)\sum_{\mu=-S+1/2}^{S-1/2}\sum_i X_i(\mu-,\mu-). \tag{I.3}$$

The one-electron operators are expressed in terms of X-operators as

$$c_{i\sigma}^\dagger = \sum_\alpha (g_{i\sigma\alpha}^\dagger + h_{i\sigma\alpha}^\dagger), \tag{I.4}$$

$$g_{i\sigma+}^\dagger = \sum_M \{(S+\sigma M+1)/(2S+1)\}^{1/2} X_i\left(M+\frac{\sigma}{2},+;M\right),$$

$$g_{i\sigma-}^\dagger = \sum_M \sigma\{(S-\sigma M)/(2S+1)\}^{1/2} X_i\left(M+\frac{\sigma}{2},-;M\right),$$

$$h^{\dagger}_{i\sigma+} = \sum_{M} \{(S+\sigma M)/(2S+1)\}^{1/2} X_i\left(M2; M-\frac{\sigma}{2},-\right),$$

$$h^{\dagger}_{i\sigma-} = \sum_{M} \sigma\{(S-\sigma M+1)/(2S+1)\}^{1/2} X_i\left(M2; M-\frac{\sigma}{2},+\right).$$

In the limit $I \to \alpha\infty$ and for conduction electron concentration $n<1$ one has to retain in (I.4) only the terms with $g_{i\alpha}$ and omit the Hamiltonian H_{sd}, which yields a constant energy shift, to obtain

$$\mathsf{H} = \sum_{\mathbf{k}\sigma} t_{\mathbf{k}} g^{\dagger}_{\mathbf{k}\sigma\alpha} g_{\mathbf{k}\sigma\alpha} + \mathsf{H}_d, \quad \alpha = \operatorname{sign} I. \tag{I.5}$$

For $n>1$, we have to retain only terms with $h_{i\alpha}$ and pass to the 'hole' representation by introducing new localized spins $\tilde{S} = S \pm 1/2$. Then the Hamiltonian takes the same form (I.6) with the replacement [700]

$$t_{\mathbf{k}} \to -t_{\mathbf{k}}([\tilde{S}]/[S]). \tag{I.6}$$

At theoretical consideration of highly-correlated compounds, e.g. copper-oxide high-T_c superconductors the *t–J* model (the *s*-band Hubbard model with $U \to \infty$ and Heisenberg exchange included) is now widely used. Its Hamiltonian in ME representation reads

$$\mathsf{H} = -\sum_{ij\sigma} t_{ij} X_i(0\sigma) X_j(\sigma 0) -$$

$$-\sum_{ij} J_{ij}\left\{X_i(+-)X_j(-+) + \frac{1}{4}[X_i(++) - X_i(--)][X_j(++) - X_j(--)]\right\}. \tag{I.7}$$

At derivation of the *t–J* model from the large-*U* Hubbard model, $J = -4t^2/U$ is the antiferromagnetic kinetic exchange integral. However, it is convenient sometimes *J* as an independent variable. In particular, one treats sometimes the 'supersymmetrical' case with $t = J$ [701] which permits to use non-trivial mathematical methods.

One can see that the model (I.7) is a particular case of the *s–d* exchange model, corresponding to $I \to -\infty$, $S = 1/2$, $t_{\mathbf{k}}$ in (I.5) being replaced by $2t_{\mathbf{k}}$ (the factor of 2 occurs because of equivalence of both electrons in the Hubbard model). The *s–d* model with arbitrary *S* is sometimes more convenient since this enables one to use at calculations, besides the small parameter $1/z$ (*z* is the nearest-neighbour number), the quasiclassical parameter $1/2S$. The quasiclassical *s–d* model with $S \gg 1$ in the atomic representation was used to investigate the metal-insulator transition [695] (Appendix H).

The Hamiltonian (I.5), similar to (C.33), may be expressed in terms of the Fermi and localized spin operators. To this end, we pick out a conduction electron operator from X-operators with the use of (A.21), (A.11):

$$g^{\dagger}_{i\sigma\alpha} = \sum_{\sigma'} c^{\dagger}_{i\sigma'}(1-n_{i,-\sigma'})\left[P_{\alpha}\delta_{\sigma\sigma'} + \frac{\alpha}{2S+1}(\mathbf{S}_i\sigma_{\sigma'\sigma})\right], \tag{I.8}$$

where

$$P_{+} = \frac{S+1}{2S+1}, \qquad P_{-} = \frac{S}{2S+1}. \tag{I.9}$$

This result was obtained by Kubo and Ohata [702] by a canonical transformation. Using properties of the Pauli matrices we get

$$\mathsf{H} = \sum_{ij\sigma\sigma'} t_{ij}\left\{\left[P_{\alpha}^2 + \frac{\mathbf{S}_i\mathbf{S}_j}{(2S+1)^2}\right]\delta_{\sigma\sigma'} + \frac{\alpha}{(2S+1)^2}P_{\alpha}(\mathbf{S}_i+\mathbf{S}_j)\sigma_{\sigma\sigma'} + \right.$$
$$\left. + \frac{i}{(2S+1)^2}\sigma_{\sigma\sigma'}[\mathbf{S}_i\times\mathbf{S}_j]\right\} c^{\dagger}_{i\sigma}(1-n_{i,-\sigma})(1-n_{j,-\sigma'})c_{j\sigma'} + \mathsf{H}_d. \tag{I.10}$$

The terms with vector products (cf. (K.8)) describe anisotropic electron scattering and may be important at considering transport phenomena in narrow bands, e.g., the extraordinary Hall effect. The Hamiltonian in the form (I.10) may be also useful at treating the states with the anomalous 'chiral' order parameters, which are now extensively investigated within the two-dimensional Heisenberg and t–J models (see, e.g., [703]); however, problems which are connected with projection onto physical space should be taken into account.

The Hamiltonian (I.5) is more convenient at considering simple approximations within $1/z$-expansion [694]. Performing the simplest decoupling we obtain the electron spectrum in the ferromagnetic phase

$$E_{\mathbf{k}\sigma\alpha} = \left(P_{\alpha} - \frac{\alpha\sigma}{2S+1}\langle S^z\rangle\right)t_{\mathbf{k}}. \tag{I.11}$$

This expression demonstrates a strong dependence of electron spectrum on magnetic ordering. The rigorous expression for spin-down Green's function at $\alpha=+$, $T=0$ [699] has the non-quasiparticle form

$$\langle\langle g_{\mathbf{k}\downarrow+} | g^{\dagger}_{\mathbf{k}\downarrow+}\rangle\rangle_E = \left[E - t_{\mathbf{k}} + 2S\left(\sum_{\mathbf{q}}\frac{1}{E-t_{\mathbf{q}}}\right)^{-1}\right]^{-1}. \tag{I.12}$$

The magnon spectrum in the model (I.5) was calculated in [79, 80, 83]. The result to the leading order in $1/z$ reads

$$\omega_{\mathbf{q}} = \frac{1}{2S}\sum_{\mathbf{k}}(t_{\mathbf{k}-\mathbf{q}} - t_{\mathbf{k}})f(t_{\mathbf{k}}), \quad \alpha = +,$$
$$\omega_{\mathbf{q}} = \frac{1}{2S+1}\sum_{\mathbf{k}}(t_{\mathbf{k}-\mathbf{q}} - t_{\mathbf{k}})f\left(\frac{2S}{2S+1}t_{\mathbf{k}}\right), \quad \alpha = -. \tag{I.13}$$

In an antiferromagnet with the spiral magnetic structure, corresponding to the wavevector **Q**, we have to pass in the *s–d* Hamiltonian to the local coordinate system by using (E.8), (G.90). Then, passing from the operators $d^{\dagger}_{i\sigma}$ to the ME operators, we obtain instead of (I.5)

$$\mathsf{H} = \sum_{\mathbf{k}\sigma}(\theta_{\mathbf{k}} g_{\mathbf{k}\sigma\alpha} g_{\mathbf{k}\sigma\alpha} + \tau_{\mathbf{k}} g_{\mathbf{k}\sigma\alpha} g_{\mathbf{k},-\sigma,\alpha}) + \mathcal{H}_d, \tag{I.14}$$

$$\tau_{\mathbf{k}} = \frac{1}{2}(t_{\mathbf{k}+\mathbf{Q}/2} - t_{\mathbf{k}-\mathbf{Q}/2}), \qquad \theta_{\mathbf{k}} = \frac{1}{2}(t_{\mathbf{k}+\mathbf{Q}/2} + t_{\mathbf{k}-\mathbf{Q}/2}).$$

Performing the 'Hubbard-I' decoupling, we obtain for the electron spectrum [687]

$$E^{1,2}_{\mathbf{k}} = P_{\alpha}\theta_{\mathbf{k}} \pm \left\{\left(\frac{\bar{S}}{2S+1}\theta_{\mathbf{k}}\right)^2 + \left[P^2_{\alpha} - \left(\frac{\bar{S}}{2S+1}\right)^2\right]\tau_{\mathbf{k}}\right\}^{1/2}, \tag{I.15}$$

with $\bar{S}$ being the sublattice magnetization. In the nearest-neighbor approximation ($\theta_{\mathbf{q}} = 0$) for $I > 0$ the band at $T = 0$ is narrowed by $(2S+1)^{1/2}$ times. At the same same time, for $I < 0$ (and also in the *t–J* model) the electron may not pass to neighbour sites in the approximation under consideration, and its motion is possible due to quantum effects only. This problem is discussed in Sect.6.7.

Fluctuation corrections to the spectrum (I.15) are discussed in [687, 620]. The result for the magnon spectrum obtained in the ME representation turn out to coincide with that of the generalized Hartree–Fock approximation (G.95).

Appendix J

ELECTRON STATES AND SPIN WAVES IN THE NARROW-BAND HUBBARD FERROMAGNET

The electron and magnon spectra of a Hubbard ferromagnet with strong correlations ($U \to \infty$), which is described by the Hamiltonian

$$\mathcal{H} = \sum_{\mathbf{k}\sigma} \varepsilon_{\mathbf{k}} X_{-\mathbf{k}}(0\sigma) X_{\mathbf{k}}(\sigma 0), \tag{J.1}$$

may be investigated rigorously in the case of small concentrations of holes $c = 1 - N_e/N$ (almost half-filled band) and low temperatures. This is formally achieved by expansion in hole and magnon occupation numbers [333, 699, 700]. Consider the one-particle Green's functions

$$G_{\mathbf{k}\sigma}(E) = \langle\langle X_{\mathbf{k}}(\sigma 0) \mid X_{-\mathbf{k}}(0\sigma) \rangle\rangle_E . \tag{J.2}$$

Using the commutation relation

$$[X_{\mathbf{k}}(+0), \mathcal{H}] = \sum_{\mathbf{p}} \varepsilon_{\mathbf{k}-\mathbf{p}} \{ [X_{\mathbf{p}}(00) + X_{\mathbf{p}}(++)] X_{\mathbf{k}-\mathbf{p}}(+0) + X_{\mathbf{p}}(+-) X_{\mathbf{k}-\mathbf{p}}(-0) \},$$

we write down for $\sigma = \uparrow$ the equation of motion (E.16a)

$$\begin{aligned}(E - \varepsilon_{\mathbf{k}}) G_{\mathbf{k}\uparrow}(E) &= 1 - n_{\downarrow} + \\ &+ \sum_{\mathbf{pq}} (\varepsilon_{\mathbf{k}-\mathbf{p}} - \varepsilon_{\mathbf{k}+\mathbf{q}-\mathbf{p}}) \langle\langle X_{-\mathbf{q}}(-+) X_{\mathbf{p}}(+-) X_{\mathbf{k}+\mathbf{q}-\mathbf{p}}(+0) \mid X_{-\mathbf{k}}(0+) \rangle\rangle_E .\end{aligned} \tag{J.3}$$

Here we have taken into account the kinematical relations

$$X_{\mathbf{p}}(++) + X_{\mathbf{p}}(--) = \delta_{\mathbf{p}0} - X_{\mathbf{p}}(--), \qquad X_{\mathbf{p}}(--) = \sum_{\mathbf{q}} X_{-\mathbf{q}}(-+) X_{\mathbf{p}+\mathbf{q}}(+-),$$

$$X_{\mathbf{k}-\mathbf{p}}(-0) = \sum_{\mathbf{q}} X_{-\mathbf{q}}(-+) X_{\mathbf{k}+\mathbf{q}-\mathbf{p}}(+0), \tag{J.5}$$

which follow from (A.28), (A.25), reduced the operator products to the 'normal form' where all the operators $X(-+)$ stand at the left, and retained only the terms, which are linear in spin deviations and neglected the terms which are proportional to the hole concentrations. Introducing the function

$$\varphi_{\mathbf{kqp}}(E) = \langle\langle X_{-\mathbf{q}}(-+)X_{\mathbf{p}}(+-)X_{\mathbf{k+q-p}}(+0) | X_{-\mathbf{k}}(0+)\rangle\rangle_E / [(E-\varepsilon_{\mathbf{k}})N_{\mathbf{q}}], \quad (J.6)$$

with

$$N_{\mathbf{q}} = \langle X_{-\mathbf{q}}(-+)X_{\mathbf{q}}(+-)\rangle,$$

we derive in the same way the closed integral equation

$$(E-\varepsilon_{\mathbf{k+q-p}})\varphi_{\mathbf{kqp}}(E) = (E-\varepsilon_{\mathbf{k}})(\delta_{\mathbf{pq}}-1) + \sum_r (\varepsilon_{\mathbf{k+q-p-r}} - \varepsilon_{\mathbf{k+q-r}})\varphi_{\mathbf{kqr}}(E), \quad (J.7)$$

which describes the hole-magnon scattering. Writing down the Dyson equation

$$G_{\mathbf{k}\uparrow}(E) = \frac{1-n_\downarrow}{E-\varepsilon_{\mathbf{k}}}\left[1 - \frac{1}{E-\varepsilon_{\mathbf{k}}}\Sigma_{\mathbf{k}}(E)\right] \simeq (1-n_\downarrow)\left[E-\varepsilon_{\mathbf{k}} - \Sigma_{\mathbf{k}}(E)\right]^{-1}, \quad (J.8)$$

we obtain for the self-energy

$$\Sigma_{\mathbf{k}}(E) = \sum_{\mathbf{pq}} (\varepsilon_{\mathbf{k-p}} - \varepsilon_{\mathbf{k+q-p}})\varphi_{\mathbf{kqp}}(E). \quad (J.9)$$

The magnon spectrum is obtained from the spin Green's function. The equation of motion for this has the form (cf.(H.15))

$$\omega G_{\mathbf{q}}(\omega) = 1 - c + \\ + \sum_{\mathbf{kp}} (\varepsilon_{\mathbf{k-p}} - \varepsilon_{\mathbf{k}})\langle\langle X_{-\mathbf{k}}(0+)X_{\mathbf{q-k+p}}(+-)X_{\mathbf{k}}(+0) | X_{-\mathbf{q}}(-+)\rangle\rangle_\omega. \quad (J.10)$$

Performing calculations at $T=0$ to lowest order in magnon occupation numbers

$$n_{\mathbf{k}} = \langle X_{-\mathbf{k}}(0+)X_{\mathbf{k}}(+0)\rangle = f(\varepsilon_{\mathbf{k}}),$$

we obtain the magnon self-energy. This turns out to be determined by the same function φ

$$G_{\mathbf{q}}(\omega) = (1-c)\left[\omega - \sum_{\mathbf{kp}} (\varepsilon_{\mathbf{k-p}} - \varepsilon_{\mathbf{k+q-p}})\varphi_{\mathbf{kqp}}(\omega + \varepsilon_{\mathbf{k}})\right]^{-1}. \quad (J.11)$$

To lowest order in the small parameter $1/z$ (each order in $1/z$ corresponds to a summation over a wavevector) we obtain for the temperature correction to the the electron spectrum and for the magnon frequency

$$\delta\varepsilon_{\mathbf{k}}(T) = \sum_{\mathbf{q}} (\varepsilon_{\mathbf{k-q}} - \varepsilon_{\mathbf{k}})N_{\mathbf{q}}, \quad (J.12a)$$

$$\omega_{\mathbf{q}} = \sum_{\mathbf{k}} (\varepsilon_{\mathbf{k-q}} - \varepsilon_{\mathbf{k}})n_{\mathbf{k}}. \quad (J.12b)$$

The equation (J.7) may be solved exactly for concrete lattices. For the simple cubic lattice in the nearest-neighbour approximation we have for the band bottom shift

$$\delta\varepsilon_0 = \kappa \frac{3\zeta(5/2)v_0}{32\pi^{3/2}m^*}\left(\frac{T}{D}\right)^{5/2} \tag{J.13}$$

and for the spin-wave frequency at small q

$$\omega_{\mathbf{q}} = Dq^2, \qquad D = \kappa c|t|, \tag{J.14}$$

where κ is expressed in terms of the lattice Green's function,

$$\kappa = \frac{1-A}{1+A} \approx 0.656,$$

$$A = \sum_{\mathbf{q}} \frac{\cos q_x}{3-\cos q_x - \cos q_y - \cos q_z} \approx 0.208.$$

It is convenient to calculate more complicated characteristics at finite temperatures with the use of the expansion in $1/z$ [700]. The magnon damping is given by (cf.(G.23)-(G.25))

$$\gamma_{\mathbf{q}} = \pi\sum_{\mathbf{kp}}(\varepsilon_{\mathbf{k-q}} - \varepsilon_{\mathbf{k}})^2[n_{\mathbf{k+q-p}}(1-n_{\mathbf{k}}) + \\ + (n_{\mathbf{k+q-p}} - n_{\mathbf{k}})N_{\mathbf{p}}]\delta(\omega + \varepsilon_{\mathbf{k}} - \varepsilon_{\mathbf{k+q-p}} - \omega_{\mathbf{p}}) \tag{J.15}$$

and turn out to be finite at $T=0$, unlike the case of Heisenberg ferromagnet. The temperature dependence of spin-wave stiffness reads (cf.(G.27))

$$\delta D(T) = \frac{\pi^2}{12}T^2 N(E_{\mathrm{F}})\frac{d}{dE_{\mathrm{F}}}\left(\frac{\partial^2\varepsilon_{\mathbf{k}}}{\partial k^2}\right)_{k=k_{\mathrm{F}}} - \\ - \frac{5\pi^{1/2}}{12}\zeta(5/2)\left(\frac{v_0}{4\pi^2}\right)^2 \frac{k_{\mathrm{F}}}{m^*}\left(\frac{T}{D}\right)^{5/2} - \frac{v_0^2 k_{\mathrm{F}}^4}{144\pi D k_{\mathrm{F}}^2}T^2 \ln\frac{4Dk_{\mathrm{F}}^2}{T}. \tag{J.16}$$

The magnon Green's function permits to calculate the spin-wave correction to magnetization

$$\langle S^z\rangle = \frac{1}{2}(n_\uparrow - n_\downarrow) = \frac{1-c}{2} - n_\downarrow = \\ = \frac{1-c}{2} + \frac{1}{\pi}\sum_{\mathbf{q}}\int N_{\mathrm{B}}(\omega)\,\mathrm{Im}\, G_{\mathbf{q}}(\omega)\,d\omega. \tag{J.17}$$

The corrections owing to the damping turn out to be cancel the factor $1-c$ in the denominator of (J.11) and we obtain to terms of order of $1/z^2$

$$\langle S^z \rangle = \frac{1-c}{2} - \sum_{\mathbf{p}} N_{\mathbf{p}}. \tag{J.18}$$

Thus the ground state is really saturated ferromagnetic, and magnetization at low temperatures obeys the usual Bloch $T^{3/2}$-law.

As follows from (J.9), spin-up states propagate freely on the background of the saturated ferromagnetic ordering, and corrections to the spectrum at low temperatures are proportional to $T^{5/2}$. The situation is more interesting for spin-down states. Using again the kinematical relation (A.25) we obtain

$$G_{\mathbf{k}\downarrow}(E) = \sum_{\mathbf{q}} \langle\langle X_{-\mathbf{q}}(-+) X_{\mathbf{k}+\mathbf{q}}(+0) | X_{-\mathbf{k}}(0+) \rangle\rangle_E. \tag{J.19}$$

The simplest decoupling in the equation of motion for the Green's function in the right-hand side of (J.19) yields

$$G^0_{\mathbf{k}\downarrow}(E) = \sum_{\mathbf{q}} \frac{N_{\mathbf{q}} + n_{\mathbf{k}+\mathbf{q}}}{E - \varepsilon_{\mathbf{k}+\mathbf{q}} + \omega_{\mathbf{q}}}. \tag{J.20}$$

The Green's function (J.20) has a purely non-quasiparticle nature. Because of the weak **k**-dependence of the corresponding distribution function at $T \to 0$,

$$\langle X_{-\mathbf{k}}(0-) X_{\mathbf{k}}(-0) \rangle = -\frac{1}{\pi} \int f(E) \operatorname{Im} G^0_{\mathbf{k}\downarrow}(E)\, dE \simeq c. \tag{J.21}$$

the non-quasiparticle states possess a small mobility and do not carry current [333, 338]. Thus the 'spin-down' excitations are reminiscent of Anderson's spinons (Sect.6.8) which are also described by the Green's function with zero residue [631]. The non-quasiparticle contribution to the density of states turns out to be appreciable:

$$N_{\downarrow}(E, T=0) = \sum_{\mathbf{k}\sigma} n_{\mathbf{k}+\mathbf{q}} \delta(E - \varepsilon_{\mathbf{k}+\mathbf{q}} + \omega_{\mathbf{q}}) = \begin{cases} N_{\uparrow}(E), & E_{\mathrm{F}} - E \gg \omega_{\max}, \\ 0, & E > E_{\mathrm{F}}. \end{cases} \tag{J.22}$$

The sense of this result is as follows. The states well below the Fermi level (of holes) do not possess spin polarization, since electrons with any spin projection may be put into these states (the holes are spinless). However, from states above the Fermi level, only spin-up electrons may be extracted in the saturated ferromagnetic state under consideration. The large non-quasiparticle T-linear contribution to specific heat may be also picked up [338]. However, it has more complicated origin in comparison with the spinon contribution [631] since the non-quasiparticle states are absent at the Fermi level.

A more advanced decoupling yields the result [338]

$$G_{\mathbf{k}\downarrow}(E)=\left\{E-\varepsilon_{\mathbf{k}}+\left[G^{0}_{\mathbf{k}\downarrow}(E)\right]^{-1}\right\}^{-1}. \tag{J.23}$$

At small c, the Green's function (J.23) has no poles below the Fermi level, so that the above conclusions are not changed qualitatively. However, with increasing c, it acquires a spin-polaron pole below E_{F}, and the saturated ferromagnetism is destroyed [332].

The expression (J.23) should be compared with the corresponding result for the paramagnetic phase in the 'Hubbard-III' approximation (cf.(H.16))

$$G_{\mathbf{k}\downarrow}(E)=\left\{E-\varepsilon_{\mathbf{k}}+\frac{1-c}{2}\left[G^{0}_{\mathbf{k}\downarrow}(E)\right]^{-1}\right\}^{-1}. \tag{J.24}$$

Unlike (J.23), equation (J.24) does not contain Fermi functions, so that the incoherent (non-quasiparticle) states do not vanish at E_{F}.

One may expect that in the non-saturated ferromagnetic state (or at high temperatures) the expression (J.20) should be replaced by the Hubbard-I result

$$G_{\mathbf{k}\downarrow}(E)=\frac{c+n_{\downarrow}}{E-\varepsilon_{\mathbf{k}}(c+n_{\downarrow})}, \tag{J.25}$$

which describes usual quasiparticle states with a narrowed band. Note that the expression (J.25) may be obtained by averaging nominator and denominator in (J.20) over $\mathbf{q}$, which corresponds to the large z-approximation. As follows from (J.22), the expressions (J.8), (J.21), unlike (J.25), enable one to satisfy the sum rule (H.13) since

$$\sum_{\mathbf{k}}\langle X_{-\mathbf{k}}(0-)X_{\mathbf{k}}(-0)\rangle=\int_{-\infty}^{E_{\mathrm{F}}}N_{\sigma}(E)dE=c=n_{0}. \tag{J.26}$$

Thus the non-quasiparticle nature of spin-down current carriers is intimately related to the description of the saturated ferromagnetic state.

Appendix K

s–f EXCHANGE MODEL AND INDIRECT EXCHANGE INTERACTION IN RARE EARTHS

For rare-earth metals, where 4*f*-electrons are well localized, the *s–f* exchange model may provide a basis of a quantitative theory of magnetic properties. In particular, the main mechanism of exchange between 4*f*-shells in rare earths and their conducting compounds is the Ruderman-Kittel-Kasuya-Yosida (RKKY) indirect interaction via conduction electrons, which occurs in the second order of perturbation theory in the *s–f* exchange parameter. Excluding from the simplest Hamiltonian of the *s–f* model (G.2) the *s–f* exchange interaction by a canonical transformation [265], we obtain the effective Heisenberg Hamiltonian

$$\mathsf{H}_f = -\sum_{\mathbf{q}} J_{\mathbf{q}}^{\mathrm{RKKY}} \mathbf{S}_{-\mathbf{q}} \mathbf{S}_{\mathbf{q}}, \qquad J_{\mathbf{q}}^{\mathrm{RKKY}} = I^2 \sum_{\mathbf{k}} \frac{n_{\mathbf{k}} - n_{\mathbf{k}+\mathbf{q}}}{t_{\mathbf{k}+\mathbf{q}} - t_{\mathbf{k}}}. \tag{K.1}$$

In real space, the RKKY exchange integrals have an oscillating and slowly decreasing dependence on distance. Performing integration for free electrons we obtain

$$J_{ij}^{\mathrm{RKKY}} = \frac{9\pi n^2 I^2}{2v_0^2 E_{\mathrm{F}}} F(2k_{\mathrm{F}} |\mathbf{R}_i - \mathbf{R}_j|), \qquad F(x) = \frac{x\cos x - \sin x}{x^4}. \tag{K.2}$$

Now we consider a more realistic model of 4*f*-metals. For most rare-earths (except for Eu and Sm), the matrix elements of intersite exchange interactions are small in comparison with the distances between *LSJ*-multiplets, and the Russell–Saunders coupling scheme is a good approximation. Using for simplicity the representation of *s*-type plane waves for conduction electrons, we derive for the *s–f* Hamiltonian (cf.(D.20))

$$\mathsf{H}_{sf} = \sum_{\mathbf{kk}'\sigma\sigma'} \sum_{\nu\Gamma_1\Gamma_2\gamma_1\gamma_2} e^{i(\mathbf{k}-\mathbf{k}')\mathbf{R}_\nu} \left\langle \gamma_1, \mathbf{k}\sigma \left| \sum_{ic} \frac{e^2}{|\mathbf{r}_i - \mathbf{r}_c|}(1-P_{ic}) \right| \gamma_2, \mathbf{k}'\sigma' \right\rangle \times$$

$$\times \langle \Gamma_1 | a^\dagger_{\nu\gamma_1} a_{\nu\gamma_2} | \Gamma_2 \rangle X_\nu(\Gamma_1, \Gamma_2) c^\dagger_{\mathbf{k}\sigma} c_{\mathbf{k}'\sigma'}, \quad \text{(K.3)}$$

where $c^\dagger_{\mathbf{k}\sigma}$ are creation operators for conduction electrons, $\gamma_i = \{lm_i\}$, $\Gamma_i = \{SLJM_i\}$; P_{ic} are operators of permutation of conduction and localized electrons. Expanding the plane waves according to (C.28) and using (C.7) we obtain the series in λ, λ' with the 'Slater integrals'

$$F^{(p)}_{\lambda\lambda'}(\mathbf{kk}') = e^2 \int r_1^2 r_2^2 R_l^2(r_1) \frac{r_<^p}{r_>^{p+1}} R_l(r_2) j_\lambda(kr_2) j_{\lambda'}(k'r_1) dr_1 dr_2; \quad \text{(K.4)}$$

$$G^{(p)}_{\lambda\lambda'}(\mathbf{kk}') = e^2 \int r_1^2 r_2^2 R_l^2(r_1) j_\lambda(kr_2) \frac{r_<^p}{r_>^{p+1}} R_l(r_2) j_{\lambda'}(k'r_1) dr_1 dr_2, \quad \text{(K.5)}$$

where $l = 3$ for f-electrons. The small parameter of the expansion is $k_\mathrm{F} r_f \sim 0.2$ where r_f is the radius of the $4f$-shell. The matrix elements that arise may be calculated by the method of double irreducible tensor operators and expressed in terms of matrix elements of the total angular momentum J, as demonstrated in Appendix D. In particular, for the zeroth-order term we obtain from (B.19)

$$\mathsf{H}_{sf}(00) = -\frac{4\pi}{[l]} \sum_{\nu\sigma\sigma'} G^{(0)}_{00} \left[\frac{n}{2} \delta_{\sigma\sigma'} + (g-1)(\boldsymbol{\sigma}_{\sigma\sigma'} \mathbf{J}_\nu) \right] c^\dagger_{\nu\sigma} c_{\nu\sigma'}. \quad \text{(K.6)}$$

Higher-order terms of the expansion are anisotropic and have the structure

$$\mathsf{H}^{\mathrm{coul}}_{sf} = \sum_{\nu\mathbf{kk}'\sigma\sigma'} e^{i(\mathbf{k}-\mathbf{k}')\mathbf{R}_\nu} c^\dagger_{\mathbf{k}\sigma} c_{\mathbf{k}'\sigma'} \times$$

$$\times (B_0 + B_1[3\{(\mathbf{kJ}_\nu), (\mathbf{k}'\mathbf{J}_\nu)\} - 2(\mathbf{kk}')J(J+1)] + \ldots); \quad \text{(K.7)}$$

$$\mathsf{H}^{\mathrm{exch}}_{sf} = \sum_{\nu\mathbf{kk}'\sigma\sigma'} (A_0 \delta_{\sigma\sigma'} + A_1(\boldsymbol{\sigma}_{\sigma\sigma'} \mathbf{J}_\nu) + iA_2([\mathbf{kk}']\mathbf{J}_\nu)\delta_{\sigma\sigma'} +$$

$$+ A_3\{(\mathbf{kJ}_\nu), (\mathbf{k}'\mathbf{J}_\nu)\} + A_4[(\mathbf{k}\boldsymbol{\sigma}_{\sigma\sigma'})(\mathbf{k}'\mathbf{J}_\nu) + (\mathbf{k}'\boldsymbol{\sigma}_{\sigma\sigma'})(\mathbf{kJ}_\nu)] +$$

$$+ A_5[(\mathbf{k}\boldsymbol{\sigma}_{\sigma\sigma'})(\mathbf{kJ}_\nu) + (\mathbf{k}'\boldsymbol{\sigma}_{\sigma\sigma'})(\mathbf{k}'\mathbf{J}_\nu)] +$$

$$+ A_6[(\mathbf{kJ}_\nu)^2 + (\mathbf{k}'\mathbf{J}_\nu)^2] + iA_7\{(\boldsymbol{\sigma}_{\sigma\sigma'}\mathbf{J}_\nu), ([\mathbf{kk}']\mathbf{J}_\nu)\} + \ldots), \quad \text{(K.8)}$$

where { , } is anticommutator. The maximum power q of momentum operator **J** is determined by the maximum value of λ, retained in the expansion (C.28), $q = \min\{2J, 2\lambda + 1\}$. The terms with the vector products $[\mathbf{kk}']$ describe the anisotropic electron scattering and are important in the theory of the anomalous Hall effect. The coefficients of the expansion (K.8) are given by [704]

$$A_1 = \frac{2}{7}(g-1)\left[G_{00}^{(3)} + (\mathbf{kk}')\eta_1\right] - D_1\left[\frac{9}{35}(\mathbf{kk}')\eta_2 + \frac{2}{7}\sqrt{6}G_{02}^{(3)}\right],$$

$$A_2 = \frac{1}{28}(g-2)\eta_3 - \frac{9\sqrt{5}}{70}\eta_2 D_2, \qquad A_3 = \frac{9\sqrt{5}}{35}D_2\eta_2,$$

$$A_4 = \frac{27}{70}\eta_2 D_1, \qquad A_5\frac{3\sqrt{6}}{7}G_{02}^{(3)}D_1, \qquad A_6 = -\frac{3\sqrt{5}}{7}G_{02}^{(3)}D_2,$$

$$A_7 = A_8 = \frac{9\sqrt{15}}{14}\eta_3(2J+1)\left[\frac{(2J-2)!}{(2J+3)!}\right]^{1/2}\begin{Bmatrix} L & J & S \\ L & J & S \\ 1 & 2 & 1 \end{Bmatrix}\langle SL \| W^{(11)} \| SL\rangle, \quad \text{(K.9)}$$

where the irreducible matrix elements are determined by (B.27),

$$\eta_1 = \frac{9}{5}G_{11}^{(2)} + \frac{4}{3}G_{11}^{(4)}, \qquad \eta_2 = \frac{9}{5}G_{11}^{(2)} + \frac{5}{9}G_{11}^{(4)}, \qquad \eta_3 = \frac{9}{5}G_{11}^{(2)} - G_{11}^{(4)},$$

$$D_1 = \left[\frac{2J+1}{J(J+1)}\right]^{1/2}\begin{Bmatrix} L & J & S \\ L & J & S \\ 2 & 1 & 1 \end{Bmatrix}\langle SL \| W^{(12)} \| SL\rangle,$$

$$D_2 = (-1)^{S+L+J}\frac{2}{\sqrt{3}}\frac{2J+1}{(2S+1)^{1/2}}\langle SL \| W^{(02)} \| SL\rangle\begin{Bmatrix} L & J & S \\ J & L & 2 \end{Bmatrix}. \quad \text{(K.10)}$$

The Hamiltonian of indirect f–f interaction is obtained in the second order in H_{sf} and has the same structure as (D.22). The main contributions may be written down in the form [389]

$$\begin{aligned} \mathsf{H}_{ff}(\nu_1\nu_2) = &-I_1(g-1)^2(\mathbf{J}_1\mathbf{J}_2) - \\ &- I_2 D_1(g-1)\left[(\mathbf{J}_1\mathbf{J}_2) - 3(\rho_{12}\mathbf{J}_1)(\rho_{12}\mathbf{J}_2)/\rho_{12}^2\right] - \\ &- I_3 n D_3\left[(\mathbf{J}_1\mathbf{J}_2) - 3(\rho_{12}\mathbf{J}_2)^2/\rho_{12}^2\right], \quad \text{(K.11)} \end{aligned}$$

where I_i are linear combinations of the integrals of the type

$$B_{\mu'\lambda''\lambda'''p'}^{\mu\lambda\lambda'p} = \int_0^\infty dk\, k^2 n_k \int_{-\infty}^{\infty} dk' \frac{k'^2}{k^2 - k'^2} j_\mu(k\rho_{12}) j_{\mu'}(k'\rho_{12}) \times$$
$$\times G_{\lambda\lambda'}^{(p)}(kk') G_{\lambda''\lambda'''}^{(p')}(kk'). \quad \text{(K.12)}$$

The largest term of this expansion, which is proportional to $(g-1)^2$, corresponds to the usual exchange between spins according to the de Gennes formula (B.19). The dependence of the *f–f* exchange parameter $J_{\text{eff}} \sim (g-1)^2$ is in a good agreement with experimental data on the paramagnetic Curie temperatures in the series of rare-earth metals. The orbital contributions to the *f–f* interaction, which are proportional to D_1 and D_2 and vanish at $L=0$, are considerably smaller. A still smaller term of the purely orbital exchange is obtained in the second order in A_2:

$$\mathsf{H}'_{ff}(\nu_1\nu_2) = -I_4(g-2)^2(\mathbf{J}_1\mathbf{J}_2) = -I_4(\mathbf{L}_1\mathbf{L}_2). \qquad \text{(K.13)}$$

The orbital terms may give an appreciable contribution to the exchange anisotropy of crystal magnetization (Sect.4.8).

Appendix L

SPIN-ORBITAL INTERACTION

Besides exchange interactions, an important role in magnetic crystals belongs to the relativistic spin-orbital interaction (SOI). The latter, although being weak, results in a partial unquenching of orbital momenta and is responsible for anisotropy of magnetic and other properties. SOI is especially important for transport phenomena, e.g., anomalous galvanomagnetic and thermomagnetic properties.

The operator of SOI for an electron with the quasimomentum $\mathbf{p}$ and spin $\mathbf{s}$ in the potential $V(\mathbf{r})$ has the form

$$\mathsf{H}_{so} = \frac{\hbar}{2m^2c^2}[\nabla V, \mathbf{p}]\mathbf{s}. \tag{L.1}$$

For the Coulomb interaction

$$V(\mathbf{r}) = -Ze^2/r,$$

we obtain

$$\mathsf{H}_{so} = \lambda(r)(\mathbf{ls}), \tag{L.2}$$

where

$$l = \frac{1}{\hbar}[\mathbf{r},\mathbf{p}], \qquad \lambda(r) = \frac{Ze^2\hbar^2}{2m^2c^2r^3}.$$

For estimating the value of $\lambda(r)$ one can use hydrogen-like wavefunctions to obtain

$$\lambda_{nl} = \frac{Z^4\hbar c\alpha^2}{n^3 l(l+1/2)(l+1)}\mathrm{Ry}, \tag{L.3}$$

with Ry = 13.6 eV being the Rydberg constant, $\alpha = 1/137$. SOI increases rapidly with increasing atomic number. We have $\lambda \sim 10^{-14}$ erg for $3d$-electrons and $\lambda \sim 10^{-13}$ erg for $4f$-electrons in rare earths. The quantity λ may be also estimated from the data of atomic spectroscopy, which turn out to agree qualitatively with the theory, provided that effective values of Z are used.

At considering SOI, the question about degeneracy of electron states is essential. Evidently, for non-degenerate wavefunctions (e.g., for Bloch electrons in a crystal) diagonal matrix elements of H_{so} (L.2) vanish, so that corrections to the energy, which are linear in H_{so}, are absent. On the other hand, in the degenerate case (e.g, for a free atom) a splitting of levels in the first order in H_{so} takes place.

Besides the proper SOI (orbital electron current in the magnetic field of its own spin momentum), there exists also the interaction of orbital current with spins of other electrons

$$\mathsf{H}'_{so} = \frac{\hbar}{m^2c^2}\sum_{i\neq j}[\nabla V_{ij}, \mathbf{p}_i]\mathbf{s}_j, \tag{L.4}$$

where i, j are electron numbers,

$$V_{ij} = \frac{e^2}{|\mathbf{r}_{ij}|}, \qquad \nabla V_{ij} = -\frac{e^2\mathbf{r}_{ij}}{|\mathbf{r}_{ij}|^3}, \qquad \mathbf{r}_{ij} = \mathbf{r}_i - \mathbf{r}_j. \tag{L.5}$$

In the case of two electrons, one of which has zero orbital momentum and moves closely to the nucleus, we can put $\mathbf{l}_1 = 0$, $\mathbf{r}_{12} = \mathbf{r}_2$. Then (L.5) takes the form [20]

$$\mathsf{H}'_{so} = -\lambda'(\mathbf{l}_2\mathbf{s}_1), \tag{L.6}$$

with $\lambda' > 0$ being proportional to Z^3 rather than Z^4 (as in (L.3)). The latter fact results in that the interaction between spin and foreign orbit is more important for light elements. As a rule, for $3d$-electrons λ' is smaller than λ by two-three orders of magnitude, but, as we shall demonstrate below, the corresponding interaction may play an important role because of its singular $\mathbf{k}$-dependence.

For a single electron, the 'proper' SOI makes favourable the antiparallel interactions of its spin and orbital momenta, but the interaction (L.6) orientates its orbital momentum parallel to spin of other electrons. However, for more than half-filled shell the sign of λ changes and the state with the total momentum $\mathbf{J} = \mathbf{L} + \mathbf{S}$ has the lowest energy.

Now we discuss SOI in periodic crystals. In the localized Heisenberg model the treatment is close to that for isolated atoms. On the other hand, for crystals containing $3d$-elements the situation changes drastically. As discussed in Sect.1.3, the local crystal potential may quench orbital momenta in the case of crystals with low symmetry only. At the same time, in real d-systems with cubic or hexagonal local symmetry, which possess degenerate irreducible representations of the point group, the degeneracy is lifted by the periodic potential in the band picture (Sect.4.8). Therefore the case of quenched orbital momenta is of interest. Then only off-diagonal matrix elements of SOI are non-zero and the perturbed wavefunctions read

$$\Psi_\gamma = \Psi_\gamma^{(0)} + \sum_{\gamma' \neq \gamma} \frac{\langle \gamma' | \mathsf{H}_{so} | \gamma \rangle}{E_\gamma - E_{\gamma'}} \Psi_{\gamma'}^{(0)}, \tag{L.7}$$

with $\gamma = \{\mathbf{k}m\sigma\}$ are the states of magnetic d-electrons in the degenerate d-band.

The wavefunctions (L.7) may be used to calculate corrections to various physical quantities, owing to SOI. In particular, such corrections to matrix elements of electrostatic interaction between conduction and localized electrons will be anisotropic. It is these corrections which cause anomalous transport phenomena in magnetic crystals. The role of the proper SOI and the interaction (L.6) is different for different situations and concrete effects. For galvanomagnetic effects in d-magnets one can consider two cases

1. The mobility of d-electrons is large, so that they determine galvanomagnetic effects directly. Then the proper SOI for itinerant d-electrons plays the dominant role.
2. There exist two electron groups – conduction s-electrons with small magnetization and "magnetic" d-electrons with small mobility. Then, among four possible types of SOI (s–s, d–d, s–d and d–s), the proper d–d interaction and the s–d interaction of s-electron orbit with d-electron spin are most important. The s–s and d–s interactions yield small contributions because of smallness of magnetization for s-electrons.

Finally, we derive the Hamiltonian of the s–d model with account of SOI of d–d and s–d type in the second quantization representation. We consider the case of the strong crystal field which destroys total orbital magnetic momenta of d-electrons. We have

$$\begin{aligned}\mathsf{H} = &\sum_{\mathbf{k}\sigma} t_\mathbf{k} c_{\mathbf{k}\sigma}^\dagger c_{\mathbf{k}\sigma} + \sum_\gamma E_\gamma a_\gamma^\dagger a_\gamma + \\ &+ \sum_{\substack{\mathbf{k}\sigma\mathbf{k}'\sigma' \\ \gamma_1\gamma_2}} [I(\mathbf{k}\sigma, \gamma_1, \mathbf{k}'\sigma', \gamma_2) - I(\mathbf{k}\sigma, \gamma_1, \gamma_2, \mathbf{k}'\sigma') + \\ &+ L(\mathbf{k}\sigma, \gamma_1, \mathbf{k}'\sigma', \gamma_2) - L(\mathbf{k}\sigma, \gamma_1, \gamma_2, \mathbf{k}'\sigma')] a_{\gamma_1}^\dagger a_{\gamma_2} c_{\mathbf{k}\sigma}^\dagger c_{\mathbf{k}'\sigma'},\end{aligned} \tag{L.8}$$

where the Coulomb and exchange matrix elements

$$I(1,2,3,4) = \int \psi_1^*(x)\psi_2^*(x') \frac{e^2}{|\mathbf{r}-\mathbf{r}'|} \psi_3(x)\psi_4(x') dx dx' \tag{L.9}$$

($x = \{\mathbf{r}, s\}$) are to be calculated for the wavefunctions with account of SOI, and the spin-orbital matrix elements are

$$L(1,2,3,4) = -\frac{e^2\hbar}{m^2c^2}\int \psi_1^*(x)\psi_2^*(x')\frac{[\mathbf{r}-\mathbf{r}',\mathbf{p}]\mathbf{s}'}{|\mathbf{r}-\mathbf{r}'|^3}\psi_3(x)\psi_4(x')dxdx'. \quad \text{(L.10)}$$

In the representation $\gamma=\{\mathbf{k}m\sigma\}$ the diagonal matrix elements of $\mathsf{H}_{so}(dd)$ vanish and the off-diagonal ones are obtained within perturbation theory with the use of (L.7). Substituting (L.7) into the exchange part of (L.9) we derive to linear approximation in H_{so} the correction

$$\mathsf{H}_{sd}^{(1)} = -\lambda\sum\frac{\langle\mathbf{k}_1 m_1\sigma_1|\mathbf{ls}|\mathbf{k}_1 m_1'\sigma_2\rangle}{E_{\mathbf{k}_1 m_1\sigma_1} - E_{\mathbf{k}_1 m_1'\sigma_2}}\times$$
$$\times I^{(0)}(\mathbf{k}\mathbf{k}_1 m_1', \mathbf{k}_2 m_2\mathbf{k}')a^\dagger_{\mathbf{k}_1 m_1\sigma_1}a_{\mathbf{k}_2 m_2\sigma}c^\dagger_{\mathbf{k}\sigma}c_{\mathbf{k}'\sigma_2} + \text{h. c.} \quad \text{(L.11)}$$

The matrix elements in (L.11) are calculated, similar to Appendix K, with the use of the plane-wave representation for conduction electrons. It is convenient to use the m-representation for d-electron functions (the functions, corresponding to irreducible representations of a point group are expressed as their linear combinations). This representation enables one to introduce in a simple way spin operators for localized d-electrons:

$$a^\dagger_{m\sigma}a_{m\sigma} = (1+\sigma 2s^z)\varphi_m(l^z), \qquad a^\dagger_{m\pm}a_{m\mp} = s^\pm\varphi_m(l^z), \quad \text{(L.12)}$$

where

$$\begin{aligned}\varphi_0 &= 1+\frac{1}{4}(l^z)^2[(l^z)^2-5],\\ \varphi_{\pm1} &= \frac{1}{6}l^z(l^z\pm1)[4-(l^z)^2],\\ \varphi_{\pm2} &= \frac{1}{24}l^z(l^z\pm1)[(l^z)^2-1], \end{aligned} \quad \text{(L.13)}$$

and the off-diagonal products $a^\dagger_{m\sigma}a_{m'\sigma}$ are expressed in terms of operators $l^\pm$. Then we obtain for a cubic crystal

$$\mathsf{H}_{sd}^{(1)} = \frac{\lambda}{\Delta E}\sum_{ll'}\sum_{\nu\mathbf{k}\mathbf{k}'\sigma}F_{ll'}(\mathbf{l}_\nu,\mathbf{s}_\nu)e^{i(\mathbf{k}-\mathbf{k}')\mathbf{R}_\nu}c^\dagger_{\mathbf{k}\sigma}c_{\mathbf{k}'\sigma}. \quad \text{(L.14)}$$

where we have put for simplicity $\Delta E_{\gamma\gamma'}(\mathbf{k}) = \Delta E = \text{const}$. The expansion term with $l = l' = 1$ reads

$$\mathsf{H}_{sd}^{(1)}(11) = -i\frac{\lambda}{\Delta E}\frac{6}{5}\pi e^2\sum_{\nu\mathbf{k}\mathbf{k}'\sigma}G_{11}^{(1)}(\mathbf{k},\mathbf{k}')[\mathbf{k}\mathbf{k}']\mathbf{s}_\nu\langle\varphi(l_\nu)\rangle e^{i(\mathbf{k}-\mathbf{k}')\mathbf{R}_\nu}c^\dagger_{\mathbf{k}'\sigma}c_{\mathbf{k}\sigma} \quad \text{(L.15)}$$

(where the radial integral G is given by (K.5)) and describes anisotropic scattering.

Now we calculate matrix elements of the 'improper' SOI (L.10). Replacing for simplicity squares of wavefunctions of d-electrons by δ-functions (which is possible since orbital momenta of d-electrons do not enter) we derive

$$L(\mathbf{kk}') = i\lambda \sum_{\nu} e^{i(\mathbf{k}-\mathbf{k}')\mathbf{R}_\nu} \frac{[\mathbf{kk}']}{(\mathbf{k}-\mathbf{k}')^2} \mathbf{s}_\nu, \qquad \text{(L.16)}$$

with $\lambda' \sim 10^{-16}$ erg. The expression (L.16) does not yield a dependence on orientation of localized spin in the crystal. However, such a dependence will occur for more complicated d-wavefunctions.

Combining (L.15) with (L.16) we write down the correction to the Hamiltonian of the s–d model owing to SOI as

$$\mathsf{H}'_{sd} = \sum_{\nu\mathbf{kk}'\sigma} e^{i(\mathbf{k}-\mathbf{k}')\mathbf{R}_\nu} (\mathbf{s}_\nu \Lambda_{\mathbf{kk}'}) c^\dagger_{\mathbf{k}\sigma} c_{\mathbf{k}'\sigma}, \qquad \text{(L.17)}$$

where

$$\Lambda^z_{\mathbf{kk}'} = i\lambda \bar{l} \frac{I^{(1)}_{\mathbf{kk}'}}{\Delta E} \frac{[\mathbf{kk}']_z}{kk'} + i\lambda' \frac{[\mathbf{kk}']_z}{(\mathbf{k}-\mathbf{k}')^2}, \qquad \text{(L.18)}$$

$$\bar{l} = \frac{2}{9} \left\langle (l^z)^2 [4-(l^z)^2] \left\{ \frac{1}{4} (l^z)^2 [(l^z)^2 - 1] + 2(l^z)^2 [4-(l^z)^2] \right\} \right\rangle, \qquad \text{(L.19)}$$

$I^{(1)}$ is defined by (L.15). Although $|\lambda'| \ll |\lambda|$, the role of the second term in (L.17) may be important provided that small values of $|\mathbf{k}-\mathbf{k}'|$ yield the dominant contribution, as it takes place for extraordinary transport phenomena at low temperatures.

Appendix M

THE DENSITY MATRIX TECHNIQUE FOR DERIVATION OF TRANSPORT EQUATIONS AND THE THEORY OF THE ANOMALOUS HALL EFFECT

Mathematical description of transport phenomena deals with an balance equation for the distribution function of current carriers in external electric, magnetic and thermal fields (5.14). In simple cases such equations may be obtained from a simple physical consideration of the electron motion in the **k**-space due to external fields and collisions. However, in more complicated situations, where higher orders in the scattering amplitude become important, such simple arguments are not sufficient. Since often we do meet with this situation for transport phenomena in magnetic crystals, constructing a general method for derivation of transport equations is needed. Most convenient is the approach which uses the density matrix. The equations of motion for the latter quantity have usual quantum mechanics form and may be simply reduced to transport equations.

Define the density matrix operator

$$\hat{\rho} = e^{-\beta \mathsf{H}} / \mathrm{Sp}\, e^{-\beta \mathsf{H}}, \qquad \beta \equiv 1/T, \tag{M.1}$$

where H is the total Hamiltonian of the system including current carriers, scattering system and external fields. The coefficient in (M.1) is determined by the normalization condition

$$\mathrm{Sp}\, \hat{\rho} = 1.$$

The operator $\hat{\rho}$, as well as H, may be written down in any quantum representation. In the solid state theory, it is convenient to use the second quantization representation. Then the symbol Sp means the summation over all the possible occupation numbers of quasiparticles in the system. For example, in the case of electron-phonon system

$$\mathrm{Sp}\ldots = \sum_{\{n_{\mathbf{k}}\}\{N_{\mathbf{q}}\}} \ldots, \tag{M.2}$$

where $n_{\mathbf{k}} = 0$, 1 are the electron occupation numbers and $N_{\mathbf{q}}$ are the phonon ones.

Average value of a physical quantity A is obtained as

$$\langle A\rangle = \mathrm{Sp}\,(\hat{\rho}A). \tag{M.3}$$

Calculate for example the average occupation numbers for non-interacting conduction electrons. We find

$$\langle n_{\mathbf{k}}\rangle = \langle c_{\mathbf{k}}^{\dagger}c_{\mathbf{k}}\rangle = \sum_{\{n_{\mathbf{k}'}\}}\langle\ldots n_{\mathbf{k}}\ldots|\prod_{\mathbf{k}'}\exp(-\beta\varepsilon_{\mathbf{k}'}c_{\mathbf{k}'}^{\dagger}c_{\mathbf{k}'})c_{\mathbf{k}}^{\dagger}c_{\mathbf{k}}|\ldots n_{\mathbf{k}}\ldots\rangle \Big/$$

$$\Big/\sum_{\{n_{\mathbf{k}'}\}}\langle\ldots n_{\mathbf{k}}\ldots|\prod_{\mathbf{k}'}\exp(-\beta\varepsilon_{\mathbf{k}'}c_{\mathbf{k}'}^{\dagger}c_{\mathbf{k}'})|\ldots n_{\mathbf{k}}\ldots\rangle =$$

$$= \sum_{n_{\mathbf{k}}}\exp(-\beta\varepsilon_{\mathbf{k}})n_{\mathbf{k}}\Big/\sum_{n_{\mathbf{k}}}\exp(-\beta\varepsilon_{\mathbf{k}}) = (\exp(\beta\varepsilon_{\mathbf{k}})+1)^{-1},$$

i.e. we derive the Fermi distribution function.

Provided that the Hamiltonian of the system is written in the form

$$\mathsf{H} = \mathsf{H}_0 + \mathsf{H}',$$

with H' being a perturbation, the equilibrium density matrix $\hat{\rho}(\mathsf{H})$ may be expanded in the powers of H', which is required at calculating the field term in the transport equation. To this end one may use the theorem about expansion of exponential operators [706]:

$$\begin{aligned}\langle n|\exp(-\beta(\mathsf{H}_0+\mathsf{H}'))|n'\rangle &= \delta_{nn'}\exp(-\beta E_n^0) - \\ &-\frac{\exp(-\beta E_n^0)-\exp(-\beta E_n^0)}{E_n^0-E_{n'}^0}\langle n|\mathsf{H}'|n'\rangle + \sum_{n''}\frac{\langle n|\mathsf{H}'|n''\rangle\langle n''|\mathsf{H}'|n'\rangle}{E_{n''}^0-E_{n'}^0}\times \\ &\times\left[\frac{\exp(-\beta E_{n'}^0)-\exp(-\beta E_n^0)}{E_n^0-E_{n'}^0}-\frac{\exp(-\beta E_{n''}^0)-\exp(-\beta E_n^0)}{E_n^0-E_{n''}^0}\right],\end{aligned} \tag{M.4}$$

where E_n^0 are the eigenvalues of H_0. In the second quantization representation, the numbers n mean the occupation number sets, and summing over them yields an expression of $\hat{\rho}(\mathsf{H})$ in terms of powers of H' and $\hat{\rho}(\mathsf{H}) = \hat{\rho}_0$. In an representation of the wavefunctions $|l\rangle$ we have

$$\hat{\rho} = \hat{\rho}^{(0)} + \hat{\rho}^{(1)} + \hat{\rho}^{(2)} + \ldots,$$

where

$$\hat{\rho}_{ll'}^{(0)} = \rho_l\delta_{ll'}, \qquad \hat{\rho}_{ll'}^{(1)} = \frac{\rho_l-\rho_{l'}}{\varepsilon_l-\varepsilon_{l'}}\mathsf{H}_{ll'}', \tag{M.6a}$$

$$\hat{\rho}_{ll'}^{(2)} = \gamma\rho_l\delta_{ll'} + \sum_{l''} \frac{\mathsf{H}_{ll''}'\mathsf{H}_{l''l'}'}{\varepsilon_l - \varepsilon_{l'}}\left(\frac{\rho_l - \rho_{l'}}{\varepsilon_l - \varepsilon_{l'}} - \frac{\rho_l - \rho_{l''}}{\varepsilon_l - \varepsilon_{l''}}\right); \tag{M.6b}$$

$$\gamma = \sum_{ll'} \frac{|\mathsf{H}_{ll'}'|^2}{\varepsilon_l - \varepsilon_{l'}}\left[\frac{\rho_l - \rho_{l'}}{\varepsilon_l - \varepsilon_{l'}} - \frac{\partial\rho_l}{\partial\varepsilon_l}\right]. \tag{M.6c}$$

The equation of motion for the density matrix operator has the usual form

$$i\frac{\partial\hat{\rho}}{\partial t} = [\mathsf{H}, \hat{\rho}]. \tag{M.7}$$

In any representation $|l\rangle$, we may write down the system of equations for diagonal and off-diagonal matrix elements. Under certain conditions this system may be reduced to transport equations of the lowest Born and next-order approximations. In this way, transport equations were obtained for elastic scattering by impurities [458], and further for scattering by phonons [460] and spin inhomogeneities [466]. The theory in the case of arbitrarily strong scattering amplitude, but small impurity concentration was developed by Luttinger [707]. However, for a complicated H_0, calculations in the matrix form are rather cumbersome even for impurity scatterng in the second Born approximation. Therefore it is very convenient for practical calculations to derive transport equations in an operator form without concretizing the Hamiltonian. Now we consider this technique constructed in [471, 472].

The only requirement to the Hamiltonian is the possibility of the representation

$$\mathsf{H}_t = \mathsf{H}_0 + \mathsf{H}_E e^{st} + \mathsf{H}', \tag{M.8}$$

where H_0 has a diagonal form in the n-representation, H_E is the energy of the system in the adiabatically included ($s \to 0$) electric field $\mathbf{E}$, H' is an off-diagonal part. The total density matrix operator is represented as

$$\hat{\rho}_t = \hat{\rho} + (\hat{f}_a + \hat{f}_b)e^{st}, \tag{M.9}$$

where $\hat{\rho}$ is the equilibrium density matrix in the absence of electric field, f_a and f_b are the diagonal and off-diagonal components of the correction. Taking into account the relations

$$[\mathsf{H}_0, \hat{f}_a] = [\mathsf{H}', \hat{f}_a]_a = [\mathsf{H}_0, \hat{f}_b]_a,$$

we obtain to linear approximation in E

$$s\hat{f}_a = [\mathsf{H}_E, \hat{\rho}]_a + [\mathsf{H}', \hat{f}_b]_a, \tag{M.10}$$

$$\hat{f}_b = \hat{L}^{-1}\{[\mathsf{H}_E, \hat{\rho}]_b + [\mathsf{H}', \hat{f}_a + \hat{f}_b]_b\}, \tag{M.11}$$

where

$$\hat{L}^{-1} = \frac{1}{is - \Delta}, \qquad \Delta\hat{f}_b = [\mathsf{H}_0, \hat{f}_b], \tag{M.12}$$

Δ is the difference of the eigenvalues of H in the corresponding states. The system (M.10), (M.11) may be solved by the iteration method. Substituting (M.11) into (M.10) we derive

$$is\hat{f}_b = [\mathsf{H}_E, \hat{\rho}]_a + [\mathsf{H}', \hat{L}^{-1}\{[\mathsf{H}_E, \hat{\rho}]_b + [\mathsf{H}', \hat{f}_a + \hat{f}_b]_b\}]_a. \tag{M.13}$$

Repeating the procedure we have after the $(n-1)$-th iteration

$$\begin{aligned} is\hat{f}_a = [\mathsf{H}_E, \hat{\rho}]_a + [\mathsf{H}', \hat{L}^{-1}\{[\mathsf{H}_E, \hat{\rho}]_b + [\mathsf{H}', \hat{f}_a]_b + \\ + [\mathsf{H}', \hat{L}^{-1}\{[\mathsf{H}_E, \hat{\rho}]_b + [\mathsf{H}', \hat{f}_a]_b + \\ + [\mathsf{H}', \hat{L}^{-1}\{\{[\mathsf{H}_E, \hat{\rho}]_b + [\mathsf{H}', \hat{f}_a]_b + \ldots\} + \\ + [\mathsf{H}', \hat{L}^{-1}\{\ldots\{[\mathsf{H}_E, \hat{\rho}]_b + [\mathsf{H}', \hat{f}_a + \hat{f}_b]_b\}\ldots\}\}\}]_a, \end{aligned} \tag{M.14}$$

where $\{\ldots\{$ stands for n curly brackets. The iteration procedure allows one to obtain the solution to any desired order in H' provided that the expansion of f_a in H' starts from a lower degree of H' than that of f_b. Then, to accuracy of $(\mathsf{H}')^{n+1}$, we may neglect the term $[\mathsf{H}', f_b]$ in (M.14) and solve for f_a, and further find f_b from (M.10).

Consider the first iteration. To lowest order in H' we have

$$[\mathsf{H}_E, \hat{\rho}^{(0)}]_b + [\mathsf{H}', \hat{L}^{-1}[\mathsf{H}', \hat{f}_a]_b]_a = 0. \tag{M.15}$$

Taking into account the identity

$$\hat{L}^{-1}_{nn'} + \hat{L}^{-1}_{n'n} = -\left(\frac{1}{\Delta_{nn'}} + \frac{1}{\Delta_{n'n}}\right) - i\pi[\delta(\Delta_{nn'}) + \delta(\Delta_{n'n})], \tag{M.16}$$

multiplying by $n_\lambda = a^\dagger_\lambda a_\lambda$ and summing over the occupation numbers we obtain from (M.15)

$$C^{(0)}_\lambda + T^{(0)}_\lambda = 0, \tag{M.17}$$

where

$$C^{(0)}_\lambda = \mathrm{Sp}\,\{[\mathsf{H}_E, \hat{\rho}^{(0)}]n_\lambda\}, \tag{M.18}$$

is the field term,

$$T_\lambda^{(0)} = \mathrm{Sp}\{[\mathsf{H}', \hat{L}^{-1}[\mathsf{H}', \hat{f}_a]_b]_a n_\lambda\} = \\ = 2\pi i\ \mathrm{Sp}\{\mathsf{H}'\hat{f}_a\underline{\mathsf{H}}' - \mathsf{H}'\hat{f}\underline{\mathsf{H}}_a'\}\underline{\delta(\Delta)}n_\lambda \tag{M.19}$$

is the collision term (the δ-functions correspond to underlined operator H'). One can see that $f_a = f^{(-2)} \sim (\mathsf{H}')^{-2}$. Thus we obtain an usual Boltzmann-type transport equation.

In the next order approximation with respect to H' we have to take into account terms with $f^{(-1)} \sim (\mathsf{H}')^{-1}$. One obtains

$$C_\lambda^{(1)} + C_\lambda'^{(1)} = T_\lambda^{(0)}(f^{(-1)}) + T_\lambda^{(0)}(f^{(-2)}) = 0, \tag{M.20}$$

where

$$C_\lambda^{(1)} = \mathrm{Sp}\{[\mathsf{H}_E, \hat{\rho}^{(1)}]n_\lambda\}, \tag{M.21}$$

$$C_\lambda'^{(1)} = 2\pi i\ \mathrm{Sp}\ \{[\mathsf{H}', \hat{\rho}^{(0)}\underline{\mathsf{H}_E} - \mathsf{H}'\underline{\mathsf{H}_E}\hat{\rho}^{(0)}]n_\lambda\delta(\Delta)\}, \tag{M.22}$$

and, in the case of purely imaginary product $\mathsf{H}'\mathsf{H}'\mathsf{H}'$ (e.g., for the spin-orbital interaction),

$$T_\lambda^{(1)} = 6\pi^2\ \mathrm{Sp}\{\hat{f}_a\underline{\delta(\Delta)\mathsf{H}'}n_\lambda\mathsf{H}'\underline{\mathsf{H}\,\delta(\Delta)}\}. \tag{M.23}$$

We see that one of important advantages of the method is a simple formation of δ-function terms.

To second order in H'

$$C_\lambda^{(2)} + C_\lambda'^{(2)} + C_\lambda''^{(2)} + T_\lambda^{(0)}(f^{(0)}) + T_\lambda^{(1)}(f^{(-)}) + T_\lambda^{(2)}(f^{(-2)}) = 0, \tag{M.24}$$

where

$$C_\lambda^{(2)} = \mathrm{Sp}\{[\mathsf{H}_E, \hat{\rho}^{(2)}]n_\lambda\}; \tag{M.25}$$

$$C_\lambda'^{(2)} = \mathrm{Sp}\{[\mathsf{H}', \hat{L}^{-1}[\mathsf{H}_E, \hat{\rho}^{(1)}]]n_\lambda\}; \tag{M.26}$$

$$C_\lambda''^{(2)} = \mathrm{Sp}\ \{[\mathsf{H}', \hat{L}^{-1}[\mathsf{H}', \hat{L}^{-1}[\mathsf{H}_E, \hat{\rho}^{(0)}]]]n_\lambda\}; \tag{M.27}$$

$$T_\lambda^{(2)} = \mathrm{Sp}\{[\mathsf{H}', \underline{\hat{L}^{-1}[\mathsf{H}', \hat{f}_a]_b}\,\underline{[[\mathsf{H}', \hat{n}_\lambda]\hat{L}^{-1}, \mathsf{H}']_b\hat{L}^{-1}}\}. \tag{M.28}$$

In the matrix form this result was derived by Kohn and Luttinger [478]. Further we consider the transport equations for concrete scattering mechanisms.

M.1 IMPURITY SCATTERING

The Hamiltonian of the electron–impurity system is given by

$$\mathsf{H}_0 = \sum_l \varepsilon_l c_l^\dagger c_l, \qquad \mathsf{H}_E = e\mathbf{E}\sum_{ll'}\mathbf{r}_{ll'}c_l^\dagger c_{l'}, \qquad \mathsf{H}' = \sum_{ll'} V_{ll'}c_l^\dagger c_{l'}, \tag{M.29}$$

where $l = \{n\mathbf{k}\}$, n is the band index

$$V_{ll'} = \sum_{i=1}^{n_i} e^{i(\mathbf{k}-\mathbf{k}')\mathbf{R}_i} \varphi_{ll'}, \tag{M.30}$$

$$\varphi_{ll'} = \int u_l^*(\mathbf{r}) u_{l'}(\mathbf{r}) \varphi(\mathbf{r}) e^{i(\mathbf{k}-\mathbf{k}')\mathbf{r}} d\mathbf{r}, \tag{M.31}$$

$\varphi(\mathbf{r})$ is the one-impurity potential, n_i is the number of impurities, u_l are the Bloch factors in (2.1). Averaging over random distribution of the impurities yields

$$\langle |V_{ll'}|^2 \rangle = n_i |\varphi_{ll'}|^2 \quad (l \neq l'). \tag{M.32}$$

The diagonal matrix elements of the impurity potential may be included into and are not important at derivation of the transport equation. Consider the lowest order equation (M.17). Substituting H_E and H' into (M.18), (M.19) we obtain

$$C_l^{(0)} = eE^\alpha \sum_{l'} (\rho_{ll'}^{(0)} r_{l'l}^\alpha - r_{ll'}^\alpha \rho_{l'l}^{(0)}) = e\mathbf{E}[\hat{\rho}^{(0)}, \mathbf{r}]_l = ie\mathbf{E}\frac{\partial n_l}{\partial \mathbf{k}} \tag{M.33}$$

(hereafter we use the notation $n_1 = \rho_1^{(0)} = f(\varepsilon_1)$ for the equilibrium Fermi distribution function),

$$T_l^{(0)} = 2\pi i \sum_{l'} |\varphi_{ll'}|^2 (f_{l'} - f_l)\delta(\varepsilon_{l'} - \varepsilon_l), \tag{M.34}$$

with

$$f_l = \mathrm{Sp}\,(\hat{f}_a c_l^\dagger c_l). \tag{M.35}$$

The expressions (M.33), (M.34) yield the field and collision terms of the transport equation for the elastic scattering by impurities in the Born approximation. The results in the two next approximation were obtained by Luttinger [459]. The second Born approximation transport equation for impurity scattering reads (first-order corrections to the field term vanish)

$$T_l^{(1)}(f^{(-2)}) + T_l^{(0)}(f^{(-1)}) = 0, \tag{M.36}$$

where

$$T_l^{(1)} = -2\pi \sum_{l'l''} \delta(\varepsilon_l - \varepsilon_{l'})[(L^{(-1)})_{ll''} \langle V_{ll''} V_{l''l'} V_{l'l} \rangle +$$
$$+ (L^{(-1)})^*_{ll''} \langle V_{ll'} V_{l'l''} V_{l''l} \rangle (f_l - f_{l'})] \tag{M.37}$$

and $T_1^{(0)}$ is given by (M.34).

Consider the case where the spin-orbital interaction is present. Picking out in (M.37) the imaginary part, which is linear in SOI, and averaging over impurities we obtain

$$T_l^{(2)}(f^{(-2)}) = -(2\pi)^2 n_i \sum_{l'l''} \delta(\varepsilon_l - \varepsilon_{l'})\delta(\varepsilon_l - \varepsilon_{l''}) \times$$
$$\times (f_l^{(-2)} - f_{l'}^{(-2)}) \operatorname{Im}(\varphi_{ll'}\varphi_{l'l''}\varphi_{l''l}). \quad \text{(M.38)}$$

Assuming that the effective radius r_0 of the potential $\varphi(r)$ is small ($k_F r_0 \ll 1$) we may put

$$\varphi_{\mathbf{kk}'} \approx \int \varphi(\mathbf{r}) d\mathbf{r} = \overline{\varphi}. \quad \text{(M.39)}$$

Then, using the effective mass approximation we obtain the solution to the first Born approximation (only intraband transitions should be taken into account due to the delta functions)

$$f_l^{(-2)} = -\tau_0 e E \frac{\partial n_l}{\partial \varepsilon_l} v_l, \quad \text{(M.40)}$$

$$\tau_0^{-1} = \frac{n_i}{2\pi} \overline{\varphi}^{-2} \frac{(2m^*)^{3/2}}{\hbar^4} E_F^{1/2}. \quad \text{(M.41)}$$

To calculate corrections owing to SOI we have to expand the matrix elements (M.31) in small $|\mathbf{k}-\mathbf{k}'|$:

$$\operatorname{Im}(\varphi_{ll'}\varphi_{l'l''}\varphi_{l''l}) = \frac{i}{2} \varphi_{kk'}\varphi_{k'k''}\varphi_{k''k}(k'_\alpha - k''_\alpha)(k'_\beta - k_\beta)\left(\frac{\partial J^l_\alpha}{\partial k_\beta} - \frac{\partial J^l_\beta}{\partial k_\alpha}\right), \quad \text{(M.42)}$$

where

$$J^l_\alpha = J^{nn}_\alpha(\mathbf{k}), \qquad J^{nn'}_\alpha(\mathbf{k}) = \int u^*_{n\mathbf{k}}(\mathbf{r}) \frac{\partial u_{n'\mathbf{k}}(\mathbf{r})}{\partial k_\alpha} d\mathbf{r} \quad \text{(M.43)}$$

is proportional to SOI and purely imaginary.

Substituting (M.40), (M.42) into (M.38) we derive

$$\frac{i n_i \overline{\varphi}^{3} E_F^2 \tau_0 (2m^*)^3}{12\pi\hbar^7} \frac{\partial n_l}{\partial \varepsilon_l} e E^\alpha k_\beta \left(\frac{\partial J^l_\alpha}{\partial k_\beta} - \frac{\partial J^l_\beta}{\partial k_\alpha}\right) +$$
$$+ 2\pi n_i \overline{\varphi}^2 \sum_{\mathbf{k}} \delta(\varepsilon_{n\mathbf{k}} - \varepsilon_{n\mathbf{k}'})(f^{(-1)}_{n\mathbf{k}} - f^{(-1)}_{n\mathbf{k}'}) = 0. \quad \text{(M.44)}$$

This equation has the structure of an usual Born equation with modified field term. Its solution has the form

$$f_l^{(-1)} = -\frac{iE_F}{3n_i\overline{\varphi}} \frac{\partial n_l}{\partial \varepsilon_l} e E^\alpha k_\beta \left(\frac{\partial J^l_\alpha}{\partial k_\beta} - \frac{\partial J^l_\beta}{\partial k_\alpha}\right). \quad \text{(M.45)}$$

Now we estimate the quantities J. To first order of perturbation theory in SOI we obtain

$$u_{n\mathbf{k}} = u_{n\mathbf{k}}^{(0)} + \sum_{n' \neq n} \frac{\langle n' | \mathsf{H}_{\mathrm{so}} | n \rangle}{\varepsilon_{n\mathbf{k}} - \varepsilon_{n\mathbf{k}'}} u_{n'\mathbf{k}}; \tag{M.46}$$

$$J_{\alpha}^{l} = -2i \sum_{n' \neq n} \frac{r_{nn'}^{\alpha} \langle n' | \mathcal{H}_{\mathrm{so}} | n \rangle}{\varepsilon_{n\mathbf{k}} - \varepsilon_{n\mathbf{k}'}}, \tag{M.47}$$

where matrix elements of coordinate are connected with those of quasimomentum

$$r_{nn'}^{\alpha}(\mathbf{k}) = \frac{-iP_{nn'}^{\alpha}(\mathbf{k})}{m^{*}(\varepsilon_{n\mathbf{k}} - \varepsilon_{n\mathbf{k}'})}. \tag{M.48}$$

Putting for simplicity $\varepsilon_{n\mathbf{k}} - \varepsilon_{n'\mathbf{k}} = \Delta = \text{const}$ and substituting the expression for H_{so} we derive

$$J_{\alpha}^{l} = \frac{1}{m^{*}\Delta^{2}} \sum_{n' \neq n} ((\mathsf{H}_{\mathrm{so}})_{nn'} P_{n'n}^{\alpha} - P_{nn'}^{\alpha} (\mathsf{H}_{\mathrm{so}})_{n'n}) = \frac{\hbar}{m^{*}\Delta^{2}} [\mathsf{H}_{\mathrm{so}}, p^{\alpha}]_{n'n}. \tag{M.49}$$

Using the Poisson equation

$$\Delta V(\mathbf{r}) = -4\pi e^{2} \rho(\mathbf{r}) \tag{M.50}$$

($\rho(\mathbf{r})$ is the charge density) and (L.1) we obtain

$$J_{\alpha}^{l} = -i \frac{2\pi \hbar^{4} e^{2} \rho_{\mathrm{eff}}}{3m^{2}c^{2}m^{*}\Delta^{2}M(0)} [\mathbf{kM}]^{\alpha}, \tag{M.51}$$

where $\mathbf{M}$ is the magnetization,

$$\rho_{\mathrm{eff}} = \int |u_{n\mathbf{k}}(\mathbf{r})|^{2} \rho(\mathbf{r}) d\mathbf{r}. \tag{M.52}$$

Substituting (M.51) into (M.45) we find the Hall conductivity

$$\sigma_{yx} = \frac{e\bar{v}_{y}}{a_{0}^{3}E_{x}} = \frac{e}{a_{0}^{3}E_{x}} \sum_{l} v_{y}^{l} f_{l}^{(-1)} = t \frac{e^{2}n}{m^{*}} \frac{2\pi}{3\Delta^{2}} \mu_{\mathrm{B}}^{2} \hbar \rho_{\mathrm{eff}} \frac{M}{M(0)} \frac{E_{\mathrm{F}}}{3n_{i}\bar{\varphi}}, \tag{M.53}$$

where t is the number of bands, $n = k_{\mathrm{F}}^{3}/6\pi^{2}$ is the electron concentration. Taking into account the expression for the diagonal component of the conductivity tensor

$$\sigma_{xx} = \frac{e\bar{v}_{x}}{a_{0}^{3}E_{x}} = \frac{e}{a_{0}^{3}E_{x}} \sum_{l} v_{x}^{l} f_{l}^{(-2)} = -\tau_{0} e^{2} \sum_{l} (v_{l})^{2} \frac{\partial n_{l}}{\partial \varepsilon_{l}} = t \frac{e^{2} n \tau_{0}}{m^{*}}, \tag{M.54}$$

we obtain for the Hall coefficient the result (5.124).

M.2 SCATTERING BY PHONONS

In the case of electron–phonon interaction we have

$$\mathrm{H}_0 = \sum_l \varepsilon_l c_l^\dagger c_l + \sum_{\mathbf{q}} \omega_{\mathbf{q}} b_{\mathbf{q}}^\dagger b_{\mathbf{q}}, \tag{M.55}$$

$$\mathrm{H}' = \sum_{ll'\mathbf{q}} (Q_{ll'\mathbf{q}} c_l^\dagger c_{l'} b_{\mathbf{q}} + Q_{ll'\mathbf{q}}^* c_{l'}^\dagger c_l b_{\mathbf{q}}^\dagger), \tag{M.56}$$

where

$$Q_{ll'\mathbf{q}} = \frac{2}{3} i (2M\omega_{\mathbf{q}})^{-1/2} q C_{nn'} \delta_{\mathbf{k}-\mathbf{k}',\mathbf{q}},$$

$\omega_{\mathbf{q}}$ is the phonon frequency, $b_{\mathbf{q}}^\dagger$ and $b_{\mathbf{q}}$ are the Bose creation and annihilation operators for phonons, C is the Bloch constant in the conductivity theory [1]. Inclusion of the second term in H_0 takes into account the inelastic character of electron-phonon scattering.

Consider the lowest-order approximation (M.17). Then the field term is determined by (M.33), and the collision term is calculated from (M.19):

$$T_l^{(0)} = 2\pi i \sum_{l'\mathbf{q}} |Q_{ll'\mathbf{q}}|^2 \{[(f_{l'} - f_l)N_{\mathbf{q}} - f_l(1-n_{l'}) + f_{l'}n_{l'}]\delta(\varepsilon_{l'} - \varepsilon_l + \omega_{\mathbf{q}}) +$$
$$+[(f_{l'} - f_l)(1+N_{\mathbf{q}}) + f_l(1-n_{l'}) + f_{l'}n_{l'}]\delta(\varepsilon_{l'} - \varepsilon_l - \omega_{\mathbf{q}})\}. \tag{M.57}$$

Here we have carried out decouplings of many-particle density matrices, so that $N_{\mathbf{q}}$ and $n_{n\mathbf{k}}$ are the equilibrium Bose and Fermi functions.

The lowest-order contribution to AHE owing to the phonon scattering is described by the equation of the second order in H. The corresponding corrections to the distribution function, which are linear in SOI, occur both from field and collision terms. To simplify the consideration, we follow to paper [460] and do not take into account all such corrections, but restrict ourselves to the equation

$$C_l'^{(2)} + T_l^{(0)}(f^{(0)}) = 0 \tag{M.58}$$

(see (M.24)) and neglect $C^{(2)}$, $C''^{(2)}$ and $T^{(2)}$. The contributions neglected apparently do not influence qualitative results.

To calculate $C_1'^{(2)}$ we have to expand equilibrium distribution functions up to the first order in the scattering amplitude. Performing the decoupling of many-particle averages with the use of the theorem about expansion of exponential operators (M.6) we obtain

$$\sum_{l'\mathbf{q}} |Q_{ll'\mathbf{q}}|^2 [\delta(\varepsilon_{l'} - \varepsilon_l + \omega_{\mathbf{q}})\varphi_{ll'\mathbf{q}}^{(0)} - \delta(\varepsilon_{l'} - \varepsilon_l - \omega_{\mathbf{q}})\varphi_{ll'\mathbf{q}}^{(0)}] +$$
$$+ e\mathbf{E} \sum_{l'\mathbf{q}} |Q_{ll'\mathbf{q}}|^2 \beta n_{l'}(1-n_l)(\mathbf{r}_{l'} - \mathbf{r}_l)[\delta(\varepsilon_{l'} - \varepsilon_l + \omega_{\mathbf{q}})N_{\mathbf{q}} +$$
$$+ \delta(\varepsilon_{l'} - \varepsilon_l - \omega_{\mathbf{q}})(1+N_{\mathbf{q}})] = 0, \tag{M.59}$$

where

$$\varphi^{(i)}_{ll'\mathbf{q}} = N_{\mathbf{q}}(n_{l'} - n_l) + f^{(i)}_l (1 - n_{l'}) + f^{(i)}_{l'}. \tag{M.60}$$

For high temperatures $T \gg \omega_{\mathbf{q}}$ we have

$$\beta n_{l'}(1 - n_{l'}) = -\frac{\partial n_l}{\partial \varepsilon_l},$$

so that the solution, which is linear in SOI, has the form

$$f^{(0)}_l = ieE^{\alpha} J^l_{\alpha} \frac{\partial n_l}{\partial \varepsilon_l}, \tag{M.61}$$

where the quantity J (see (M.43)) is determined by the diagonal part of the coordinate, which is due to SOI [459]:

$$r^{\alpha}_{ll'} = i\delta_{ll'} \frac{\partial}{\partial k_{\alpha}} + iJ^{nn'}_{\alpha}(\mathbf{k})\delta_{\mathbf{kk}'}. \tag{M.62}$$

The off-diagonal part of $f^{(0)}$ is obtained from the equation (M.11):

$$\begin{aligned} f^{(0)}_{l'l} = \frac{1}{\Delta_{l'l}} \Bigg\{ & eE^{\alpha}[\hat{\rho}^{(0)}, \mathbf{r}^{\alpha}]_{l'l} - \\ & - \sum_{l''\mathbf{q}} \Bigg[Q_{l'l''\mathbf{q}} Q_{ll''\mathbf{q}} \frac{\varphi^{(-2)}_{l''l\mathbf{q}}}{\Delta^{+}_{l''l\mathbf{q}}} - Q^{*}_{l'l''\mathbf{q}} Q_{l''l\mathbf{q}} \frac{\varphi^{(-2)}_{ll''\mathbf{q}}}{\Delta^{-}_{l''l\mathbf{q}}} - \\ & - Q_{ll''\mathbf{q}} Q^{*}_{l''l'\mathbf{q}} \frac{\varphi^{(-2)}_{l'l''\mathbf{q}}}{\Delta^{+}_{l'l''\mathbf{q}}} + Q_{ll''\mathbf{q}} Q_{l'l''\mathbf{q}} \frac{\varphi^{(-2)}_{ll''\mathbf{q}}}{\Delta^{-}_{l'l''\mathbf{q}}} \Bigg] \Bigg\}, \end{aligned} \tag{M.63}$$

where

$$\Delta_{l'l} = \varepsilon_l - \varepsilon_{l'} - is, \qquad \Delta^{\pm}_{l'l\mathbf{q}} = \varepsilon_l - \varepsilon_{l'} \pm \omega_{\mathbf{q}} - is, \quad s \to 0.$$

Using (M.61) and (M.63) we can calculate average velocity of current carriers to zeroth order in Q:

$$\overline{\mathbf{v}} = \sum_l f^{(0)}_l \mathbf{v}^l + \sum_{l' \neq l} f^{(0)}_{l'l} \mathbf{v}^{ll'} \equiv \overline{\mathbf{v}}_a + \overline{\mathbf{v}}_b, \tag{M.64}$$

$$\mathbf{v}^l = \frac{\partial \varepsilon_l}{\partial \mathbf{k}}, \qquad v^{ll'}_{\alpha} = -(\varepsilon_l - \varepsilon_{l'}) J^{nn'}_{\alpha} \delta_{\mathbf{kk}'}. \tag{M.65}$$

Taking into account (M.62) and the relation

$$\sum_{n'} (J^{nn'}_{\beta} J^{n'n}_{\alpha} - J^{nn'}_{\alpha} J^{n'n}_{\beta}) = \frac{\partial J^l_{\beta}}{\partial k_{\alpha}} - \frac{\partial J^l_{\alpha}}{\partial k_{\beta}}, \tag{M.66}$$

we obtain

$$\overline{\mathbf{v}}_a = ieE^{\alpha} \sum_l \frac{\partial n_l}{\partial \varepsilon_l} \mathbf{v}^l J^l_{\alpha}, \tag{M.67}$$

$$\overline{v}_{b\beta} = ieE^{\alpha}\sum_l n_l \left(\frac{\partial J^l_{\beta}}{\partial k_{\alpha}} - \frac{\partial J^l_{\alpha}}{\partial k_{\beta}} \right), \tag{M.68}$$

where we have neglected interband transitions. Integrating (M.68) by parts and combining with (M.67) we get the final expression for the average velocity

$$\overline{v}_{\alpha} = ie\mathbf{E}\sum_l \frac{\partial n_l}{\partial \varepsilon_l} \mathbf{v}^l J^l_{\alpha}. \tag{M.69}$$

Substituting the expression (M.51) into (M.69) we obtain

$$\sigma^{\mathrm{ph}}_{yx} = -\frac{2\pi}{3}\frac{e^4\hbar^4}{m^*(mc\Delta)^2}\frac{M^z}{M(0)}\rho_{\mathrm{eff}}\sum_l \frac{\partial n_l}{\partial \varepsilon_l} v^l_x k_x. \tag{M.70}$$

Integrating over $\mathbf{k}$ and averaging over subbands yields, similar to (M.53)

$$\sigma^{\mathrm{ph}}_{yx} = \frac{2\pi}{3}\frac{e^4 n\hbar^4}{(mc\Delta)^2}\rho_{\mathrm{eff}}t\left\langle \frac{1}{m^*} \right\rangle \frac{M^z}{M(0)}. \tag{M.71}$$

Then the spontaneous Hall coefficient reads

$$R^{\mathrm{ph}}_{\mathrm{s}} = -\frac{\sigma_{yx}}{4\pi M^z}\rho^2 = -\frac{2}{3}\frac{\mu^2_{\mathrm{B}}e^2 n\hbar}{\Delta^2}\rho_{\mathrm{eff}}t\left\langle \frac{1}{m^*} \right\rangle \frac{\rho^2}{M(0)}. \tag{M.72}$$

M.3 SCATTERING BY SPIN INHOMOGENEITIES

We describe the interaction of conduction electrons with magnetic moments within the s–d exchange model with inclusion of spin-orbital interaction (L.17). In the case of high temperatures we include into the interaction of electrons with the mean field:

$$\mathsf{H}_d = -W_M \sum_i S^z_i, \qquad W_M = 2J_0\langle S^z \rangle. \tag{M.73}$$

In the lowest Born approximation we derive

$$\begin{aligned} -eE^{\alpha}\frac{\partial n_{l\pm}}{\partial k_{\alpha}} = 2\pi\sum_{l'}\{ &| I_{ll'} \pm \Lambda^z_{ll'} |^2 (f^{(-2)}_{l'\pm} - f^{(-2)}_{l\pm})K^{zz}\delta(\varepsilon_{l\pm} - \varepsilon_{l'\pm}) + \\ &+ | I_{ll'} |^2 (\Phi^{(-2)}_{l\pm,l'\pm}K^{\pm\mp} - \Phi^{(-2)}_{l'\pm l\pm}K^{\mp\pm})\delta(\varepsilon_{l\pm} - \varepsilon_{l'\pm} \pm W_M) + \\ &+ \Lambda^+_{l'l}\Lambda^-_{l'l}[(\Phi^{(-2)}_{l\pm,l'\pm}K^{+-} - \Phi^{(-2)}_{l'\pm l\pm}K^{-+})\delta(\varepsilon_{l\pm} - \varepsilon_{l'\pm} + W_M) + \\ &+ (\Phi^{(-2)}_{l\pm,l'\pm}K^{-+} - \Phi^{(-2)}_{l'\pm l\pm}K^{+-})\delta(\varepsilon_{l\pm} - \varepsilon_{l'\pm} - W_M)]\}, \end{aligned} \tag{M.74}$$

where

$$\Phi^{(i)}_{\lambda'\lambda} = f^{(i)}_{\lambda}(1-n_{\lambda'}) - f^{(i)}_{\lambda'} n_{\lambda}, \tag{M.75}$$

$$\varepsilon_{l\pm} = \varepsilon_l \pm I_{ll}\langle S^z \rangle,$$

and we have introduced the on-site averages

$$K^{zz} = \langle (S^z)^2 \rangle - \langle S^z \rangle^2, \qquad K^{\pm\mp} = \langle S^{\pm} S^{\mp} \rangle = \langle (S \pm S^z)(S \mp S^z + 1) \rangle, \tag{M.76}$$

neglecting intersite spin correlators in spirit of mean-field approximation. The equation (M.74) was first derived in [466] by the matrix method. Omitting spin-orbital terms we obtain in the one-band approximation ($l = \mathbf{k}$) for $I_{\mathbf{kk}'}$= const the result by Kasuya [422]

$$-eE^{\alpha}\frac{\partial n_{k\pm}}{\partial k_{\alpha}} = 2\pi I^2 \sum_{\mathbf{k}'} [(f_{\mathbf{k}'\pm} - f_{\mathbf{k}\pm}) K^{zz} \delta(\varepsilon_{\mathbf{k}\pm} - \varepsilon_{\mathbf{k}'\pm}) +$$

$$+ (\Phi^{(-2)}_{\mathbf{k}\pm,\mathbf{k}'\mp} K^{\pm\mp} - \Phi^{(-2)}_{\mathbf{k}'\pm,\mathbf{k}\pm} K^{\mp\pm}) \delta(\varepsilon_{\mathbf{k}\pm} - \varepsilon_{\mathbf{k}'\pm} \pm W_M)]. \tag{M.77}$$

The trial solution to the equation (M.77) has the standard form

$$f^{(-2)}_{\mathbf{k}\pm} = -eE^{\alpha} \tau_0^{\pm}(\varepsilon_{\mathbf{k}\pm}) \frac{\partial n_{\mathbf{k}\pm}}{\partial \varepsilon_{\mathbf{k}\pm}} v_x. \tag{M.78}$$

On substituting (M.78) into (M.77) and performing integration over angles we derive

$$\tau_0^{\pm}(\varepsilon) = \frac{2\pi}{v_0} \frac{\hbar^4}{I^2 \varepsilon} (2m^*)^{-3/2} \varphi^{\pm}\left(\frac{\varepsilon}{T}, \frac{W_M}{T}\right), \tag{M.79}$$

where

$$\varphi^{\pm}(\varepsilon, \eta) = \left[K^{zz} + K^{\mp\pm} \frac{1+e^{-\varepsilon}}{1+e^{-\varepsilon \pm \eta}} \right]. \tag{M.80}$$

In particular,

$$\varphi^{+}(0,\eta) = \varphi^{-}(0,\eta) \equiv \varphi(\eta) = [K^{zz} + 2K^{-+}(1+e^{\eta})]^{-1}, \tag{M.81}$$

since

$$K^{+-} = e^{W_M/T} K^{-+}.$$

For $S = 1/2$ one has

$$[\varphi(\eta)]^{-1} = 3\left(\frac{1}{4} - \langle S^z \rangle^2\right). \tag{M.82}$$

Now we may introduce the transport relaxation time

$$\tau_0 = -\frac{1}{3\pi}\frac{\hbar}{I^2}\int(\varphi^- + \varphi^+)\frac{\partial f(\varepsilon)}{\partial\varepsilon}d\varepsilon \simeq \frac{2}{3\pi}\frac{\hbar}{I^2}E_F\varphi\left(\frac{W_M}{T}\right). \tag{M.83}$$

Then the magnetic resistivity reads

$$\rho_{\text{mag}} = \left(\frac{e^2 n\tau_0}{m^*}\right)^{-1} = \frac{3\pi}{2}\frac{m^*}{e^2 n}\frac{I^2}{\hbar E_F}\left[\varphi\left(\frac{W_M}{T}\right)\right]^{-1}. \tag{M.84}$$

Now we consider the case of low temperatures $T \ll T_C$ where one can use the spin-wave approximation. Using the Holstein-Primakoff representation for spin operators we derive

$$\begin{aligned} \mathsf{H}' = &-(2S)^{1/2} I\sum_{\mathbf{kq}}(c^\dagger_{\mathbf{k-q}\uparrow}c_{\mathbf{k}\downarrow}b^\dagger_{\mathbf{q}} + c^\dagger_{\mathbf{k}\downarrow}c_{\mathbf{k-q}\uparrow}b_{\mathbf{q}}) - \\ &- I\sum_{\mathbf{kpq}\sigma}\sigma(I + \sigma\Lambda^z_{\mathbf{k,k+q-p}})c^\dagger_{\mathbf{k}\sigma}c_{\mathbf{k+q-p}\sigma}b^\dagger_{\mathbf{q}}b_{\mathbf{p}}. \end{aligned} \tag{M.85}$$

The lowest-order collision term is calculated similar tò case of electron–phonon interaction. We have

$$\begin{aligned} T^{(0)}_{\mathbf{k}\pm} = 2\pi I^2 S\sum_{\mathbf{q}} N_{\mathbf{q}}\Bigg[&\frac{f(\pm\varepsilon_{\mathbf{k}\pm})\exp(\beta\omega_{\mathbf{q}})}{f(\pm\varepsilon_{\mathbf{k}\pm} - \omega_{\mathbf{q}})} f_{\mathbf{k+q}\pm} - \\ &- \frac{f(\pm\varepsilon_{\mathbf{k}\pm} - \omega_{\mathbf{q}})}{f(\pm\varepsilon_{\mathbf{k}\pm})} f_{\mathbf{k+q}\pm}\Bigg]\delta(\varepsilon_{\mathbf{k+q}\mp} - \varepsilon_{\mathbf{k}\pm} \pm \omega_{\mathbf{q}}). \end{aligned} \tag{M.86}$$

The solution to the corresponding transport equation (M.17) which describes scattering by spin waves has the form

$$f_{\mathbf{k}\sigma} = \frac{1}{T}\frac{\partial n_{\mathbf{k}\sigma}}{\partial\varepsilon_{\mathbf{k}\sigma}}\chi_{\alpha\sigma}(\varepsilon_{\mathbf{k}\sigma})k_\alpha, \tag{M.87}$$

$$\chi_{\alpha\sigma}(\varepsilon) = C_{\alpha\sigma} + \psi_{\alpha\sigma}(\varepsilon).$$

The energy dependence of $\chi(\varepsilon)$ is needed to satisfy the integral equation. However, it turns out that, as well as in the case of phonon scattering [1], this dependence results in temperature corrections to resistivity of higher orders only; however, it is important at considering the anomalous Hall effect. Substituting (M.87) into the transport equation yields for the isotropic electron spectrum

$$-eE^x \frac{\partial n_{\mathbf{k}}}{\partial \varepsilon_{\mathbf{k}}} = 2I^2 S \frac{ka_0}{4\pi}\left(\frac{\partial \varepsilon_{\mathbf{k}}}{\partial \mathbf{k}}\right)^{-1} \frac{T}{T_C}\int_\delta^\infty N_B(Tx) f(\pm\varepsilon_{\mathbf{k}}) f(\pm\varepsilon_{\mathbf{k}} - Tx)\times$$

$$\times\{\chi_{x\pm}(\varepsilon_{\mathbf{k}} \mp Tx)(1-\alpha x) - \chi_{x\pm}(\varepsilon_{\mathbf{k}})\}e^{\pm\beta\varepsilon_{\mathbf{k}}}\, dx, \qquad \text{(M.88)}$$

where we have neglected the small spin splitting,

$$\delta = \frac{T}{T_0}, \qquad T_0 \sim \left(\frac{I}{E_F}\right)^2 T_C(k_F a_0) \to 0, \qquad \alpha = \frac{T}{2T_C}(ka_0)^{-2}.$$

Integrating (M.88) over ε we evaluate the constant C

$$C_x = -eE^x \frac{4\pi ka_0}{I^2 S\gamma}\left(\frac{\partial \varepsilon_{\mathbf{k}}}{\partial \mathbf{k}}\right)^2 \left(\frac{T_C}{T}\right)^2, \qquad \text{(M.89)}$$

with

$$\gamma = \int_{-\infty}^{\infty} \bar{\gamma}(y)dy, \qquad \bar{\gamma}(y) = \int_\delta^\infty N_B(Tx) f(Ty) f(T(x+y)) e^y\, dx.$$

Then the equation for ψ takes the form

$$\int_0^\infty N_B(Tx) f(\varepsilon) f(\varepsilon - Tx)[\psi_x(\varepsilon - Tx) - \psi_x(\varepsilon)]e^{\beta\varepsilon}\, dx =$$

$$= -\frac{2\pi eE^x}{I^2 Ska_0}\left(\frac{\partial \varepsilon_{\mathbf{k}}}{\partial \mathbf{k}}\right)^2 \frac{T_C}{T}\frac{\partial f(\varepsilon)}{\partial \varepsilon}. \qquad \text{(M.90)}$$

One can see that $\psi_x \sim T_C/T$, so that the contribution of the function ψ to resistivity is proportional to $(T/T_C)^3$ and may be neglected. Integrating over $\mathbf{k}$ we derive for the conductivity

$$\sigma_{xx} = \frac{4}{3\pi\gamma}\frac{e^2 n}{I^2 Sm^*}\frac{a_0}{v_F}(k_F a_0)^3 \left(\mathbf{k}\frac{\partial \varepsilon_{\mathbf{k}}}{\partial \mathbf{k}}\right)_{k_F} \left(\frac{T_C}{T}\right)^2, \qquad \text{(M.91)}$$

which agrees with the result (5.62).

The next-order approximation transport equation is important for consideration of the contribution to the anomalous Hall effect in ferromagnets owing to magnetic scattering. At high temperatures we obtain [466]

$$\sum_{l'} |I_{ll'}|^2 K^{zz}(f_{l\pm}^{(-1)} - f_{l'\pm})\delta(\varepsilon_{l\pm} - \varepsilon_{l'\pm}) +$$
$$+ (K^{\pm\mp}\Phi_{l\pm,l'\pm}^{(-1)} - K^{\mp\pm}\Phi_{l'\pm,l\pm}^{(-1)})\delta(\varepsilon_{l\pm} - \varepsilon_{l'\pm} \pm W_M)\times$$
$$\times 2\pi i \sum_{l'} W_{l\pm}^{(1)} \Lambda_{ll'}^{z} (f_{l\pm}^{(-2)} - f_{l'\pm}^{(-2)})\delta(\varepsilon_{l\pm} - \varepsilon_{l'\pm}) = 0, \qquad \text{(M.92)}$$

where $f^{(-2)}$ is the solution to the equation (M.74),

$$W_{l\pm}^{(1)} = \sum_{l''} |I_{ll''}|^2 \{\langle (S^z - \langle S^z \rangle)^3 \rangle \delta(\varepsilon_{l\pm} - \varepsilon_{l''\pm}) +$$
$$+ [\langle (S^z - \langle S^z \rangle) S^{\pm} S^{\mp} \rangle \pm (2K^{zz} - K^{\pm\mp}) n_{l''\mp} \pm K^{\mp\pm} n_{l\pm} +$$
$$+ 2\langle S^z \rangle n_{l''\mp} n_{l\pm}] \delta(\varepsilon_{l\pm} - \varepsilon_{l''\pm} \pm W_M)\}. \qquad \text{(M.93)}$$

In the one-band approximation, on substituting into (M.92) the solution to the lowest-order transport equation (M.78) and the matrix elements of SOI (L.18) we obtain

$$-eE^x \frac{\partial n_{\mathbf{k}\pm}}{\partial \varepsilon_{\mathbf{k}\pm}} v_y W_{\pm}^{(1)} \tau_0^{\pm}(\varepsilon_{\mathbf{k}\pm}) \frac{a_0^3 k m^*}{4\pi\hbar^3} \left(\frac{64}{35} \bar{l} \frac{I^{(1)}}{\Delta E} \lambda + \frac{1}{2} \lambda' \right) +$$
$$+ I^2 \sum_{\mathbf{k}} [K^{zz}(f_{\mathbf{k}\pm}^{(-1)} - f_{\mathbf{k}'\pm}^{(-1)})\delta(\varepsilon_{\mathbf{k}\pm} - \varepsilon_{\mathbf{k}'\pm}) +$$
$$+ (K^{\pm\mp}\Phi_{\mathbf{k}\pm,\mathbf{k}'\pm}^{(-1)} - K^{\mp\pm}\Phi_{\mathbf{k}'\pm,\mathbf{k}\pm}^{(-1)})\delta(\varepsilon_{\mathbf{k}\pm} - \varepsilon_{\mathbf{k}'\pm})] = 0. \qquad \text{(M.94)}$$

The solution to (M.94) is searched as

$$f_{\mathbf{k}\sigma}^{(-1)} = -eE^x \tau^{\sigma}(\varepsilon_{\mathbf{k}\sigma}) \frac{\partial n_{\mathbf{k}\sigma}}{\partial \varepsilon_{\mathbf{k}\sigma}} v_y. \qquad \text{(M.95)}$$

Then we derive

$$\tau^{\sigma}(\varepsilon_{\mathbf{k}\sigma}) = \frac{m^* k}{2\hbar} a_0^3 \lambda_{\text{eff}} W_{\sigma}^{(1)} [\tau_0^{\sigma}(\varepsilon_{\mathbf{k}})]^2, \qquad \text{(M.96)}$$

with

$$\lambda_{\text{eff}} = \frac{32}{35} \frac{I^{(1)}}{\Delta E} \bar{l} \lambda + \lambda'. \qquad \text{(M.97)}$$

Now we can calculate the Hall current

$$j_y = 2e \sum_{\mathbf{k}\sigma} f_{\mathbf{k}\sigma}^{(-1)} v_y.$$

The result for the Hall coefficient reads

$$R_{\mathrm{s}}^{\mathrm{mag}}=\frac{9\pi}{32}\left(\frac{I}{E_{\mathrm{F}}}\right)^2\frac{m^*\lambda_{\mathrm{eff}}}{e^2n\hbar}\frac{S}{M(0)}\left\{K^{zz}+(4\langle S^z\rangle)^{-1}\left[2K^{zz}-\right.\right.$$

$$\left.\left.-\frac{1}{2}(K^{+-}+K^{-+})+\langle S^z\rangle\coth\frac{\beta W_M}{2}\right]\frac{\sinh\beta W_M-\beta W_M}{\cosh\beta W_M-1}\right\}. \qquad (M.98)$$

In particular, using the equation for magnetization (4.14) we obtain for $S=1/2$

$$R_{\mathrm{s}}^{\mathrm{mag}}=\frac{9\pi}{64}\left(\frac{I}{E_{\mathrm{F}}}\right)^2\frac{m^*\lambda_{\mathrm{eff}}}{e^2n\hbar}\frac{1/4-\langle S^z\rangle^2}{M(0)}\left[1+\coth\frac{\beta W_M}{2}\frac{\sinh\beta W_M-\beta W_M}{\cosh\beta W_M-1}\right]. \qquad (M.99)$$

At neglecting the weak temperature dependence of the function in the square brackets the result (M.99) may be represented in the form (5.130).

The transport equation of the next Born approximation with account of linear corrections in the spin-orbital interaction, which describes electron-magnon scattering at low temperatures, has the form

$$T_\lambda^{(0)}(f^{(-1)})+T_\lambda^{(1)}(f^{(-2)})=0, \qquad (M.100)$$

where the field term $C^{(1)}$ is zero (as well as for impurity scattering), $T^{(0)}$ is given by (M.86), the solution to the lowest Born equation is given by (M.87). To calculate T we write down for (M.85)

$$\mathsf{H}'=\mathsf{H}''+\mathsf{H}''',$$

where H'' contains the exchange terms and H''' the spin-orbital ones. Using (M.23) and the properties

$$\mathsf{H}''_{n_1n_2}=\mathsf{H}''_{n_2n_1},\qquad \mathsf{H}'''_{n_1n_2}=-\mathsf{H}'''_{n_2n_1}, \qquad (M.101)$$

calculating the commutators and performing the decouplings, which are valid to the order under consideration, we obtain:

$$T_{\mathbf{k}\pm}^{(1)}=3\pi I^2S\sum_{\mathbf{pq}}N_{\mathbf{q}}N_{\mathbf{p}}\times$$

$$\times\left\{\Lambda_{\mathbf{k},\mathbf{k}+\mathbf{q}-\mathbf{p}}^z\delta(\varepsilon_{\mathbf{k}-\mathbf{p}\mp}-\varepsilon_{\mathbf{k}\pm}\pm\omega_{\mathbf{q}})\delta(\varepsilon_{\mathbf{k}-\mathbf{p}\mp}-\varepsilon_{\mathbf{k}\pm}\mp\omega_{\mathbf{q}}\pm\omega_{\mathbf{p}})\times\right.$$

$$\times\left[\frac{f(\pm\varepsilon_{\mathbf{k}\pm})f(\pm\varepsilon_{\mathbf{k}\pm}+\omega_{\mathbf{q}}-\omega_{\mathbf{p}})}{[f(\pm\varepsilon_{\mathbf{k}\pm}-\omega_{\mathbf{p}})]^2}e^{\beta(\omega_{\mathbf{q}}+\omega_{\mathbf{p}})}f_{\mathbf{k}-\mathbf{p}\mp}-\right. \qquad (M.102)$$

$$\left.-\frac{f(\pm\varepsilon_{\mathbf{k}\pm})}{f(\pm\varepsilon_{\mathbf{k}\pm}+\omega_{\mathbf{q}}-\omega_{\mathbf{p}})}e^{\beta\omega_{\mathbf{p}}}f_{\mathbf{k}+\mathbf{q}-\mathbf{p}\mp}\right]+$$

$$+\Lambda^{z}_{\mathbf{k-q,k-p}}\delta(\varepsilon_{\mathbf{k-p}\mp}-\varepsilon_{\mathbf{k}\pm}\pm\omega_{\mathbf{p}})\delta(\varepsilon_{\mathbf{k-q}\mp}-\varepsilon_{\mathbf{k}\pm}\pm\omega_{\mathbf{p}})\times$$

$$\left.\times\frac{f(\pm\varepsilon_{\mathbf{k}\pm}-\omega_{\mathbf{q}})}{f(\pm\varepsilon_{\mathbf{k}\pm}-\omega_{\mathbf{p}})}e^{\beta\omega_{\mathbf{p}}}f_{\mathbf{k-p}\mp}\right\}.$$

An interesting feature of the equation (M.100) should be noted. Substituting $f^{(2)}\sim C_x=\text{const}$ into (M.102) yields zero after integration over ε in the lowest order in q/k, and higher orders terms in q result in higher powers of T/T_{C}. Therefore one has to take into account the energy dependence of the function (M.87). Then the solution of (M.100) yields [472]

$$f^{(-1)}_{\mathbf{k}\sigma}=\frac{3a_0}{16\pi}\left(\frac{\partial\varepsilon_{\mathbf{k}\sigma}}{\partial\mathbf{k}}\right)^{-1}\frac{T}{T_{\mathrm{C}}}\left[4,4\lambda\overline{I}\frac{I^{(1)}}{\Delta E}\frac{T}{T_{\mathrm{C}}}-0,24\lambda'-\nu(\mp\varepsilon)\right]\frac{\partial n_{\mathbf{k}\sigma}}{\partial\varepsilon_{\mathbf{k}\sigma}}k_y, \quad \text{(M.103)}$$

with

$$\nu(\varepsilon)=\lambda\overline{I}\frac{I^{(1)}}{\Delta E}\int_{\delta}^{\infty}\left(N_{\mathrm{B}}(\omega)\omega e^{\beta\omega}\frac{f(\varepsilon+\omega)}{f(\varepsilon)}-\lambda'\ln(1+e^{\beta\omega})\right)d\omega.$$

Calculating σ_{yx} with the use of (M.103) and using the expression for the diagonal conductivity (M.91) we find the expression for the Hall coefficient

$$R^{\mathrm{mag}}_{\mathrm{s}}=-\frac{3\pi}{512}\frac{2\hbar I^2 S}{e^2 k_{\mathrm{F}} M(0)}\left(\mathbf{k}\frac{d\varepsilon_{\mathbf{k}}}{d\mathbf{k}}\right)^{-3}_{k_{\mathrm{F}}}\left[1,1\lambda\overline{I}\frac{I^{(1)}}{\Delta E}\left(\frac{T}{T_{\mathrm{C}}}\right)^4+0,8\lambda'\left(\frac{T}{T_{\mathrm{C}}}\right)^3\right], \quad \text{(M.104)}$$

where the largest T^3-term occurs due to the energy dependence of $\psi(\varepsilon)$.

Appendix N

DEGENERATE ANDERSON MODEL

The periodic Anderson model describes the situation where highly correlated $d(f)$-electrons do not participate directly in the band motion, but are hybridized with the conduction band states. Such a situation takes place in a number of rare-earth and actinide compounds (Chapter 6). The hybridization (many-configuration) picture is sometimes useful also for discussing some properties of transition d-metals and other d-electron systems. For example, strong p–d hybridization takes place in copper-oxide high-T_c superconductors (Sect.6.7). At neglecting spin-orbit coupling, which is appropriate for transition metals and their compounds, we write down the Anderson-lattice Hamiltonian in the form

$$\mathsf{H} = \mathsf{H}_0 + \sum_{\mathbf{k}\sigma} t_{\mathbf{k}} c^{\dagger}_{\mathbf{k}\sigma} c_{\mathbf{k}\sigma} + \sum_{\mathbf{k}lm\sigma} (V_{\mathbf{k}lm} c^{\dagger}_{\mathbf{k}\sigma} a_{\mathbf{k}lm\sigma} + \text{h. c.}), \tag{N.1}$$

where H_0 is the Hamiltonian of intrasite interaction between d-electrons. A symmetry analysis of hybridization mixing in various situations is performed in the reviews [708, 565]. To simplify the model consideration, we describe the states of conduction electrons by plane waves. Using the expansion in spherical harmonics (C.28) we obtain for the matrix element of hybridization

$$V_{\mathbf{k}lm} = i^l Y^*_{lm}(\hat{\mathbf{k}}) v_l(k), \tag{N.2}$$

where

$$v_l(k) = 4\pi \int r^2 R_l(r) v(r) j_l(kr) dr \tag{N.3}$$

and $v(r)$ is a spherically symmetric potential of a given site. In the limit of jj-coupling (actinide compounds) one has to replace in (N.1) $lm\sigma \to j\mu$ with $j = l \pm 1/2$ being the total electron momentum and m its projection.

In the case of strong correlations for d-electrons it is convenient to pass to the representation of the Hubbard operators which reduces H_0 to the diagonal form (H.2). Retaining two lowest terms $\Gamma_n = \{SL\}$, $\Gamma_{n-1} = \{S'L'\}$ for the configurations d^n and d^{n-1} and defining new conduction electron operators

$$d^{\dagger}_{\mathbf{k}lm\sigma} = \sum_{\mu\mu' MM'} C^{S\mu}_{S'\mu', \frac{1}{2}\sigma} C^{LM}_{L'M', lm} X_{\mathbf{k}}(SL\mu M, S'L'\mu' M'), \tag{N.4}$$

$$c^{\dagger}_{\mathbf{k}lm\sigma} = i^l Y^*_{lm}(\hat{\mathbf{k}}) c^{\dagger}_{\mathbf{k}\sigma},$$

we present the Hamiltonian (N.1) in the form

$$\mathsf{H} = \mathsf{H}_0 + \sum_{\mathbf{k}\sigma} \left[t_{\mathbf{k}} c^{\dagger}_{\mathbf{k}\sigma} c_{\mathbf{k}\sigma} + \tilde{v}_l(k)(c^{\dagger}_{\mathbf{k}lm\sigma} d_{\mathbf{k}lm\sigma} + \text{h. c.}) \right], \tag{N.5}$$

where

$$\mathsf{H}_0 = \Delta \sum_{\mathbf{k}lm\sigma} d^{\dagger}_{\mathbf{k}lm\sigma} d_{\mathbf{k}lm\sigma} + \text{const}, \tag{N.6}$$

$$\Delta = E_{SL} - E_{S'L'} - \zeta$$

(we have passed to the grand canonical ensemble by introducing the chemical potential ζ), effective hybridization parameters are given by

$$\tilde{v}_l(k) = n^{1/2} G^{S_n L_n}_{S_{n-1} L_{n-1}} v_l(k). \tag{N.7}$$

Now we discuss rare-earth systems. Because of strong Coulomb interaction between $4f$-electrons, formation of the f-bands, containing 14 electron states, is non-realistic. Thus one has to use the model with two configurations f^n and $s(d)f^{n-1}$, which corresponds to delocalization of only one f-electron per atom. In the Russel-Saunders scheme we may confine ourselves to two lower multiplets of the $4f$-ion, $\Gamma_n = SLJ$ and $\Gamma_{n-1} = S'L'J'$. Passing in (N.1), (A.31) to the J-representation with the use of (B.5) and summing the product of Clebsh–Gordan coefficients

$$\sum_{\mu\mu' MM'} C^{JM_J}_{S\mu, LM} C^{J'M'_J}_{S'\mu', L'M'} C^{LM}_{L'M', lm} C^{S\mu}_{S'\mu', \frac{1}{2}\sigma} =$$

$$= \sum_{j\mu} \begin{Bmatrix} S & L & J \\ S' & L' & J' \\ 1/2 & l & j \end{Bmatrix} ([j][j'][L])^{1/2} C^{j\mu}_{\frac{1}{2}\sigma, lm} C^{JM_J}_{J'M'_J, j\mu}, \tag{N.8}$$

we derive the Hamiltonian, which is formally similar to that in the case of jj-coupling,

$$\mathsf{H} = \sum_{\mathbf{k}j\mu} \left[\Delta f^{\dagger}_{\mathbf{k}j\mu} f_{\mathbf{k}j\mu} + t_{\mathbf{k}} c^{\dagger}_{\mathbf{k}j\mu} c_{\mathbf{k}j\mu} + \tilde{v}_j(k)(c^{\dagger}_{\mathbf{k}j\mu} f_{\mathbf{k}j\mu} + \text{h. c.}) \right]. \tag{N.9}$$

where we have introduced new electron operators

$$f_{\mathbf{k}j\mu}^{\dagger} = \sum_{M_J M_{J'}} C_{J'M',j\mu}^{JM_J} X_{\mathbf{k}}(SLJM_J, S'L'J'M'_J),$$

$$c_{\mathbf{k}j\mu}^{\dagger} = i^l \sum_{m\sigma} C_{\frac{1}{2}\sigma,lm}^{j\mu} Y_{lm}^{*}(\hat{\mathbf{k}}) c_{\mathbf{k}\sigma}^{\dagger}, \tag{N.10}$$

and the effective hybridization parameters are expressed in terms of a $9j$-symbol

$$\tilde{v}_j(k) = \begin{Bmatrix} S & L & J \\ S' & L' & J' \\ 1/2 & l & j \end{Bmatrix} ([j][j'][L])^{1/2} G_{S'L'}^{SL} v_l(k). \tag{N.11}$$

Thus hybridization effects in ME systems depend strongly on ME quantum numbers S, L, J and, consequently, on the atomic number [709]. Such a dependence in the rare-earth element series is similar to de Gennes correlation for the s–f exchange parameter and paramagnetic Curie temperature. Its experimental investigation, e.g., in spectroscopic data, is of great interest.

In the case where $|\Delta|$ is large in comparison with the width of the $d(f)$ level we may exclude the hybridization term from the Hamiltonians (N.5), (N.9) by a canonical transformation to obtain, respectively,

$$\begin{aligned} \mathcal{H} = -\frac{1}{\Delta} \sum v_l(k) v_l(k') C_{L'M',lm}^{LM} C_{S'\mu',\frac{1}{2}\sigma}^{S\mu} C_{L'M'',lm'}^{LM''} C_{S'\mu'',\frac{1}{2}\sigma'}^{S\mu''} \times \\ \times [X_{\mathbf{k}-\mathbf{k}'}(SL\mu''M'', SL\mu M) \delta_{\mu'\mu''} \delta_{M'M} + \\ + X_{\mathbf{k}-\mathbf{k}'}(S'L'\mu'M', S'L'\mu'''M''') \delta_{\mu\mu''} \delta_{MM''}] c_{\mathbf{k}lm\sigma}^{\dagger} c_{\mathbf{k}'lm'\sigma'}, \end{aligned} \tag{N.12a}$$

$$\begin{aligned} \mathcal{H} = -\frac{1}{\Delta} \sum v_j(k) v_{j'}(k') C_{J'M',j\mu}^{JM} C_{J'M'',j'\mu'}^{JM''} \times \\ \times [X_{\mathbf{k}-\mathbf{k}'}(JM'', JM) \delta_{M'M''} + \\ + X_{\mathbf{k}-\mathbf{k}'}(J'M', J'M''') \delta_{MM''}] c_{\mathbf{k}j\mu}^{\dagger} c_{\mathbf{k}'j'\mu'}. \end{aligned} \tag{N.12b}$$

For $\Delta < 0$ ($\Delta > 0$) the filling of the level is n ($n-1$) and only the first (second) term in the brackets of (N.12) should be retained.

The expressions (N.12) describe the exchange interaction of conduction electrons with $d(f)$-electrons. It should be noted that in the case under consideration the interaction is highly anisotropic because of spherical harmonics which enter (N.4), (N.10). This should result in a strong anisotropy of indirect RKKY-type f–f interaction which is obtained in the second order in the s–f exchange. Such an anisotropy is observed in a number of rare-earth and actinide compounds. Using the identity

$$\sum_{M'} C^{LM}_{L'M',lm} C^{LM''}_{L'M',lm'} =$$

$$= \sum_{pq} (-1)^{p-q} [p] \left(\frac{[L]}{[l]} \right)^{1/2} \begin{Bmatrix} l & l & p \\ L & L' & L' \end{Bmatrix} C^{LM''}_{LM,pq} C^{lm}_{lm',p-q}, \qquad \text{(N.13)}$$

the Hamiltonians (N.12) may be decomposed into sum of terms which correspond to interaction of conduction electrons with different multipole components of orbital and spin (or total momentum) degrees of freedom.

As well as in the usual s–f exchange model [552], perturbation theory expansion in the models (N.12) yields logarithmic corrections to various physical quantities which indicates reconstruction of the system state at low temperatures. In particular, such a correction to electron self-energy and resistivity occurs in the third order in I. Unfortunately, the complicated tensor structure of the Hamiltonians (N.12) prevents the calculation of the unique energy scale for the infrared divergences (the Kondo temperature). However, such a calculation may be performed in the case where the energy of the $d(f)$ level Δ does not depend on the many-electron term and is determined by the number of electrons only (see Sect.6.2).

Consider the anticommutator retarded Green's function of localized d-electrons (H.3) in the non-magnetic phase of the model (N.1). The simplest decoupling yields

$$G_{\mathbf{k}lm}(E) = \left[\Phi(E) - \frac{|V_{\mathbf{k}lm}|^2}{E - t_{\mathbf{k}}} \right]^{-1}, \qquad \text{(N.14)}$$

where the function Φ is defined by (H.5). The corresponding energy spectrum contains a system of subbands, separated by hybridization gaps (or pseudogaps, provided that $V(\mathbf{k})$ vanishes for some $\mathbf{k}$) which are surrounded by density-of-states peaks. In the model with strong correlations (N.7), we have

$$E^{1,2}_{\mathbf{k}} = \frac{1}{2}(t_{\mathbf{k}} + \Delta) \pm \left[\frac{1}{4}(t_{\mathbf{k}} - \Delta)^2 + |\tilde{V}_{\mathbf{k}lm}|^2 \right], \qquad \text{(N.15)}$$

with

$$\tilde{V}_{\mathbf{k}lm} = i^l Y^*_{lm}(\hat{\mathbf{k}})\tilde{v}_l(k)\left\{\frac{[S][L]}{2[l]}(N_{SL} + N_{S'L'})\right\}^{1/2}. \tag{N.16}$$

One can see that the width of hybridization gap depends appreciably on the many-electron occupation numbers (in particular, on the position of d-level). The approximation (N.14) does not take into account spin-flip processes, which result in the Kondo effect and can change substantially the structure of electron spectrum near the Fermi level. To take into account the Kondo anomalies, we perform a more accurate calculation of the Green's function. For brevity we consider the model (N.9); in the model (N.5)

$$[J] \to [S][L], \qquad \tilde{v}_j \to \tilde{v}_l.$$

It is convenient to use the operators, which are averaged over angles of the vector $\mathbf{k}$:

$$f^\dagger_{kjm} = \int f^\dagger_{\mathbf{k}jm} d\hat{\mathbf{k}}, \qquad c^\dagger_{kjm} = \int c^\dagger_{\mathbf{k}jm} d\hat{\mathbf{k}}. \tag{N.17}$$

Following to the consideration of the SU(N) Anderson model in Sect.6.4, we write down the equation of motion

$$\begin{aligned}(E-\Delta)\langle\langle f_{\mathbf{k}jm} \mid f^\dagger_{kjm}\rangle\rangle_E &= R_j(1+\tilde{v}_j(k)\langle\langle c_{\mathbf{k}jm} \mid f^\dagger_{kjm}\rangle\rangle_E) + \\ &+ \sum_{j'm'\mu M\mathbf{q}} \tilde{v}_{j'}(q) C^{J\mu}_{J'M',jm} \left\langle\left\langle \left[\sum_{M'} C^{J\mu}_{J'M',jm'} X_{\mathbf{k}-\mathbf{q}}(J'M, J'M') + \right.\right.\right. \\ &\left.\left.\left. + \sum_{\mu'} C^{J\mu'}_{J'M,j'm'} X_{\mathbf{k}-\mathbf{q}}(J\mu', J\mu)\right] c_{\mathbf{q}j'm'} \mid f^\dagger_{kjm} \right\rangle\right\rangle_E,\end{aligned} \tag{N.18}$$

where we have carried out a decoupling for the term, which describes the processes without changing m,

$$R_j = \frac{1}{[j]}\left\{\frac{[J]}{[J']} - \left(\frac{[J]}{[J']} - 1\right)\sum_M \langle X(JM, JM)\rangle\right\}. \tag{N.19}$$

Further we neglect for simplicity the above-discussed influence of hybridization gap, which is possible provided that the latter lies far below the Fermi level (note that the corresponding contributions are formally small in the inverse degeneracy of f-level, $1/N$). Carrying out decouplings in the equations for the Green's function in the right-hand side of (N.18) we obtain

$$(E-t_q)\langle\langle X_{\mathbf{k-q}}(J\mu',J\mu)c_{\mathbf{q}j'm'} \mid f^\dagger_{kjm}\rangle\rangle_E = \\ = \tilde{v}_{j'}(q)n_q\sum_{M'} C^{J\mu'}_{J'M',j'm'}\langle\langle X_{\mathbf{k-q}}(J'M,J\mu'') \mid f^\dagger_{kjm}\rangle\rangle_E,$$

$$(E-t_q)\langle\langle X_{k-q}(J'M,J'M')c_{\mathbf{q}j'm'} \mid f^\dagger_{kjm}\rangle\rangle_E = \\ = -\tilde{v}_{j'}(q)n_q\sum_{\mu''} C^{J\mu''}_{J'M',j'm'}\langle\langle X_{\mathbf{k-q}}(J'M,J\mu'') \mid f^\dagger_{kjm}\rangle\rangle_E, \quad \text{(N.20)}$$

with $n_k = \langle c^\dagger_{kjm}c_{kjm}\rangle = f(t_k)$ being the Fermi distribution functions. Substituting (N.20) into (N.19), averaging over angles and using the orthogonality relations for the Clebsh–Gordan coefficients, we find

$$\langle\langle f_{kjm} \mid f^\dagger_{kjm}\rangle\rangle_E = R_J[E-\Delta-\Sigma_f(E)]^{-1}, \quad \text{(N.21)}$$

$$\Sigma_f(E) = 2\rho\sum_j \tilde{v}^2_j(k_F)\frac{J-J'}{[J']}\ln\left|\frac{W}{E}\right|. \quad \text{(N.22)}$$

where we have used the appıoximation (6.5). At $J > J'$ the Green's function (N.21) has the pole

$$|\Delta^*| = T_K \approx W\exp\left\{-\left(\frac{[J]}{[J']}-1\right)^{-1}|\Delta|\left[\rho\sum_j \tilde{v}^2_j(k_F)\right]^{-1}\right\}. \quad \text{(N.23)}$$

The usual Kondo effect corresponds to the total compensation of magnetic moment ($J'=0$). At $J' > J$ the pole (N.23) is absent (the strong coupling regime does not occur) since the model under consideration is mapped by a canonical transformation into a Coqblin–Schrieffer model with a positive exchange parameter. An analogue of the result (N.23) for d-impurities has the form

$$T_K = W\exp\left\{-\left(\frac{[S][L]}{[S'][L']}-1\right)^{-1}\frac{|\Delta|}{\rho\tilde{v}^2_l(k_F)}\right\}. \quad \text{(N.24)}$$

Note that the formula (N.24) satisfies the condition of the particle-hole symmetry ($n \to n' = 2[l]+1-n$, $\Delta \to -\Delta$) owing to the relation for the fractional parentage coefficients

$$\left\{G^{SL}_{S'L'}(n',n'-1)\right\}^2 = \frac{n[S'][L']}{n'[S][L]}\left\{G^{S'L'}_{SL}(n,n-1)\right\}^2. \quad \text{(N.25)}$$

At neglecting the dependence $\Delta(LS)$, all the ME terms of configurations d^n and d^{n-1} contribute equally to spin-flip processes, and, consequently, to the Kondo temperature. Then the coefficients G which enter (N.7) may be summed up in the equations of motion with the use of orthogonality relations (A.8), (A.9) to obtain for the Kondo temperature

$$T_K \approx W \exp\left[-(2[l]+1-2n)^{-1}(|I|\rho)^{-1}\right], \qquad I = \frac{\tilde{v}_l^2(k_F)}{\Delta}. \qquad \text{(N.26)}$$

Note that the expression (N.26) may be represented in the the form, which is similar to (N.24),

$$T_K \approx W \exp\left[-\left(\frac{2[l]+1-2n}{n}-1\right)^{-1}\frac{1}{n|I|\rho}\right], \qquad \text{(N.27)}$$

with the factor $(2[l]+1-n)/n$ being again the ratio of statistical weights for the configurations d^n and d^{n-1}:

$$\frac{2[l]+1-n}{n} = \frac{(2[l])!}{n!(2[l]-n)!} \Big/ \frac{(2[l])!}{(n-1)!(2[l]-n+1)!}.$$

The result (N.26) differs from the result of high-temperature perturbation theory (6.22) by an unity in the denominator of the exponential only. Such a difference is typical for the calculation of the Kondo temperature in the degenerate Anderson model [565, 574] and is explained by that this approach is justified, strictly speaking, only in the large-N limit.

Appendix O

MEAN-FIELD APPROXIMATION FOR THE GROUND STATE OF MAGNETIC KONDO LATTICES

To construct the mean-field approximation describing the ground state of Kondo lattices we use the Abrikosov pseudofermion representation for localized spins $S = 1/2$ (which coincides with the Dirac representation (B.1))

$$\mathbf{S}_i = \frac{1}{2}\sum_{\sigma\sigma'} f^\dagger_{i\sigma}\sigma_{\sigma\sigma'} f_{i\sigma'} \tag{O.1}$$

with the subsidiary condition

$$f^\dagger_{i\uparrow}f_{i\uparrow} + f^\dagger_{i\downarrow}f_{i\downarrow} = 1.$$

Making the saddle-point approximation for the path integral describing the spin-fermion interacting system [711] one can reduce the Hamiltonian of the s–f exchange interaction to the effective hybridization model:

$$-I\sum_{\sigma\sigma'} c^\dagger_{i\sigma}c_{i\sigma}\left(\sigma_{\sigma\sigma'}\mathbf{S}_i - \frac{1}{2}\delta_{\sigma\sigma'}\right) \rightarrow f_i^\dagger V_i c_i + c_i^\dagger V_i^\dagger f_i - \frac{1}{2I}\mathrm{Sp}\,(V_i V_i^\dagger), \tag{O.2}$$

where the vector notations are used

$$f_i^\dagger = (f^\dagger_{i\uparrow}, f^\dagger_{i\downarrow}), \qquad c_i^\dagger = (c^\dagger_{i\uparrow}, c^\dagger_{i\downarrow}),$$

V is the effective hybridization matrix which is determined from a minimum of the free energy. Coleman and Andrei [711] considered the formation of a spin-liquid state in the two-dimensional situation. Here we treat the more simple case of ferromagnetic ordering following to [608]. The Heisenberg Hamiltonian of the f-subsystem H_f is considered in the mean-field approximation. For a ferromagnet we have $(\bar{S} = \langle S^z \rangle)$

$$\mathsf{H}_f = -J_0\bar{S}\sum_{i\sigma}\sigma f^\dagger_{i\sigma}f_{i\sigma}, V_{\sigma\sigma'} = V_\sigma\delta_{\sigma\sigma'} \tag{O.3}$$

and the Hamiltonian of the s–f model takes the form

$$\mathsf{H} - \zeta n = \sum_{\mathbf{k}\sigma}\left[(t_{\mathbf{k}} - \zeta)c^{\dagger}_{\mathbf{k}\sigma}c_{\mathbf{k}\sigma} + w_{\sigma}f^{\dagger}_{\mathbf{k}\sigma}f_{\mathbf{k}\sigma} + V_{\sigma}(c^{\dagger}_{\mathbf{k}\sigma}f_{\mathbf{k}\sigma} + f^{\dagger}_{\mathbf{k}\sigma}c_{\mathbf{k}\sigma})\right] + \text{const}, \tag{O.4}$$

where

$$w_{\sigma} = w - \sigma J_0 \overline{S},$$

w being the energy of 'f-level'. The equations for w, the chemical potential ζ and $\overline{S}$ read

$$n^{\dagger}_{\sigma} \equiv \sum_{\mathbf{k}}\langle f^{\dagger}_{\mathbf{k}\sigma}f_{\mathbf{k}\sigma}\rangle = \frac{1}{2} + \sigma\overline{S}; \tag{O.5}$$

$$n = \sum_{\mathbf{k}}\langle c^{\dagger}_{\mathbf{k}\sigma}c_{\mathbf{k}\sigma}\rangle. \tag{O.6}$$

One can see that $-w$ plays the role of the chemical potential for pseudofermions (note that the numbers of electrons and pseudofermions are conserved separately). After the minimization one obtains the equation for V_{σ}

$$V_{\sigma} = 2I\sum_{\mathbf{k}}\langle f^{\dagger}_{\mathbf{k}\sigma}c_{\mathbf{k}\sigma}\rangle. \tag{O.7}$$

Diagonalizing the Hamiltonian (O.4) by a canonical transformation

$$c^{\dagger}_{\mathbf{k}\sigma} = u_{\mathbf{k}\sigma}\alpha^{\dagger}_{\mathbf{k}\sigma} - v_{\mathbf{k}\sigma}\beta^{\dagger}_{\mathbf{k}\sigma}, \qquad f^{\dagger}_{\mathbf{k}\sigma} = u_{\mathbf{k}\sigma}\beta^{\dagger}_{\mathbf{k}\sigma} - v_{\mathbf{k}\sigma}\alpha^{\dagger}_{\mathbf{k}\sigma}, \tag{O.8}$$

$$u_{\mathbf{k}\sigma} = \cos(\theta_{\mathbf{k}\sigma}/2), \qquad v_{\mathbf{k}\sigma} = \sin(\theta_{\mathbf{k}\sigma}/2),$$

with

$$\sin\theta_{\mathbf{k}\sigma} = \frac{2V_{\sigma}}{E_{\mathbf{k}\sigma}}, \qquad \cos\theta_{\mathbf{k}\sigma} = \frac{t_{\mathbf{k}} - \zeta - w_{\sigma}}{E_{\mathbf{k}\sigma}}, \tag{O.9}$$

we obtain the energy spectrum of a hybridization form

$$E^{\alpha,\beta}_{\mathbf{k}\sigma} = \frac{1}{2}(t_{\mathbf{k}} - \zeta + w_{\sigma} \pm E_{\mathbf{k}\sigma}). \tag{O.10}$$

Then the equations (O.5)–(O.7) take the form

$$n^{f}_{\sigma} = \frac{1}{2}\sum_{\mathbf{k}}\left[(1 - \cos\Theta_{\mathbf{k}\sigma})n^{\alpha}_{\mathbf{k}\sigma} + (1 + \cos\Theta_{\mathbf{k}\sigma})n^{\beta}_{\mathbf{k}\sigma}\right], \tag{O.11}$$

$$n = \frac{1}{2}\sum_{\mathbf{k}\sigma}\left[(1 + \cos\Theta_{\mathbf{k}\sigma})n^{\alpha}_{\mathbf{k}\sigma} + (1 - \cos\Theta_{\mathbf{k}\sigma})n^{\beta}_{\mathbf{k}\sigma}\right], \tag{O.12}$$

$$1 = -2I\sum_{\mathbf{k}}(n^{\beta}_{\mathbf{k}\sigma} - n^{\alpha}_{\mathbf{k}\sigma})/E_{\mathbf{k}\sigma}. \tag{O.13}$$

At small $|V_{\sigma}|$, $|w_{\sigma}|$ and $T = 0$ we have

$$\cos\Theta_{\mathbf{k}\sigma} \approx \text{sign}(t_{\mathbf{k}} - \zeta - w_{\sigma}), \tag{O.14}$$

so that the equations (O.11) and (O.12) are simplified. The edges of the hybridization gaps in spin subbands are given by

$$E^{\alpha}_{\mathbf{k}\sigma} > w_\sigma + V_\sigma^2/\zeta, \qquad E^{\beta}_{\mathbf{k}\sigma} < w_\sigma - V_\sigma^2/(W-\zeta). \tag{O.15}$$

Further we consider different types of ferromagnetic solutions. We confine ourselves to the case where the conduction electron concentration $n<1$ (the results for $n>1$ are obtained after the particle-hole transformation). As follows from (O.15), at not too large $\bar{S}$ for both σ the condition

$$w_\sigma > V_\sigma^2/(W-\zeta), \tag{O.16}$$

holds, i.e. the chemical potential lies below the energy gap, as well as in the non-magnetic case. Define the function $\zeta(c)$ by

$$c = 2\int_0^{\zeta(c)} \rho(E)dE, \tag{O.17}$$

with $\rho(E)$ $(0<E<W)$ being the bare density of states. Then the equation (O.12) takes the form $\zeta(n)=\zeta$, and (O.11) and (O.13) yield

$$\lambda_\sigma \equiv V_\sigma^2/w_\sigma = \zeta(n+2n_\sigma^f) - \zeta(n), \tag{O.18}$$

$$1 = -2I\int_0^{\zeta+\lambda_\sigma} \frac{\rho(E)}{[(E-\zeta-w_\sigma)^2 + 4V_\sigma^2]^{1/2}}dE. \tag{O.19}$$

Calculating the integral in (O.19) to leading and next-order terms in $1/\ln|W/V_\sigma|$ we obtain

$$1 = -4I\rho\ln\left|\frac{C}{V_\sigma}\right| - 2I\left[\int_C^{\zeta}\frac{\rho(E+\zeta)}{E}dE + \int_{C+\zeta}^{\zeta(n+2n_\sigma^f)}\frac{\rho(E)}{E-\zeta}dE\right], \tag{O.20}$$

where C is a cutoff parameter which does not enter final results. In the leading approximation V_σ does not depend on σ and we have

$$|V_\sigma| \sim (WT_{\mathrm{K}})^{1/2}, \qquad T_{\mathrm{K}} \equiv W\exp\frac{1}{2I\rho}. \tag{O.21}$$

Retaining the next order terms, we calculate the ratio and obtain the self-consistent equation for magnetization

$$\tanh\left(\frac{1}{4\rho}\int_{\zeta(n+1-2\bar{S})}^{\zeta(n+1+2\bar{S})}\frac{\rho(E)-\rho}{E-\zeta}dE\right) = \frac{J_0\bar{S}}{w}, \tag{O.22}$$

where

$$w \approx V^2/[\zeta(n+1)-\zeta] \sim T_{\mathrm{K}}. \tag{O.23}$$

The equation (O.22) has no non-trivial solutions for the bare DOS

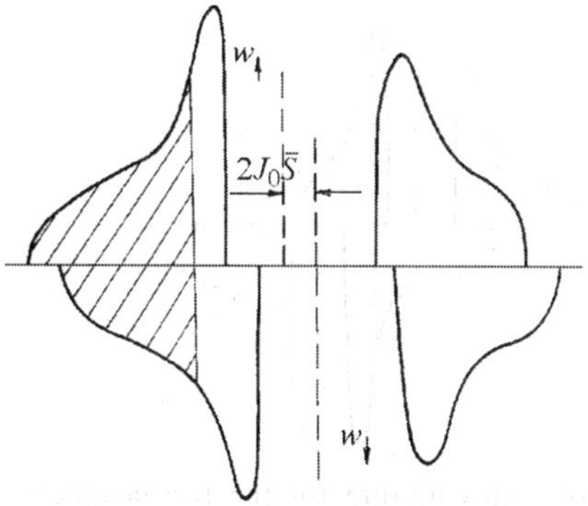

Fig.O.1. The density of states picture for the ferromagnetic solution with a small spin splitting.

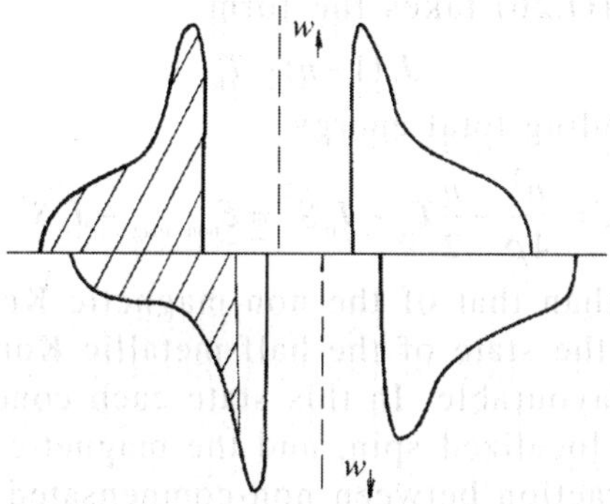

Fig.O.2. The density of states picture for the saturated (half-metallic) ferromagnetic solution.

$\rho(E) = \text{const}$. However, solutions with $\bar{S} \neq 0$ may occur for some $\rho(E)$ if both the left and right-hand side of (O.22) are of order unity, i.e. $J_0 \sim T_K$ (see Fig.O.1).

Provided that

$$w_\downarrow > V_\downarrow^2/(W-\zeta), \qquad w_\uparrow < -V_\uparrow^2/\zeta, \tag{O.24}$$

i.e. ζ lies in the energy gap for $\sigma = \uparrow$, we obtain the 'half-metallic' ferromagnetic solution with

$$n_\uparrow^f = 1 - n/2, \qquad n_\downarrow^f = n/2, \qquad \bar{S} = (1-n)/2 \tag{O.25}$$

(see Fig.O.2). Such a solution exists under the condition

$$-\varphi/\zeta < [\zeta(2n) - \zeta]^{-1} - J_0(1-n)/V_\downarrow^2 < \varphi/(W-\zeta), \tag{O.26}$$

$$\varphi \equiv \left|\frac{V_\uparrow}{V_\downarrow}\right|^2 = \exp\left(\frac{1}{\rho}\int_{\zeta(2n)}^{W} \frac{\rho(E)}{E-\zeta} dE\right). \tag{O.27}$$

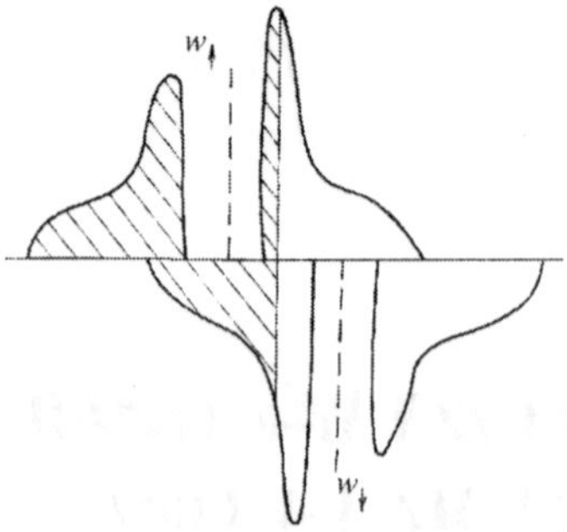

Fig.O.3. The density of states picture for the ferromagnetic solution where the spin splitting exceeds the energy gap.

For $\rho(E) = \text{const}$ (O.26) takes the form

$$J_0(1-n) < T_K \tag{O.28}$$

and the corresponding total energy

$$\mathcal{E} = \frac{n^2}{4\rho} - \frac{n}{2}T_K - J_0\overline{S}^2 = \mathcal{E}_{\text{non-mag}} - J_0\overline{S}^2 \tag{O.29}$$

is always lower than that of the non-magnetic Kondo state. Thus the formation of the state of the half-metallic Kondo ferromagnet is energetically favourable. In this state each conduction electron compensates one localized spin, and the magnetic ordering is due to exchange interaction between non-compensated moments. Such a picture is reminiscent of the situation in the narrow-band ferromagnet in the Hubbard or *s–d* exchange model with large intrasite interaction (Sect.4.6). In our case the bare interaction is small, but effective interaction is large in the strong coupling regime.

The expression (O.29) should be compared with the energy of the usual ferromagnetic state with the Kondo effect suppressed:

$$\mathcal{E} = \frac{n^2}{4\rho} - J_0/4 \quad (V_\sigma = 0, \quad \overline{S} = 1/2), \tag{O.30}$$

We see that the latter state becomes energetically favourable at

$$J_0\left(1 - \frac{n}{2}\right) > T_K. \tag{O.31}$$

At the critical point, a first-order transition takes place.

Third type of possible ferromagnetic solutions correspond to the situation of large energy splitting where ζ lies in the lower hybridization subband for $\sigma = \downarrow$ and in the upper one for $\sigma = \downarrow$ (Fig.O.3). However, such solutions (at least for $\rho(E) = \text{const}$) are energetically unfavourable [608].

Appendix P

SCHWINGER AND DYSON–MALEEV REPRESENTATIONS IN THE THEORY OF TWO-DIMENSIONAL HEISENBERG ANTIFERROMAGNETS

The Schwinger boson representation for spin operators has the form, which is similar to the Abrikosov representation (O.1), the Fermi operators being replaced by Bose operators $b_{i\sigma}$:

$$\mathbf{S}_i = \frac{1}{2}\sum_{\sigma\sigma'} b^\dagger_{i\sigma}\sigma_{\sigma\sigma'}b_{i\sigma'},$$

$$S_i^+ = b^\dagger_{i\uparrow}b_{i\downarrow}, \qquad S_i^- = b^\dagger_{i\downarrow}b_{i\uparrow}, \qquad S_i^z = \frac{1}{2}(b^\dagger_{i\uparrow}b_{i\uparrow} - b^\dagger_{i\downarrow}b_{i\downarrow}). \tag{P.1}$$

For the given localized spin S, these operators should satisfy at each site i the subsidiary condition (constraint)

$$\sum_\sigma b^\dagger_{i\sigma}b_{i\sigma} = 2S. \tag{P.2}$$

It should be noted that the 'hyperbolic' operators (B.14) which change the value of momentum may be easily represented in terms of the Schwinger bosons [656]:

$$K^+ = b^\dagger_\uparrow b^\dagger_\downarrow, \qquad K^- = b_\downarrow b_\uparrow, \qquad K^z = \frac{1}{2}(b^\dagger_\uparrow b_\uparrow + b^\dagger_\downarrow b_\downarrow + 1). \tag{P.3}$$

The representation (P.1) turns out to be convenient at considering low-dimensional systems which do not possess long-range ordering at finite temperatures, but demonstrate developed spin fluctuations (short-range order).

Consider the 'mean field' approximation for the two-dimensional Heisenberg model within the self-consistent spin-wave theory [623]. (This differs drastically from the usual mean field approximation of Sect 4.1 since permits to describe the strong short-range order above the ordering point.) The constraint (P.2) is taken into account in the average by introducing the Lagrange multiplier λ which does not depend on i. Anomalous averages $\langle b_{i\uparrow} b_{j\downarrow} \rangle$ describe singlet pairing of bosons, i.e. the short-range order parameters. Occurence of long-range ordering with the wavevector $\mathbf{Q}$ corresponds to the Bose–Einstein condensation with the quasimomenta $\mathbf{k}=\pm\mathbf{Q}/2$. It is convenient to introduce the interaction with small external magnetic field H. Then the Bogoliubov transformation

$$b_{\mathbf{Q}/2+\mathbf{k}\uparrow} = \cosh\frac{\Theta_{\mathbf{k}}}{2}\alpha_{\mathbf{k}} - \sinh\frac{\Theta_{\mathbf{k}}}{2}\beta^{\dagger}_{-\mathbf{k}},$$

$$b_{\mathbf{Q}/2-\mathbf{k}\downarrow} = \cosh\frac{\Theta_{\mathbf{k}}}{2}\beta_{-\mathbf{k}} - \sinh\frac{\Theta_{\mathbf{k}}}{2}\alpha^{\dagger}_{\mathbf{k}}, \tag{P.4}$$

reduces the Heisenberg Hamiltonian to the diagonal form

$$\mathsf{H}_d = \sum_{\mathbf{k}} (E^{\alpha}_{\mathbf{k}} \alpha^{\dagger}_{\mathbf{k}} \alpha_{\mathbf{k}} + E^{\beta}_{\mathbf{k}} \beta^{\dagger}_{\mathbf{k}} \beta_{\mathbf{k}}), \tag{P.5}$$

where, for the square lattice with the parameter $a_0 = 1$,

$$E^{\alpha,\beta}_{\mathbf{k}} = E_{\mathbf{k}} \mp \left(\frac{1}{2}H - 2|J|\langle S^z \rangle\right), \qquad E_{\mathbf{k}} = (\lambda^2 - \gamma^2_{\mathbf{k}})^{1/2}, \tag{P.6}$$

$$\sinh\Theta_{\mathbf{k}} = \gamma_{\mathbf{k}}/E_{\mathbf{k}}, \qquad \cosh\Theta_{\mathbf{k}} = \lambda/E_{\mathbf{k}}, \tag{P.7}$$

with

$$\gamma_{\mathbf{k}} = \frac{1}{2}\gamma(\sin k_x + \sin k_y).$$

Equations for λ and γ, which are obtained from (P.7), have the form

$$2S+1 = \frac{1}{N}\sum_{\mathbf{k}} \frac{\lambda}{E_{\mathbf{k}}}(1 + N_{\mathbf{k}\alpha} + N_{\mathbf{k}\beta}), \tag{P.8}$$

$$1 = \frac{1}{2N}\sum_{\mathbf{k}} \frac{|J|}{E_{\mathbf{k}}}(\sin k_x + \sin k_y)^2 (1 + N_{\mathbf{k}\alpha} + N_{\mathbf{k}\beta}). \tag{P.9}$$

where $N_{\mathbf{k}\alpha,\beta} = N_{\mathrm{B}}(E^{\alpha,\beta}_{\mathbf{k}})$, N is the number of lattice sites which should be retained explicitly in the problem under consideration.

At $T=0$ we have $\lambda=\gamma$, and $N_{\mathbf{k}\alpha}$ (but not $N_{\mathbf{k}\beta}$) contains the condensate term at $E^{\alpha}_{\mathbf{k}}=0$, i.e.

$$\mathbf{k} = \pm\mathbf{Q}/2, \qquad E_{\mathbf{k}} = \frac{1}{2}H - 2J\langle S^z \rangle \sim H. \tag{P.10}$$

Then we obtain

$$N_{\pm Q/2} = N\langle S^z \rangle = NE_{Q/2}n_B/\lambda, \qquad (P.11)$$

with $2n_B$ being the density of condensed bosons. Equation (P.8) yields

$$n_B = S + \frac{1}{2} - \frac{1}{2N}\sum_{\mathbf{k}}\left[1 - \frac{1}{4}(\sin k_x + \sin k_y)^2\right]^{-1/2} = S - 0.197, \qquad (P.12)$$

so that n_B equals the ground state sublattice magnetization of the Neel antiferromagnet $\bar{S}(0)$ with account of zero-point spin-wave corrections.

At finite T the boson spectrum (P.6) contains the gap and the condensate is absent. Then we may put $H=0$, $N_{\mathbf{k}\alpha,\beta} = N_{\mathbf{k}}$ from the beginning.

Consider the spin spectral density

$$K_{\mathbf{q}}(\omega) = -\frac{1}{\pi} N_B(\omega) \operatorname{Im} \langle\langle \mathbf{S}_{\mathbf{q}} | \mathbf{S}_{-\mathbf{q}} \rangle\rangle_\omega \qquad (P.13)$$

in the Schwinger boson representation. The spin Green's function are expressed in terms of polarization operators of non-interacting bosons α and β, and we derive

$$\begin{aligned} K_{\mathbf{q}}(\omega) = \frac{1}{4}\sum_{\mathbf{k}} \sum_{\nu,\mu=\alpha,\beta} \Big\{ &(2-\delta_{\mu\nu})\cosh^2\frac{\Theta_{\mathbf{k}} - \Theta_{\mathbf{k+q}}}{2} \times \\ &\times N_{\mathbf{k}\mu}(1+N_{\mathbf{k+q}\nu})\delta(\omega + E^\nu_{\mathbf{k+q}} - E^\mu_{\mathbf{k}}) + \\ &+ (1+\delta_{\mu\nu})\sinh^2\frac{\Theta_{\mathbf{k}} - \Theta_{\mathbf{k+q}}}{2}[N_{\mathbf{k}\mu}N_{\mathbf{k+q}\nu}\delta(\omega - E^\nu_{\mathbf{k+q}} - E^\mu_{\mathbf{k}}) + \\ &+ (1+N_{\mathbf{k}\mu})(1+N_{\mathbf{k+q}\nu})\delta(\omega + E^\nu_{\mathbf{k+q}} + E^\mu_{\mathbf{k}})]\Big\}. \qquad (P.14) \end{aligned}$$

As follows from (P.11), (P.14), the spectral density contains at $T=0$, $H \to 0$ the delta-function contribution

$$\delta K_{\mathbf{q}}(\omega) = \frac{3}{2} n_B^2 N \delta_{\mathbf{qQ}} \delta(\omega). \qquad (P.15)$$

The factor of 3/2 in (P.15) should in fact be omitted since it is an artifact of the mean field approximation which yields, because of inaccurate account of the condition (P.2),

$$\langle \mathbf{S}_i^2 \rangle = \frac{3}{2} S(S+1),$$

violating thereby the corresponding sum rule at the site.

At finite temperatures we have

$$E_{\mathbf{k}}^2 = \frac{1}{2}\lambda^2\left[(\mathbf{k}\mp\mathbf{Q}/2)^2 + \xi^{-2}\right], \quad \mathbf{k}\to\pm\mathbf{Q}/2, \tag{P.16}$$

where the correlation length is given by

$$\xi \sim \exp(\pi\lambda n_{\mathrm{B}}/2T). \tag{P.17}$$

At $\mathbf{q}=\mathbf{Q}$ the integral in (P.14) is "almost" divergent at the points $\mathbf{k}=\pm\mathbf{Q}/2$ with the cutoff scale $|\mathbf{k}\mp\mathbf{Q}/2| \sim \xi^{-1}$. Using the expansion $N_{\mathbf{k}} \cong T/E_{\mathbf{k}}$ we present the corresponding singular contribution in the form

$$\delta K_{\mathbf{q}}(\omega) = \frac{3}{2}\left[\frac{2T}{\pi\lambda}\ln\xi\right]^2 \Delta_{\mathbf{q}}\Delta_{\omega} \approx \frac{3}{2}n_{\mathrm{B}}^2\Delta_{\mathbf{q}}\Delta_{\omega}, \tag{P.18}$$

where $\Delta_{\mathbf{q}}$ and Δ_{ω} are $\delta(\mathbf{q}-\mathbf{Q})$ and $\delta(\omega)$ like functions smoothed on the scales ξ^{-1} and $\omega_{\xi} \sim J/\xi$ respectively. At $T \ll J$ we have $\omega_{\xi} \ll T$ and we may neglect the smoothing which yields a formal description of strong short-range order. To obtain the temperature dependence of the coefficient in (P.18) ('sublattice magnetization') we estimate the intensity of the Ornstein–Cernike peak at $\mathbf{q}=\mathbf{Q}$ in the static spin correlator

$$S_{\mathrm{eff}}(T) = \left(\sum_{|\mathbf{q}-\mathbf{Q}|<q_0} \langle \mathbf{S}_{-\mathbf{q}}\mathbf{S}_{\mathbf{q}}\rangle\right)^{1/2}, \tag{P.19}$$

where $q_0 \gg \xi^{-1}$ is a cutoff wavevector. Using the result of the scaling consideration [712], which yields the correct preexponential factor

$$\xi = C_{\xi}\exp\left(\pi c\overline{S}(0)/\sqrt{2T}\right), \tag{P.20}$$

where c is the magnon velocity and $C_{\xi} \cong 0.01/(2\pi) \ll 1$ is a numerical factor. Substituting (P.20) into (P.19), (P.18) and neglecting $\ln q$ in comparison with $\ln C_{\xi}$ we obtain the linear dependence [713]

$$S_{\mathrm{eff}}(T) = \overline{S}(0) - \frac{\sqrt{2T}}{\pi c}|\ln C_{\xi}|. \tag{P.21}$$

The contributions to spectral density from antiferromagnetic spin waves are determined by the terms, which are linear in n_{B}. Putting $E_{\mathbf{k}}\to 0$, $N_{\mathbf{k}} = T/E_{\mathbf{k}}$, but retaining $E_{\mathbf{k}+\mathbf{q}}$ and vice versa, and performing integration over $\mathbf{k}$, we derive

$$\delta_1 K_{\mathbf{q}}(\omega)=\frac{3}{2}\frac{T}{E_{\mathbf{q}+\mathbf{Q}/2}}\frac{1}{N}\sum_{\mathbf{k}}\frac{\lambda^2-\gamma_{\mathbf{k}}\gamma_{\mathbf{k}+\mathbf{q}}}{E_{\mathbf{k}}^2}\times$$
$$\times[(1+N_{\mathbf{q}+\mathbf{Q}/2})\delta(\omega+E_{\mathbf{q}+\mathbf{Q}/2})+N_{\mathbf{q}+\mathbf{Q}/2}\delta(\omega-E_{\mathbf{q}+\mathbf{Q}/2})]\approx$$
$$\approx\frac{3}{2}n_{\mathrm{B}}\left(\frac{1-\varphi_{\mathbf{q}}}{1+\varphi_{\mathbf{q}}}\right)^{1/2}\{[1+N_{\mathrm{B}}(\omega_{\mathbf{q}})]\delta(\omega+\omega_{\mathbf{q}})+N_{\mathrm{B}}(\omega_{\mathbf{q}})\delta(\omega-\omega_{\mathbf{q}})\},\quad (\mathrm{P.22})$$

where, for q, $|\mathbf{q}-\mathbf{Q}|\gg\xi^{-1}$,

$$\omega_{\mathbf{q}}=\lambda\left(1-\varphi_{\mathbf{q}}^2\right)^{1.2}\approx E_{\mathbf{q}+\mathbf{Q}/2},\quad \varphi_{\mathbf{q}}\equiv\frac{1}{2}\left(\cos q_x+\cos q_y\right)\qquad(\mathrm{P.23})$$

so that $\omega_{\mathbf{q}}$ is the renormalized magnon frequency.

For comparison, we consider the application of the Dyson–Maleev representation (E.2) to the same problem. In the case of a two-sublattice antiferromagnet this has the form

$$S_l^-=(2S)^{1/2}a_l^\dagger,\qquad S_l^+=(2S)^{1/2}\left(1-\frac{1}{2S}a_l^\dagger a_l\right)a_l,$$
$$S_l^z=S-a_l^\dagger a_l,\quad l\in A,$$
$$S_m^-=(2S)^{1/2}b_m^\dagger,\qquad S_m^+=(2S)^{1/2}b_m^\dagger\left(1-\frac{1}{2S}b_m^\dagger b_m\right)b_m,$$
$$S_m^z=-S+b_m^\dagger b_m,\quad m\in B.\qquad(\mathrm{P.24})$$

In the self-consistent approach [624] one puts on each site the condition $\langle S^z\rangle=0$, i.e.

$$\langle a_l^\dagger a_l\rangle=\langle b_l^\dagger b_l\rangle=S.\qquad(\mathrm{P.25})$$

Using the Bogoliubov transformation

$$a_{\mathbf{k}}=\cosh\frac{\Theta_{\mathbf{k}}}{2}\alpha_{\mathbf{k}}-\sinh\frac{\Theta_{\mathbf{k}}}{2}\beta_{-\mathbf{k}}^\dagger,$$
$$b_{-\mathbf{k}}^\dagger=\cosh\frac{\Theta_{\mathbf{k}}}{2}\beta_{-\mathbf{k}}^\dagger-\sinh\frac{\Theta_{\mathbf{k}}}{2}\alpha_{\mathbf{k}},\qquad(\mathrm{P.26})$$

we diagonalize the Hamiltonian to obtain

$$\mathcal{H}=\sum_{\mathbf{k}}{}'\omega_{\mathbf{k}}(\alpha_{\mathbf{k}}^\dagger\alpha_{\mathbf{k}}+\beta_{\mathbf{k}}^\dagger\beta_{\mathbf{k}}),\qquad(\mathrm{P.27})$$

with

$$\langle\alpha_{\mathbf{k}}^\dagger\alpha_{\mathbf{k}}\rangle=\langle\beta_{\mathbf{k}}^\dagger\beta_{\mathbf{k}}\rangle=N_{\mathbf{k}}=N_{\mathrm{B}}(\omega_{\mathbf{k}}),$$
$$\omega_{\mathbf{k}}=(\lambda^2-\gamma^2\varphi_{\mathbf{k}}^2)^{1/2},\qquad \tanh 2\Theta_{\mathbf{k}}=\gamma\varphi_{\mathbf{k}}/\lambda.\qquad(\mathrm{P.28})$$

Here Σ' stands for the sum over the reduced Brillouin zone. The equations for λ and γ read

$$S+\frac{1}{2}=\sum_{\mathbf{k}}{}'\frac{\lambda}{\omega_{\mathbf{k}}}(1+2N_{\mathbf{k}}), \tag{P.29}$$

$$1=\sum_{\mathbf{k}}{}'\frac{|J|}{\omega_{\mathbf{k}}}(\cos k_x+\cos k_y)(1+2N_{\mathbf{k}}). \tag{P.30}$$

At $T=0$ (in more general case, in the presence of the long-range ordering), $\lambda=\gamma+O(1/N)$, so that the spectrum is gapless. The $1/N$-corrections describe the Bose condensation:

$$\left(1-\gamma^2/\lambda^2\right)^{-1/2}=Nn_{\mathrm{B}}. \tag{P.31}$$

As follows from the constraint (P.25) and the structure of the Dyson–Maleev transformation, the transverse spin correlation function $\langle S^+_{\mathbf{q}}S^-_{-\mathbf{q}}\rangle$ vanishes. Thus the corresponding spectral density reads

$$\begin{aligned} K_{\mathbf{q}}(\omega)=K^{zz}_{\mathbf{q}}(\omega)= \\ =\frac{1}{N}\sum_{\mathbf{k}}{}'\Bigg\{\cosh\frac{\Theta_{\mathbf{k}}-\Theta_{\mathbf{k+q}}}{2}N_{\mathbf{k}}(1+N_{\mathbf{k+q}})\delta(\omega+\omega_{\mathbf{k+q}}-\omega_{\mathbf{k}})+ \\ +\frac{1}{2}\sinh^2\frac{\Theta_{\mathbf{k}}-\Theta_{\mathbf{k+q}}}{2}[N_{\mathbf{k}}N_{\mathbf{k+q}}\delta(\omega-\omega_{\mathbf{k+q}}-\omega_{\mathbf{k}})+ \\ +(1+N_{\mathbf{k}})(1+N_{\mathbf{k+q}})\delta(\omega+\omega_{\mathbf{k+q}}+\omega_{\mathbf{k}})]\Bigg\}. \end{aligned} \tag{P.32}$$

The delta-function contribution is

$$\delta K_{\mathbf{q}}(\omega)=Nn_{\mathrm{B}}^2\delta_{\mathbf{qQ}}\delta(\omega). \tag{P.33}$$

Unlike (P.15), the expression (P.33) does not contain the superfluous factor 3/2. On the other hand, in contrast with the 'isotropic' Schwinger representation, the Dyson–Maleev representation violates the rotational invariance even in the paramagnetic phase.

The versions of self-consistent spin-wave theory (SSWT), based on non-linear boson representations, yield a gap in the spin-wave spectrum in the absence of long-range order. This contradicts, e.g., to exact results for one-dimensional lattice with half-integer spin S where the spectrum is gapless (for integer spin, the spectrum contains so-called Haldane gap [714] which is reproduced qualitatively by SSWT). However, as discussed in [622], such approaches are satisfactory in the cases where the ground state possesses long-range order.

It should be noted that both approaches under consideration describe the long-range order in terms of delta-function singularity of the spin correlation function, the sublattice magnetization being always zero. The possibility of such a description in the three-dimensional situation for a wide range of physical properties, including local characteristics, is demonstrated in [715]. In particular, in such an antiferromagnetic state without sublattices the Green's function of a nuclear spin at each site turns out to have both poles $\pm AS$ (A is the hyperfine interaction parameter).

Applications of the non-linear boson representations to systems with a weak interlayer coupling and magnetic anisotropy are discussed in [713].

CONCLUSIONS

In the present book we have attempted to consider the whole variety of physical properties of transition metals. The characteristics of TM are much more complicated and interesting than those of simple metals. Simultaneous consideration of all the properties permits to establish some regularities with filling of *d*- and *f*-shells and their connection with the electronic structure. One might expect similarity of the properties within a given transition series, since, with some exceptions, the total number of external *sp*-electrons does not change.

In fact, chemical and most physical properties (crystal structure etc.) of rare earths are very close. At the same time, magnetic characteristics of 4*f*-metals change appreciably with the number of *f*-electrons. Moreover, as demonstrated in Sect.4.8.2, there exists some periodicity within the 4*f*-series, so that the latter forms a miniature "periodic table". This regularity is connected with the many-electron term structure of the 4*f*-shell.

On the other hand, non only magnetic, but also other properties of *d*-metals depend strongly on the number of *d*-electrons n_d, and they play an important role in the formation of a crystal. An appreciable delocalization of *d*-states is directly observed in the experimental investigations of the Fermi surface.

In some anomalous rare earth and actinide systems, *f*-electrons also take part in formation of the Fermi surface, due to both one-electron mechanisms (hybridization) and many-electron effects (as in Kondo lattices). This may lead to occurrence of states with rather exotic properties, e.g. to greatly enhanced electronic effective masses (Chapter 6). In metallic cerium, direct *f–f* overlap plays an important role.

The "dual" (localized vs. itinerant) nature of *d*-electrons requires using various approaches for analyzing physical properties of TM and their compounds. One can pick out two main approaches. The first one starts from first-principle one-electron band calculations. According to the Hohenberg-Kohn theorem, these calculations can provide an accurate description of some ground state characteristics. At the same time, standard band calculations, which use the local density approximation, are often insufficient for narrow *d*-, 5*f*- and especially 4*f*-bands. Besides that, the density functional approach is, generally speaking, unable to describe the whole excitation spectrum and thermodynamic properties.

The band approach can explain anomalies in TM properties, which are connected with singularities of electron density of states. Presence of such singularities leads sometimes to considerable modifications of standard formulas of solid state theory, e.g., for electronic specific heat and paramagnetic susceptibility. It should be also noted that for "flat" regions of spectrum with small values of grad $E(\mathbf{k})$, i.e. for density of states peaks, many-electron perturbation corrections become large and the role of correlation effects increases.

The second approach takes into account electron correlations in a microscopic way starting from the atomic picture. The adequacy of this picture is evident for strongly localized $4f$-states. However, atomic features are retained to some extent for d-states too. In particular, this is confirmed by a characteristic "toothed" n_d-dependence of electronic properties in transition metals and by many-electron term effects in spectral measurements.

In this connection, the problem of strong (ten-fold) degeneracy of d-states seems to be important. Although the degeneracy is lifted at broadening of the atomic levels into energy bands by the periodic lattice potential, it survives in some points of the Brillouin zone. This degeneracy is important, e.g., for the orbital momenta which determine anisotropy of a number of properties. It should be stressed that such a scheme of degeneracy lift takes place in the one-electron picture only, and the classification of electron states changes in the many-electron representation where additional quantum numbers occur. The corresponding new quasiparticles may possess different degree of localization and mobility. This may change essentially the results of the standard band theory.

From the point of view of the qualitative microscopic description, investigation of simple theoretical models, which include effects of strong intraatomic electron correlations, turns out to be very useful. Such effects turn out to be especially bright for some d- and f-compounds. In the case of narrow bands (large Coulomb interaction) the correlations result in a radical rearrangement of electron spectrum formation of the Hubbard subbands. A convenient tool for describing atomic statistics of excitations in such a situation is the formalism of many-electron Hubbard operators. On the other hand, even small interaction among localized and itinerant electrons may result in a reconstruction of electron spectrum at low temperatures owing to peculiarities of resonance scattering in many-particle systems (the Kondo effect).

Methods which combine band structure calculations and model considerations seem to be promising. As discussed in the book, such approaches were developed, e.g, for transition metal oxides and heavy-fermion systems.

In a number of cases, correlation effects are reduced to a modification of electron spectrum and density of states (e.g., formation of hybridization gap or Abrikosov-Suhl resonance in intermediate valence and Kondo systems), so that electron properties may be further calculated in a phenomenological way with the use of one-electron theory results. However, the modifications of electron spectrum parameters themselves may not be obtained in the standard band theory. In particular, the parameters are often strongly temperature dependent owing to many-electron renormalizations.

On the other hand, sometimes the excitation spectrum is not described within the usual quasiparticle picture and has an essentially incoherent nature. Simple examples are provided by the electron spectrum of the Hubbard ferromagnet (Appendix J) and two-dimensional conducting antiferromagnet (Sect.6.7).

The spectrum of highly-correlated systems is often described in terms of auxiliary (slave) Fermi and Bose operators, which correspond to quasiparticles with exotic properties (neutral fermions, charged bosons etc.). Last time such ideas have been extensively applied in connection with the unusual spectra of high-T_c superconductors and heavy-fermion systems. Investigation of these problems leads to complicated mathematics, which uses the whole variety of modern quantum field theory methods, and very beautiful physics. For example, description of the Fermi liquid state in terms of Bose excitations becomes possible. These concepts change essentially classical notions of the solid state theory.

BIBLIOGRAPHY

1. A.Sommerfeld and H.Bethe, Electronentheorie der Metalle, Handb.Phys, Vol.XXIV, ed. H.Geiger and K.Scheel, Springer, Berlin, 1933.
2. A.H.Wilson, Theory of Metals, Cambridge, 1953.
3. F.Seitz, Modern Theory of Solids, McGraw Hill, New York, 1940.
4. R.E.Peierls, Quantum Theory of Solids, Oxford, Clarendon Press, 1955.
5. C.G.Kittel, Introduction to Solid State Physics, Wiley, New York, 1953.
6. J.M.Ziman, Principles of the Theory of Solids, London, 1964.
7. J.M.Ziman, Electrons and Phonons, Clarendon Press, Oxford, 1960.
8. F.J.Blatt, Physics of Electronic Conduction in Solids, McGraw Hill, New York, 1968.
9. W.A.Harrison, Solid State Theory, McGraw Hill, New York, 1970.
10. I.M.Lifshitz, M.Ya.Azbel and M.I.Kaganov, Electron Theory of Metals, New York, Consultants Bureau, 1973.
11. A.P.Cracknell and K.C.Wong, The Fermi Surface, Clarendon Press, Oxford, 1973.
12. N.W.Ashcroft and N.D.Mermin, Solid State Physics, Holt, Rienart and Winston, 1976.
13. W.A.Harrison, Electronic Structure and the Properties of Solids, Freeman, San Francisko, 1980.
14. A.A.Abrikosov, Theory of Metals, Moscow, Nauka, 1987.
15. B.Coqblin, The Electronic Structure of Rare Earth Metals and Alloys, Academic Press, 1977.
16. K.N.R.Taylor and M.I.Darby, Physics of Rare Earth Solids, Chapman and Hall, London, 1972.
17. S.A.Nikitin, Magnetic Properties of Rare Earth Metals and their Allcys, Moscow State University, 1989; K.P.Belov, M.A.Belyanchikova, R.Z.Levitin and S.A.Nikitin, Rare Earth Ferro- and Antiferromagnets, Moscow, Nauka, 1965.
18. D.C.Griffin, K.Andrew and R.D.Cowan, Phys.Rev.177, 62 (1969).
19. A.J.Freeman, In: Magnetic Properties of Rare Earth Metals, Ed.R.J.Elliot, Plenum Press, 1972, p.245.
20. I.I.Sobelman, Introduction to the Theory of Atomic Spectra, Pergamon Press, Oxford, 1973.
21. P.Soederlind, O.Eriksson, J.M.Wills and B.Johansson, Phys.Rev.B48, 9212 (1993).
22. U.Benedict, W.A.Grosshans and W.B.Holtzapfel, Physica B144, 14 (1986).
23. U.Benedict, J.R.Peterson, R.G.Haire and C.Dufour, J.Phys.F14, L43 (1984).
24. V.L.Moruzzi, J.F.Janak and A.R.Williams, Calculated Electronic Properties of Metals, Pergamon Press, New York, 1978.
25. N.F.Mott, Metal-Insulator Transitions, London, Taylor and Francis, 1974.
26. T.Moriya, Spin Fluctuations in Itinerant Electron Magnetism. Springer, Berlin, 1985.
27. A.Abrikosov, L.Gorkov and I.Dzyaloshinsky, Methods of Quantum Field Theory in Statistical Physics, Prentice Hall, New York, 1963.
28. J.Hubbard, Proc.Roy.Soc. A276, 238 (1963).
29. J.Hubbard, Proc.Roy.Soc. A277, 237 (1963).
30. J.Hubbard, Proc.Roy.Soc. A281, 401 (1964).
31. J.Hubbard, Proc.Roy.Soc. A285, 542 (1965).
32. B.R.Judd, Operator Techniques in Atomic Spectroscopy, Mc Graw Hill, New York, 1963; Second Quantization and Atomic Spectroscopy, John Hopkins University Press, Baltimore, 1967.
33. S.Fraga, K.Saxena and J.Karwowski, Handbook of Atomic Data, Elsevier, Amsterdam, 1976.

34. C.E.Moore, Atomic Energy Levels, US National Bureau of Standards Circular No 1463, vol.1–3, 1958.
35. R.Dagis and D.Segzda, Fiz.Tverd.Tela 36, 705 (1994).
36. S.P.Kowalczuk, R.A.Pollak, F.R.McFeely, L.Ley and D.A.Shirley, Phys.Rev.B8, 2387 (1973).
37. P.A.Bennett, J.C.Fuggle and F.U.Hillebrecht, Phys.Rev.B27, 2194 (1983).
38. E.Antonides et al, Phys.Rev.B15, 1669 (1977).
39. Yu.P.Irkhin, Usp.Fiz.Nauk 154, 321 (1988) [Sov.Phys.Usp.31, 163 (1988)].
40. R.J.H.Kappert, H.R.Borsje and J.C.Fuggle, J.Magn.Magn.Mat.100, 363 (1991).
41. S.A.Altshuler and B.M.Kozyrev, Electron paramagnetic resonance of transition group compounds (in Russian), Moscow, Nauka, 1972.
42. K.I.Kugel and D.I.Khomsky, Usp.Fiz.Nauk 136, 621 (1982).
43. R.S.Knox and A.Gold, Symmetry in the Solid State, Benjamin, New York, 1964.
44. B.H.Brandow, Adv.Phys.26, 651 (1977).
45. S.Huefner, Adv.Phys.43, 183 (1994).
46. B.Koiller and L.M.Falicov, J.Phys.C7, 299 (1974).
47. L.H.Tjeng, C.T.Chen, G.Ghijsen, R.Rudolf and F.Sette, Phys.Rev.Lett.67, 501 (1991).
48. A.Fujimori et al, Phys.Rev.B42, 7580 (1990).
49. J.van Elp, P.H.Potze, H.Eskes, R.Berger and G.A.Sawatzky, Phys.Rev.B44, 1530 (1991).
50. J.van Elp, H.Eskes, P.Kuiper and G.A.Sawatzky, Phys.Rev.B44, 5927 (1991).
51. A.Fujimori, F.Minami and S.Sugano, Phys.Rev.B29, 5225 (1984); A.Fujimori and F.Minami, Phys.Rev.B30, 957 (1984).
52. J.M.Ziman, The Calculation of Bloch Functions, In: Solid State Physics, ed.H.Ehrenreich et al., vol.26, Academic Press, 1971.
53. J.O.Dimmock, In: Solid State Physics, ed.H.Ehrenreich et al., vol.26, Academic Press, 1970, p.103.
54. T.L.Loucks, APW-Method, Benjamin, Menlo Park, California, 1967.
55. W.Heine, M.Cohen and D.Weaire, Solid State Physics, ed.H.Ehrenreich et al., vol.24, Academic Press, 1970.
56. A.R.Mackintosh and O.K.Andersen, The Electronic Structure of Transition Metals, in: Electrons at the Fermi Surface, Ed.M.Springford, Cambridge University Press, 1980, p.149.
57. V.V.Nemoshkalenko and V.N.Antonov, Methods of Calculation Physics in the Solid State Theory. Band Theory of Metals (in Russian), Kiev, Naukova Dumka, 1985.
58. J.C.Slater and G.F.Koster, Phys.Rev.94, 1498 (1954).
59. P.Hohenberg and W.Kohn, Phys.Rev.B136, 864 (1964).
60. J.A.Moriarty, Phys.Rev.B1, 1363 (1970); B5, 2066 (1972).
61. M.D.Girardeau, J.Math.Phys., 12, 165 (1971).
62. Z.A.Gursky and B.A.Gursky, Fiz.Metallov Metalloved.50, 928 (1980).
63. G.B.Bachelet, D.R.Hamann and M.Schluter, Phys.Rev.B26, 4199 (1982).
64. K.-M.Ho, S.G.Louie, J.R.Chelikowsky and M.L.Cohen, Phys.Rev.B15, 1755 (1977).
65. B.A.Greenberg, M.I.Katsnelson, V.G.Koreshkov et al., phys.stat.sol.(b) 158, 441 (1990); A.S.Ivanov, M.I.Katsnelson, A.G.Mikhin et al, Phil.Mag.69, 1183 (1994).
66. T.L.Loucks, Phys.Rev.139, A1181 (1965).
67. O.K.Andersen, Phys.Rev.B12, 864 (1975).
68. E.Wimmer, H.Krakauer, M.Weinert and A.J.Freeman, Phys.Rev.B24, 864 (1981).

69. J.C.Slater, The Self-Consistent Field for Molecules and Solids, Mc Graw Hill, New York, 1974.
70. W.Kohn and L.J.Sham, Phys.Rev.A140, 1133 (1965); L.J.Sham and W.Kohn, Phys.Rev. 145, 561 (1966).
71. O.Gunnarsson and B.I.Lundqvist, Phys.Rev.B13, 4274 (1976).
72. R.O.Jones and O.Gunnarsson, Rev.Mod.Phys, 61, 689 (1989).
73. L.Hedin and B.I.Lundqvist, J.Phys.C4, 2064 (1971).
74. U.von Barth and L.Hedin, J.Phys.C5, 1629 (1972).
75. T.Ziegler, A.Rauk and E.J.Baerends, Theor.Chim.Acta, 43, 261 (1977).
76. U.von Barth, Phys.Rev.A20, 1963 (1979).
77. A.Svane and O.Gunnarsson, Phys.Rev.B37, 9919 (1988); Phys.Rev.Lett.65, 1148 (1990); Europhys.Lett.7, 171 (1988).
78. D.A.Papaconstantopoulos, Handbook of Bandstructure of Elemental Solids, New York, Plenum Press, 1986.
79. R.J.Jelitto, J.Phys.Chem.Sol., 30, 609 (1969).
80. R.H.Swendsen and H.Callen, Phys.Rev.B6, 2860 (1972).
81. S.V.Vonsovsky, M.I.Katsnelson and A.V.Trefilov, Fiz.Metallov Metalloved.76, N3, 3 (1993).
82. D.Koelling, F.M.Mueller, A.J.Arko and J.B.Ketterson, Phys.Rev.B10, 4889 (1974).
83. J.Friedel, Canad.J.Phys.34, 1190 (1956).
84. W.A.Harrison, Phys.Rev.B28, 550 (1983).
85. N.F.Mott and K.W.H.Stevens, Phil.Mag.2, 1364 (1957).
86. W.M.Lomer and W.M.Marshall, Phil.Mag.3, 185 (1958).
87. E.O.Wollan, Phys.Rev.117, 387 (1960).
88. J.B.Goodenough, Phys.Rev.120, 67 (1960).
89. L.F.Matheiss, Phys.Rev.134, A970 (1964).
90. F.J.Himpsel and D.Eastman, Phys.Rev.B18, 5236 (1978).
91. O.K.Andersen, Phys.Rev.B2, 883 (1970).
92. N.E.Christensen, Phys.Rev.B14, 3446 (1976).
93. F.M.Mueller, A.J.Freeman, J.O.Dimmock and A.M.Furduya, Phys.Rev.B1, 4617 (1970).
94. S.G.Louie, Phys.Rev.Lett.40, 1525 (1978).
95. N.I.Mazin, E.G.Maksimov, S.I.Rashkeev and Yu.A.Uspensky, Proc. P.N.Lebedev Physical Institute of Academy of Sciences, vol.190, Moscow, Nauka, 1988, p.3.
96. K.D.Sevier, Low Energy Electron Spectrometry, New York, Wiley, 1972.
97. R.F.Davis, R.S.Williams, S.D.Kevan, P.S.Wehner and D.A.Shirley, Phys.Rev.B31, 1997 (1985).
98. P.S.Wehner, R.S.Williams, S.D.Kevan, D.Denley and D.A.Shirley, Phys.Rev.B19, 6164 (1979).
99. K.A.Mills, R.F.Davis, S.D.Kevan, G.Thornton and D.A.Shirley, Phys.Rev.B22, 581 (1980).
100. A.Svane, Phys.Rev.Lett., 68, 1900 (1992).
101. J.F.van der Veen, F.J.Himpsel and D.E.Eastman, Phys.Rev.B22, 4226 (1980).
102. W.Speier et al., Phys.Rev.B30, 6921 (1984).
103. H.Hoechst et al, Z.Phys.B42, 99 (1981).
104. D.E.Eastman, F.J.Himpsel and J.A.Knapp, Phys.Rev.Lett.44, 95 (1980).
105. E.Kisker, K.Schroeder, W.Gudat and M.Campagna, Phys.Rev.B31, 329 (1985).
106. R.Clauberg, E.M.Haines and P.Feder, J.Magn.Magn.Mater. 54–57, 622 (1986).
107. E.Kisker, J.Magn.Magn.Mater.45, 23 (1984).
108. D.E.Eastman, F.J.Himpsel and J.A.Knapp, Phys.Rev.Lett.40, 1514 (1978); F.J.Himpsel, J.A.Knapp and D.E.Eastman, Phys.Rev.B19, 2919 (1979).

109. K.Kakizaki, J.Fujii, K.Shimada et al, Phys.Rev.Lett.72, 2781 (1994).
110. M.Pessa, P.Heinmann and H.Neddermeyer, Phys.Rev.B14, 3488 (1976).
111. G.Bush, M.Campagna and H.C.Siegmann, Phys.Rev.B4, 746 (1971).
112. P.Genoval, A.A.Manuel, E.Walker and M.Peter, J.Phys.:Cond.Mat. 3, 4201 (1991).
113. C.M.Schneider, P.Schuster, M.S.Hammond and J.Kirschner, Europhys.Lett.16, 689 (1991).
114. P.Guletsky, Yu.A.Knyazev, M.M.Kirillova and L.M.Sandratsky, Fiz.Metallov Metalloved.67, 279 (1989).
115. T.Jo and G.A.Sawatzky, Phys.Rev.B43, 8771 (1991).
116. H.Martensson and P.O.Wilsson, Phys.Rev.B30, 3047 (1984).
117. A.Liebsh, Phys.Rev.B23, 5203 (1981).
118. D.R.Penn, Phys.Rev.Lett.42, 921 (1979).
119. R.H.Victora and L.M.Falicov, Phys.Rev.Lett.55, 1140 (1985).
120. E.Jensen and D.M.Wieliczka, Phys.Rev.B30, 7340 (1984).
121. O.Gunnarson and K.Schoenhammer, Phys.Rev.B28, 4315 (1983).
122. Yu.P.Irkhin, In: Electron Structure and Physical Propertties of Rare Earths and Actinides, Ural Science Centre, Sverdlovsk, 1981, p.50.
123. J.T.Waber and A.C.Switendick, Proc.V-th Rare Earth Research Conf., Iowa State University, 1965, v.B2, p.75.
124. G.Murhopadhyay and C.K.Majumdar, J.Phys.C2, 924 (1969).
125. W.E.Pickett, A.J.Freeman and D.D.Koelling, Phys.Rev.B23, 1266 (1981).
126. G.Johansen and A.R.Mackintosh, Solid State Comm.8, 121 (1970).
127. B.N.Harmon, J.de Phys.40, 65 (1979).
128. M.M.Islam and D.J.Newman, J.Phys.F5, 939 (1975) Tm.
129. T.A.Matveeva and R.F.Egorov, Fiz.Metallov Metalloved.51, 717, 950 (1981).
130. O.V.Farberovich, G.P.Nizhnikova, S.V.Vlasov and E.P.Domashevskaya, phys.stat.sol(b) 121, 241 (1984).
131. H.J.F.Jansen, A.J.Freeman and R.Monnier, Phys.Rev.B31, 4092 (1985).
132. D.M.Bylander and L.Kleinman, Phys.Rev.B49, 1608 (1994); B50, 1363 (1994).
133. J.F.Herbst, D.N.Lowy and R.E.Watson, Phys.Rev.B6, 1913 (1972).
134. J.F.Herbst, R.E.Watson and J.W.Wilson, Phys.Rev.B13, 1439 (1976); B17, 3089 (1978).
135. E.I.Zabolotsky, Yu.P.Irkhin and L.D.Finkelstein, Fiz.Tverd.Tela, 16, 1142 (1974).
136. P.O.Heden, H.Logfren and S.B.M.Hagstrom, Phys.Rev.Lett.26, 432 (1971).
137. C.Bounele, R.C.Karnatek and C.Jorgensen, Chem.Phys.Lett.14, 145 (1972).
138. A.J.Freeman, D.D.Koelling, In: The Actinides: Electron Structure and Related Properties, ed. A.J.Freeman and J.B.Darby, New York, Academic Press, 1974, p.51.
139. B.V.Karpenko, In: Electron Structure and Physical Propertties of Rare Earths and Actinides, Ural Science Centre, Sverdlovsk, 1981, p.86.
140. S.C.Keeton, T.L.Loucks, Phys.Rev.146, 429 (1966).
141. D.D.Koelling, A.J.Freeman, Solid State Comm.9, 1369 (1971); Phys.Rev.B12, 5622 (1975).
142. P.Weinberger, A.M.Boring and J.L.Smith, Phys.Rev.B31, 1964 (1985).
143. O.Eriksson and J.M.Wills, Phys.Rev.B45, 3198 (1992).
144. J.Kittel, Quantum Theory of Solids, Wiley, New York, 1963.
145. C.G.Lonzarich, Fermi Surface Studies in Ferromagnets, in: Electrons at the Fermi Surface, Ed.M.Springford, Cambridge University Press, 1980, p.225.
146. A.P.Cracknell, Band Structures and Fermi Surfaces of Metallic Elements, Landolt-Boernstein New Series, Band 13c, Springer, 1984.
147. F.S.Ham, Phys.Rev.128, 2524 (1962).

148. D.R.Jennison, Phys.Rev.B16, 5147 (1977).
149. C.B.So, K.Takegahara and S.Wang, J.Phys.F7, 105 (1977).
150. M.J.G.Lee, Phys.Rev.178, 953 (1969).
151. S.L.Altmann, A.P.Cracknell, Proc.Phys.Soc. 84, 761 (1964).
152. P.Blaha and J.Callaway, Phys.Rev.B32, 7664 (1985).
153. J.H.Condon, J.A.Marcus, Phys. Rev. 134, A 446 (1964).
154. C.S.Fleming, T.L.Loucks, Phys.Rev. 173, 685 (1968).
155. T.L.Loucks, Phys.Rev. 144, 504 (1966).
156. T.L.Loucks, Phys.Rev.A139, 1181 (1965).
157. R.C.Young, P.C.Yordan, D.W.Jones and V.J.Hems, J.Phys.F4, L84 (1984).
158. G.S.Fleming, S.H.Liu and T.E.Loucks, Phys.Rev.Lett.21, 1524 (1968).
159. O.Jepsen, Phys.Rev.B12, 2988 (1975).
160. K.M.Welch and E.H.Hygh, Phys.Rev.B9, 1993 (1974).
161. S.L.Altmann and C.J.Bradley, Proc. Phys. Soc. 92, 764 (1967).
162. A.J.Hughes and J.Callaway, Phys. Rev. 136, A 1390 (1964).
163. T.L.Loucks, Phys. Rev. 159, 544 (1967).
164. A.C.Thorsen and A.S.Joseph, Phys. Rev.131, 2078 (1963).
165. P.M.Everett, Phys.Rev.B20, 1419 (1979).
166. L.L.Boyer, D.A.Papaconstantopoulos and B.M.Klein, Phys.Rev.B15, 3685 (1977).
167. D.G.Laurent, C.S.Wang and J.Gallaway, Phys.Rev.B17, 455 (1978).
168. R.D.Parker and M.H.Halloran, Phys.Rev.B9, 4130 (1974).
169. R.A.Phillips, Phys.Lett.A36, 361 (1971).
170. A.C.Thorsen, T.G.Berlincourt, Phys. Rev. Lett. 7, 244 (1961).
171. G.B.Scott, M.Springford, J.R.Stockton, Phys.Lett.A27, 655 (1968); G.B.Scott and M.Springford, Proc.Roy.Soc.A320, 115 (1970).
172. L.F.Mattheiss, Phys. Rev. B1, 373 (1970).
173. L.L.Boyer, D.A.Papaconstantopoulos and B.M.Klein, Phys.Rev.B15, 3685 (1977).
174. M.Cardona and L.Ley, Photoemission in Solids (Berlin:Springer) 1978.
175. M.H.Halloran et al, Phys.Rev.B1, 366 (1970).
176. T.L.Loucks, Phys.Rev.139, A1181 (1965).
177. L.F.Mattheiss, Phys. Rev.139, A 1893 (1965).
178. D.G.Laurent, J.Callaway, J.L.Fry and N.E.Brener, Phys.Rev.B23, 4877 (1981).
179. L.F.Mattheiss and D.R.Hamann, Phys.Rev.B33, 823 (1986).
180. S.Asano and J.Yamashita, J.Phys.Soc.Jpn, 23, 714 (1967).
181. N.I.Kulikov and E.I.Kulatov, J.Phys.F12, 2292 (1982).
182. K.Okumara and I.M.Tepleton, Proc.R.Soc.A287, 89 (1965).
183. M.J.G.Lee, Phys.Rev.178, 953 (1969).
184. J.B.Ketterson, D.D.Koelling, J.C.Shaw and Windmiller, Phys.Rev.B11, 1447 (1975).
185. D.M.Sparlin and J.A.Marcus, Phys. Rev. 144, 484 (1966).
186. J.A.Hoekstra and J.L.Stanford, Phys.Rev.B8, 1416 (1973).
187. R.F.Girvan, A.V.Gold and R.A.Phillips, J.Phys.Chem.Sol.29, 1485 (1968).
188. G.S.Fleming, S.H.Liu and T.E.Loucks, Phys.Rev.Lett.21, 1524 (1968).
189. G.C.Fletcher, J. Phys.C2, 1440 (1969).
190. L.F.Mattheiss, Phys.Rev.151, 450 (1966).
191. A.C.Thorsen, A.S.Joseph, L.I.Valby, Phys.Rev.150, 523 (1966).
192. J.H.Wood, Phys. Rev. 126, 517 (1962).
193. R.A.Tawil and J.Callaway, Phys.Rev.B7, 4242 (1973).
194. J.Callaway and C.S.Wang, Phys.Rev.B16, 2095 (1977).
195. A.V.Gold, L.Hodges, P.T.Panousis and D.R.Stone, Int.J.Magn.2, 357 (1971).
196. D.R.Baraff, Phys.Rev.B8, 3439 (1973).
197. S.Wakoh, J.Yamashita, J. Phys Soc. Japan 38, 1151 (1970).

198. C.M.Singal and T.P.Das, Phys.Rev.B16, 5068 (1977).
199. J.R.Anderson and J.F.Shirber, J.Appl.Phys.52, 1630 (1980).
200. N.I.Kulikov and E.I.Kulatov, J.Phys.F12, 2267 (1982).
201. F.Batallan, S.Rosenman and C.Summers, Phys.Rev.B11, 545 (1975).
202. I.Rosenman and F.Batallan, Phys.Rev.B5, 1340 (1972).
203. J.R.Anderson, J.J.Hudak and D.R.Stone, AIR Conf.Proc.5, 477 (1972).
204. J.Callaway and C.S.Wang, Phys.Rev.B7, 1096 (1973); C.S.Wang and J.Callaway, Phys.Rev.B9, 4897 (1974).
205. R.Prasad, S.K.Joshi and S.Auluck, Phys.Rev.B16, 1765 (1977); S.Ahuja, S.Auluck and B.Johansson, Physica Scripta 50, 573 (1994).
206. J.R.Anderson, D.A.Papaconstantopoulos, L.L.Boyer and J.E.Schriber, Phys.Rev.B20, 3172 (1979).
207. E.I.Zornberg, Phys.Rev.B1, 244 (1970).
208. D.S.Tsui, Phys.Rev.164, 669 (1967).
209. G.N.Kamm and J.R.Anderson, Phys.Rev.B2, 2944 (1970).
210. J.F.Cooke, H.L.Davies and R.F.Wood, Phys.Rev.Lett.25, 28 (1970).
211. P.T.Coleridge and I.M.Templeton, J.Phys.F2, 643 (1972).
212. P.T.Coleridge, Phys. Lett. 22, 367 (1966); J.Low Temp.Phys.1, 577 (1969).
213. E.S.Alekseyev, V.A.Ventsel, O.A.Voronov, A.I.Likhter and M.V.Magnitskaya, Zh.Eksp.Teor.Fiz.76, 215 (1979) [Sov.Phys.JETP 49, 110 (1979)].
214. G.N.Kamm, Phys.Rev.B2, 2944 (1970).
215. O.K Andersen, A.R.Mackintosh, Solid State Commun. 6, 285 (1968).
216. P.T.Coleridge, Phys. Lett. 15, 223 (1965).
217. P.T.Coleridge, Proc. Roy. Soc. A295, 458 (1966).
218. J.J.Grodski, A.E.Dixson, Solid State Commun.7, 735 (1969).
219. S.Hornfeldt, Solid State Commun. 8, 673 (1970).
220. N.V.Volkenshtein, V.A.Novoselov, V.E.Startsev and Yu.N.Tsiovkin, Zh.Eksp.Teor.Fiz.60, 1733 (1971); Sov.Phys.JETP 33, 959 (1971).
221. G.O.Arbman and S.Hornfeldt, J.Phys.F2, 1033 (1972).
222. G.O.Arbman and S.Hornfeldt, J.Phys.F2, 1033 (1972); M.Cooper and B.Williams, Philos.Mag.26, 1441 (1972).
223. J.J.Vuillemin, Phys. Rev. 144, 396 (1966).
224. J.J.Vuillemin, M.G.Priestley, Phys. Rev. Lett. 15, 307 (1965).
225. D.H.Dye, S.A.Campbell, G.W.Grabtree, J.B.Ketterson, N.B.Sandesara and J.J.Vuillemin, Phys.Rev.B23, 462 (1981).
226. J.B.Ketterson and L.R.Windmiller, Phys.Rev.B1, 4747 (1970); B2, 4813 (1970).
227. D.H.Dye, J.B.Ketterson and G.W.Crabtree, J.Low Temp.Phys.30, 813 (1978).
228. N.F.Christensen et al, J.Magn.Magn.Mat.76–77, 23 (1988).
229. S.Tanuma, W.R.Datars, H.Doi, A.Dunsworth, Solid State Commun.8, 1107 (1970).
230. R.C.Young, R.G.Jordan and D.W.Jones, Phys.Rev.Lett.31, 1473 (1973).
231. S.C.Keeton and T.L.Loucks, Phys. Rev. 168, 672 (1968).
232. S.C.Keeton and T.L.Loucks, Phys. Rev. 146, 428 (1966).
233. A.C.Thorsen and A.S.Joseph, L.E.Valby, Phys. Rev. 162, 574 (1967).
234. R.P.Gupta and T.L.Loucks, Phys. Rev. Lett. 22, 458 (1969).
235. K.A.Gschneidner, Jr., Physical Properties and Interrelationships of Metallic and Semimetallic Elements, In: Solid State Physics, ed.H.Ehrenreich et al., vol.16, Academic Press, 1964, p.275.
236. B.Johansson and M.S.S.Brooks, Theory of Cohesion in Rare Earths and Actinides, In: Handbook Phys.Chem. Rare Earths, ed.K.A.Gschneidner,Jr. et al, vol.17, p.149.
237. Properties of Elements, ed.M.E.Dritz, Moscow, Metallurgiya, 1985.

238. Properties of Elements, ed.G.V.Samsonov, Moscow, Metallurgiya, 1976.
239. V.E.Zinov'yev, Thermophysical Properties of Metals at High Temperatures, Moscow, Metallurgiya, 1989.
240. A.M.Oles, Phys.Rev.B23, 271 (1981).
241. J.Friedel and C.M.Sayers, J.de Phys.38, 697 (1977).
242. C.D.Gelatt, H.Ehrenreich and R.E.Watson, Phys.Rev.B15, 1613 (1977).
243. V.L.Moruzzi, J.F.Janak and K.Schwarz, Phys.Rev.B37, 790 (1988).
244. H.L.Skriver and M.M.Rosengaard, Phys.Rev.B45, 9410 (1992).
245. H.L.Skriver, Phys.Rev.B31, 1909 (1985).
246. V.A.Finkel, Structure of Rare Earth Metals (in Russian), Moscow, Metallurgiya, 1978.
247. C.M.Hurd and J.E.A.Alderson, Solid State Comm., 12, 375 (1974).
248. W.Hume-Rothery, The Metallic State, New York, University Press, 1931.
249. H.Jones, Proc.Phys.Soc., 49, 423 (1937).
250. N.Engel, Trans.Amer.Soc.Metals, 57, 610 (1964).
251. G.V.Samsonov, I.F.Pryadko and L.F.Pryadko, Electron Localization in Solids (in Russian), Moscow, Nauka, 1976.
252. W.Hume-Rothery, Progr.Mater.Sci., 13, 229 (1968).
253. D.G.Pettifor, J.Phys.C3, 367 (1970).
254. J.W.Davenport, R.E.Watson and M.Weinert, Phys.Rev.B32, 4883 (1985).
255. M.Sigalas, D.Papaconstantopoulos and N.C.Bacalis, Phys.Rev.B45, 5777 (1992).
256. J.M.Wills and O.Eriksson, Phys.Rev.B45, 13879 (1992).
257. G.Gladstone, M.A.Jensen and J.R.Schrieffer, In: Superconductivity, ed.R.D.Parks, New York, 1969.
258. J.A.Moriarty, Phys.Rev.B49, 12431 (1994).
259. M.M.Steiner, R.C.Albers and L.J.Sham, Phys.Rev.B45, 13272 (1992).
260. G.Grimwall, J.Phys.Chem.Sol., 29, 1221 (1968).
261. J.Osuzu, Thesis, Nagoya University, 1968.
262. M.Shimizu, Rep.Progr.Phys., 44, 329 (1981).
263. P.Fulde and M.Loewenhaupt, Adv.Phys., 34, 589 (1985).
264. L.M.Noskova, E.V.Rosenfeld and Yu.P.Irkhin, Fiz.Nizkikh Temperatur, 11, 469 (1985).
265. S.V.Vonsovsky, Magnetism, New York, Wiley, 1974.
266. M.Dixon, F.E.Hoare, T.M.Holden and D.E.Moody, Proc.Roy.Soc.A285, 561 (1965).
267. R.E.Pawell and E.E.Stansbury, J.Phys.Chem.Sol.26, 757 (1965).
268. E.V.Rosenfeld, A.A.Siventsev, Yu.P.Irkhin and L.M.Noskova, Fiz.Tverd.Tela 33, 202 (1991).
269. S.A.Nemnonov, Fiz.Metallov Metalloved.19, 550 (1965).
270. E.V.Galoshina, Usp.Fiz.Nauk 113, 105 (1974).
271. R.Kubo and Y.Obata, J.Phys.Soc.Jpn 11, 547 (1956).
272. M.Yasui, Physica B149, 139 (1988).
273. S.Tikadzumi, Physics of Magnetism, Wiley, New York, 1964.
274. R.Huguenin and D.Baldock, Phys.Rev.Lett.Phys.Rev.Lett.16, 795 (1966); R.Huguenin, G.P.Pells and D.N.Baldock, J.Phys.F1, 281 (1971).
275. R.Peierls, Z.Phys.80, 763 (1933).
276. F.Ducastelle and F.Cyrot-Lackmann, J.de Phys.Colloque 32, C1–534 (1971).
277. M.Matsumoto, J.Staunton and P.Strange, J.Phys.:Cond.Mat.3, 1453 (1991).
278. S.G.Das, Phys.Rev.B13,3978 (1976).
279. A.H.MacDonald, K.L.Liu and S.H.Vosko, Phys.Rev.B16, 777 (1977).
280. P.Rhodes and E.P.Wohlfarth, Proc.Roy.Soc. A273, 247 (1963).
281. H.Ido, J.Magn.Magn.Mat.54–57, 937 (1986); H.Ido, S.Yasuda, J.de Phys.Colloque C8, 49, 141 (1988).

282. M.J.Otto, R.A.M. van Woerden, P.J. van der Valk et al. J.Phys.:Cond.Mat.1, 2341 (1989).
283. P.J.Webster, J.Phys.Chem.Sol., 32, 1221 (1971).
284. R.Vijayaraghavan, J.Magn.Magn.Mater.47–48, 561 (1985); M.Kasaya et al, J.Magn.Magn.Mater.76–77, 347 (1988).
285. E.S.Stoner, Proc.Roy.Soc.A165, 372 (1938); A169, 339 (1939).
286. E.P.Wohlfarth, Phil.Mag.42, 374 (1951).
287. D.M.Ceperley and B.J.Adler, Phys.Rev.Lett.45, 566 (1980).
288. V.Yu.Irkhin, M.I.Katsnelson and A.V.Trefilov, J.Magn.Magn.Mat.117, 210 (1992); J.Phys.:Cond.Mat.5, 8763 (1993).
289. G.H.O.Daalderop, M.H.Boon and F.M.Mueller, Phys.Rev.B41, 9803 (1990).
290. T.Jarlborg and A.J.Freeman, Phys.Rev.B22, 2332 (1980); T.Jarlborg, A.J.Freeman and D.D.Koelling, J.Magn.Magn.Mat.23, 291 (1981).
291. Yu.P.Irkhin, E.V.Rosenfeld and A.A.Siventsev, Fiz.Tverd.Tela 33, 1646 (1991).
292. O.Gunnarson, J.Phys.F6, 587 (1976).
293. A.M.Oles and G.Stollhoff, J.Magn.Magn.Mat.54–57, 1045 (1986).
294. E.V.Rosenfeld and Yu.P.Irkhin, Fiz.Metallov Metalloved. 57, 837 (1984); Yu.P.Irkhin and E.V.Rosenfeld, Solid State Comm., 44, 1371 (1982).
295. C.Herring, Magnetism, vol.4, New York, Academic Press, 1966.
296. T.Moriya and A.Kawabata, J.Phys.Soc.Jpn, 34, 639 (1973); 35, 669 (1973).
297. I.E.Dzyaloshinskii and P.S.Kondratenko, Sov.Phys.JETP 43, 1036 (1976).
298. J.A.Hertz and M.A.Klenin, Phys.Rev.B10, 1084 (1974); Physica 91B, 49 (1977).
299. K.K.Murata and S.Doniach, Phys.Rev.Lett.29, 285 (1982).
300. J.Hubbard, Phys.Rev.B19, 2626 (1979); B20, 4584 (1979).
301. H.Hasegawa, J.Phys.Soc.Jpn 49, 178; 963 (1980).
302. E.A.Turov and V.I.Grebennikov, Physica B149, 150 (1988).
303. A.A.Siventsev and Yu.P.Irkhin, Fiz.Tverd.Tela 35, 1965 (1993).
304. P.Mohn and E.P.Wohlfarth, J.Phys.F17, 2421 (1987); E.P.Wohlfarth and P.Mohn, Physica B149, 145 (1988).
305. G.G.Lonzarich and L.Taillefer, J.Phys.C18, 4339 (1985).
306. R.A.de Groot, F.M.Mueller, P.G.Mueller, P.G. van Engen and K.H.J.Bushow, Phys.Rev.Lett.50, 2024 (1983).
307. R.A.de Groot, F.M.Mueller, P.G. van Engen and K.H.J.Bushow, J.Appl.Phys.55(6), 2151 (1984).
308. R.A.de Groot and K.H.J.Buschow, J.Magn.Magn.Mat.54–57, 1377 (1986).
309. J.Kuebler, Physica B+C 127, 257 (1984).
310. R.A.de Groot, A.M.von der Kraan and K.H.J.Buschow, J.Magn.Magn.Mat.61, 330 (1986).
311. J.Kuebler, J.R.Williams and C.B.Sommers, Phys.Rev.B28, 1745 (1983).
312. S.Fujii, S.Sugimura, S.Ishida and S.Asano, J.Phys.:Cond.Mat. 2, 8583 (1990).
313. K.Schwarz, J.Phys.F16, L211 (1986).
314. E.Kulatov and I.Mazin, J.Phys.:Cond.Mat. 2, 343 (1990).
315. R.C.Albers, A.M.Boring, G.H.O.Daalderop and F.M.Mueller, Phys.Rev.B36, 3661 (1987).
316. S.V.Halilov and E.T.Kulatov, J.Phys.: Cond.Mat.3, 6363 (1991); Zh.Eksp.Theor.Fiz.98, 1778 (1989).
317. A.Yanase, K.Siratori, J.Phys.Soc.Jpn 53, 312 (1984) (1988).
318. V.Yu.Irkhin and M.I.Katsnelson, Usp.Fiz.Nauk 164, 705 (1994) [Sov.Phys.Uspekhi 37, 565 (1994)].
319. K.Schwarz, O.Mohn, P.Blaha, J.Kuebler, J.Phys.F14, 2659 (1984).

320. S.S.Jaswal, Phys.Rev.B41, 9697 (1990); S.S.Jaswal, W.B.Yelon, G.C.Hadjipanayis et al., Phys.Rev.Lett. 67, 644 (1991).
321. B.I.Min, J.-S.Kang, J.H.Hong et al., Phys.Rev.B48, 6317 (1993).
322. S.K.Malik, P.J.Arlinghaus, W.E.Wallace, Phys.Rev.B25, 6488 (1982).
323. J.Inoue and M.Shimizu, J.Phys.F15, 1511 (1985).
324. S.Matar, P.Mohn, G.Demazeau and B.Siberchicot, J.de Phys.49, 1761.
325. S.Fujii, S.Ishida and S.Asano, J.Phys.:Cond.Mat.4, 1575 (1992).
326. A.Sakuma, J.Phys.Soc.Jpn 60, 2007 (1991).
327. Y.Noda and Y.Ishikawa. J.Phys.Soc.Jpn 40, 690, 699 (1976).
328. K.Tajima and Y.Ishikawa, P.J.Webster et al. J.Phys.Soc.Jpn 43, 483 (1977).
329. M.I.Auslender and V.Yu.Irkhin, J.Phys.C18, 3533 (1985).
330. E.Kisker, G.Baum, A.Mahan, W.Raith and B.Reihl, Phys.Rev.B18, 2256 (1978).
331. M.J.Otto, R.A.M. van Woerden, P.J. van der Valk et al. J.Phys.: Cond.Mat.1, 2351 (1989).
332. J.A.Hertz and D.M.Edwards, J.Phys.F3, 2174 (1973); D.M.Edwards and J.A.Hertz, J.Phys.F3, 2191 (1973).
333. V.Yu.Irkhin and M.I.Katsnelson, Fiz.Tverd.Tela 25, 3383 (1983) [Sov.Phys.-Solid State 25, 1947 (1983)].
334. G.L.Bona, F.Meier, M.Taborelli, E.Bucher and P.H.Schmidt, Solid State Comm.56, 391 (1985).
335. T.Moriya, J.Phys.Soc.Jpn 19, 681 (1964).
336. E.A.Turov and M.P.Petrov, Nuclear Magnetic Resonance in Ferro- and Antiferromagnets [in Russian], Moscow, Nauka, 1969.
337. V.Yu.Irkhin and M.I.Katsnelson, Fiz.Metallov Metalloved. 66, 41 (1988).
338. V.Yu.Irkhin and M.I.Katsnelson, J.Phys.:Cond.Mat.2, 7151 (1990).
339. H.Enokiya, J.Phys.Soc.Jpn 31, 1037 (1971).
340. M.M.Matsuura, J.Phys.Soc.Jpn 21, 886 (1966).
341. K.Terakura, T.Oguchi, A.R.Williams and J.Kuebler, Phys.Rev.B30, 4734 (1984).
342. Z.Szotek, W.M.Temmerman and H.Winter, Phys.Rev.B47, 4029 (1993).
343. V.I.Anisimov, A.Zaanen and O.K.Andersen, Phys.Rev.B44, 943 (1991).
344. L.L.Chase, Phys.Rev.B10, 2226 (1974).
345. S.Lauer, A.X.Trautwein and F.E.Harris, Phys.Rev.B29, 6774 (1984).
346. G.L.Zhao, J.Callaway and M.Hayashibara, Phys.Rev.B48, 15781 (1994).
347. H.S.Jarrett, W.H.Cloud, H.J.Bouchard, S.R.Butler, C.G.Frederick and J.L.Gillson, Phys.Rev.Lett.21, 617 (1968).
348. J.Kanamori, Progr.Theor.Phys.30, 275 (1963).
349. Y.Nagaoka, Phys.Rev.147, 392 (1966).
350. D.N.Khomsky, Fiz.Metallov Metalloved.29, 31 (1970).
351. P.B.Vissher, Phys.Rev.B10, 932, 943 (1974).
352. E.L.Nagaev, Physics of Magnetic Semiconductors, Moscow, Mir, 1983.
353. M.I.Auslender, V.Yu.Irkhin and M.I.Katsnelson, J.Phys.C21, 5521 (1988).
354. J.Hubbard and K.P.Jain, J.Phys.C1, 1650 (1968).
355. A.Sakurai, Progr.Teor.Phys.39, 312 (1968).
356. J.J.Field, J.Phys C5, 664 (1972).
357. S.V.Tyablikov, Methods of Quantum Theory of Magnetism [in Russian], Moscow, Nauka, 1975.
358. G.Chrobok, M.Hofmann, G.Regenfus and P.Sizmann, Phys.Rev.B15, 429 (1975).
359. A.Vaterlaus, F.Millani and F.Meier. Phys.Rev.Lett.65, 3041 (1990).
360. S.A.Nikitin, A.S.Andreenko, A.K.Zvezdin and A.F.Popov, Zh.Eksp.Teor.Fiz.76, 2159 (1979).
361. K.Yosida and A.Watabe, Progr.Theor.Phys.28, 361 (1962).

362. W.E.Evenson and S.V.Lin, Phys.Rev.178, 783 (1969).
363. R.J.Elliott and F.A.Wedgwood, Proc.Roy.Soc.81, 846 (1963); 84, 63 (1964).
364. P.G.de Gennes, Journ.Phys.Rad.23, 630 (1932).
365. H.Miwa, Proc.Phys.Soc.85, 1197 (1965).
366. A.J.Fedro and T.Arai, Phys.Rev.B6, 911 (1972).
367. V.Yu.Irkhin and M.I.Katsnelson, Z.Phys.B75, 67 (1989).
368. I.E.Dzyaloshinsky, Zh.Eksp.Teor.Fiz.46, 1420 (1964); 47, 336, 992 (1964).
369. J.Bohr and D.Gibbs, Physica A140, 349 (1986).
370. S.K.Godovikov, Thesis, Moscow State University, 1994.
371. K.G.Gurtovoy and R.Z.Levitin, Usp.Fiz.Nauk, 153, 193 (1987).
372. P.G.Huray, S.E.Nave and R.G.Haire, J.Less-Common Met, 93, 393 (1983).
373. S.E.Nave et al, Physica B130, 225 (1985).
374. I.V.Solovyev, A.I.Liechtenstein, V.A.Gubanov, V.A.Antropov and O.K.Andersen, Phys.Rev.B43, 14414 (1991).
375. I.V.Solovyev, Thesis, Sverdlovsk, 1991.
376. E.A.Turov and A.I.Mitsek, Zh.Eksp.Teor.Fiz.37, 1127 (1959).
377. Ch.Kittel and J.H.Van Vleck, Phys.Rev.118, 1231 (1960).
378. H.B.Callen and E.R.Callen, J.Phys.Chem.Sol.27, 1271 (1966).
379. H.Brooks, Phys.Rev.58, 909 (1940).
380. E.I.Kondorsky and E.Straube, Pis'ma ZhETF 17, 41 (1973).
381. E.I.Kondorsky, Band Theory of Magnetism (in Russian), Moscow University Press, 1976 (pt.1), 1977 (pt.2).
382. R.Gersdorf, Phys.Rev.Lett.40, 344 (1978).
383. G.H.O.Daalderop, P.J.Kelly and M.F.H.Schuurmans, Phys.Rev.B41, 11919 (1990).
384. O.Eriksson, G.W.Fernando, R.C.Albers and A.M.Boring, Solid State Comm. 78, 801 (1991); O.Eriksson, A.M.Boring, R.C.Albers, G.W.Fernando and B.R.Cooper, Phys.Rev.B45, 2868 (1992).
385. H.J.F.Jansen, J.Appl.Phys.67, 4555 (1990).
386. D.C.Johnston and J.H.Cho, Phys.Rev.B42, 8710 (1990).
387. Yu.P.Irkhin, E.V.Rosenfeld and L.D.Finkelstein, Fiz.Tverd.Tela 35, 2769 (1993).
388. G.Di Castro, L.F.Feiner and M.Grilly, Phys.Rev.Lett.66, 3209 (1991).
389. Yu.P.Irkhin, V.V.Druzhinin and A.A.Kazakov, Zh.Eksper.Teor.Fiz. 54, 1183 (1968) [Sov.Phys.JETP 27, 633 (1968)].
390. S.K.Godovikov, M.G.Kozin, V.V.Turovtsev and V.S.Spinel, phys.stat.sol.(b) 78, 103 (1976).
391. Yu.P.Irkhin and E.V.Rosenfeld, Fiz.Tverd.Tela 16, 485 (1974).
392. A.S.Ermolenko et al, Zh.Eksp.Teor.Fiz.69, 1743 (1975).
393. Yu.P.Irkhin, E.I.Zabolotskii and E.V.Rosenfeld, Fiz.Metallov Metalloved.49, 1216 (1980).
394. Yu.P.Irkhin, Thesis, Sverdlovsk, 1968.
395. V.N.Novogrudsky and I.G.Fakidov, Zh.Eksp.Teor.Fiz.47, 40 (1964).
396. E.A.Turov, V.G.Shavrov and Yu.P.Irkhin, Zh.Eksp.Teor.Fiz.47, 296 (1964).
397. S.S.Levina, V.N.Novogrudsky and I.G.Fakidov, Zh.Eksp.Teor.Fiz.45, 52 (1963).
398. E.A.Turov, Transport, Optical and Acoustical Properties of Antiferromagnets (in Russian), Sverdlovsk, 1990.
399. K.B.Vlasov et al, Fiz.Metallov Metalloved.42, 513 (1976); Fiz.Tverd.Tela 22, 1656 (1980).
400. K.B.Vlasov et al, Fiz.Metallov Metalloved.44, 1005 (1977); 53, 493 (1982); Fiz.Tverd.Tela 24, 1338 (1982).
401. L.A.Boyarsky, Fiz.Nizkhikh Temperatur, 22, 912 (1996).
402. R.Kubo, J.Phys.Soc.Jpn 12, 570 (1957).

403. H.Nakano, Progr.Theor.Phys.17, 145 (1957).
404. H.Mori, Progr.Theor.Phys., 34, 399 (1965).
405. V.Yu.Irkhin, M.I.Katsnelson and A.V.Trefilov, Physica C160, 397 (1989).
406. N.V.Volkenshtein, V.P.Dyakina and V.E.Startsev, phys.stat.sol.(b) 57, 9 (1973).
407. N.V.Volkenshtein et al, Fiz.Metallov Metalloved.67, 734 (1989).
408. G.Baber, Proc.Roy.Soc.A158, 383 (1937).
409. J.Appel, Phil.Mag.90, 1071 (1963).
410. M.J.Rice, Phys.Rev.Lett.20, 1439 (1968).
411. P.A.Schroeder, Physica B+C 109–110, 1901 (1982).
412. N.V.Volkenshtein, V.A.Novoselov and V.E.Startsev, Zh.Eksp.Teor.Fiz.60, 1078 (1971); V.E.Startsev, Thesis, Sverdlovsk, 1985.
413. N.V.Volkenshtein et al, Fiz.Nizkikh Temp.10, 841 (1989).
414. F.J.Blatt and H.G.Satz, Helv.Phys.Acta, 33, 1007 (1960).
415. Yu.Kagan and A.P.Zhernov, Zh.Eksp.Teor.Fiz.50, 1107 (1966).
416. K.Ueda and T.Moriya, J.Phys.Soc.Jpn 39, 605 (1975).
417. A.H.MacDonald, Can.J.Phys.60, 710 (1982).
418. N.F.Mott, Proc.Roy.Soc.A153, 699 (1936); A157, 368 (1936); Adv.Phys.13, 325 (1964).
419. A.H.Wilson, Proc.Roy.Soc.A167, 580 (1938).
420. Yu.P.Irkhin, Fiz.Metallov Metalloved.6, 214, 586 (1958).
421. M.D.Wilding and E.W.Lee, Proc.Phys.Soc.85, 955 (1965).
422. T.Kasuya, Progr.Theor Phys. 16, 58 (1956).
423. Sh.Sh.Abelsky and E.A.Turov, Fiz.Metallov Metalloved.10, 801 (1960).
424. S.V.Vonsovsky and Yu.A.Izyumov, Usp.Fiz.Nauk, 58, 3 (1962).
425. E.A.Turov, Izv. AN SSSR (Ser.Fiz.) 19, 474 (1955).
426. T.Kasuya, Progr.Theor Phys. 22, 227 (1959).
427. M.Roessler, phys.stat.sol. 9, K27 (1965).
428. V.S.Lutovinov, M.Yu.Reizer and M.A.Savchenko, Zh.Eksp.Theor.Fiz.77, 707 (1979).
429. A.R.Mackintosh, Phys.Lett.4, 140 (1963).
430. R.V.Colvin and S.Arais, phys.stat.sol.4, 37 (1964).
431. A.A.Berdyshev and I.N.Vlasov, Fiz.Metallov Metalloved. 10, 628 (1960).
432. M.Roesler, phys.stat.sol. 8, K31 (1965).
433. V.Yu.Irkhin and M.I.Katsnelson, Phys.Rev.B52, 6181 (1995).
434. N.V.Ryzhanova, Sh.Sh.Abelsky and Yu.P.Irkhin, Fiz.Metallov Metalloved.56, 843 (1983).
435. H.Ehrenreich and L.Schwartz, In: Solid State Physics, vol.31, ed.F.Seitz and D.Turnbull, New York, Academic Press, 1976.
436. I.A.Campbell and A.Fert, In: Ferromagnetic Materials, ed.P.Wohlfarth, North-Holland, vol.3, 1982, p.747; A.Fert and I.A.Campbell, J.Phys.F6, 849 (1976).
437. Proc.First Int.Symp.on Magnetic Multilayers (Kyoto, 1993), J.Magn.Magn. Mat.126 (1993).
438. M.V.Vedernikov, Adv.Phys.18, 337 (1969).
439. L.T.Rayevskaya, E.V.Rosenfeld and Yu.P.Irkhin, Fiz.Metallov Metalloved.66, 73 (1988).
440. K.R.Krylov, A.I.Ponomarev, I.M.Tsidilkovsky et al, Phys.Lett.A131, 203 (1988).
441. R.C.Yu et al, Phys.Rev.B37, 7963 (1988).
442. V.Yu.Irkhin, M.I.Katsnelson and A.V.Trefilov, Europhys.Lett.15, 649 (1991); Zh.Eksp.Theor.Fiz.105, 1733 (1994).
443. C.M.Hurd, The Hall Effect in Metals and Alloys, Plenum, New York, 1972.

444. M.J.Ziman, Phys.Rev.121, 1320 (1961).
445. T.Matsuda, J.Phys.Chem.Sol.30, 859 (1969); A.Hasegawa and T.Kasuya, J.Phys.Soc.Jpn 28, 75 (1970).
446. N.V.Volkenshtein et al, Zh.Eksp.Teor.Fiz.89, 172 (1985).
447. E.I.Kondorsky, Sov.Phys.JETP 28, 1256 (1969).
448. E.Fawcett, Adv.Phys.13, 139 (1964).
449. E.M.Pugh, Phys.Rev.36, 1503 (1930).
450. I.K.Kikoin, Sow.Phys.9, 1 (1936).
451. N.V.Volkenshtein and G.V.Fedorov, Zh.Eksp.Teor.Fiz.38, 64 (1960).
452. T.Penney et al, J.Magn.Magn.Mat.54–57, 370 (1986).
453. F.Lapierre et al, J.Magn.Magn.Mat.63–64, 338 (1987).
454. V.Rudnitskii, Zh.Eksp.Teor.Fiz.9, 262 (1939); 10, 744 (1940).
455. R.Hulme, Proc.Roy.Soc.A135, 237 (1932).
456. A.G.Samoilovich and B.L.Kon'kov, Zh.Eksp.Teor.Fiz. 20, 783 (1950).
457. R.Karplus and J.M.Luttinger, Phys.Rev. 95, 1154 (1954).
458. W.Kohn and J.M.Luttinger, Phys.Rev.108, 590 (1957).
459. J.M.Luttinger, Phys.Rev.112, 739 (1958).
460. Yu.P.Irkhin and V.G.Shavrov, Zh.Eksp.Teor.Fiz.42, 1233 (1962) [Sov. Phys.-JETP 15, 854 (1962)].
461. H.R.Leribaux, Phys.Rev.150, 384 (1966).
462. A.B.Granovskii and E.I.Kondorskii, Fiz.Metallov Metalloved. 39, 718 (1975).
463. A.N.Voloshinsky and L.F.Savitskaya, Zh.Eksp.Theor.Fiz.61, 2018 (1971).
464. Yu.P.Irkhin and Sh.Sh.Abelskii, Fiz.Metallov Metalloved.14, 641 (1962).
465. J.Kondo, Progr.Theor.Phys.27, 772 (1962); 28,846 (1962).
466. Yu.P.Irkhin and Sh.Sh.Abelskii, Fiz.Tverd.Tela 6, 1635 (1964) [Sov. Phys. – Solid State 5, 1283 (1964)].
467. L.F.Savitskaya and A.N.Voloshinsky, Fiz.Metallov Metalloved.33, 849 (1972).
468. E.I.Kondorskii, Zh.Eksp.Teor.Fiz.48, 506 (1965).
469. L.Berger, Phys.Rev.B2, 4559 (1970); B5, 1862 (1972); B8, 2351 (1973); P.Nozieres and C.Lewiner, J.Phys.(Paris) 34, 901 (1973).
470. S.K.Lyo and T.Holstein, Phys.Rev.Lett.29, 423 (1972).
471. Yu.Kagan and L.A.Maksimov, Fiz.Tverd.Tela 7, 530 (1965) [Sov.Phys. – Solid State 7, 422 (1965)].
472. Yu.P.Irkhin, A.N.Voloshinskii and Sh.Sh.Abelskii, phys.stat.sol. 22, 309 (1967).
473. R.Huguenin and D.Rivier, Helv.Phys.Acta 38, 900 (1965).
474. E.I.Kondorskii, Zh.Eksp.Teor.Fiz.55, 558 (1968) [Sov.Phys.- JETP 28, 291 (1968)]; Zh.Eksp.Teor.Fiz.55, 2367 (1968) [Sov.Phys.- JETP 28, 1256 (1968)].
475. A.A.Abdurakhmanov, Transport effects in ferromagnetic metals (in Russian), Rostov University Press, 1978.
476. F.E.Maranzana, Phys.Rev.160, 421 (1967).
477. B.Giovannini, Proc.Conf. sur l'effet Hall extraordinaire, ed.G.Cohen et al, University of Geneva, 1973, p.82.
478. A.Fert, J.Phys.Lett.35, L107 (1974).
479. A.Fert and P.M.Levy, Phys.Rev.B36, 1907 (1987); P.M.Levy, Phys.Rev.B38, 6779 (1988).
480. N.O.Kourov, Yu.N.Tsiovkin and N.V.Volkenshtein, Fiz.Tverd.Tela 24, 1059 (1981).
481. J.Smit, Physica 17, 612 (1951); 21, 817 (1955); 24, 39 (1958).
482. V.Marsocci, Phys.Rev.137, A1842 (1965).
483. Vu Dinh Ky, phys.stat.sol.15, 739 (1966).
484. A.Kundt, Wied.Ann. 49, 257 (1893).

485. H.R.Hulme, Proc.Roy.Soc.A135, 237 (1952).
486. C.Kittel, Phys.Rev.53, A208 (1951).
487. P.N.Argyres, Phys.Rev.97, 334 (1965).
488. B.R.Cooper, Phys.Rev.139, A1504 (1965).
489. A.N.Voloshinsky and G.A.Bolotin, Fiz.Metallov Metalloved.17, 481 (1964).
490. L.E.Gurevich and I.N.Yassievich, Fiz.Tverd.Tela 6, 3341 (1964).
491. E.I.Kondorsky and A.V.Vedyaev, J.Appl.Phys.39, 559 (1968).
492. D.Weller et al, J.de Phys.Colloque C8, 49, 41 (1988).
493. A.V.Druzhinin, Yu.V.Knyazev and G.A.Bolotin, Fiz.Metallov Metalloved. 52, 1194 (1981).
494. R.A. de Groot, P.G. van Engen, P.P.J. van Engelen, K.H.J.Buschow, J.Magn.Magn.Mat.86, 326 (1990).
495. J.H.Wijngaard, C.Haas, R.A.de Groot, Phys.Rev.B40, 9318 (1989).
496. H.Fuiji, H.Kawanaka, T.Takabatake et al, J.Phys.Soc.Jpn 58, 3481 (1989).
497. M.Yethirai, R.A.Robinson, J.J.Rhyne, J.A.Gotaas and K.H.J.Bushow, J.Magn.Magn.Mat.79, 355 (1993).
498. G.S.Krinchik and M.V.Chetkin, Zh.Eksp.Teor.Fiz.36, 1924 (1959).
499. V.N.Mayevsky and G.A.Bolotin, Fiz.Metallov Metalloved. 40, 258 (1975).
500. A.V.Druzhinin, Thesis, Sverdlovsk, 1986.
501. G.S.Krinchik and R.D.Nuralieva, Fiz.Metallov Metalloved.7, 697 (1959).
502. E.I.Kondorsky and R.P.Vasilyeva, Zh.Eksp.Teor.Fiz.45, 401 (1963).
503. E.I.Kondorsky, Zh.Eksp.Teor.Fiz.45, 511 (1963); 46, 2080 (1964).
504. A.N.Voloshinsky and N.V.Ryzhanova, Fiz.Metallov Metalloved. 40, 7 (1975); 41, 903 (1976).
505. P.A.Lee, T.M.Rice, J.W.Serene, L.J.Sham and J.W.Wilkins, Comm.Cond. Mat.Phys.12, 99 (1986).
506. A.de Visser and J.J.M.Franse, J.Magn.Magn.Mat.100, 204 (1991).
507. G.R.Stewart, Rev.Mod.Phys. 56, 755 (1984).
508. S.Barth et al, Phys.Rev.B39, 11695 (1989).
509. H.Nakamura, Y.Kitaoka, K.Asayama and J.Floquet, J.Phys.Soc.Jpn 57, 2644 (1988).
510. D.Jaccard, J.Sierro, J.P.Brison and J.Floquet, J.de Phys. Coll.C8, 49, 741 (1988).
511. W.H.Wong and W.G.Clark, J.Magn.Magn.Mater.108, 175 (1992); O.Avenel, J.S.Xia, B.Andraka et al., Phys.Rev.B45, 5695 (1992).
512. J.M.Lawrence, P.S.Riseborough and R.D.Parks, Rep.Progr.Phys.44, 1 (1981).
513. H.Nakamura, Y.Kitaoka, H.Yamada and K.Asayama, J.Magn.Magn.Mater. 76–77, 517 (1988); H.Nakamura, Y.Kitaoka, T.Iwai and K.Asayama, J.Phys.Cond.Mat.4, 473 (1992).
514. Y.J.Uemura et al, Phys.Rev.B39, 4726 (1989); C.Tieng, Phys.Rev.B43, 83 (1991).
515. C.Jin et al, Bull.Am.Phys.Soc.36, 717 (1991).
516. E.Gratz, E.Bauer, B.Barbara et al, J.Phys.F15, 1975 (1985).
517. E.Bauer, E.Gratz and C.Schmitzer, J.Magn.Magn.Mater.63–64, 37 (1987).
518. F.R. de Boer et al, J.Magn.Magn.Mater.63–64, 91 (1987).
519. S.Takagi, T.Kimura, N.Sato, T.Satoh and T.Kasuya, J.Phys.Soc.Jpn, 57, 1562 (1988).
520. D.T.Adroya and S.K.Malik, J.Magn.Magn.Mat.100, 126 (1991) (see also corresponding references therein).
521. D.Gignoux et al, J.Magn.Magn.Mat.74, 1 (1988).

522. D.T.Adroya, S.K.Malik, B.T.Padalia and R.Vijayaraghavan, Sol.St.Comm.66, 1201 (1988).
523. Y.Nakazawa et al, Techn.Rep.ISSP A2663 (1993).
524. G.Aeppli, F.Bucher, C.Broholm, J.K.Kjems, J.Baumann and J.Hufnagl, Phys.Rev.Lett. 60, 615 (1988).
525. T.T.M.Palstra et al., Phys.Rev.Lett.55, 2727 (1985).
526. J.K.Kjems and C.Broholm, J.Magn.Magn.Mat.76–77, 371 (1988).
527. R.N.Kleiman, D.J.Bishop, R.H.Ott, Z.Fisk and J.L.Smith, Phys.Rev.Lett. 64,1975 (1990).
528. A.de Visser et al, Phys.Rev.B45, 2962 (1992).
529. L.Degiorgi, H.R.Ott, M.Dressel, G.Gruener and Z.Fisk, Europhys.Lett.26, 221 (1994).
530. J.D.Thompson, Z.Fisk and H.R.Ott, J.Magn.Magn.Mat.54–57, 393 (1986).
531. E.Brueck et al Phys.Rev.B49, 8562 (1994).
532. C.Geibel et al, Z.Phys.B83, 305 (1991); A.Schroeder et al, Phys.Rev.Lett.72, 136 (1994).
533. C.Geibel et al, Z.Phys.B84, 1 (1991).
534. E.Brueck, F.R. de Boer, V.Sechovsky and L.Havela, Europhys.Lett. 7, 177 (1988).
535. E.Sugiura et al, J.Magn.Magn.Mater.90–91, 65 (1990); H.Fujii et al, J.Magn.Magn.Mater.90–91, 507 (1990).
536. N.Sato et al, J.Phys.Soc.Jpn 60, 757 (1991).
537. P.Bonville et al. J.Magn.Magn.Mater.76–77, 473 (1988).
538. T.Sakon, N.Sato, A.Oyamada et al. J.Phys.Soc.Jpn 62, 2209 (1992).
539. P.Bonville et al, J.Phys.:Cond.Mat.1, 8567 (1989).
540. M.Kasaya, T.Tani, K.Kawata et al, J.Phys.Soc.Jpn 60, 3145 (1991).
541. S.K.Dhar et al, Phys.Rev.B49, 641 (1994).
542. A.Ochiai, T.Suzuki and T.Kasuya, J.Magn.Magn.Mater.52, 13 (1985).
543. E.V. Sampathkumaran et al, Solid St.Comm.78, 971 (1991).
544. V.A.Stolyarov, L.S.Danelyan and L.Yu.Prokofyeva, Fiz.Tverdogo Tela 28, 2474 (1986).
545. N.B.Brandt and V.V.Moshchalkov, Adv.Phys.33, 373 (1984); V.V.Moshchalkov and N.B.Brandt, Usp.Fiz.Nauk, 149, 585 (1986).
546. F.Steglich, J.Magn.Magn.Mat.100, 186 (1991).
547. E.Bauer, Adv.Phys.40, 417 (1991).
548. K.Miyake, T.Matsuura and C.M.Varma, Solid State Comm.71, 1149 (1989).
549. M.Springford and P.H.P.Reinders, J.Magn.Magn.Mat.76–77, 11 (1988).
550. G.Zwicknagl, Adv.Phys.41, 203 (1992).
551. P.Fulde, J.Keller and G.Zwicknagl, Solid State Physics, vol.41, ed.F.Seitz et al., New York, Academic Press, 1988, p.1.
552. J.Kondo, Solid State Physics, vol.23, ed.F.Seitz and D.Turnbull, New York, Academic Press, 1969, p.183.
553. P.Nozieres, J.Low Temp.Phys.17, 31 (1974).
554. P.W.Anderson, J.Phys.C3, 2346 (1970).
555. P.Nozieres and A.Blandin, J.Phys.(Paris) 41, 193 (1980).
556. K.G.Wilson, Rev.Mod.Phys.47, 773 (1975).
557. N.Andrei, K.Furuya and J.H.Loewenstein Rev.Mod.Phys.55, 331 (1983).
558. A.M.Tsvelick and P.B.Wiegmann, Adv.Phys.32, 745 (1983).
559. M.D.Daybell and W.A.Steyert, Rev.Mod.Phys.40, 380 (1968); J.Appl.Phys.40, 1056 (1969).
560. A.J.Heeger, Solid State Physics, vol.23, ed.F.Seitz and D.Turnbull, New York, Academic Press, 1969, p.283.

561. J.R.Schrieffer, J.Appl.Phys. 38, 1143 (1967).
562. L.L.Hirst, Z.Phys.244, 230 (1971); Int.J.Magn. 2, 213 (1972); Adv.Phys.21 295 (1972).
563. V.Yu.Irkhin and Yu.P.Irkhin, Zh.Eksp.Theor.Fiz., 107, 616 (1995).
564. J.S.Griffith, The Theory of Transition-Metal Ions, Cambridge, University Press, 1961; The Irreducible Tensor Method for Molecular Symmetry Groups, New Jersey, Prentice-Hall, Englewood Cliffs, 1962.
565. D.M.Newns and N.Read, Adv.Phys.36, 799 (1987).
566. F.G.Aliev et al, J.Magn.Magn.Mat.76–77, 295 (1988); Phys.Rev.B47, 729 (1993).
567. M.Kyogaku et al, J.Phys.Soc.Jpn 59, 1728 (1990).
568. T.Takabatake et al, J.Magn.Magn.Mat.76–77, 87 (1988).
569. S.K.Malik and D.T.Adroya, Phys.Rev.B43, 6277 (1991).
570. B.Andraka, Phys.Rev.B49, 348 (1994).
571. P.C.Canfield et al, J.Appl.Phys.70, 5800 (1991).
572. M.F.Hundley et al, Phys.Rev.B42, 6842 (1990); A.Severing, J.D.Thompson, P.C.Canfield and Z.Fisk, Phys.Rev.B44, 6832 (1991).
573. W.F.Brinkman and S.Engelsberg, Phys.Rev.169, 417 (1968).
574. V.Yu.Irkhin and M.I.Katsnelson, Z.Phys.B70, 371 (1988).
575. M.E.Fisher, J.S.Langer, Phys.Rev.Lett.20, 665 (1968).
576. J.Sakurai, J.Matsuura and Y.Komura, J.Phys.Colloque C8, 49, 783 (1988).
577. Y.Onuki et al, J.Phys.Colloque C8, 49, 481 (1988).
578. T.Ohyama, J.Sakurai and Y.Komura, Sol.St.Comm.60, 975 (1986).
579. G.Sparn et al, J.Magn.Magn.Mat.47–48, 521 (1985).
580. U.Gottwick et al, J.Magn.Magn.Mat.63–64, 341 (1987).
581. P.Coleman, Phys.Rev.B29, 3035 (1984).
582. Y.Ono, T.Matsuura and Y.Kuroda, Physica C159, 878 (1989).
583. F.Marabelly and P.Wachter, J.Magn.Magn.Mat.70, 364 (1989).
584. D.I.Khomsky, Usp.Fiz.Nauk, 129, 443 (1979).
585. B.Johansson, Phil.Mag.34, 469 (1974).
586. D.Wielizka, I.H.Wesver, D.W.Lynch and C.G.Olson, Phys.Rev.B26, 7056 (1983).
587. R.Podloucky and D.Gloetzel, Phys.Rev.B27, 3390 (1983).
588. L.D.Finkelstein, Fiz.Metallov Metalloved., 57, 402 (1984).
589. B.Johansson, J.Magn.Magn.Mat.47–48, 231 (1985).
590. V.Yu.Irkhin and M.I.Katsnelson, Zh.Eksp.Teor.Fiz.90, 1080 (1986) [Sov.Phys.JETP 63, 631 (1986)]; Solid State Comm. 58, 881 (1986).
591. G.Travaglini and P.Wachter, Phys.Rev.B30, 893, 5877 (1984); P.Wachter and G.Travaglini, J.Magn.Magn.Mat.47–48, 423 (1985).
592. A.Amato and J.Sierro, J.Magn.Magn.Mat.47–48, 475 (1985).
593. P.Strange, J.Phys.C17, 4273 (1984); F.Lopez-Aguilar and J.Costa-Quintana, J.Phys.C19, 2185 (1986).
594. B.Bucher, P.Steiner and P.Wachter, Phys.Rev.Lett.67, 2717 (1991).
595. S.S.Bader, N.E.Phillips and D.B.McWhan, Phys.Rev.B7, 4686 (1973).
596. T.S.Altshuler, V.N.Mironov, G.G.Khaliullin and D.I.Khomsky, Pis'ma ZhETF 40, 28 (1984) [JETP Lett. 40, 754 (1984)].
597. V.Yu.Irkhin and M.I.Katsnelson, Fiz.Tverd.Tela 29, 1461 (1987).
598. Yu.P.Irkhin, Pis'ma ZhETF 32, 205 (1980).
599. K.Sugiyama et al, J.Magn.Magn.Mat.52, 283 (1985).
600. M.Kasaya, F.Ida, N.Takigawa and T.Kasuya, J.Magn.Magn.Mat.47–48, 429 (1985).
601. V.Yu.Irkhin and M.I.Katsnelson, Fiz.Metallov Metalloved.N1, 16 (1991).

602. P.Frings, B.Renker and C.Vettier, J.Magn.Magn.Mater. 63–64, 202 (1987).
603. Y.Onuki et al, J.Magn.Magn.Mater.76–77, 119 (1989); B.Luthi, P.Thalmeier, G.Bruls and D.Weber, J.Magn.Magn.Mater.90–91, 37 (1990).
604. M.A.Continentino, Phys.Rev.B47, 11587 (1993).
605. C.Lacroix, J.Magn.Magn.Mater.100, 90 (1991).
606. S.M.M.Evans, J.Phys.:Cond.Mat.3, 8441 (1991).
607. S.V.Vonsovsky, V.Yu.Irkhin and M.I.Katsnelson, Physica B163, 321 (1990).
608. V.Yu.Irkhin and M.I.Katsnelson, Z.Phys.B82, 77 (1991).
609. S.V.Vonsovsky, V.Yu.Irkhin and M.I.Katsnelson, Physica B171, 135 (1991).
610. P.Coleman, Physica B171, 3 (1991).
611. P.Fazekas and E.Mueller-Hartmann, Z.Phys.B85, 285 (1991).
612. V.Yu.Irkhin and M.I.Katsnelson, J.Phys.:Cond.Mat.4, 9661 (1992); Phys.Rev.B56, 8109 (1997).
613. H.Yashima, N.Sato, H.Mori and T.Satoh, Solid State Comm.43, 595 (1982); S.K.Dhar, K.A.Gschneidner (Jr.), W.H.Lee, P.Klavins and R.N.Shelton, Phys.Rev.B36, 341 (1987).
614. S.V.Vonsovsky, Yu.P.Irkhin, V.Yu.Irkhin and M.I.Katsnelson, J.Phys.Colloque C8, 49, 253 (1988).
615. M.Kyogaku, Y.Kitaoka, K.Asayama, T.Takabatake and K.Fujii, J.Phys.Soc.Jpn 61, 43 (1992); A.Kratzner et al, Europhys.Lett.19, 649 (1992).
616. R.Kuentzler, R.Claud, G.Schmerber and Y.Dossmann, J.Magn.Magn.Mat. 104–107, 1976 (1992).
617. J.G.Bednorz and K.A.Mueller, Z.Phys.B64, 189 (1986).
618. M.K.Wu et al, Phys.Rev.Lett.58, 908 (1987).
619. F.C.Zhang, T.M.Rice, Phys.Rev.B37, 3759 (1988).
620. V.Yu.Irkhin and M.I.Katsnelson, J.Phys.:Cond.Mat.3, 6439 (1991).
621. R.J.Birgeneau, H.J.Guggenheim and G.Shirane, Phys.Rev.B1, 2211 (1970); R.J.Birgeneau, D.R.Gabbe, H.P.Jenssen et al, Phys.Rev.B38, 6614 (1988).
622. P.Arovas and A.Auerbach, Phys. Rev.B38, 316 (1988).
623. D.Yoshioka, J.Phys.Soc.Jpn 58, 3393 (1989).
624. M.Takahashi, Phys.Rev.B40, 2494 (1989).
625. C.L.Kane, P.A.Lee and N.Read, Phys.Rev.B39, 6880 (1989).
626. K.J. von Szczepanski, P.Horsch, W.Stephan and M.Ziegler, Phys.Rev.B41, 2017 (1990).
627. A.Kebede, C.S.See, J.Schwegler et al, Phys.Rev.B40, 4453 (1989).
628. J.L.Peng, R.N.Shelton and H.B. Radonsky, Phys.Rev.B41, 187 (1989).
629. T.Brugger, T.Schreiner, G.Roth, P.Adelman and G.Czjzek, Phys.Rev.Lett. 71, 2481 (1991).
630. C.C.Yu and P.W.Anderson, Phys.Rev.B29, 6165 (1984).
631. P.W.Anderson, in Frontiers and Borderlines in Many Particle Physics, ed.J.Schrieffer and R.Broglia, North-Holland, 1988, p.1.
632. S.Kivelson, Phys.Rev.B36, 7237 (1987).
633. P.W.Anderson, Int.J.Mod.Phys.4, 181 (1990); Yu.A.Izymov, M.I.Katsnelson and Yu.N.Skryabin, Itinerant Electron Magnetism [in Russian], Moscow, Nauka, 1994.
634. V.Yu.Irkhin, A.A.Katanin and M.I.Katsnelson, J.Phys.:Cond.Mat. 4, 5227 (1992).
635. V.Yu.Irkhin and M.I.Katsnelson, Pis'ma ZhETF 49, 500 (1989); Phys.Lett.A150, 47 (1990).
636. J.Deportes, B.Ouladdiaf and K.R.A.Ziebeck, J.Magn.Magn.Mat.70, 14 (1987).
637. H.Wada et al, J.Magn.Magn.Mat.70, 17 (1987); M.Shiga, J.Magn.Magn.Mat.129, 17 (1994).

638. G.Krill et al, J.Phys.C9, 761 (1976); S.Sudo, J.Magn.Magn.Mat.114, 57 (1992).
639. T.Furuno, K.Ando, S.Kunii et al, J.Magn.Magn.Mat.76–77, 117 (1988); K.Fraas, U.Ahlheim, P.H.P.Reinders et al, J.Magn.Magn.Mater.108, 220 (1992).
640. O.Nakamura, A.Oyamada, A.Ochiai, S.Kunii, T.Takeda, T.Suzuki and T.Kasuya, J.Magn.Magn.Mater.76–77, 293 (1988).
641. Z.Fisk et al, Phys.Rev.Lett.67, 3310 (1991).
642. J.Rossat-Mignod et al, Phys.Rev.B16, 440 (1977); P.Fisher et al, J.Phys.C11, 345 (1978).
643. M.Sera, T.Fujita, T.Suzuki and T.Kasuya, In: Valence Instabilities, North-Holland, 1982, p.435.
644. P.W.Anderson, Phys.Rev.Lett., 64, 239 (1990); 65, 2306 (1990); 66, 3226 (1991); 71, 1220 (1993).
645. C.M.Varma, P.B.Littlewood, A.Schmitt-Rink, E.Abrahams and A.E.Ruckenstein, Phys.Rev.Lett.63, 1996 (1989); P.B.Littlewood and C.M.Varma, Phys.Rev.B46, 405 (1992).
646. M.B.Maple et al, J.Low Temp.Phys.95, 225 (1994); G.R.Stewart, Rev.Mod.Phys. 73, 797 (2001); 78, 743 (2006).
647. A.A.Romanyukha, Yu.N.Shvachko, V.Yu.Irkhin, M.I.Katsnelson, A.A.Koshta and V.V.Ustinov, Physica C171, 276 (1990).
648. J.M.Luttinger, J.Math.Phys.15, 609 (1963); G.Mahan, Many-Particle Physics, New York, Plenum Press (1981).
649. F.D.M.Haldane, J.Phys.C12, 4791 (1979).
650. A.H.Castro Neto and E.Fradkin, Phys.Rev.B49, 10877 (1994) A.Houghton, H.-J.Kwon and J.B.Marston, Phys.Rev.B50, 1351 (1994).
651. N.N.Bogoliubov, Lectures in Quantum Statistics. In: Collected works, vol.2, Kiev, Naukova Dumka, 1970 (in Russian).
652. Yu.P.Irkhin, Zh.Eksper.Teor.Fiz.50, 379 (1966) [Sov.Phys.JETP 23, 253 (1966)].
653. V.Yu.Irkhin and Yu.P.Irkhin, Fiz.Metallov Metalloved. 76, N4, 49 (1993).
654. V.Yu.Irkhin and Yu.P.Irkhin, phys.stat.sol.(b) 183, 9 (1994).
655. Yu.P.Irkhin, Zh.Eksper.Teor.Fiz.66, 1005 (1974).
656. D.C.Mattic, The Theory of Magnetism, New York, Harper and Row, 1965.
657. V.A.Popov and A.A.Loginov, phys.stat.sol.(b) 84, 83 (1977).
658. C.W.Nielson and G.F.Koster, Spectroscopic coefficients for the p, d, f-configurations, Cambridge, 1963.
659. B.G.Wybourne, Symmetry Principles and Atomic Spectroscopy, Wiley, New York, 1970.
660. V.Yu.Irkhin and Yu.P.Irkhin, Zh.Eksper.Teor.Fiz.104, 3868 (1993) [Sov.Phys.JETP 77, 858 (1993)].
661. P.W.Anderson, Solid State Physics, vol.14, ed.F.Seitz and D.Turnbull, New York, Academic Press, 1963, p.99.
662. S.P.Shubin and S.V.Vonsovsky, Proc. Roy.Soc.A145, 159 (1934).
663. S.V.Vonsovsky and M.I.Katsnelson, J.Phys.C12, 2043 (1979).
664. H.Bethe, Intermediate Quantum Mechanics, Benjamin, New York, 1964.
665. A.B.Bolotin and V.K.Shugurov, Zh.Vych.Mat.Mat.Fiziki, 3, 560 (1963); H.J.Silverstone, J.Chem.Phys. 47, 537 (1967).
666. P.M.Levy, Phys.Rev. 177, 509 (1969).
667. V.V.Druzhinin and A.S.Moskvin, Fiz.Metallov Metalloved. 26, 415 (1968).
668. E.L.Nagaev, Usp.Fiz.Nauk, 136, 61 (1989).
669. T.Moriya, Phys.Rev.120, 91 (1960).
670. R.J.Elliott and M.F.Torpe, J.Appl.Phys. 39, 802 (1968).

671. D.N.Zubarev, Usp.Fiz.Nauk 71, 71 (1960).
672. V.G.Bar'yakhtar, V.N.Krivoruchko and D.A.Yablonsky, Solid St.Comm.46, 613 (1983); Zh.Eksp.Teor.Fiz.85, 602 (1983).
673. R.O.Zaitsev, Zh.Eksp.Teor.Fiz.68, 207 (1975).
674. V.V.Valkov and S.G.Ovchinnikov, Teor.Mat.Fiz. 50, 466 (1982).
675. V.V.Valkov and T.A.Valkova, Teor.Mat.Fiz. 59, 453 (1984).
676. V.V.Valkov, T.A.Valkova and S.G.Ovchinnikov, Zh.Eksp.Teor.Fiz. 88, 550 (1985); phys.stat.sol.(b) 142, 255 (1987).
677. F.P.Onufrieva, Zh.Eksp.Teor.Fiz.80, 2372 (1981); 86, 1691 (1984); 89, 2270 (1985); 94, 232 (1988).
678. I.Affleck, J.Phys.:Cond.Mat. 1, 3047 (1989).
679. E.V.Rosenfeld, Pis'ma ZhETF, 24, 60 (1976).
680. M.I.Kaganov and A.V.Chubukov, Usp.Fiz.Nauk, 153, 537 (1987).
681. K. von Klitzing, G.Dorda and M.Pepper, Phys.Rev.Lett.45, 494 (1980).
682. V.P.Silin and A.Z.Solontsov, Fiz.Metallov Metalloved.58, 1080 (1984).
683. M.I.Auslender, V.Yu.Irkhin, Z.Phys.B56, 301 (1984); B61, 129 (1985).
684. H.Wada, M.Nishigori and M.Shiga, J.Phys.:Cond.Mat.3, 2083 (1991).
685. G.M.Carneiro and C.J.Pethick, Phys.Rev.B11, 1106 (1975).
686. F.Rys, J.C.Helman, W.Baltensperger, Phys.Kond.Mat.6, 105 (1967).
687. V.Yu.Irkhin, Fiz.Tverd.Tela 28, 3066 (1986).
688. S.Jin et al, Science 264, 413 (1994).
689. V.Yu.Irkhin and M.I.Katsnelson, Fiz.Metallov Metalloved.65, 446 (1988).
690. N.A.Cade and W.Young, Adv.Phys.26, 393 (1977).
691. M.C.K.Wiltshire, M.M.Elcombe and C.J.Howard, J.Phys.F15, 1595 (1985).
692. M.I.Katsnelson and V.Yu.Irkhin, J.Phys.C17, 4291 (1984).
693. V.Yu.Irkhin and A.M.Entelis, J.Phys.:Cond.Mat.1, 4111 (1989); V.Yu.Irkhin and M.I.Katsnelson, Phys.Rev.B62,5647 (2000).
694. A.O.Anokhin and V.Yu.Irkhin, phys.stat.sol.(b) 165, 129 (1991).
695. A.O.Anokhin, V.Yu.Irkhin and M.I.Katsnelson, J.Phys.: Cond.Mat.3, 1475 (1991).
696. L.A.Maksimov and K.A.Kikoin, Fiz.Metallov Metalloved.28, 43 (1969).
697. R.O.Zaitsev, Zh.Eksp.Teor.Fiz. 75, 2362 (1978).
698. M.Sh.Erukhimov and S.G.Ovchinnikov, phys.stat.sol.(b) 123, 105 (1984).
699. V.Yu.Irkhin and M.I.Katsnelson, Zh.Eksp.Teor.Fiz.88, 522 (1985).
700. V.Yu.Irkhin and M.I.Katsnelson, J.Phys.C18, 4173 (1985).
701. P.B.Wiegmann, Phys.Rev.Lett.60, 821 (1988); Y.Chen, D.Foerster and P.Larkin, Phys.Rev.B46, 5370 (1992).
702. K.Kubo and N.Ohata, J.Phys.Soc.Jpn, 33, 21 (1972).
703. O.Narikiyo, K.Kuboki and H.Fukuyama, J.Phys.Soc.Jpn 59, 2443 (1990).
704. V.V.Druzhinin and Yu.P.Irkhin, Zh.Eksp.Teor.Fiz.51, 1856 (1966) [Sov.Phys.JETP 24, 1250 (1967)].
705. A.Georges and G.Kotliar, Phys.Rev.B45, 6479 (1993); X.Y.Zhang and C.M.Zhang, Phys.Rev.B49, 7929 (1994); M.J.Rozenberg, G.Kotliar and X.Y.Zhang, Phys.Rev.B49, 10181 (1994); F.J.Ohkawa, J.Phys.Soc.Jpn 61, 1615 (1992); A. Georges et al., Rev. Mod. Phys. 68, 13 (1996).
706. R.Karplus and J.Schwinger, Phys.Rev.73, 1020 (1948).
707. J.M.Luttinger and W.Kohn, Phys.Rev.109, 1892 (1958).
708. L.L.Hirst, Adv.Phys. 27, 231 (1978).
709. Yu.P.Irkhin, Fiz.Tverd.Tela 30, 1202 (1988).
710. K.A.Kikoin and L.A.Maksimov, Zh.Eksp.Teor.Fiz. 58, 2184 (1970).
711. P.Coleman and N.Andrei, J.Phys.:Cond.Mat.1, 4057 (1989).

712. S.Chakravarty, B.I.Halperin and D.R.Nelson, Phys.Rev.B39, 2344 (1989).
713. V.Yu.Irkhin, A.A.Katanin and M.I.Katsnelson, Fiz.Metallov Metalloved., 79, N1, 65 (1995); Phys.Lett.A157, 295 (1991).
714. F.D.M.Haldane, Phys.Rev.Lett.50, 1153 (1983).
715. V.Yu.Irkhin and M.I.Katsnelson, Z.Phys.B62, 201 (1986).
716. V.Yu.Irkhin and M.I.Katsnelson, Phys.Rev.B53, 14008 (1996).
717. V.Yu.Irkhin and Yu.P.Irkhin, J.Magn.Magn.Mater.164, 119 (1996).
718. M.Akai, H.Akai and J.Kanamori, J.Phys.Soc.Jpn 54, 4246,4257 (1985).
719. R.Zeller, J.Phys.F17, 2123 (1987).
720. N.Stefanou, A.Oswald, R.Zeller and P.H.Dederichs, Phys.Rev.B35, 6911 (1987).
721. B.Drittler, N.Stefanou, S.Bluegel, R.Zeller and P.H.Dederichs, Phys.Rev.B40, 8203 (1989).
722. P.H.Dederichs, R.Zeller, H.Akai, H.Ebert, J.Magn.Magn.Mater.100, 241 (1991).
723. I.Mertig, E.Mrosan and R.Schoepke, J.Phys.F12, 1689 (1982); I.Mertig, E.Mrosan, R.Zeller and P.H.Dederichs, phys.stat.sol.(b) 117, 619 (1983).
724. I.Mertig, E.Mrosan, R.Zeller and P.H.Dederichs, Proc.Int.Conf. on the Physics of Transition Metals (Darmstadt, July 1992), World Scientific, 1992, p.778.
725. M.D.Daybell, In: Magnetism, vol.5, ed.H.Suhl, New York, Academic Press, 1973.
726. A.R.Miedema, J.F.Dorleijn, J.Phys.F5, 487,1543 (1975); O.Jaoul, I.A.Campbell and A.Fert, J.Magn.Magn.Mater.5, 23 (1977); T.Farrell and D.Greig, J.Phys.C3, 138 (1970); F.Mauri, A.Lucasson, P.Lucasson, P.Moser, F.Faudot, J.Phys.:Cond.Mat.2, 3269 (1990).
727. V.Yu.Irkhin, Phys.Rev.B57, 13375 (1998).
728. V.Yu.Irkhin, A.A.Katanin, Phys.Rev.B55, 12318 (1997); Phys.Rev.B57, 379 (1998); V.Yu.Irkhin, A.A.Katanin, M.I. Katsnelson, Phys.Rev.B60, 1082 (1999).
729. V.Yu.Irkhin, A.V.Zarubin, Phys.Rev.B70, 035116 (2004); J.Magn.Magn.Mat. 300, 246 (2006).
730. V.Yu.Irkhin, A.V.Zarubin, Eur.Phys.Journal B38, 563 (2004).
731. V.Yu.Irkhin and M.I.Katsnelson, Eur.Phys.Journal B19, 401 (2001).
732. V.Yu.Irkhin and M.I.Katsnelson, JETP Lett. 80, 312 (2004).
733. V.Yu.Irkhin and M.I.Katsnelson, Phys.Rev.B61, 14640 (2000).
734. V.Yu.Irkhin and M.I.Katsnelson, Eur.Phys.Journal B43, 479 (2005).
735. V.Yu.Irkhin and M.I.Katsnelson, Eur.Phys.Journal B30, 481 (2002).
736. V.Yu.Irkhin and M.I.Katsnelson, Phys.Rev.B73, 104429 (2006).

Index

I

J

K

L

M